Outer Magnetospheric Boundaries: Cluster Results

Space Sciences Series of ISSI

Volume 20

The International Space Science Institute is organized as a foundation under Swiss law. It is funded through recurrent contributions from the European Space Agency, the Swiss Confederation, the Swiss National Science Foundation, and the University of Bern. For more information, see the homepage at http://www.issi.unibe.ch/.

OUTER MAGNETOSPHERIC BOUNDARIES: CLUSTER RESULTS

Edited by

G. PASCHMANN
ISSI, Bern, Switzerland

S.J. SCHWARTZ
Imperial College, London, UK

C.P. ESCOUBET
ESTEC, Noordwijk, The Netherlands

S. HAALAND
MPE, Garching, Germany

Reprinted from *Space Science Reviews*, Volume 118, Nos. 1–4, 2005

A.C.I.P. Catalogue record for this book is available from the Library of Congress

ISBN: 1-4020-3488-1

Published by Springer
P.O. Box 990, 3300 AZ Dordrecht, The Netherlands

Sold and distributed in North, Central and South America
by Springer,
101 Philip Drive, Norwell, MA 02061, U.S.A.

In all other countries, sold and distributed
by Springer,
P.O. Box 322, 3300 AH Dordrecht, The Netherlands

Printed on acid-free paper

Printed in the Netherlands

Contents

Foreword

The present 20th volume of the ISSI Space Science Series is devoted to the results obtained by ESA's Cluster mission regarding the outer boundaries of Earth's magnetopshere. These boundaries and their environs are rich in exciting processes in collisionless plasmas, such as shock formation, magnetic reconnection, particle acceleration and wave-particle interactions. They are interesting in their own right, but also because they are prototypes of similar structures and phenomena that are ubiquitous in the universe, having the unique advantage that they are accessible to in situ measurements. This book represents the first synthesis of new results on this topic obtained in the first few years of the Cluster mission.

ISSI's involvement with Cluster began in March 1996 with the first meeting of a working group that wrote what is now considered the reference book on multi-spacecraft data analysis methods. It was natural then that Cluster would become the focus of an ISSI workshop series, with the working title Magnetospheric Boundaries and Turbulence: Cluster Results, once the data analysis had progressed beyond a certain basic stage. The first workshop, held in March 2003 with nearly 40 invited participants, defined the objectives of the project and provided a first outline of the book, based on inputs from three Working Groups (solar wind/foreshock/magnetosheath; bow shock; magnetopause/cusp). A second workshop in November 2003, and a final meeting by the main authors in June 2004, reviewed the available material, defined the remaining work, and started the editing process. The meetings were organized by a coregroup consisting of A. Balogh, Imperial College, London; P. Décréau, LPCE/CNRS, Orléans; M. Goldstein, Goddard Space Flight Center, Greenbelt; C. P. Escoubet, ESTEC, Noordwijk; M. Scholer, MPE, Garching; S. J. Schwartz, Queen Mary College, London. G. Paschmann, ISSI, Bern, was the project leader. A total of of 45 authors from 9 different countries finally contributed to the book. G. Paschmann, S. J. Schwartz, P. Escoubet, and S. Haaland served as the Editors.

ISSI's involvement in Cluster was also personal. The founding directors, Johannes Geiss and Bengt Hultqvist, played key roles in making Cluster part of the first cornerstone of ESA's Horizons 2000 program. B.H. was also one of the proposers of the mission. Of the present directors, G.P. was among the mission proposers, and later became a PI of one of the instruments. R.M.B., as ESA's Director of Science, accompanied Cluster through its long history, including its resurrection after the 1996 catastrophic loss.

Bern in January 2005 R.M. Bonnet, R.von Steiger, G. Paschmann

Introduction

G. Paschmann[1], C. P. Escoubet[2], S. J. Schwartz[3], and S. E. Haaland[4]

When the stream of plasma emitted from the Sun (the solar wind) encounters Earth's magnetic field, it slows down and flows around it, leaving behind a cavity, the magnetosphere. The magnetopause is the surface that separates the solar wind on the outside from the Earth's magnetic field on the inside, thus its name. Because the solar wind moves at supersonic speed, a bow shock must form ahead of the magnetopause that acts to slow the solar wind to subsonic speeds. Interest in these boundaries and their environs arises from the fact that they are the site of many fascinating plasma phenomena and processes, such as particle acceleration, magnetic reconnection, wave excitation, and turbulence, that are thought to be ubiquitous in the universe, but must be studied in situ because they are not accessible by remote sensing techniques.

The boundaries of the magnetosphere have been the target of direct in-situ measurements since the beginning of the space age. But because they are constantly moving, changing their orientation, and undergoing evolution, the interpretation of single-spacecraft measurements has been plagued by the fundamental inability of a single observer to unambiguously distinguish spatial from temporal changes. The boundaries are thus a prime target for the study by a closely spaced fleet of spacecraft. There have been previous attempts with two spacecraft missions (ISEE, AMPTE, Interball), but with two stations one can only infer variations along the line between the spacecraft, and three stations always form a plane. Thus Cluster, with its four spacecraft in a three-dimensional configuration at variable separation distances, represents a giant step forward.

Cluster's history goes back to 1982, when a group of scientists responded to a call for proposals by the European Space Agency, ESA. In February 1986 ESA chose Cluster and the SOlar and Heliospheric Observatory (SOHO) as the first 'Cornerstone' in its Horizons 2000 Science Programme. Ready for launch in the fall of 1995, the four spacecraft were destroyed when the Ariane 5 rocket exploded during its maiden flight on 4 June, 1996. The spacecraft and instruments were rebuilt and then successfully launched pairwise on two Soyuz-Fregate rockets from

[1]International Space Science Institute, Bern, Switzerland
[2]ESTEC, Noordwijk, The Netherlands
[3]Astronomy Unit, Queen Mary, University of London, London, UK
[4]Max-Planck-Institut für extraterrestrische Physik, Garching, Germany

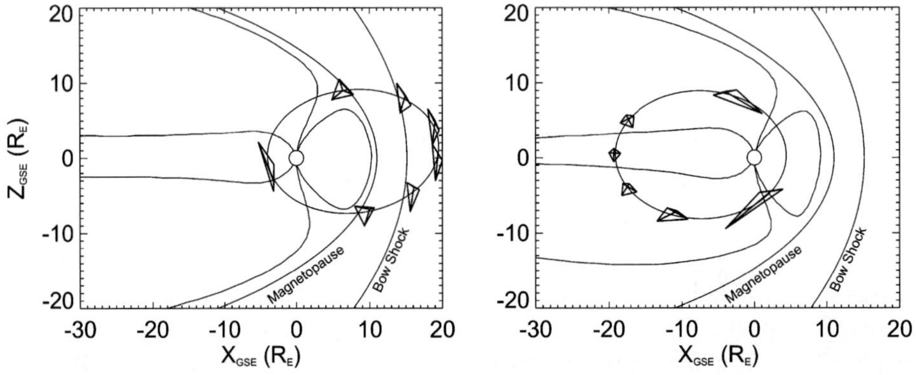

Figure i. Cluster orbit in spring (left) and in fall (right), as it cuts through the key regions of the magnetosphere and its boundaries. The spacecraft separations are exaggerated for clarity.

Baikonur on 9 July and 12 August 2000, respectively. After extensive commissioning of the payload, the science phase officially began on 1 February 2001. Originally intended for a two year lifetime, the mission was first extended until December 2005, and is likely to be extended even further.

Following launch, the Cluster spacecraft were placed in nearly identical, highly eccentric polar orbits, with an apogee of 19.6 R_E and a perigee of $\sim$4 R_E. Figure i shows the orbit superimposed on a cut of the magnetosphere, and illustrates that Cluster moves outbound over the northern polar cap, crosses the magnetopause and bow shock into the solar wind, before recrossing those boundaries in reverse order and moving over the southern polar cap back towards perigee. Figure i(left) applies to Spring of each year. As the orbit is inertially fixed, it rotates around the Earth once a year, as the latter revolves around the Sun. As a result, the apogee of the orbit will be located in the geomagnetic tail half a year later, as shown in Figure i(right). In the course of the mission Cluster will thus encounter the regions of interest several times.

The orbits are tuned so that the four spacecraft are located at the vertices of a nearly regular tetrahedron when crossing the major boundaries. The separation distances between the spacecraft are adjusted from $\sim$100 km to $\sim$10,000 km during the mission. This allows to study the phenomena at a number of important spatial scales. Figure ii shows the sequence of separation distances achieved so far and planned for the extended mission. This separation strategy is possible because each Cluster spacecraft carried 650 kg of fuel at launch, of which 75% was expended in order to reach the operational orbit. The four Cluster spacecraft carry identical sets of 11 scientific instruments, designed to measure the ambient electromagnetic fields and particle populations over a wide range of frequencies and energies. The instruments, as well as the spaceraft and mission, are described in Escoubet et al.

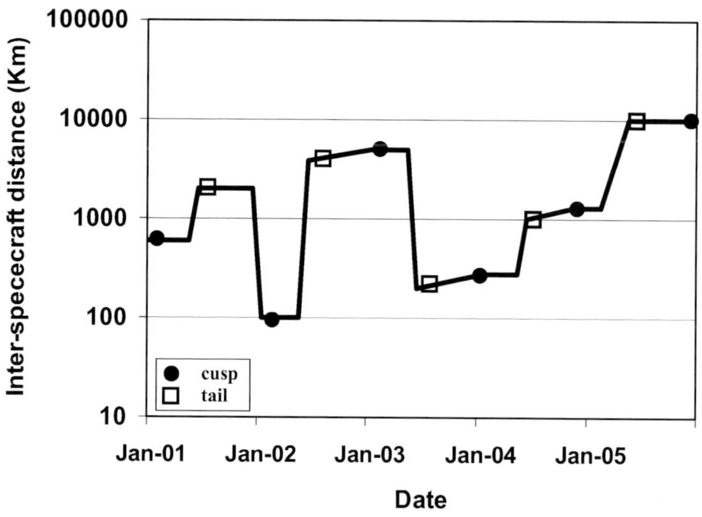

Figure ii. Spacecraft separation distances as a function of time.

(1997). The truly international scale of the endeavour is evident from the fact that more than 200 investigators from ESA member states, the United States, Canada, China, the Czech Republic, Hungary, India, Israel, Japan and Russia are involved in analysis of the Cluster data.

Cluster data analysis involves a number of methods that were not previously available or proven. Among them are methods to infer the orientation, motion and thickness of the boundaries from four-spacecraft timing; the ability to infer spatial derivatives of scalar and vector quantities from the differences measured at the four locations; and the technique to extract the three-dimensional wave properties, another unique Cluster capability. These methods were described in a book published before launch (Paschmann and Daly, 1998). (A second edition of the book is available in electric form on the ISSI web site.) Naturally, application, validation, and further development of these methods and techniques have played a major role in the Cluster data analysis, as will become apparent throughout this book.

The purpose of this book is to provide a synopsis of the Cluster results pertaining to the magnetospheric boundaries. Written only four years after launch, this can only be a snapshot of a field in motion, but should nevertheless give an impression of the Cluster achievements and prospects. Cluster's main impact is that it provides a magnifying glass for small scale phenomena and processes. By the same token there often is a need to put the Cluster observations into a more global perspective. Fortunately, observations from the ground and from other spacecraft can fill in the global context.

The book is organized as follows. Part I deals with the environs of the boundaries, starting with the unperturbed solar wind (Chapter 1), turning to the foreshock region (Chapter 2) that is strongly modified by the presence of waves and particles originating from the bow shock, and finally to the magnetosheath (Chapter 3), which is the often highly turbulent region between the bow shock and the magnetopause. Part II deals with the bow shock itself. After an introduction (Chapter 4), Chapter 5 deals with the bow shock when it is in its quasi-perpendicular geometry, i.e., when the shock normal and the upstream magnetic field form angles greater than about 45°. Chapter 6 deals with the bow shock under quasi-parallel geometry. The two geometries have enormous consequences for the appearance and properties of the shock and are therefore addressed in separate chapters. The bow shock results are summmarized in Chapter 7. Part III, finally, focusses on the magnetopause and cusp. Chapter 8 describes the properties of the magnetopause, including the boundary layer of solar-wind plasma that often is formed on the inside of the magnetopause. Chapter 9 deals with the polar cusps, the unique funnel-shaped feature that represents a weak spot in the defense of the magnetosphere against the solar wind. Chapter 10, finally, reviews what we have learned from Cluster with regard to the processes occurring at the magnetopause, including their microphysical aspects. Because of their importance for the results, methodological issues are interspersed throughout the book.

References

Escoubet, C. P., C. T. Russell, and R. Schmidt (eds.): 1997, *The Cluster and Phoenix missions*. Dordrecht: Kluwer.

Paschmann, G. and P. W. Daly (eds.): 1998, *Analysis Methods for Multi-Spacecraft Data*. ESA Publications Division.

I

SOLAR WIND, FORESHOCK, MAGNETOSHEATH

Chapter 1

The Near-Earth Solar Wind

M. L. Goldstein[1], J. P. Eastwood[1], R. A. Treumann[2], E. A. Lucek[3], J. Pickett[4], and P. Décréau[5]

1.1 The Solar Wind at 1 AU

Observational evidence for a continuous stream of plasma filling interplanetary space was deduced from the properties of cometary tails by Biermann (1951). Parker (1958) demonstrated that solutions of the fluid equations describing the solar atmosphere necessitated the existence of a continuous supersonic wind. The first *in situ* measurements of this wind were made by Gringauz et al. (1960), and Snyder and Neugebauer (1966).

At 1 Astronomical Unit (AU) the solar wind is a tenuous ionised gas that carries with it magnetic fields reaching back into the solar atmosphere. The density of this gas at 1 AU is typically about 5 particles cm^{-3}. In composition it is about 95% protons, 5% He, with a small fraction of heavy ions. (Although we will not discuss the heavy ions further here, they are of profound importance to studies of the solar atmosphere and composition of the Sun.) The embedded magnetic field has a typical value (at 1 AU) of 5 nT. The wind flows at speeds ranging from a couple of hundred km s^{-1} to 1000 km s^{-1}, or more. The speed of sound in this plasma is about 40 km s^{-1}. In addition, the speed of the most characteristic plasma wave, i.e., Alfvén waves (discussed below), is also about 40 km s^{-1} at 1 AU. Consequently, the flow is both supersonic and super-Alfvénic with a plasma $\beta \sim 1$ where β is the ratio of thermal to magnetic pressure $2\mu_0 n k_B T/B^2$ and k_B is Boltzmann's constant.

[1]NASA Goddard Space Flight Center, Greenbelt, MD, USA
[2]Max-Planck-Institut für extraterrestrische Physik, Garching, Germany
[3]Space and Atmospheric Physics, The Blackett Laboratory, Imperial College London, London, UK
[4]Department of Physics and Astronomy, The University of Iowa, Iowa City, IA, USA
[5]LPCE/CNRS and Université d' Orléans, France

Space Science Reviews 118: 7–39, 2005.
DOI: 10.1007/s11214-005-3823-4

The proton temperature (at 1 AU) is of order 1.2×10^5 K, while the electrons tend to be somewhat hotter ($\approx 1.4 \times 10^5$ K).

Matthaeus et al. (1986) showed that magnetic fluctuations in the solar wind obtained from ISEE 3 constituted a stationary ensemble. They used the ensemble to determine a variety of average solar wind properties. For example, they found that the ensemble-average correlation length of the magnetic fluctuations was 4.9×10^9 m and that the ratios of the average value of the variances of the magnetic field components were 8:9:10, where the third component is the local mean field direction. Solar wind speeds for the intervals included in the ensemble ranged from 295 km s^{-1} to 700 km s^{-1} with the modal value lying near 360 km s^{-1} (no attempt was made to separate the data into fast and slow flow regimes).

Although the solar wind flows radially from the Sun, the rotation of the Sun causes the magnetic field line to form an Archimedean (or 'Parker') spiral in space (Parker, 1958). Given the equatorial rotation rate of the Sun (25.4 days), the magnetic field is makes an angle of $\approx 45°$ to the flow at 1 AU. Because the polarity of solar magnetic fields tends to be of one sign or the other over large polar regions of the Sun, especially during the minimum of solar activity, the interplanetary magnetic field may point sunward or anti-sunward for extended periods of time as measured by single spacecraft. During each solar rotation near solar minimum, 2 or 4 'sector boundaries', where the field orientation rapidly reversed, are observed in the magnetic field time series (Wilcox and Ness, 1965). The three-dimensional structure of this boundary is referred to as the Heliospheric Current Sheet (HCS).

The fairly simple picture described above applies primarily to the minimum activity levels during the 22-year solar activity cycle. That is the time period during which one observes fast solar wind coming from coronal holes at relatively high speed ($\approx$ 750 km s^{-1}) and from high latitudes. Slow solar wind, which is significantly more variable ($300 - 450$ km s^{-1}) is observed at low latitudes. The origin of slow solar wind is less clear. It may arise from the boundary between open and closed magnetic field, or the source of slow wind may be plasma leaking from the helmet streamers that arise from plasma transiently leaking from open magnetic field lines not associated with coronal holes (e.g., Axford and McKenzie, 1996). Models of the global structure of the solar wind have become increasingly detailed and sophisticated. Recent models have incorporated physical sources of heat to the corona and thus relax the (unphysical) isothermal assumption that was used in the earliest treatments (see, e.g., Israelevich et al., 2001; Gombosi et al., 2004; Riley et al., 2002; Usmanov et al., 2000; Usmanov and Goldstein, 2003, and references therein).

One of the curiosities of solar wind observations is the apparent success (as implied in the discussion above) of the fluid or magnetofluid equations in describing many of the macroscopic properties of the wind. This despite the fact that the proton-proton collision time is $\approx 4 \times 10^6$ sec, which implies that the solar wind is essentially collisionless (the electrons in the core of the distribution function are not as collisionless, but have proton-electron collision times that correspond to a few collisions between the solar corona and 1 AU). Cluster, with its ability to determine three-dimensional structure and with its full complement of wave and particle instruments, is capable of investigating the relationship between kinetic and fluid properties, especially at boundaries of structures whose scale is of order the spacecraft separation. These structures include the bow shock and interplanetary shock waves, the HCS, interplanetary shocks, flux tubes in magnetic clouds and interplanetary coronal mass ejections (ICMEs), magnetic holes, discontinuities, and weak double layers (bipolar and tripolar electrostatic structures). Investigations of these structures has begun and some results are described below. In 2005, when the spacecraft separation is increased to $\approx 10,000$ km, the range of phenomena that can be studied with Cluster will expand considerably. In particular, the three-dimensional symmetries of solar wind turbulence in the inertial and dissipation ranges will be amenable to intensive investigation.

For a recent compendium of all aspects of solar wind research, the reader is referred to Velli et al. (2002); a detailed discussion of global solar wind properties can be found in Hundhausen (1995).

1.1.1 Waves and turbulence

One of the most striking features of the solar wind is the presence of large fluctuations in both the magnetic field and plasma velocity. These fluctuations were conjectured by Coleman (1966) to be evidence of magnetofluid turbulence. In making that conjecture, Coleman was primarily motivated by two properties of the wind: The first was that the power spectrum of the magnetic and velocity fluctuations was a power law with an index of approximately $-5/3$, the value characterising the inertial range of three-dimensional incompressible fluid turbulence (Kolmogorov, 1941). The second was the existence of large velocity shears at the boundaries between fast and slow solar wind, being significant sources of free energy that would stir the medium and generate turbulence.

The correlation between the fluctuating magnetic, $\delta\mathbf{b}$, and velocity, $\delta\mathbf{v}$, fields suggested that the fluctuations were nearly pure Alfvén waves (Belcher and Davis, 1971; Unti and Neugebauer, 1968) propagating outward from the solar corona, defined by the relation $\delta\mathbf{v} = \pm\delta\mathbf{b}/\sqrt{\mu_0\rho}$. Here ρ is the mass density of the plasma, and the + and - signs indicate propagation parallel or anti-parallel to the background magnetic field, respectively. Pure Alfvén waves are exact solutions of the incompressible equations of magnetohydrodynamics (MHD). This led to questioning whether or not the solar wind was a dynamically turbulent medium, or merely reflected remnants of coronal processes of no dynamical importance.

The resolution that emerged from analyses of data from the Helios, Voyager, ISEE, and IMP 8 spacecraft, among others, combined with numerical solutions of the MHD equations, was that the fluctuations represented both a source of 'waves' emanating from the solar corona as well as a consequence of turbulent evolution

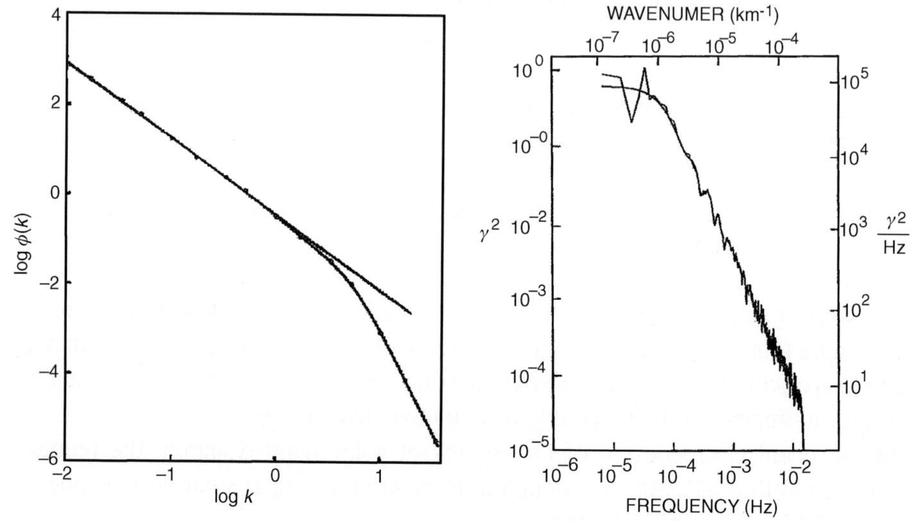

Figure 1.1. Two examples of turbulence. Left: A one-dimensional spectrum from a tidal channel collected on March 10, 1959. The straight line has a slope of 5/3 (Grant et al., 1962). Right: Solar wind magnetic field fluctuations from Voyager 1 at 1 AU (Matthaeus and Goldstein, 1982a). In the inertial range, the spectral index is almost exactly $-5/3$. The time resolution of these data was 28.8 s. In general, at 1 AU, the -5/3-spectrum continues almost another decade before steepening as dissipation sets in near 0.1 Hz.

driven by the free energy contained in velocity shears that bound fast and slow solar wind (Matthaeus and Goldstein, 1982a; Roberts et al., 1987a,b). For a review, see Tu and Marsch (1995).

Understanding the nature of magnetohydrodynamic turbulence is even more challenging than is the problem of understanding fluid turbulence. On one level there are many similarities between magnetic and velocity fluctuations observed in the solar wind and fluctuations observed in fluids. For example, a classic example of fluid turbulence is the nearly perfect power-law power spectrum of velocity fluctuations obtained from flow in a tidal channel. That spectrum spans more than two-decades in frequency (see Figure 1.1, left panel) with a power law index of $-5/3$ (Grant et al., 1962). A similar spectrum of magnetic fluctuations derived from Voyager observations of solar wind magnetic field fluctuations also spans more than two decades (Figure 1.1, right panel) and also has a power law index of $-5/3$ (Matthaeus and Goldstein, 1982a; Tu and Marsch, 1995). However, fluid turbulence is generally isotropic and homogeneous. The presence of the large-scale solar wind magnetic field implies that the solar wind is not isotropic. The Voyager spectra shown above are one-dimensional spectra – the three-dimensional properties of those fluctuations are not well understood and Cluster, at large spacecraft

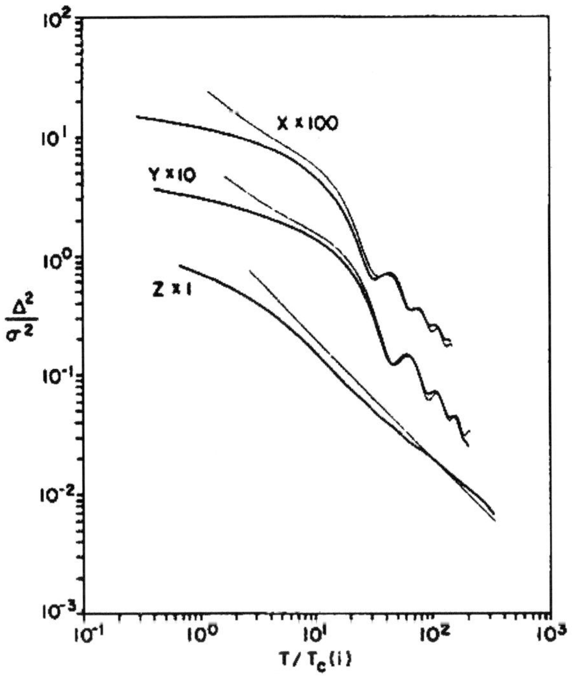

Figure 1.2. Comparison of observations (heavy lines) determined from 621 days of IMP data with predictions of the ergodic theorem modified to include the effects of solar rotation (for the x- and y-components). From Matthaeus and Goldstein (1982b).

separations, should have the capability to elucidate the nature of the turbulence in an anisotropic magnetofluid.

As a first step in the investigation of solar wind fluctuations as an example of a turbulent magnetofluid, one must justify the use of Fourier transforms and other techniques in the construction of power spectra of solar wind magnetic and velocity fluctuations. For such power spectra to be meaningful, it is necessary that the fluctuations represent a stationary random process. The assumption of stationarity is implicit in theoretical derivations of spatial mean-free-paths for 180° scattering of energetic solar and galactic particles in the solar wind in resonance with the Fourier components of the magnetic field. Energetic protons 'resonate' with wave numbers $k_{\|} r_{ci} = \cos\theta$, where r_{ci} is the Larmor radius of the energetic particles and $\cos\theta = \mu$ is the pitch-angle. Matthaeus and Goldstein (1982b) analysed two years of magnetic field data and constructed a zero parameter fit to the predictions of the ergodic theorem (Panchev, 1971; Monin and Yaglom, 1975) for stationary random data. When the effects of solar rotation were included, the solar wind was shown to satisfy the conditions of 'weak' stationarity (cf. Figure 1.2).

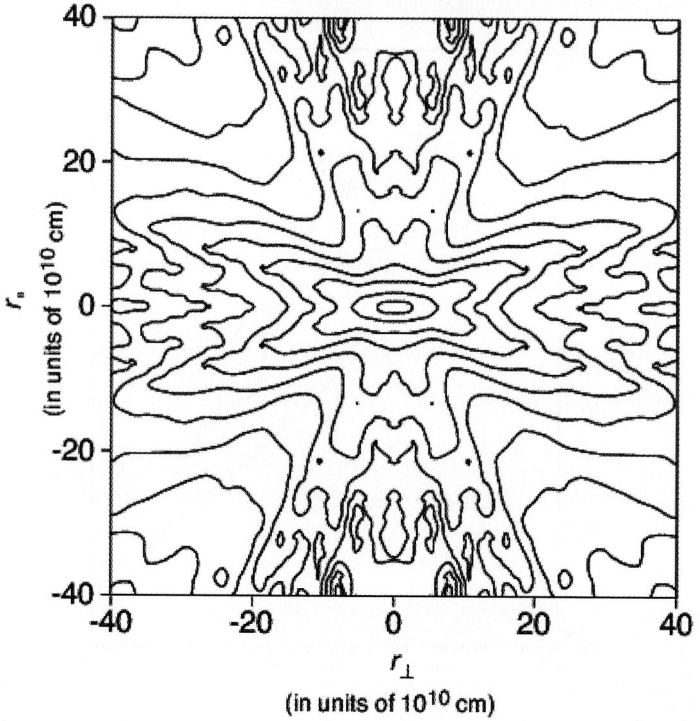

Figure 1.3. Contour plot of the two-dimensional correlation function of solar wind fluctuations as a function of distance parallel and perpendicular to the mean magnetic field. The four quadrant plot was produced by reflecting the data across the axes from the first quadrant. From Matthaeus et al. (1990).

As mentioned above, the power spectra constructed from Voyager data are one-dimensional. Construction of such spectra is possible because the solar wind is super-Alfvénic. Thus, one can invoke the frozen-in-flow hypothesis (Taylor, 1938), which states that for supersonic flows, the fluctuations are essentially frozen into the flow and can be treated as being approximately static. Consequently, the power spectrum of the time series is equivalent to a power spectrum of wave numbers using the relation $k = \omega/V_{sw}$, where V_{sw} is the super-Alfvénic flow speed of the solar wind. Analyses of the time series produce 'reduced' one-dimensional power spectra (see, e.g., Batchelor, 1970; Moffatt, 1978) in the sense that they are equivalent to taking the full three-dimensional spectrum (which one cannot obtain directly from single spacecraft measurements) and integrating it over the two-dimensions perpendicular to the supersonic flow vector.

With only a single spacecraft, one needs additional assumptions to make progress on investigating two- and three-dimensional symmetries of solar wind turbulence. The first comprehensive effort to deduce the two-dimensional structure of the mag-

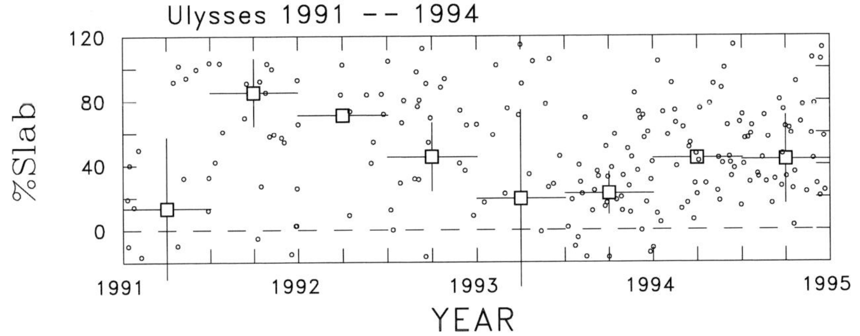

Figure 1.4. The percentage of slab (planar, parallel propagating) Alfvén waves observed in the solar wind. Square symbols represent averages over the time intervals represented by horizontal lines. Vertical lines give the variance of the underlying subset. Adapted from Smith (2002).

netic fluctuations was carried out using the statistical ensemble mentioned above by Matthaeus et al. (1990). They showed that solar wind fluctuations consisted of two components. The first component was comprised of slab (planar, parallel propagating) Alfvén waves; the second component consisted of fluctuations with wave vectors nearly transverse to the mean magnetic field. The two-dimensional correlation function from that analysis is shown in Figure 1.3.

Subsequent single spacecraft analyses of Bieber et al. (1996) and more recently from Ulysses data Smith (2002) found that typically 50% of the magnetic fluctuation energy resides in the quasi-two-dimensional component and 50% in the slab component (see Figure 1.4). There was no obvious way to tell, however, whether the quasi-2D component reflected a nearly-incompressible population of fluctuations or merely one that contained nearly perpendicular **k**-vectors.

1.1.2 Studies using Cluster

The apogee of the Cluster orbit precesses so that the spacecraft are outside the magnetosphere and magnetosheath for at most 12 – 24 hours/orbit during a significant part of the year – from late fall to early spring. To date, the separation of the four spacecraft when in the solar wind has ranged from about 100 – 5000 km. At those relatively small separations, most of the studies using Cluster solar wind data have focused on sharp gradients, structures, and discontinuities, i.e., as mentioned above, the microphysics or kinetic physics, of the wind. The studies to date include:

- *The properties of the heliospheric current sheet, Section 1.2.1.* Both timing and minimum variance analyses have been used to determine the orientation and speed of the current sheet crossing. A change in the sign of the electron heat flux, seen by the PEACE instrument, is consistent with ACE observations of the same HCS crossing. Using the curlometer technique (Dunlop et al., 1988, 2002b) to analyse the crossing, the main component of the current was independently determined to be confined to the plane of the discontinuity. This current sheet crossing also provided an opportunity to look for magnetic holes, which are often associated with crossings of the HCS (see discussion below in Section 1.2.1).

- *The signatures of magnetic clouds and interplanetary coronal mass ejections, Section 1.2.2.* The orientations of discontinuities and shocks associated with interplanetary coronal mass ejections (ICMEs) provide information as to the large-scale morphology of the ICME (see Jones et al., 2002, among others). Small-scale flux ropes are associated with the HCS (Moldwin et al., 1995). In particular, using minimum variance analysis, Cluster data has been used to determine the change in the orientation of a small flux rope as it passed through Earth's bow shock.
- *Properties of discontinuities, Section 1.2.3.* For the first time, with four spacecraft, it has become possible to remove the ambiguities in determining the normals of interplanetary discontinuities and thus better ascertain the distribution between rotational (RDs) and tangential discontinuities (TDs). Such analyses are important in understanding the general morphology and dynamics of the solar wind in that TDs represent the boundaries between separate flow regions or flux tubes while RDs may be related to the dissipation of Alfvénic fluctuations (Tsurutani et al., 2005). Thus far, it has been difficult to find unambiguous examples of RDs (Tsurutani and Ho, 1999). For very well defined directions of minimum variance (i.e., with the ratio of intermediate to minimum eigenvalues exceeding 10), the discontinuities analysed were found to be planar on the scale of the Cluster tetrahedron (600 – 900 km).
- *Isolated magnetic holes, Section 1.3.1.* Isolated magnetic drop outs have been found to convect with the solar wind, but sufficient uncertainty remains so as to allow for propagation at $\sim V_A$. The observed holes have scale sizes consistent with excitation via a mirror mode instability in which cold electrons are important (Fränz et al., 2003; Treumann et al., 2004).
- *Weak double layers, Section 1.3.2.* The Wide Band Data experiment on Cluster has demonstrated sensitivity sufficient to detect both bipolar and tripolar electrostatic structures in the solar wind. While the bipolar structures exhibit small potential jumps, the tripolar structures show virtually no change in potential. The solar wind observations have been used in a statistical study that shows that the amplitudes of the bi- and tripolar structures are related to the strength of the background magnetic field.

1.2 Solar Wind Structures

1.2.1 The heliospheric current sheet

The heliospheric current sheet (HCS) 'represents the magnetic equator of the global heliosphere' (Smith, 2001) and is a fundamental component of the heliosphere. It has been widely studied in its own right, and also in the context of other problems; for example, in studies of cosmic rays dynamics (Jokipii et al., 1977; Smith and Lockwood, 1990). Recent research has concentrated on the heliospheric plasma sheet (HPS), the denser layer of solar wind plasma that envelops the HCS, and how the HPS is related to coronal streamers. Research has also examined the transient replacement of the HCS by magnetic clouds, and folding of the HCS (e.g., Crooker et al., 2004a,b). This has important implications for the structure of the heliosphere at solar maximum.

The majority of HCS observations have been made by single spacecraft. Assuming that HCS structure is convected over the spacecraft faster than any intrinsic variations occur (the Taylor frozen-in-flow hypothesis), such measurements have been used together with minimum variance analysis and a number of simplifying assumptions to estimate the thickness of the HCS and HPS (Winterhalter et al., 1994). In addition, when applying a single spacecraft analysis, one arrives at a final answer only by considering the properties of the plasma over the entire encounter; particularly when integrating the magnetic field time series to characterise the current density.

The multi-spacecraft observations afforded by the Cluster mission, allow a fundamentally different approach to be used. Multipoint discontinuity and current density analysis techniques (Dunlop et al., 2002a; Dunlop et al., 2002b) can be used to calculate boundary motions and currents with fewer assumptions. In this section we review a published Cluster case study of the HCS (Eastwood et al., 2002b), to illustrate how Cluster has contributed to our understanding of the HCS.

1.2.1.1 Cluster observations of the HCS

Figure 1.5 shows the magnetic field observed by the Cluster 1 FGM experiment (Balogh et al., 2001) from 18:00 UT February 12, 2001 to 06:00 UT February 13, 2001. The spacecraft separation was ~600 km. At the start of the interval, $\Phi_B =$ 135° and the field is aligned to an anti-sunward spiral. At 00:50 UT, $\Phi_B = 315°$, indicating a sector boundary and crossing of the HCS. This current sheet was also observed by ACE at the Lagrange 1 point upstream of the Earth, approximately 50 min beforehand (Eastwood et al., 2002b). Observations of the superthermal electron 'strahl' at ACE reveal that the direction of the strahl switches from 0° to 180° when the current sheet crosses ACE (J. Gosling, private communication), as expected for a consistently outward directed heat flux. Figure 1.6 plots the electron pitch angle distributions of the PEACE instrument from Cluster 3 (left) and Cluster 1 (right). Before the HCS crossing at 00:50 UT, the distribution is domi-

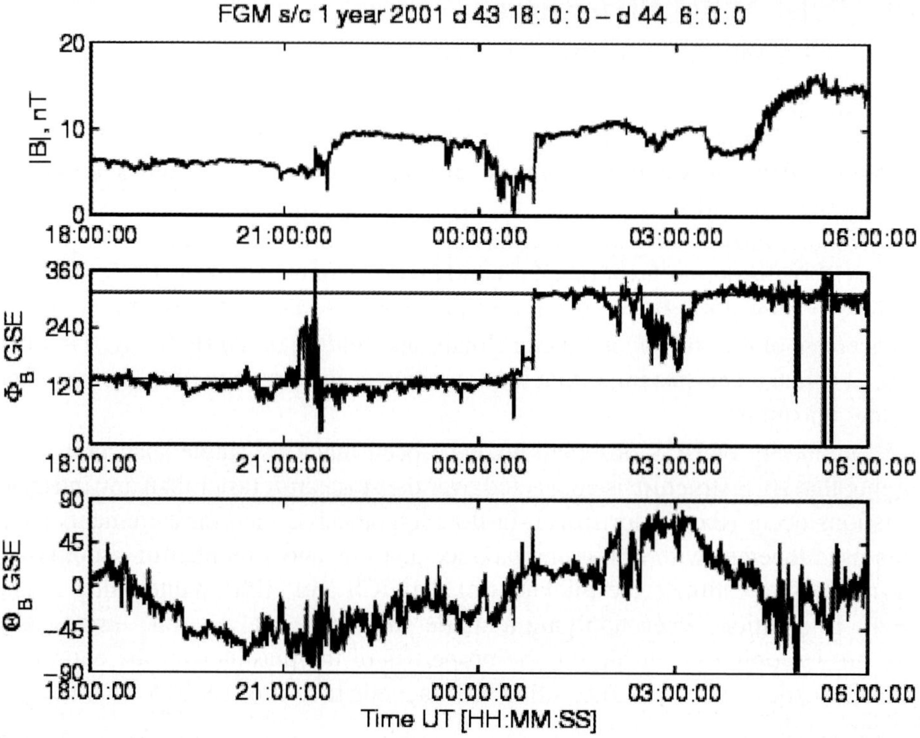

Figure 1.5. Magnetic field observed by Cluster 1 between 18:00 UT on February 12, 2001 and 06:00 UT on February 13, 2001. The top panel shows the magnetic field magnitude, the middle panel Φ_B, the in-ecliptic field angle, measured counter-clockwise from the sunward direction. Horizontal gray lines at $\Phi_B = 135°$ and (315° show where the magnetic field is aligned to either the anti-sunward or sunward Parker spiral direction. The bottom panel shows Θ_B, the elevation of the magnetic field vector out of the ecliptic plane. The data is at spin resolution. On this scale, differences between the magnetic field time series from each Cluster spacecraft are indistinguishable. (Figure provided by J. Eastwood.)

nated by 0° particles, although there is a significant return flux at 180°. After the crossing, the electrons have pitch angles near 180° with virtually no flux at 0°. Thus, the Cluster observations are consistent with the findings from ACE.

Timing analysis. To determine the orientation and normal speed of the HCS, a timing analysis was applied to the Cluster observations, assuming a planar discontinuity (on the scale of the spacecraft separation) that is moving at constant speed over the spacecraft array (Schwartz, 1998). The results of this analysis are shown in Table 1.1. Magnetic field data averaged to 12 vectors s^{-1} was used. The uncertainty in the time at which each spacecraft encountered the HCS is 0.2 s. These results may be compared with the normal orientation obtained from minimum variance

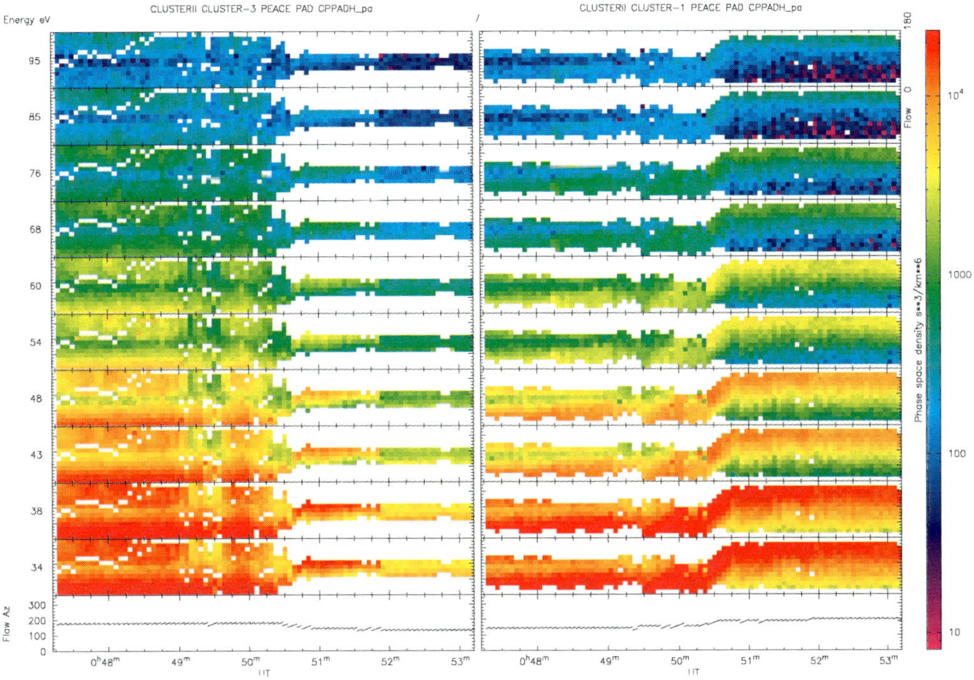

Figure 1.6. Electron pitch angle distributions from the PEACE/HEAA sensors on Cluster 3 (left panel) and Cluster 1 (right panel) on February 13, 2001. Each panel represents an energy slice and within each slice the pitch angles range from 0° to 180° (white indicates pitch angles that were not sampled). The crossing of the HCS occurred at about 00:50. Because the spacecraft were within ≈ 600 km of each other and crossed the HCS at essentially the same time, it is possible to use the four spacecraft together to get a clear picture of how the heat flux changed as the HCS was crossed. During this (early) time in the mission, the pitch angle sampling on each spacecraft was often not optimal. For this crossing of the HCS, the pitch angle sampling was good on Cluster 3 before the crossing, and on Cluster 1 after the crossing. (Figure provided by M. Goldstein.)

analysis. In the time interval 00:45 UT–00:55 UT, a minimum variance direction (-0.36, -0.69, -0.62) was obtained. In the interval 00:00 UT–02:00 UT the minimum variance direction was (0.44, 0.66, 0.62).

The mean field in the minimum variance direction, $\langle \mathbf{B} \cdot \mathbf{n} \rangle$ was 0.1 ± 0.4 nT, identifying the HCS as a tangential discontinuity since no magnetic field threads the HCS. The solar wind velocity (-450, 10, -2) km s^{-1} measured from the CIS/HIA instrument is fairly constant during the HCS boundary crossing. Its projection onto the discontinuity normal yields ~ 190 km s^{-1}. If compared to the normal speed in Table 1.1 one concludes that the HCS is not propagating in the solar wind rest frame, which is consistent with the conclusion that the discontinuity is tangential. The main change in field angle occurs over ~ 22 s, corresponding to a thickness of a few 1000 km.

normal speed (kms^{-1})	170 ± 20
normal orientation	(-0.45, -0.67, -0.59)
uncertainty (°)	17

Table 1.1. Heliospheric current sheet timing analysis. Directions are with respect to the GSE co-ordinate system.

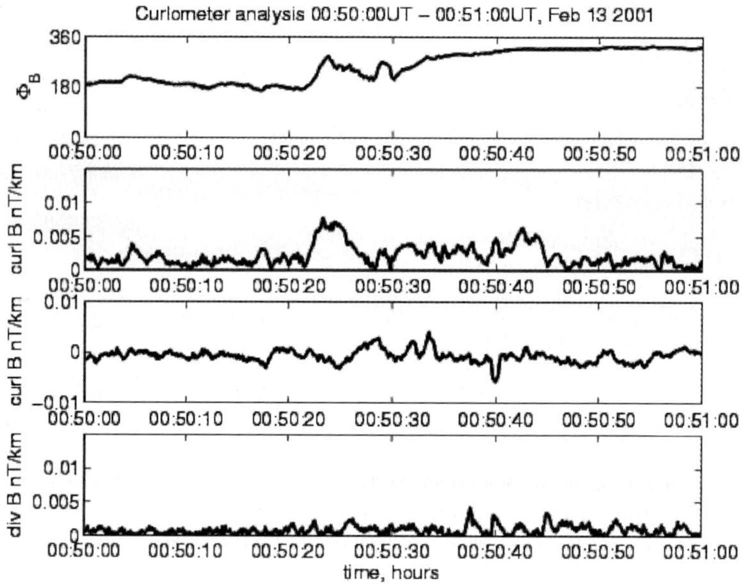

Figure 1.7. Curlometer analysis for the duration of the HCS encounter by Cluster. Top panel: Change in Φ_B indicating the slow rotation in the magnetic field. Initially $\Phi_B = 180°$. Second panel: $\nabla \times \mathbf{B}$ in the plane of the HCS. Third panel: Component of the $\nabla \times \mathbf{B}$ parallel to the HCS normal. Bottom panel: Linear estimate of $\nabla \cdot \mathbf{B}$. The current is confined to the plane of the HCS and is structured. (From Eastwood et al., 2002b).

Current analysis One of Cluster's unique capabilities is that it allows the current density to be estimated directly from $\nabla \times \mathbf{B}$ (Dunlop et al., 1988; Chanteur and Mottez, 1993; Khurana et al., 1996). This technique is reliable only when all of the spacecraft are embedded within the current layer. From the timing analysis the thickness of the HCS is significantly larger than the tetrahedron scale size. Thus the curlometer analysis can be used. It was applied to the interval 00:50 UT–00:51 UT, yielding the results in Figure 1.7.

The main current component is confined to the plane of the HCS discontinuity. Note that the orientation of the HCS is calculated independently of the current vector. Moreover, the current structure in the second panel displays substructure. The current density is bifurcated, peaking at both sides of the layer at $\nabla \times \mathbf{B} = 0.01\,\mathrm{nT\,km^{-1}}$, corresponding to a current density of $10^{-8}\,\mathrm{A\,m^{-2}}$. The peak current may be compared with the $0.5\text{-}1.5\times10^{-8}\,\mathrm{A\,m^{-2}}$ of the single spacecraft analysis.

Magnetic holes For completeness we mention that crossings of the HCS frequently show the presence of 'magnetic holes (see, Burlaga and Lemaire, 1978; Fitzenreiter and Burlaga, 1978; Turner et al., 1977), sometimes referred to as D Sheets. These are dropouts in the magnitude of the main magnetic field. There appear to be several 'species of magnetic holes in the solar wind; relatively short duration holes are thought to be associated with mirror modes (cf. Section 1.3.1). Longer-lasting magnetic holes were described first by Burlaga and Lemaire (1978). Recent theoretical work and simulations (Roberts et al., 2002) suggest that magnetic holes may naturally accompany the HCS and arise from dynamical forces associated with the necessity of maintaining pressure balance between the HCS, HPS and the surrounding fast solar wind.

Cluster should, in principle, be capable of investigating the very structure of such magnetic holes during passages like the one in Figure 1.8 which shows the rotation of the magnetic field on February 13, 2001, near 00:50 UT. The magnetic field direction is indicated by the arrows and the magnitude is indicated both by the (rainbow) colour and by the length of the arrows. A systematic search for magnetic holes using Cluster observations is underway.

Conclusion We have used multi-spacecraft techniques made possible with Cluster to investigate and describe properties of the HCS. In particular, the four spacecraft allow for estimates of the current density using the curlometer technique that employ fewer assumptions than were necessary using data form single spacecraft. From the case study described above, we were able to conclude that the HCS was a tangential discontinuity with a bifurcated current structure that had a peak current density of $10^{-8}\,\mathrm{Am^{-2}}$. This observation of the HCS indicates that it is a slowly moving planar transverse current layer that is several 1000 km thick. Clearly more statistics are required before one can conclude that the HCS is generally a tangential rather than a rotational discontinuity. Which has important implications for stability with respect to magnetic reconnection (Smith, 2001).

1.2.2 CMEs and magnetic clouds

Some of the most interesting phenomena observed in the solar wind are the interplanetary manifestations (ICMEs) of coronal mass ejections (CMEs), the large eruptions (10^{11}-10^{12} kg) of plasma (see, e.g., Webb and Howard, 1994; Low, 2001) from the solar corona. As CMEs propagate into interplanetary space, they can be observed *in situ*. Their radial thickness at Earth is $\sim$0.2 AU. It thus takes a day or two for an ICME to pass over an interplanetary spacecraft (e.g., Gosling et al.,

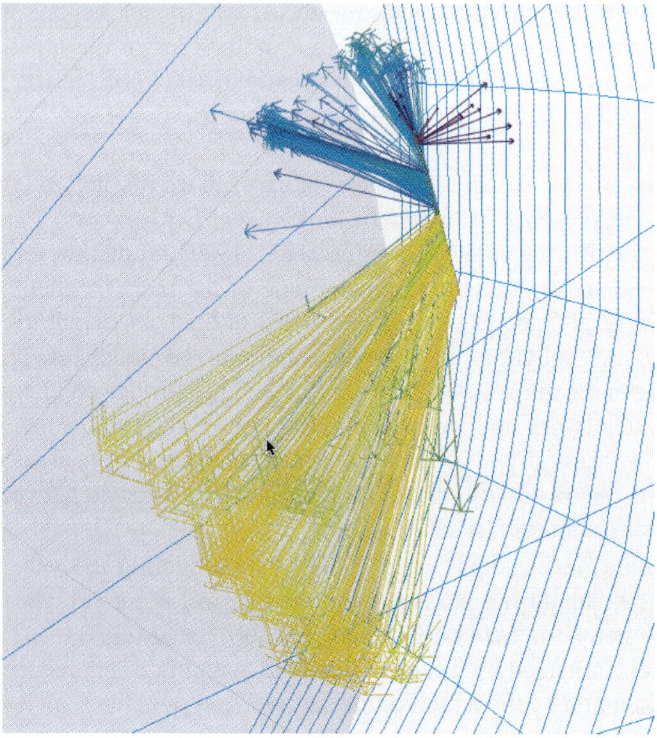

Figure 1.8. A view of the HCS crossing on February 13, 2001 showing the magnetic field (coloured by $|B|$). Note the rapid rotation of the magnetic field followed by further rotations and a decrease in magnitude (dark blue). The picture was made using the visualisation software package 'Visbard'. Figure provided by M. Goldstein and D. A. Roberts.

1990) when it travels into the heliosphere at super-magnetosonic speeds relative to the ambient solar wind, driving an ICME-associated bow shock that is typically observed shortly before onset of the ICME itself. In general, about $1/3$ of all ICMEs have the large-scale rotation and other properties of magnetic clouds. The relationship between ICMEs and magnetic clouds has been investigated by Gopalswamy et al. (1998).

The orbit of Cluster is not in the solar wind long enough to observe an ICME or magnetic cloud in its entirety, and the separation of the Cluster spacecraft has been too small so far to allow for a multipoint analysis of their geometric properties. However, Cluster can be used to study the properties of the shock and sheath regions of an ICME. Furthermore, the properties of small-scale magnetic flux ropes (Moldwin et al., 1995, 2000) with durations of tens of minutes, can be analysed by Cluster.

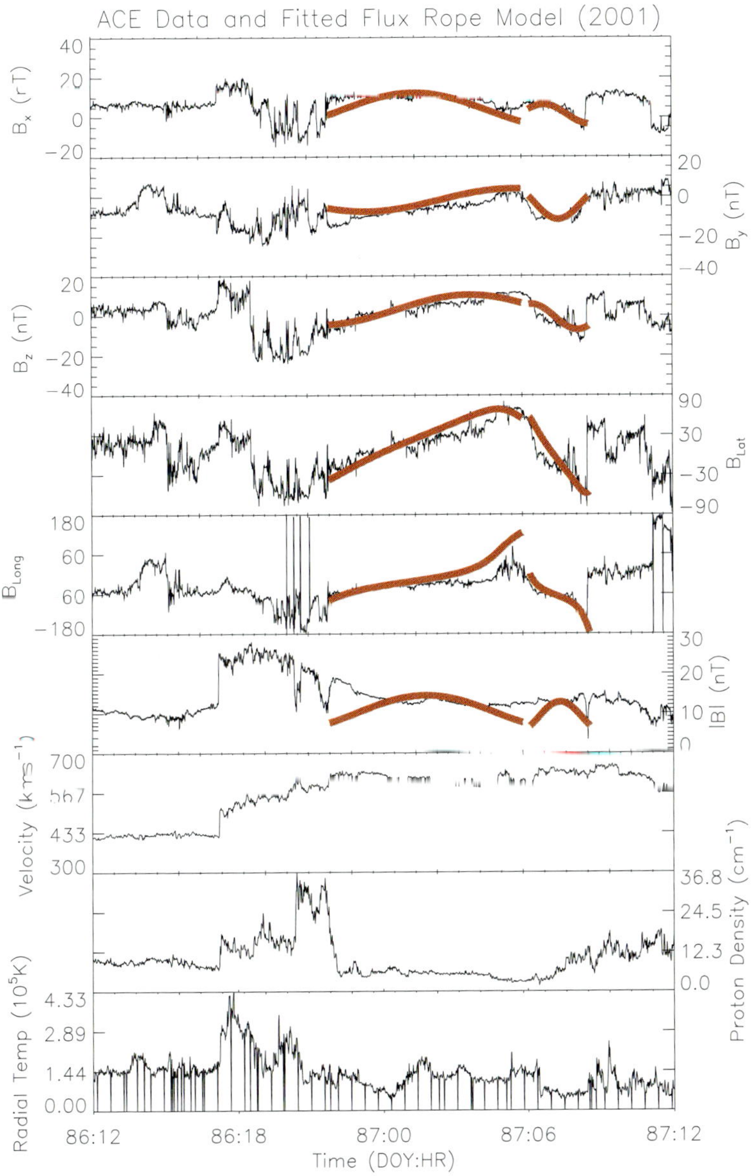

Figure 1.9. Flux rope observed by ACE in the solar wind between 12:00 UT on March 27, 2001 and 12:00 UT, on March 28, 2001 (day 86 on March 27). The top three panels show the Cartesian components of the magnetic field in GSE. The middle three panels show a the same data in polar coordinates. Overplotted in red is the magnetic field profile of a double flux rope model. The bottom three panels show plasma velocity, density and radial temperature. (From Rees, 2002).

1.2.2.1 Planar magnetic structures in ICME sheath regions

ICMEs (and magnetic clouds) are common large-scale interplanetary structures. During solar minimum they are often isolated, but during solar maximum they frequently overlap. To understand the macroscopic structure, evolution, and dynamics of the solar corona and heliosphere, it is important to understand how ICMEs are structured and orientated. ICMEs are also important triggers of geomagnetic activity. The mechanisms for their interaction with the magnetosphere depend critically on their large- and small-scale magnetic structure. For example, the prolonged intervals of southward IMF at Earth that cause large geomagnetic disturbances (Richardson et al., 2001) are often found to be related with the passage of ICMEs. Approximately 1/3 of ICMEs are associated with a magnetic flux rope structures that can be modeled as a force-free helical magnetic field configuration (e.g., Burlaga, 1988)–these are commonly referred to as magnetic clouds. This magnetic flux rope model provides the orientation axis of magnetic clouds, but for those ICMEs that are not magnetic clouds, alternative methods must be used. One approach is to use the fact that ICMEs move supersonically relative to the ambient solar wind. In the ICME frame of reference, the solar wind flows across the ICME bow shock, where it is slowed, deflected, heated and compressed. The sheath plasma is then draped over the ICME, which is thought to lead to the formation of planar magnetic structures consisting of ordered sheets of magnetic field. These may be visualised as a stack of plasma sheets, separated by tangential discontinuities (Jones et al., 2002). The normal of each individual planar structure is aligned with the normal of the ICME's leading edge. The orientation of the leading edge, which itself is perpendicular to the axis of the ICME, can be inferred by analysing the orientation of the planar structures.

This alternative analysis method can be tested by looking at ICMEs that contain a flux rope. Such a magnetic cloud was observed by ACE on March 27 – 28, 2001. Figure 1.9 shows the properties of the solar wind plasma and magnetic field as measure by ACE [Rees, Ph.D. Thesis, University College London, 2002]. Initially, ACE observed the relatively unperturbed solar wind. The ICME-driven shock was observed after 17:00 UT on day 86, indicated by the simultaneous jump in magnetic field strength, density and velocity. The ICME itself, defined by the relatively low density and smooth magnetic field rotation, starts after 22:30 UT. The interval between these two times, characterised by particularly high plasma densities, corresponds to the upstream sheath region of the ICME. In Figure 1.9, the magnetic field profile of a flux rope model is plotted in red, assuming cylindrical symmetry. Although it is a single structure, it is best modeled as a double flux rope. The axis of the trailing rotation was found to be (0.32, 0.93, 0.21) in GSE coordinates and the axis of the leading rotation to be (0.93, 0.08, 0.37). The axis of the trailing rotation is consistent with properties of typical flux ropes, however, the axis of the leading rotation is primarily aligned to the solar wind flow, which is unusual (A. Rees, private communication).

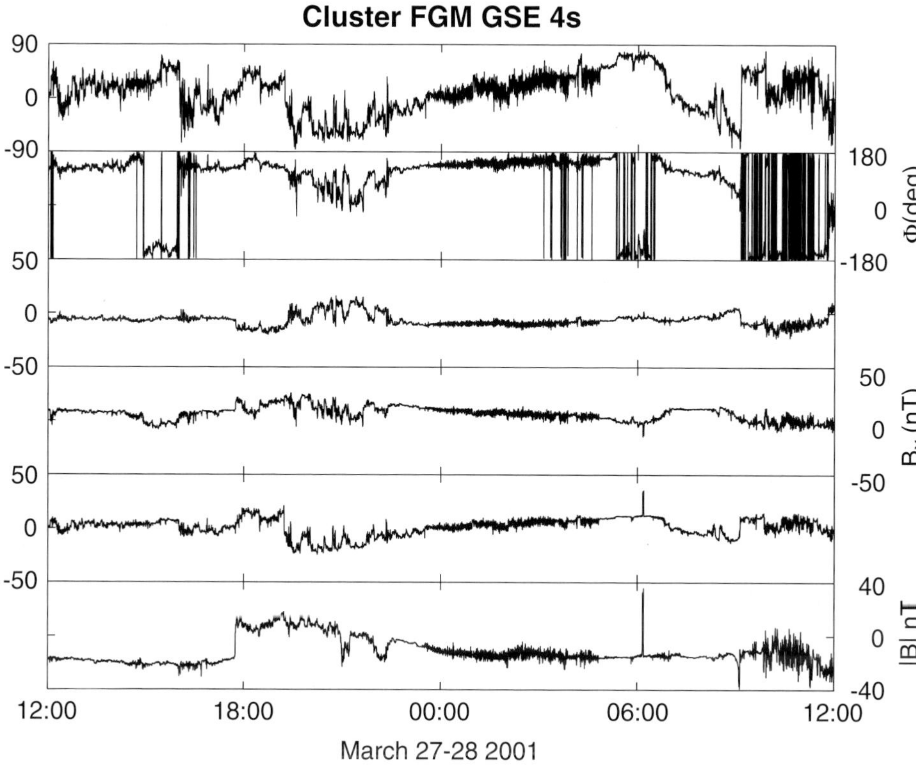

Figure 1.10. Magnetic field observed by Cluster 1 in the solar wind for the same time interval as Figure 1.9. The spike after 06:00 UT is a shock encounter. Enhanced fluctuations are foreshock-related. (Figure provided by T. Horbury).

Cluster encountered this flux rope as well. The Cluster data are shown in Figure 1.10. There was a brief bow shock encounter just after 06:00 UT on March 28, causing the spike in *B*. Significant foreshock wave activity was observed from 00:00-05:00 UT on March 28, and, again, after 09:00 UT. Soon after the end of this interval the Cluster spacecraft moved into the magnetosheath. Ten discontinuities were identified between the shock and the edge of the first flux rope. The orientations of these discontinuities and the CME driven shock were determined at 17:45 UT on March 27 using the Cluster timing analysis technique. The discontinuity normals lay within 25° of the shock normal. There was a spread of 17° in the orientation of the 10 normals (Rees, 2002). The angle between the axis of the leading rope and the average PMS/shock normal was $\sim$120°, greater than the 90° difference in the ideal case. The angle between the trailing flux rope axis and the average PMS/shock normal was $\sim$83°, suggesting that the flux rope did not have

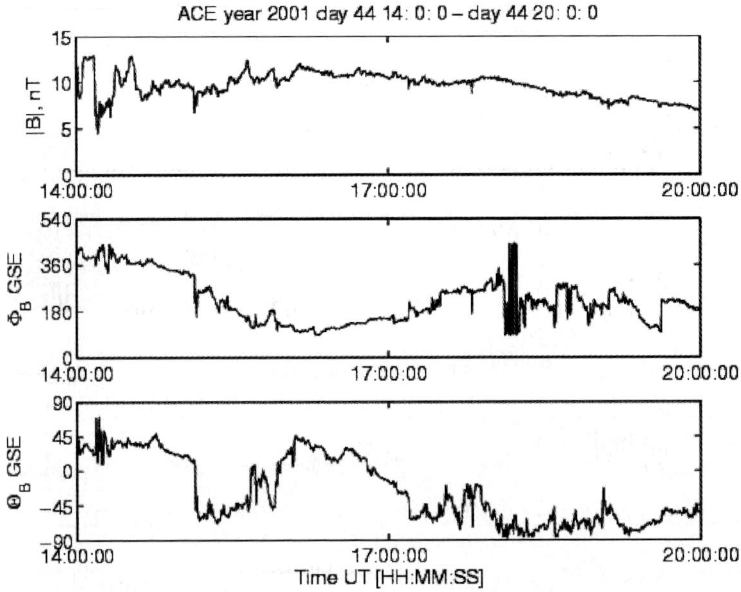

Figure 1.11. Small-scale magnetic flux rope observed by ACE on February 13 (day 44), 2001. The magnetic field is shown in polar (GSE) coordinates. Note that Φ_B is wrapped over above 360° at the start of the interval. (Adapted from Eastwood et al., 2002b).

a circular cross section. While the exact interpretation relevant to this example has not been resolved fully, it illustrates how Cluster can be targeted to certain problems and can be used to augment single spacecraft analyses.

1.2.2.2 Interaction of small-scale magnetic flux rope with the bow shock

A second class of flux rope structure with much smaller transit times (a few hours) has also been identified, after occurring in the vicinity of HCS crossings. These ropes have been termed 'small scale magnetic flux ropes (Moldwin et al., 1995, 2000). Moldwin et al. (1995) explored the stability of the HCS with respect to reconnection across the HCS generating small scale magnetic flux ropes. However, structure in the vicinity of the HCS may also be due to transients (particularly ICMEs) distorting, or locally replacing, the HCS (cf., e.g., Crooker et al., 1998).

Eastwood et al. (2002b) used Cluster and ACE observations to examine how the properties of a small scale magnetic flux rope change when interacting with the terrestrial bow shock. The observations were made on February 13, 2001, shortly after the HCS encounter discussed previously. A few hours after the HCS, both Cluster (by that time located in the magnetosheath) and ACE observed a 'small scale magnetic field rotation, lasting for a few hours.

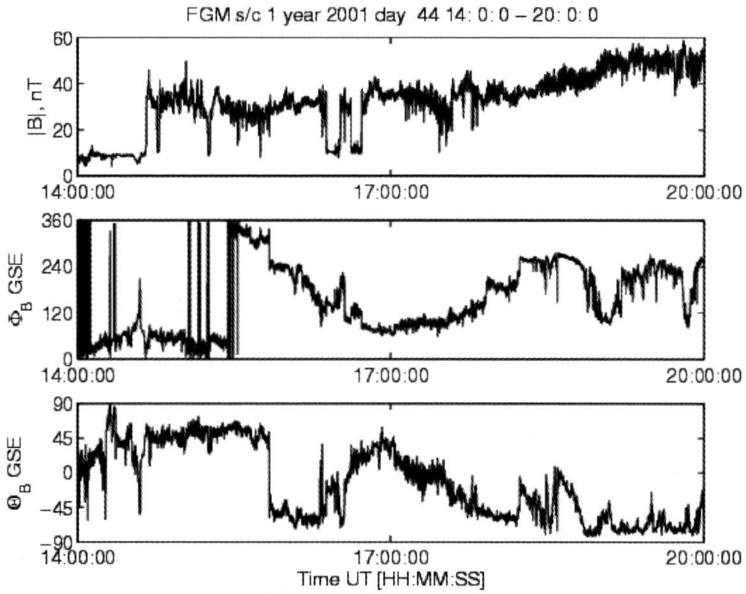

Figure 1.12. Small-scale magnetic flux rope observed by Cluster 1 on February 13 (day 44), 2001. On this scale, differences between the Cluster spacecraft are indistinguishable. For the majority of the interval, the Cluster spacecraft were located in the magnetosheath, although they were initially in the solar wind. The two apparent drops outs in the magnetic field strength correspond to brief transitions back into the solar wind caused by movement of the shock. (Adapted from Eastwood et al., 2002b).

Figure 1.11 shows the magnetic field observed by ACE between 14:00 UT and 20:00 UT on February 13, 2001. The flux rope begins shortly after 15:00 UT. The middle panel, showing Φ_B, is wrapped round at the start of the interval above 360°, showing that the rotation in Φ_B is continuous and smooth through the first part of the interval. A discontinuous change occurs in Θ_B at approximately 15:00 UT.

Figure 1.12 shows the magnetic field observed by Cluster 1. Only Cluster 1 is shown because differences between the Cluster spacecraft are indistinguishable. Φ_B is constrained to lie between $0 - 360°$ (the data are not wrapped over 360° as in Figure 1.11). At 14:30 UT Cluster 1 crossed the bow shock into the magnetosheath staying there for the majority of the interval. The two drop-outs in magnetic field strength brief before 17:00 UT correspond to transitions back into the solar wind, caused by fluctuations in the standoff distance of the bow shock. The rotation in the field persists. By advecting the ACE data using the solar wind velocity measured at ACE, it is clear that the structure was convected directly to Cluster. Although the solar wind plasma is compressed in the magnetosheath, the flow velocity is accordingly reduced so that mass flux through the shock is conserved. Consequently, the rotation of the magnetic field vector appears to proceed at the same rate.

The orientation of the flux rope is altered by its passage through the shock; as can be quantified by a MVA. The relationship of the MVA coordinates to a model magnetic flux rope has been examined by Lepping et al. (1990). If the structure is force-free and cylindrically symmetric, the intermediate variance direction is aligned with the rope axis. The closer the spacecraft trajectory approaches the axis, the better this alignment is likely to be.

MVA was applied to the the interval 15:30 UT–19:30 UT in the Cluster data, and to the same interval in the advected ACE data. The left hand panel of Figure 1.13 shows the orientation of the Cluster MVA axes projected onto the ecliptic plane at Earth. A cut through a model bow shock in the ecliptic plane is also shown. The right hand panel gives the orientation of the ACE MVA axes projected into the ecliptic plane. The intermediate variance direction at ACE is aligned to the Parker spiral, whereas the intermediate direction at Cluster is more aligned with Y_{GSE}. In transit from ACE to Cluster, the axis of the flux rope has been rotated away from the Sun-Earth line. This is most likely interpreted in terms of processes at the bow shock. The shock amplified transverse magnetic field causes the magnetic field vector to rotate away from the shock normal, curving the field around the magnetosphere in a plane similar to that occupied by the shock. This effect, together with field line draping deeper in the magnetosheath would cause a field line parallel to the Parker spiral to become more aligned with Y_{GSE} behind the nose of the bow shock. At the center of the simple flux rope model, the magnetic field is aligned along the axis. Consequently, the axis of the rope is rotated away from the Sun-Earth line, as is seen in the observations.

1.2.3 Solar wind discontinuities

Interplanetary magnetic field discontinuities are an intrinsic property of the solar wind (e.g., Ness et al., 1966). In the MHD approximation, discontinuities can either be rotational (RD) or tangential (TD). TDs separate unrelated regions of plasma; there is no flow of plasma across the discontinuity plane. A TD does not propagate in the plasma rest frame, and no magnetic field threads the discontinuity (see Burlaga, 1995, for general background). The magnetic field direction, magnitude, plasma density and even composition can change across a TD, but since the plasma characteristics on either side are unrelated, changes need not necessarily occur. In contrast, however, RDs represent a rotation in the magnetic field, possess a component of magnetic field across the discontinuity plane, and plasma transport can take place across the boundary. RDs propagate in the plasma frame.

1.2.3.1 Previous discontinuity observations

The relative proportion of RDs and TDs in the solar wind contains information on the source of the solar wind and the contribution of dynamical effects to solar wind properties. Methods to reliably distinguish RDs and TDs have therefore been of interest. Smith (1973) classified discontinuities by the fraction of magnetic field

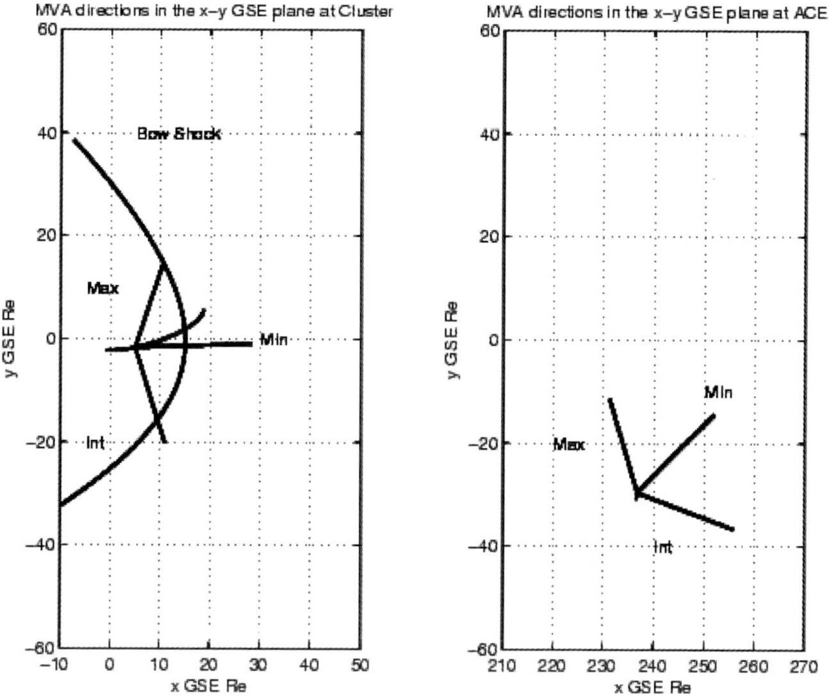

Figure 1.13. Comparison of the MVA axes derived from Cluster and ACE observations of the same small-scale magnetic flux rope. The axis of the flux rope is best identified as the intermediate variance direction. It can be seen that in transit from ACE to Cluster, the intermediate variance direction, and therefore the presumed axis of the flux rope, has rotated away from the Sun-Earth line. (From Eastwood et al., 2002b).

threading the discontinuity ($\mathbf{B} \cdot \mathbf{n}/B$) together with the size of magnetic field rotation across the discontinuity ($\Delta|\mathbf{B}|/B$). He defined four classes of discontinuity: rotational (RDs), tangential (TDs), 'either' (EDs) and 'neither' (NDs).

Essential to all methods of classification is a reliable estimate of the normal to the discontinuity. Single spacecraft studies typically use minimum variance analysis (MVA) (Sonnerup and Cahill, 1967) to estimate normals (e.g., Smith, 1973; Lepping and Behannon, 1980; Neugebauer et al., 1984). MVA can be influenced by the presence of waves on the discontinuity surface or in the discontinuity environment. Burlaga and Ness (1969) analysed a small sample of six discontinuities seen by Explorer 33, 34 and 35 separated by 10 – 135 R_E, and concluded that each was consistent with being a TD. Using Geotail, Wind and IMP-8 observations for determining the normal, the largest class is not TD but ED (Horbury et al., 2001).

Recalculation of the normals using MVA showed, in contrast, a predominance of RDs. In determining the discontinuity normal with less than four spacecraft it is necessary to assume that the discontinuity is convected in the solar wind flow and

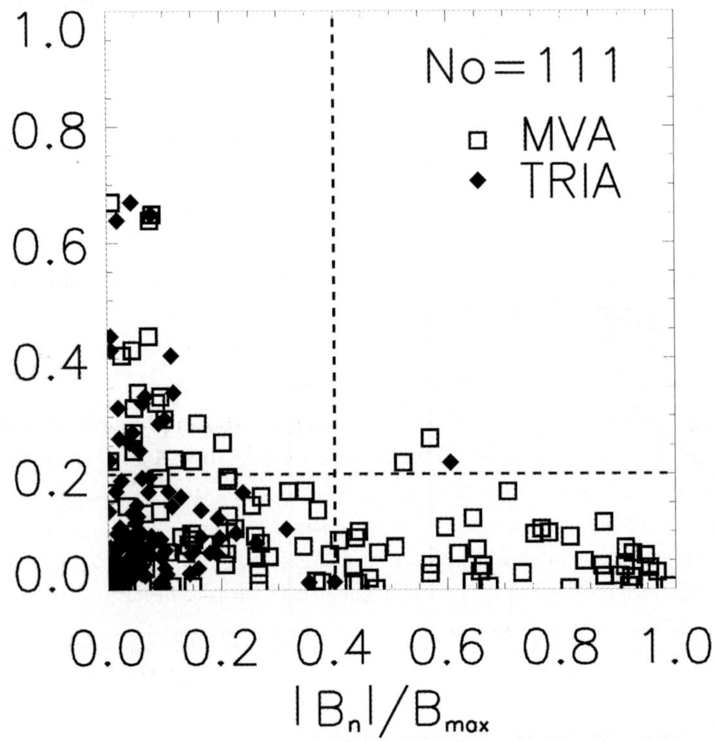

Figure 1.14. Scatterplot based on normal estimates from four spacecraft timings (squares) and Cluster 4 magnetic field MVA (diamonds). (From Knetter et al., 2004).

that the discontinuity velocity could be approximated by the solar wind velocity measured at one of the spacecraft. Four point observations are required to remove this ambiguity.

1.2.3.2 Cluster observations

Knetter et al. (2004) used a four spacecraft triangulation method (Schwartz, 1998), together with the Cluster data set to make a comprehensive survey of the properties of a large set of interplanetary discontinuities. They compared the results to analysis of the same data set using MVA. The results are summarised in Figure 1.14. The data set consisted of 129 interplanetary discontinuities observed in the undisturbed solar wind, without visible foreshock activity, between February 1 and May 13, 2001 when the Cluster tetrahedron scale was approximately 600-900 km. Even when the ambiguity of the discontinuity speed was removed, results consistent with Horbury et al. (2001) were recovered. The number of unambiguous RDs identified using triangulation normals was a few %, while the largest class (83%)

was EDs, i.e., RDs with a small magnetic field rotation, or TDs with a small change in magnetic field magnitude across them.

Using MVA defined normals, the number of unambiguous RDs was found to depend on how well the minimum variance direction was defined, as measured by the ratio between the intermediate and minimum eigenvalues, λ_2/λ_3. Knetter et al. (2004) concluded that this effect arose from the influence of wave fields in the plasma containing the discontinuity. The threshold for λ_2/λ_3 was typically set at 2, but the comparison of MVA and triangulation results was much improved when only MVA normals with $\lambda_2/\lambda_3 > 10$ were used. Analysis of the subset of MVA normals with $\lambda_2/\lambda_3 > 10$ showed that they typically were co-linear between the four spacecraft, suggesting that the interplanetary discontinuities were planar on the particular scale of the spacecraft tetrahedron.

The conclusions of this work not only demonstrate the planarity of interplanetary discontinuities on 600-900 km scales and that the largest group of interplanetary discontinuities using the Smith (1973) criteria is the ED group, but that MVA applied to data from a single spacecraft can be used reliably to estimate the normal direction of a discontinuity only if the eigenvalue ratio λ_2/λ_3 exceeds 10: a result consistent with findings from observations of preshock waves (e.g., Eastwood et al., 2002), and the analysis of magnetopause orientation (e.g., Sonnerup and Scheible, 1998).

1.3 Solar Wind Microstructures

1.3.1 Isolated magnetic holes

Isolated magnetic holes, which are not associated with current sheet crossings, occur in the solar wind (e.g., Winterhalter et al., 1994) at a rate such that they might have a significant effect on energetic particle diffusion. Whether these structures are solitary MHD waves (e.g., Baumjohann et al., 1999), or whether they arise from a kinetic instability such as the mirror mode instability is an open question. As described elsewhere in this volume, the mirror mode instability is a non-propagating mode, generated under conditions of high β, and anisotropic proton temperature where $T_{\parallel} > T_{\perp}$ (e.g., Gary, 1993). For the first time, the four Cluster spacecraft observations allow the motion of such magnetic dropouts to be measured. Mirror mode holes should convect with the solar wind plasma, while solitary MHD waves would be expected to have a finite propagation speed in the solar wind frame. Knowledge of the solar wind rest frame velocity, together with observations of the associated ion and electron distributions is essential to establishing the source of these structures.

Observations of two sets of isolated magnetic dropouts in the solar wind were presented by Fränz et al. (2003). One of the intervals analysed is shown in Figure 1.15. The estimated velocities of these holes were consistent with convection by the solar wind, but timing uncertainties arising from the small tetrahedron scale

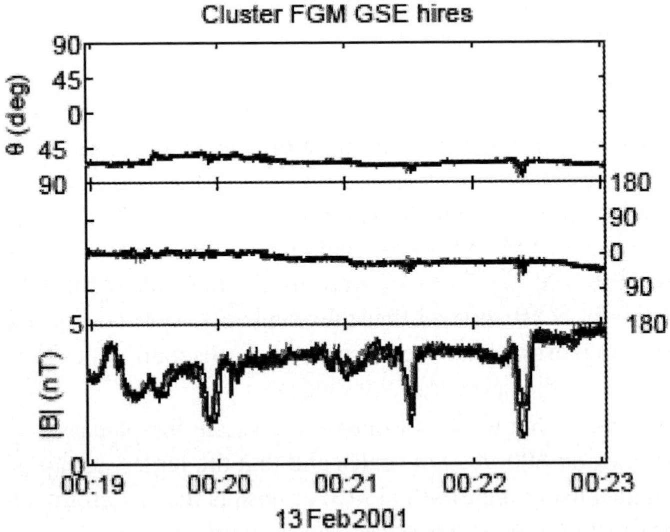

Figure 1.15. The first set of solitary linear waves observed by Cluster upstream of the Earth's bow shock at 19.5 R_E on 12.6 MLT on 13 Feb 2001. Shown are magnetic field GSE polar and azimuth angles and total magnitude at 22 vectors s^{-1} for all four spacecraft (grey tones). The extension of the structures in time is about 10 s, corresponding to a size of 5000 km at a solar wind speed of $V_p = 450\,\mathrm{km\,s^{-1}}$. (From Fränz et al., 2003).

size of ~600 km made it difficult to rule out that some structures were propagating at MHD wave speeds.

Detailed examination of the ion and electron distributions requires that the holes have a minimum duration of about 30 s implying that they are few R_E in lateral width. Such holes must thus be large structures on the solar wind scale. Linear theory of mirror modes suggests that the growth rate maximises at perpendicular wavelengths $\lambda_\perp \sim 2\pi r_{ci} \approx 1\,R_E$ (Pokhotelov et al., 2004). The preference is therefore to observe magnetic holes under slow solar wind conditions. Fränz et al. (2003) presented one event for which particle distributions were available. The transverse scale of this event (~ 5000 km) is in excellent agreement with the above condition for the maximum growth rate. In this case the strongest signature was found in the electron data, which showed a drop in electron temperature and fluctuations in electron density. The ion data showed neither the strong proton anisotropy expected for a mirror mode, nor the density increase or velocity perturbations expected for an MHD soliton, suggesting that these structures are seen in their nonlinear state and that the electrons might play a crucial role (Fränz et al., 2003) in generation of the holes. Interestingly, this is in agreement with recent theoretical considerations and full particle numerical simulations of mirror modes (Treumann et al., 2004) showing the presence of cold electrons in the nonlinear state of mirror mode evolution.

1.3.2 Weak double layers

Mangeney et al. (1999) discovered the presence of Isolated Electrostatic Structures (IES) in the free solar wind near the Lagrange point L1 using waveform data obtained by the Time Domain Sampler (TDS) which is part of the WAVES experiment on the Wind spacecraft. These IES appear as isolated spikes in the electric field waveforms and are similar in shape and duration to the electric pulses parallel to the magnetic field observed in the auroral acceleration region identified as double layers (Temerin et al., 1982; Bostrom et al., 1988). IES also resemble electrostatic solitary waves (ESW) observed in various regions of the magnetosphere, (cf., e.g., Matsumoto et al., 1994; Franz et al., 1998), although the ESW are typically reported as being bipolar pulses (two successive peaks of roughly equal amplitude but of opposite sign), whereas the IES are often referred to as tripolar pulses, consisting of three successive peaks, the first and last being of about equal magnitude and sign; the middle peak has both larger magnitude and opposite sign.

Mangeney et al. (1999) analysed 75 IES finding that the tripolar IES had finite potential jumps with typical values of $q\Delta\Phi/(k_B T_e) \simeq 3 \times 10^{-4}$. In contrast, bipolar IES have no potential jump. Due to the small but finite potential jumps found in the tripolar IES, it was concluded that they were weak double layers, consisting of two main layers of approximately opposite charge; the positively charged one usually preceding the negatively charged one. The potential usually drops in the anti-Sunward direction, varying in the same sense as the interplanetary electric potential which tends to decelerate outward propagating electrons and accelerate outward propagating protons.

Similar tripolar pulses, or weak double layers, have been routinely detected by the Cluster Wideband (WBD) plasma wave receiver when Cluster is in the solar wind. The characteristic electric field amplitudes and time durations of these pulses are shown from a Cluster survey in Figure 1.16a (Pickett et al., 2004a). Figure 1.16a illustrates that the amplitudes of the tripolar pulses observed in the solar wind are some of the smallest observed in near Earth space (the largest are seen in the the auroral zones at $4.5 - 6.5\,\mathrm{R_E}$). The difference is due largely to the difference in the magnetic field strength in the two regions. The observations show a broad range of amplitudes at any magnetic field strength. However, there is a general trend for the amplitude of the tripolar pulses to increase as the strength of the magnetic field increases. (This topic is discussed more fully in the Section: 'Electrostatic waves:Broadband structures' in Lucek et al. (2005)). The time durations of the tripolar pulses observed in the solar wind, Figure 1.16b, were found to be of the same order of magnitude as those in all other regions (few milliseconds) with the exception of the magnetosheath, where there is a tendency for the tripolar pulses to be much shorter.

Tripolar pulses have also been observed within SLAMS (Short Large-Amplitude Magnetic Structures), using the waveform data obtained by the EFW experiment on Cluster (R. Behlke, private communication). Figure 1.17 shows an example of one

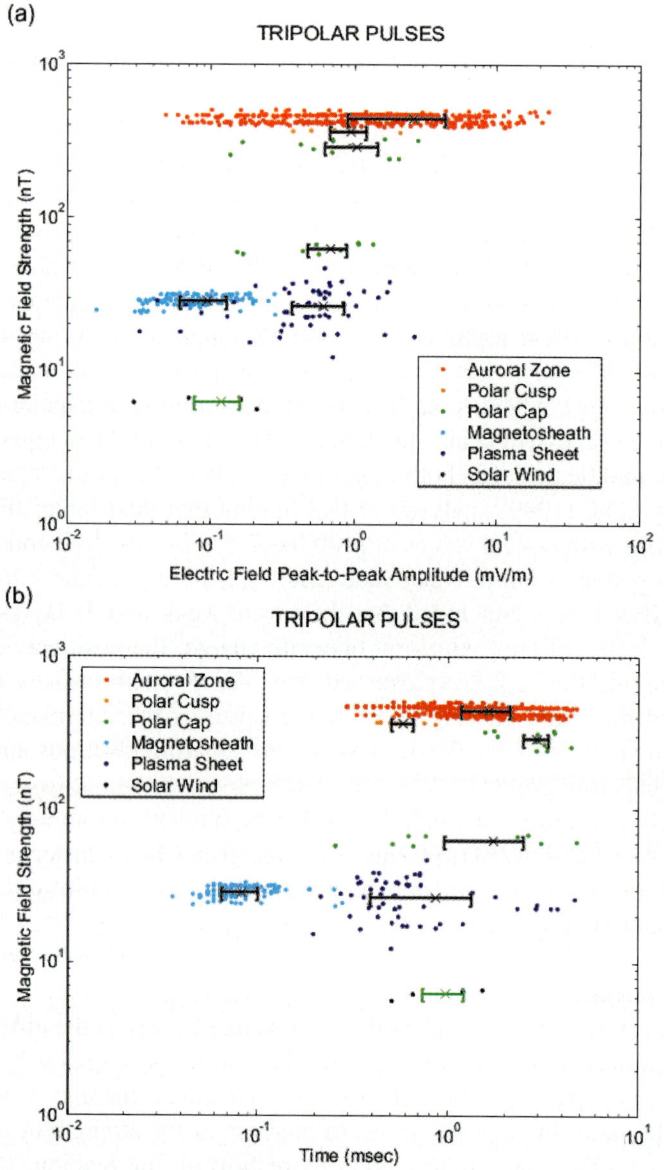

Figure 1.16. Survey of the tripolar pulses observed by Cluster WBD over a two-year period. (a) Electric field amplitude vs. magnetic field strength showing a trend of increasing electric field amplitude with increasing magnetic field strength, just as for the bipolar pulses. (b) Pulse duration vs. magnetic field strength showing no trend between the two, but pointing out the obvious difference of the time durations of the magnetosheath pulses to pulses in all other regions, just as for the bipolar pulses. (From Pickett et al., 2004a).

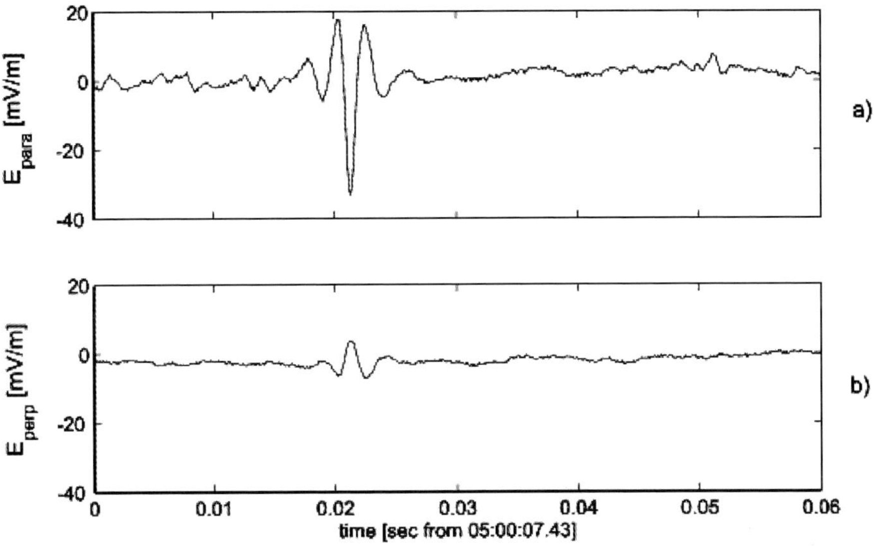

Figure 1.17. Example of a tripolar solitary structure observed in a Short Large-Amplitude Magnetic Structure on February 3, 2002 on Cluster 2. The parallel and perpendicular (to **B**) components are shown in panels (a) and (b), respectively. (Figure provided by R. Behlke).

of these tripolar pulses obtained within a SLAMS structure on February 3, 2002. The amplitudes of the electric field parallel and perpendicular to the magnetic field are shown in panels 1 and 2, respectively. The time duration of this tripolar pulse is ~ 6 ms with a peak-to-peak amplitude in the parallel electric field of $\sim 55\,\mathrm{mV\,m^{-1}}$. The time duration of this pulse is consistent with those found by Pickett et al. (2004) in the solar wind outside SLAMS, however, the amplitude is somewhat larger within the SLAMS structure. Behlke et al. (2004) also observed bipolar type structures within SLAMS, with velocities on the order of $400 - 1200\,\mathrm{km\,s^{-1}}$ and peak-to-peak amplitudes in the parallel electric field of up to $\sim 65\,\mathrm{mV\,m^{-1}}$, suggesting they may be ion depletion structures, since they have negative potentials moving at velocities above the typical ion thermal speed.

Currently none of the theories commonly used to describe these types of structures are well suited to the observations. Future work will thus concentrate on trying to understand better their nature by carrying out statistical studies both within and outside of SLAMS in the solar wind and comparing these results to theories of of solitary waves and phase space holes.

1.4 Outlook

As the focus of Cluster science turns more to the solar wind when the separation distances have become larger, contributions will expand in several critical scientific areas, including:

- Further characterising the properties of the HCS and the large-scale magnetic holes sometimes seen near the HCS.
- Determine the orientation of ICMEs and small-scale flux ropes (and the effect of the Earth's bow shock) on such structures.
- Determine the normals of interplanetary discontinuities using the four spacecraft and better understand the relative distribution of RDs and TDs.
- Determine the origin of isolated magnetic holes.
- Detect low amplitude short duration electrostatic wave forms which have either a bipolar or tripolar structure.
- Further define the physics of solar wind turbulence by characterising the dynamical dissipation of sharply crested fluctuations and by defining the physical properties of fluctuations in the dissipation range of the Kolmogorov spectrum.

One of the goals of the Cluster mission is to use the four spacecraft to determine various properties of the turbulence that are difficult, if not impossible, to study with a single spacecraft. Even at the projected maximum separation of 10^4 km, however, the transverse separation will be much less than the average correlation length which is $\sim 5 \times 10^6$ km (Matthaeus et al., 1986). Nonetheless, at those separations, properties of the dissipation range can be investigated. However, by taking into account the orientation of the interplanetary magnetic field, as was done with the ISEE 3 data, one should be able to use the tetrahedral formation to fill in much more of the correlation function with better statistics than was done previously (see, e.g., Horbury, 1999). Combining Cluster with ACE and Wind, will in addition enable even more of the correlation function to be estimated.

Acknowledgements

MLG and JP acknowledge support from NASA. JPE holds a National Research Council Resident Research Associateship Award at NASA GSFC. Cluster research in the UK and France is supported by PPARC and CNES, respectively.

References

Axford, W. I. and J. F. McKenzie: 1996, 'Implications of observations of the solar wind and corona for solar wind models'. *Astrophysics and Space Science* **243**(1), 1–3.

Balogh, A., C. M. Carr, M. H. Acuña, M. W. Dunlop, T. J. Beek, P. Brown, K.-H. Fornaçon, E. Georgescu, K.-H. Glassmeier, J. Harris, G. Musmann, T. Oddy, and K. Schwingenschuh: 2001, 'The Cluster Magnetic Field Investigation: overview of in-flight performance and initial results'. *Ann. Geophys.* **19**, 1207–1217.

Batchelor, G. K.: 1970, *Theory of Homogeneous Turbulence*. New York: Cambridge University Press.

Baumjohann, W., R. A. Treumann, E. Georgescu, G. Haerendel, K.-H. Fornaçon, and U. Auster: 1999, 'Waveform and packet structure of lion roars'. *Ann. Geophys.* **17**, 1528–1534.

Behlke, R., M. Andre, S. D. Bale, J. S. Pickett, C. A. Cattell, E. Lucek, and A. Balogh: 2004, 'Solitary structures associated with Short Large-Amplitude Magnetic Structures (SLAMS) upstream of the Earth's quasi-parallel bow shock'. *Geophys. Res. Lett.* **31**, L16805, doi:10.1029/2004GL019524.

Belcher, J. W. and L. Davis: 1971, 'Large-amplitude Alfvén waves in the interplanetary medium, 2'. *J. Geophys. Res.* **76**, 3534.

Bieber, J. W., W. Wanner, and W. H. Matthaeus: 1996, 'Dominant two-dimensional solar wind turbulence with implications for cosmic ray transport'. *J. Geophys. Res.* **101**(A2), 2511–2522.

Biermann, L.: 1951, 'Kometenschweife und solare Korpuskularstrahlung'. *Zeitschrift für Astrophysik* **29**, 274.

Bostrom, R., G. Gustafsson, B. Holback, G. Holmgren, H. Koskinen, and P. Kintner: 1988, 'Characteristics of solitary waves and weak double layers in the magnetospheric plasma'. *Phys. Res. Lett.* **61**, 82.

Burlaga, L. F.: 1988, 'Magnetic clouds and force-free fields with constant alpha'. *J. Geophys. Res.* **93**, 7212–7224.

Burlaga, L. F.: 1995, *Interplanetary Magnetohydrodynamics*. New York: Oxford University Press.

Burlaga, L. F. and J. F. Lemaire: 1978, 'Inter-planetary magnetic holes - Theory'. *J. Geophys. Res.* **83**(NA11), 5157–5160.

Burlaga, L. F. and N. F. Ness: 1969, 'Tangential discontinuities in the solar wind'. *Sol. Phys.* **9**, 467–477.

Chanteur, G. and F. Mottez: 1993, 'Geometrical Tools for Cluster Data Analysis'. In: *Spatio-temporal analysis for resolving plasma turbulence (START), ESA WWP-047*. European Space Agency, pp. 341–344.

Coleman, P. J.: 1966, 'Hydromagnetic waves in the interplanetary medium'. *Phys. Rev. Lett.* **17**, 207.

Crooker, N. U., J. T. Gosling, and S. W. Kahler: 1998, 'Magnetic clouds at sector boundaries'. *J. Geophys. Res.* **103**(1), 301–306.

Crooker, N. U., C. L. Huang, S. M. Lamassa, D. E. Larson, S. W. Kahler, and H. E. Spence: 2004a, 'Heliospheric plasma sheets'. *J. Geophys. Res.* **109**(A3), A03107, doi:10.1029/2003JA010170.

Crooker, N. U., S. W. Kahler, D. E. Larson, and R. P. Lin: 2004b, 'Large-scale magnetic field inversions at sector boundaries'. *J. Geophys. Res.* **109**(A3), A03108, doi:10.1029/2003JA010278.

Dunlop, M. W., A. Balogh, and K. H. Glassmeier: 2002a, 'Four-point Cluster application of magnetic field analysis tools: The discontinuity analyzer'. *J. Geophys. Res.* **107**(A11), 1385, doi:10.1029/2001JA0050089.

Dunlop, M. W., A. Balogh, K. H. Glassmeier, and P. Robert: 2002b, 'Four-point Cluster application of magnetic field analysis tools: The Curlometer'. *J. Geophys. Res.* **107**(A11), 1384, doi:10.1029/2001JA0050088.

Dunlop, M. W., D. J. Southwood, K.-H. Glassmeier, and F. M. Neubauer: 1988, 'Analysis of multipoint magnetometer data'. *Adv. Space Res.* **8**(9-10), 273–277.

Eastwood, J. P., A. Balogh, M. W. Dunlop, T. S. Horbury, and I. Dandouras: 2002a, 'Cluster observations of fast magnetosonic waves in the terrestrial foreshock'. *Geophys. Res. Lett.* **29**, 2046, doi: 10.1029/2002GL015582.

Eastwood, J. P., A. Balogh, M. W. Dunlop, and C. W. Smith: 2002b, 'Cluster observations of the heliospheric current sheet and an associated magnetic flux rope and comparisons with ACE'. *J. Geophys. Res.* **107**(A11), 1365, doi:10.1029/2001JA0009158.

Fitzenreiter, R. J. and L. F. Burlaga: 1978, 'Structure of current sheets in magnetic holes at 1-AU'. *J. Geophys. Res.* **83**, 5579–5585.

Franz, J. A., P. M. Kintner, and J. S. Pickett: 1998, 'POLAR observations of coherent electric field structures'. *Geophys. Res. Lett.* **25**, 1277–1280.

Fränz, M., T. S. Horbury, V. Génot, O. Moullard, H. Rème, I. Dandouras, A. N. Fazakerley, A. Korth, and F. Frutis-Alfaro: 2003, 'Solitary waves observed by Cluster in the solar wind'. In: M. Velli, R. Bruno, and F. Malara (eds.): *Solar Wind Ten: Proceedings of the Tenth International Solar Wind Conference*. American Institute of Physics, pp. 562–565.

Gary, S.: 1993, *Theory of Space Plasma Microinstabilities*. Cambridge: Cambridge Atmos. Space Science Series.

Gombosi, T. I., K. G. Powell, D. L. De Zeeuw, C. R. Clauer, K. C. Hansen, W. B. Manchester, A. J. Ridley, I. I. Roussev, I. V. Sokolov, Q. F. Stout, and G. Toth: 2004, 'Solution-adaptive magnetohydrodynamics for space plasmas: Sun-to-Earth simulations'. *Computing in Science and Engineering* **6**(2), 14–35.

Gopalswamy, N., Y. Hanaoka, T. Kosugi, R. P. Lepping, J. T. Steinberg, S. Plunkett, R. A. Howard, B. J. Thompson, J. Gurman, G. Ho, N. Nitta, and H. S. Hudson: 1998, 'On the relationship between coronal mass ejections and magnetic clouds'. *Geophys. Res. Lett.* **25**(14), 2485–2488.

Gosling, J. T., M. F. Thomsen, S. J. Bame, T. G. Onsager, and C. T. Russell: 1990, 'The electron edge of the low latitude boundary layer during accelerated flow events'. *Geophys. Res. Lett.* **17**, 1833–1836.

Grant, H. L., R. W. Stewart, and A. Moilliet: 1962, 'Turbulence spectra from a tidal channel'. *J. Fluid Mech.* **12**, 241.

Gringauz, K. I., V. V. Bezrukikh, V. D. Ozerov, and R. E. Ribchinsky: 1960, 'Study of the interplanetary high energy electrons and solar corpuscular radiation by means of three electrode traps for charged particles on the second Soviet cosmic rocket'. *Soviet Phys. Doklady* **53**, 61.

Horbury, T.: 1999, 'Cluster-II Analysis of Turbulence using Correlation Functions'. In: *Cluster-II Workshop Multiscale/Multipoint Plasma Measurements*, Vol. SP-449. Imperial College, London, UK, p. 98.

Horbury, T. S., D. Burgess, M. Fränz, and C. J. Owen: 2001, 'Three spacecraft observations of solar wind discontinuities'. *Geophys. Res. Lett.* **28**, 677–680.

Hundhausen, A. J.: 1995, 'The Solar Wind'. In: M. G. Kivelson and C. T. Russell (eds.): *Introduction to Space Physics*. Cambridge University Press, pp. 91–128.

Israelevich, P. L., T. I. Gombosi, A. I. Ershkovich, K. C. Hansen, C. P. T. Groth, D. L. DeZeeuw, and A. G. Powell: 2001, 'MHD simulation of the three-dimensional structure of the heliospheric current sheet'. *Astronomy & Astrophysics* **376**(1), 288–291.

Jokipii, J. R., E. H. Levy, and W. B. Hubbard: 1977, 'Effects of particle drift on cosmic-ray transport. I. General properties, application to solar modulation'. *Astrophys. J.* **213**, 861–868.

Jones, G. H., A. Rees, A. Balogh, and R. J. Forsyth: 2002, 'The draping of heliospheric magnetic fields upstream of coronal mass ejecta'. *Geophys. Res. Lett.* **29**, 10.1029/2001GL014110.

Khurana, K. K., E. L. Kepko, M. G. Kivelson, and R. C. Elphic: 1996, 'Accurate determination of magnetic field gradients from four-point vector measurements - II: Use of natural constraints on vector data obtained from four spinning spacecraft'. *IEEE Trans. Mag.* **32**(5), 5193–5205.

Knetter, T., F. M. Neubauer, T. Horbury, and A. Balogh: 2004, 'Four-point discontinuity observations using Cluster magnetic field data: A statistical survey'. *J. Geophys. Res.* **109**(A18), A06102, doi:10.1029/2003JA010099.

Kolmogorov, A.: 1941, 'The local structure of turbulence in incompressible viscous fluid for very large Reynolds' numbers'. *Compte rendus (Dokl.) de l'Academie des sciences de l'URSS* **30**, 301–305.

Lepping, R. P. and K. W. Behannon: 1980, 'Magnetic field directional discontinuities: 1. Minimum variance errors'. *J. Geophys. Res.* **85**, 4695–4703.

Lepping, R. P., J. A. Jones, and L. F. Burlaga: 1990, 'Magnetic field structure of interplanetary magnetic clouds at 1 AU'. *J. Geophys. Res.* **95**(8), 11957–11965.

Low, B. C.: 2001, 'Coronal mass ejections, magnetic flux ropes, and solar magnetism'. *J. Geophys. Res.* **106**, 25141–25163.

Lucek, E. A., D. Constantinescu, M. L. Goldstein, J. Pickett, J. L. Pinçon, F. Sahraoui, R. Treumann, S. N. Walker, and P. Décréau: 2005, 'The Magnetosheath'. *Space Sci. Rev.* **this issue**.

Mangeney, C., C. Salem, C. Lacombe, J.-L. Bougeret, C. Perche, R. Manning, P.-J. Kellogg, K. Goetz, J. Monson, and J.-M. Bosqued: 1999, 'WIND observations of coherent electrostatic waves in the solar wind'. *Ann. Geophys.* **17**, 307–320.

Matsumoto, H., H. Kojima, T. Miyatake, Y. Omura, M. Okada, I. Nagano, and M. Tsutsui: 1994, 'Electrostatic solitary waves (ESW) in the magnetotail: BEN wave forms observed by GEOTAIL'. *Geophys. Res. Lett.* **21**, 2915–2918.

Matthaeus, W. H. and M. L. Goldstein: 1982a, 'Measurement of the rugged invariants of magnetohydrodynamic turbulence'. *J. Geophys. Res.* **87**, 6011.

Matthaeus, W. H. and M. L. Goldstein: 1982b, 'Stationarity of magnetohydrodynamic fluctuations in the solar wind'. *J. Geophys. Res.* **87**, 10347.

Matthaeus, W. H., M. L. Goldstein, and J. H. King: 1986, 'An interplanetary field ensemble at 1 AU'. *J. Geophys. Res.* **91**, 59.

Matthaeus, W. H., M. L. Goldstein, and D. A. Roberts: 1990, 'Evidence for the presence of quasi-two-dimensional nearly incompressible fluctuations in the solar wind'. *J. Geophys. Res.* **91**, 20673–20683.

Moffatt, H.: 1978, *Magnetic Field Generation in Electrically Conducting Fluids.* New York: Cambridge University Press.

Moldwin, M. B., S. Ford, R. P. Lepping, J. A. Slavin, and A. Szabo: 2000, 'Small-scale magnetic flux ropes in the solar wind'. *Geophys. Res. Lett.* **27**(1), 57–60.

Moldwin, M. B., J. L. Phillips, J. T. Gosling, E. E. Scime, D. J. Mccomas, a. Balogh, and R. J. Forsyth: 1995, 'Ulysses Observation of a Noncoronal Mass Ejection Flux Rope - Evidence of Interplanetary Magnetic Reconnection'. *J. Geophys. Res.* **100**(A10), 19903–19910.

Monin, A. S. and A. M. Yaglom: 1975, *Statistical Fluid Mechanics: Mechanics of Turbulence*, Vol. 2. Cambridge, Mass.: MIT Press.

Ness, N. F., C. S. Scearce, and S. Canarano: 1966, 'Preliminary results from Pioneer 6 magnetic field experiment'. *J. Geophys. Res.* **71**, 3305–3313.

Neugebauer, M., D. R. Clay, B. E. Goldstein, B. T. Tsurutani, and R. D. Zwicki: 1984, 'A reexamination of rotational and tangential discontinuities in the solar wind'. *J. Geophys. Res.* **89**, 5395–5408.

Panchev, S.: 1971, *Random Functions and Turbulence*. New York: Pergammon.

Parker, E. N.: 1958, 'Dynamics of the interplanetary gas and magnetic fields'. *Astrophys. J.* **128**, 664.

Pickett, J. S., L.-J. Chen, S. W. Kahler, O. Santolik, D. A. Gurnett, B. T. Tsurutani, and A. Balogh: 2004, 'Isolated electrostatic structures observed through the Cluster orbit: Relationship to magnetic field strength'. *Ann. Geophys.* **22**, 2515–2523.

Pokhotelov, O. A., R. Z. Sagdeev, M. A. Balikhin, and R. A. Treumann: 2004, 'The mirror instability at finite ion-Larmor radius wavelengths'. *J. Geophys. Res.* **109**, A09213, doi:10.1029/2004JA010568.

Rees, A.: 2002, 'Ulysees observations of magnetic clouds in the 3-D heliosphere'. Thesis/dissertation, Imperial College, University of London.

Richardson, I. G., E. W. Cliver, and H. V. Cane: 2001, 'Sources of geomagnetic storms for solar minimum and maximum conditions during 1972-2000'. *Geophys. Res. Lett.* **28**, 2569–2572.

Riley, P., J. A. Linker, and Z. Mikic: 2002, 'Modeling the heliospheric current sheet: Solar cycle variations'. *J. Geophys. Res.* **107**(A7).

Roberts, D. A., M. L. Goldstein, and A. Deane: 2002, 'Three-Dimensional MHD Simulation of Solar Wind Structure'. In: M. Velli, R. Bruno, and F. Malara (eds.): *Solar Wind 10*, Vol. 679. Pisa, Italy, p. 133.

Roberts, D. A., M. L. Goldstein, L. W. Klein, and W. H. Matthaeus: 1987a, 'Origin and evolution of fluctuations in the solar wind: Helios observations and Helios-Voyager comparisons'. *J. Geophys. Res.* **92**, 12,023.

Roberts, D. A., L. W. Klein, M. L. Goldstein, and W. H. Matthaeus: 1987b, 'The nature and evolution of magnetohydrodynamic fluctuations in the solar wind: Voyager observations'. *J. Geophys. Res.* **92**, 11021.

Schwartz, S.: 1998, 'Shock and discontinuity normals, Mach numbers, and related parameters'. In: G. Paschmann and P. W. Daly (eds.): *Analysis Methods for Multi-Spacecraft Data*, ISSI SR-001. ESA Publications Division, pp. 249–270.

Smith, C. W.: 2002, 'The geometry of turbulent magnetic fluctuations at high heliographic latitudes'. In: M. Velli, R. Bruno, and F. Malara (eds.): *Solar Wind 10*, Vol. 679. Pisa, Italy, p. 413.

Smith, E. J.: 1973, 'Identification of interplanetary tangential and rotational discontinuities'. *J. Geophys. Res.* **78**, 2054–2063.

Smith, E. J.: 1990, 'The Heliospheric Current Sheet and modulation of galactic cosmic rays'. *J. Geophys. Res.* **95**, 18731–15743.

Smith, E. J.: 2001, 'The heliospheric current sheet'. *J. Geophys. Res.* **106**, 15819–15831.

Snyder, C. W. and M. Neugebauer: 1966, 'The Relation of Mariner-2 Plasma Data to Solar Phenomena'. In: *The Solar Wind*. pp. 25–32.

Sonnerup, B. U. Ö. and L. J. Cahill: 1967, 'Magnetopause structure and attitude from Explorer 122 observations'. *J. Geophys. Res.* **72**, 171–183.

Sonnerup, B. U. Ö. and M. Scheible: 1998, 'Minimum and Maximum Variance Analysis'. In: G. Paschmann and P. W. Daly (eds.): *Analysis Methods for Multi-Spacecraft Data*, ISSI SR-001. ESA Publications Division, pp. 185–220.

Taylor, G. I.: 1938, 'The spectrum of turbulence'. *Proc. R. Soc. London Ser. A* **164**, 476.

Temerin, M., K. Cerny, W. Lotko, and F. S. Mozer: 1982, 'Observations of double layers and solitary waves in the auroral plasma'. *Phys. Rev. Lett.* **48**, 1175–1179.

Treumann, R. A., C. H. Jaroschek, O. D. Constantinescu, R. Nakamura, O. A. Pokhotelov, and E. Georgescu: 2004, 'The strange physics of low frequency mirror mode turbulence in the high temperature plasma of the magnetosheath'. *Nonlin. Proc. Geophys.* **11**, 647–657.

Tsurutani, B., G. Lakhina, J. Pickett, F. Guarnieri, N. Lin, and B. Goldstein: 2005, 'Nonlinear Alfvn Waves, Discontinuities, Proton Perpendicular Acceleration, and Magnetic Holes/Decreases in Interplanetary Space and the Magnetosphere: Intermediate Shocks?'. *Nonlinear Proc. Geophys.* in press.

Tsurutani, B. T. and C. M. Ho: 1999, 'A review of discontinuities and Alfven waves in interplanetary space: Ulysses results'. *Rev. Geophys.* **37**(4), 517–541.

Tu, C.-Y. and E. Marsch: 1995, 'MHD structures, waves and turbulence in the solar wind: Obsevations and theories'. *Space Sci. Rev.* **73**, 1–210.

Turner, J. M., L. F. Burlaga, N. F. Ness, and J. F. Lemaire: 1977, 'Magnetic holes in solar-wind'. *J. Geophys. Res.* **82**(13), 1921–1924.

Unti, T. W. and M. Neugebauer: 1968, 'Alfvén waves in the solar wind'. *Phys. Fluids* **11**, 563.

Usmanov, A. V. and M. L. Goldstein: 2003, 'A tilted-dipole MHD model of the solar corona and solar wind'. *J. Geophys. Res.* **108**(A9).

Usmanov, A. V., M. L. Goldstein, B. P. Besser, and J. M. Fritzer: 2000, 'A global MHD solar wind model with WKB Alfven waves: Comparison with Ulysses data'. *J. Geophys. Res.* **105**(A6), 12675–12695.

Velli, M., R. Bruno, and F. Malara (eds.): 2002, *Solar Wind 10*, Vol. 679. Pisa, Italy: AIP.

Webb, D. F. and R. A. Howard: 1994, 'The solar cycle variation of coronal mass ejections and the solar wind mass flux'. *J. Geophys. Res.* **99**, 4201–4220.

Wilcox, J. M. and N. F. Ness: 1965, 'Extension of Photospheric Magnetic Field into Interplanetary Space'. *Astronomical Journal* **70**(5), 333.

Winterhalter, D., M. Neugebauer, B. Goldstein, and E. J. Smith: 1994, 'Ulysses field and plasma observations of magnetic holes in the solar wind and their relation to mirror-mode structures'. *J. Geophys. Res.* **99**, 23371–23381.

Chapter 2

The Foreshock

J. P. Eastwood[1], E. A. Lucek[2], C. Mazelle[3], K. Meziane[4], Y. Narita[5], J. Pickett[6], and R. A. Treumann[7]

2.1 Introduction

Although collisionless shocks primarily exist to mediate the flow of supermagnetosonic plasma, they also act as sites for particle acceleration. It is now well known that for certain magnetic field geometries, a portion of the inflowing plasma returns to the upstream region rather than being processed by the shock and passing irreversibly downstream. The combination of the inflowing plasma and this counterstreaming component upstream of the shock is subject to a number of plasma instabilities, leading to the generation of waves. These waves interact in a highly complex manner with the ions and electrons making up the plasma and can cause part of the plasma distribution to reach high energies.

The region of space upstream of the bow shock, magnetically connected to the shock and filled with particles backstreaming from the shock is known as the foreshock. As discussed in Balogh et al. (2005), the bow shock can be classified into quasi-perpendicular and quasi-parallel shock regions according to the angle θ_{Bn} between the shock normal **n** and the direction of the solar wind magnetic field **B**. For the quasi-perpendicular bow shock ($\theta_{Bn} > 45°$), the foreshock is restricted to the shock foot, while in the quasi-parallel part of the bow shock ($\theta_{Bn} < 45°$), it

[1]NASA Goddard Space Flight Center, Greenbelt, MD, USA
[2]Space and Atmospheric Physics, The Blackett Laboratory, Imperial College London, London, UK
[3]Centre d'Etude Spatiale de Rayonnement, CNRS/UPS/OMP, Toulouse, France
[4]Physics Department, University of New Brunswick, Fredericton, Canada
[5]Institut für Geophysik und extraterrestrische Physik, Technische Universität, Braunschweig, Germany
[6]Department of Physics and Astronomy, The University of Iowa, Iowa City, IA, USA
[7]Max-Planck-Institut für extraterrestrische Physik, Garching, Germany

Space Science Reviews 118: 41–94, 2005.
DOI: 10.1007/s11214-005-3824-3

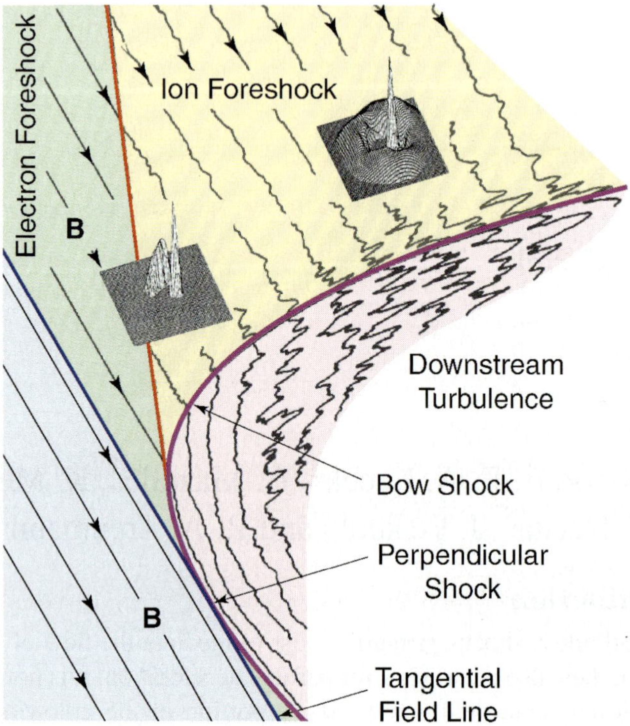

Figure 2.1. Schematic view of the terrestrial shock/foreshock system. In this picture, the solar wind flows from the left hand side; the bow shock, which sits upstream of the magnetopause, is represented by the curved line (note the magnetopause is not shown here). The structure of the shock and the nature of the downstream flow depend on the angle between the shock normal and the magnetic field, θ_{Bn}, with the quasi-parallel shock being much more extended and time variable. The foreshock, largely upstream of the quasi-parallel shock and confined to the region of space behind the tangent field line, exhibits significant spatial structure. Just behind the tangent field line is the electron foreshock; here, only backstreaming electrons are observed. Behind the ion foreshock boundary, field-aligned backstreaming ion distributions are typically observed. Deeper in the foreshock, close to the quasi-parallel shock, diffuse backstreaming ion distributions are observed. Two-dimensional velocity space relief plots are used to represent the field-aligned (close to the ion foreshock boundary) and the diffuse (close to the quasi-parallel shock) ion distributions. In these two dimensional relief plots, the sharp peak corresponds to the solar wind. (From Treumann and Scholer, 2001).

covers a much larger upstream domain and becomes a proper foreshock. Figure 2.1 illustrates the basic structure of the Earth's foreshock.

Much of our understanding of foreshock phenomena has been derived from the large number of in-situ measurements made of the terrestrial foreshock/ bow shock system. Early single spacecraft observations indicated the existence of ion distributions upstream of the bow shock that could not be classified as being either solar wind like or magnetosheath like (Asbridge et al., 1968), together with enhanced d.c. magnetic field fluctuations (Greenstadt et al., 1968). It was shown that these fluctuations were in fact well defined quasi-monochromatic waves with periods of about 30 seconds and typically left-handed in the spacecraft frame (Fairfield, 1969). Also observed were what appeared to be linearly polarised steepened waves, termed shocklets, associated with discrete wave packets (Russell et al., 1971), and so called 1Hz waves (Fairfield, 1974). It was proposed that the backstreaming ions were responsible for the generation of the low frequency waves (Barnes, 1970). Further observations indicated the existence of a high energy ion population upstream of the shock (Lin et al., 1974) and the existence of backstreaming electrons and associated electric field waves near the electron plasma frequency (Scarf et al., 1971; Fredricks et al., 1971; Feldman et al., 1973).

Observations from the ISEE-1 and ISEE-2 spacecraft (Ogilvie et al., 1977) were instrumental in improving our understanding of the foreshock. Advances made included the identification of different types of ion distributions (Gosling et al., 1978); 'reflected', 'intermediate' and 'diffuse' (Paschmann et al., 1981). Ultra Low Frequency (ULF) waves were shown to be associated with the diffuse ions, but not with reflected backstreaming ion distributions (Paschmann et al., 1979); the association of different wave types with different particle distributions was examined by (Hoppe and Russell, 1983). Furthermore, the dual spacecraft ISEE data were used in a number of case studies to show that ULF waves observed to be left handed in the spacecraft frame were propagating sunwards in the solar wind rest frame, away from the shock and in the direction of the backstreaming ion beams (Hoppe et al., 1981). Consequently, they were identified as being intrinsically right handed and therefore qualitatively consistent with generation by backstreaming ions through the ion-ion right hand resonant beam instability (e.g., Gary, 1993). This was also tested using single spacecraft data, comparing the spacecraft frame wave period with the solar wind speed via the proposed beam formation and wave generation theories (Watanabe and Terasawa, 1984). It was also found that the solar wind was slowed and deflected in the presence of diffuse ions (Bame et al., 1980).

To explain these observations, it was thought that the 'reflected' distributions, now called 'field-aligned' (to avoid any implicit assumption as to their origin), generated the 30s waves and evolved into intermediate distributions. It was then suggested that further wave-particle interaction would result in hot diffuse distributions associated with shocklets and discrete wave packets. Compositional evidence from ISEE showed that diffuse distributions contained a significant He^{++} component (Ipavich et al., 1984), whereas field aligned distributions did not (e.g., Fuselier and Thomsen, 1992, and references therein); consequently, field aligned beams are now not thought to be the major seed population for the diffuse population. The diffuse distributions are significantly hotter than the other backstreaming ion distributions. Although their density is similar, most of the energy they contain is thermal (Paschmann et al., 1981). The distribution energy extends to at least 150 keV/e.

Diffuse distributions are typically observed upstream of the quasi-parallel shock (Bonifazi and Moreno, 1981). Considerable attention has been paid to the mechanisms by which ions might reach such high energies through interactions with the shock and the upstream foreshock wave field (e.g., the reviews by Drury, 1983; Jones and Ellison, 1991). Energetic foreshock particles can be sourced from the magnetosphere directly (Sarris et al., 1987), but diffuse ions are closely associated with quasi-parallel shock processes (e.g., Gosling et al., 1989).

The quasi-parallel shock (Burgess et al., 2005), is strongly influenced by the foreshock. The foreshock is always upstream of the quasi-parallel shock, as can be seen in Figure 2.1, and the perturbations introduced into the solar wind by the foreshock significantly affect the shock. For example, it is thought that the foreshock ULF wave field can introduce significant period variations in θ_{Bn}, effectively controlling the thermalisation process (Greenstadt, 1985). Such effects have been observed by Cluster (see Section 2.3.2.2). The existence of Short Large Amplitude Magnetic Structures (SLAMS) has also been reported. These localised increases in magnetic field strength appear in (and presumably grow out of) the upstream wave field, and it has been suggested that they coalesce together to form the quasi-parallel shock front (Schwartz and Burgess, 1991), although not all quasi-parallel shocks fit this paradigm (Burgess, 1995). A third proposal is of cyclical shock formation (Burgess, 1989a). Here the quasi-parallel geometry allows a large number of particles to escape upstream; the upstream reflected ions are so dense that they constitute a pressure pulse strong enough to relaunch the shock from a position upstream, as the old shock fails to stand in the flow and convects downstream (Burgess, 1995).

We do not discuss the quasi-parallel shock and the role it plays in foreshock dynamics further here. The interested reader is referred to Burgess et al. (2005), which treats the quasi-parallel shock in detail. The work reported by Burgess et al. includes a discussion of SLAMS, which are closely related to the foreshock wave field, and also a discussion of diffuse particle spectra, demonstrating that diffusive transport plays a role in their production (Kis et al., 2004).

Further results from ISEE demonstrated the existence of gyrophase bunched backstreaming ion distributions in the foreshock; these distributions can resemble intermediate distributions in energy-time spectrograms (Fuselier et al., 1986b). Also identified were gyrotropic ion distributions (Gurgiolo et al., 1983). The production of gyrophase bunched ions and associated ULF waves has been studied numerically (Hoshino and Terasawa, 1985) and theoretically (Mazelle et al., 2000), as well as experimentally using data from the AMPTE (Fazakerley et al., 1995) and WIND spacecraft (Meziane et al., 2001).

Other work based on the ISEE dataset has helped to establish the basic morphology of the foreshock (Le and Russell, 1992a,b, and references therein), and the correlation length of the low frequency waves (Le and Russell, 1990). The existence of a new category of waves, called '3s waves' (Le et al., 1992), has also been reported. These waves are circularly polarised, right handed and have periods of 2 - 5 s in the spacecraft frame. They may be intrinsically left handed and generated by a non resonant fire hose type instability, on the basis of comparisons with the predictions of kinetic theory (Blanco-Cano et al., 1999).

A number of different sources have been proposed for the 1 Hz waves, including the shock (Fairfield, 1974), the backstreaming electrons (Sentman et al., 1983) and the backstreaming ions (Wong and Goldstein, 1987). On the basis of observations of both Venus' and the Earth's foreshock, the balance of evidence indicates that the shock is the most likely source of these waves (Orlowski et al., 1995, and references therein). In particular, it has been found that in order to correctly identify wave properties, a kinetic rather than Hall MHD approach should be used (Russell, 1994; Greenstadt et al., 1995).

Compared to the ions, there is less variability in the properties of the backstreaming electrons. On and just behind field lines tangent to the shock, one observes more energetic electrons ($>$ 1keV), whereas deeper in the foreshock less energetic electron beams are encountered. This may be understood in terms of drift in the solar wind convection electric field, which is more important for the slower population (e.g., Fitzenreiter, 1995). The backstreaming electrons exist as a high energy tail on the main solar wind distribution, which results in the production of Langmuir waves at the electron plasma frequency, particularly at the leading edge of the electron foreshock where the backstreaming beams are most energetic. Deeper in the electron foreshock, the wave frequency is shifted away from the electron plasma frequency (e.g., Burgess, 1997). Harmonics of the electron plasma frequency are also observed, corresponding to electromagnetic waves again sourced from the electron foreshock (Reiner et al., 1996).

The evolution of our understanding of the foreshock can be traced through different reviews and special publications. In particular, we direct the interested reader to the reviews by Burgess (1997), Le and Russell (1994), Greenstadt et al. (1995), Fitzenreiter (1995), Fuselier (1994, 1995), as well as volume 15(8/9) of Advances in Space Research and volume 81(A6) of the Journal of Geophysical Research. The AGU Geophysical Monographs 34 and 35 (Stone and Tsurutani, 1985; Tsurutani and Stone, 1985) summarise the knowledge of the bow shock and foreshock following the first set of major results from ISEE.

In the remainder of this introduction, we concentrate on the questions relating to the foreshock that remained unanswered at the launch of Cluster. Here, we group problems into different types.

- **Large scale morphology**

  The magnetic field, the solar wind velocity and the shock normal at the point the magnetic field intersects the shock are not a priori co-planar. Consequently, diagrams of the foreshock that are two dimensional in nature are implicitly simplified. Whilst the upstream boundary of the ion and electron foreshock is easily accounted for theoretically, it is instructive to map these boundaries in three dimensions. How stable are the locations of the ion and electron foreshock boundaries? With multipoint observations, such problems are easily addressed. Furthermore, the plasma properties on either side of these boundaries can be measured simultaneously. The existence of a boundary to the upstream ULF wave field has also been a subject of research. The location of this boundary relative to the ion foreshock boundary is presumably a function of the wave growth time, and therefore a clear understanding of its location will give extra information as to the wave generation mechanisms. When at large separations, (>1 R_E), it will be possible to use the Cluster spacecraft to study the properties of the plasma either side of these boundaries simultaneously. This will, for example, allow the deceleration of the solar wind in the foreshock to be characterised more accurately.

- **ULF wave properties**

To establish the properties of the waves in the solar wind rest frame, multipoint observations are required. The ISEE mission used two spacecraft to investigate upstream waves. However, a dual spacecraft analysis relies on minimum variance analysis of the magnetic field (MVA - see e.g., Sonnerup and Scheible, 1998), which is restricted to the analysis of circularly/elliptically polarised waves, and is also constrained geometrically; if the spacecraft separation vector is perpendicular to the wave fronts, then the speed of the waves cannot be established. Cluster offers the opportunity to test independently the results of ISEE concerning intrinsic wave properties, and can also investigate linearly polarised waves, without suffering dual spacecraft geometrical restrictions. The multi-point capability of Cluster can also be used to study the coherence lengths of waves directly.

A key question at the launch of Cluster concerned the problem of oblique wave propagation in the foreshock. The growth rates of the instabilities proposed for the generation of upstream ULF waves are maximised for parallel propagation, at least in the linear regime. However, the waves themselves are observed to propagate obliquely to the field. Whilst theories have been put forward to explain this in terms of non-local effects (Hada et al., 1987), there is no full explanation, particularly since this mechanism only works for certain geometries; oblique wave propagation is a persistent feature of the foreshock for all geometries (e.g., Burgess, 1997). A second question concerned the minority of upstream ULF waves observed to be right handed in the spacecraft frame. These waves could either be intrinsically right handed and propagating towards the shock (i.e. anti parallel to the beam), in which case they would be generated through a non-resonant firehose type instability, or they could be intrinsically left handed and propagating parallel to the beam, in which case they would be generated by the left hand resonant ion-ion beam instability (e.g., Le and

Russell, 1994). This has been investigated by comparing observations with predictions of kinetic theory (Blanco-Cano and Schwartz, 1997), but by using the multipoint Cluster data, the properties of these waves can be established directly, thus allowing a better understanding of wave particle processes in the foreshock.

- **Ion distributions**

  The division of backstreaming ion populations into reflected (i.e., field-aligned), intermediate and diffuse distributions was made on the basis of 2D velocity distribution functions and energy-time spectrograms. Further observations of ion distributions showed the existence of gyrotropic and gyrophase bunched distributions. Since intermediate, gyrophase bunched and gyrotropic distributions appear similar in such spectrograms, it is important to ascertain the extent to which gyrophase bunched ions may have been misidentified as intermediate type.

  Following the launch of ISEE, the beam disruption hypothesis was used to efficiently and simply explain foreshock dynamics. This theory proposed that the field aligned distributions generated the 30s waves, which then disrupted the beam into an intermediate type distribution. Further wave particle interactions then thermalized the beam (into a diffuse distribution), and caused the waves to steepen into shocklets with associated discrete wave packets. However, observations have shown that whilst diffuse ion distributions contain solar wind concentrations of He^{++}, field aligned distributions are largely composed of protons. Consequently, field aligned distributions do not evolve directly into diffuse distributions. It is of interest to better understand the extent to which these different populations interact. The composition of the intermediate, gyrophase bunched and gyrotropic distributions remains unclear.

- **Wave particle interactions**

  Data from ISEE were used in a small number of case studies to establish that ULF waves observed to be left handed in the spacecraft frame were in fact right handed and attempting to propagate away from the shock, parallel to the ion beams (Hoppe and Russell, 1983). This showed that, qualitatively, they had to be generated by the backstreaming ion beams through the right hand ion-ion beam instability. However, the properties of the waves were not used to compute the expected beam speed, although it was expected that the field aligned distributions were responsible for their generation. With Cluster, it is expected that quantitative tests of the wave particle interaction can be made.

- **Foreshock transients: cavities and hot flow anomalies**

  Many pictures of the foreshock assume a dynamical equilibrium. However, the bow shock - foreshock system is highly non-linear, with many complicated feedback mechanisms, and it is by no means clear that changes to the upstream conditions, particularly the magnetic field orientation, cause proportional changes to large-scale foreshock structure. Nor are the system response time scales particularly clear. Work has been published examining the effects of brief connection to the foreshock and the way in which these connections can highly perturb the solar wind - known as foreshock cavities (Sibeck et al., 2002). Also, if an interplanetary discontinuity intersects the bow shock, a hot flow anomaly may be generated - causing a massive (in the context of shock dynamics) disruption to solar wind flow upstream of the shock (Schwartz et al., 2000). Identifying the structure, dynamics and evolution of hot flow anomalies is problem highly suited to multi-spacecraft observations.

Since its launch in 2000, data from the Cluster mission has been used to address many of these questions. Here, we attempt to gather together the main results that have been derived from Cluster thus far. In particular, results from Cluster have contributed to our understanding of:

- the morphology of particle and wave boundaries in the foreshock,
- the nature of upstream waves and the origin of cyrating ions,
- the three-dimensional structure of hot-flow anomalies.

The remainder of this chapter is divided according to these three topics. In Section 2.2, work relating to the large scale structure of the foreshock and its boundaries is presented. In Section 2.3, work relating to the 'microphysics' of the foreshock, the basic wave particle interactions, is presented. In Section 2.4, work relating to hot flow anomalies (HFAs) is presented. The chapter is summarised in Section 2.5.

2.2 Foreshock Boundaries

2.2.1 Electron foreshock boundary

The electron foreshock, as sketched in Figure 2.1 occupies the region from the bow-shock to just downstream of the tangent magnetic field line. Like the ion foreshock, the electron foreshock is a very dynamic region in which bow shock reflected electrons are convected downstream toward the bow shock by the $\mathbf{v} \times \mathbf{B}$ electric field in the solar wind. Similar to the ions in the ion foreshock, the highest energy electrons are observed close to the foreshock boundary, while the lower energy electrons are observed further downstream due to time-of-flight effects and the solar wind electric field (Filbert and Kellogg, 1979). This results in electron beams producing a bump-on-tail distribution function (Fitzenreiter et al., 1984). The waves observed in this region, basically Langmuir waves, are typically at the plasma frequency f_{pe} and its second harmonic, but far away from the foreshock boundary deep in the electron foreshock, these waves are often seen shifted to frequencies above and below f_{pe} (Fuselier et al., 1985).

The Langmuir wave amplitudes in the electron foreshock were shown by Filbert and Kellogg (1979) to be largest near the tangent magnetic field lines. Etcheto and Faucheaux (1984), using data from ISEE 1, reported maximum amplitudes of a few $\mathrm{mV\,m^{-1}}$ at the edge of the foreshock and only a few tens to a few hundreds of $\mu\mathrm{V\,m^{-1}}$ further inside the electron foreshock. Cairns et al. (1997) also used ISEE 1 data to show that there is a slight offset of the large wave amplitude region from the boundary of the foreshock and that the largest amplitude waves are observed in a relatively narrow region with the Langmuir wave amplitudes falling off slowly at larger distances away from the foreshock.

Cluster multi-point measurements allow for the comparison of Langmuir wave amplitudes at various positions in the foreshock and near the boundary of the electron foreshock. Another important aspect of the Cluster measurements is the capture of waveforms by the WBD plasma wave receiver (Gurnett et al., 1997) in this region since only the Wind spacecraft up to now has had the advantage of looking at Langmuir waveforms in the foreshock (Bale et al., 1997). All previous studies were carried out using spectral density measurements, which underestimate the wave amplitudes due to temporal and spectral averaging (Robinson et al., 1993). Using the WBD measurements, some initial studies of the Langmuir waves observed in Earth's electron foreshock have been carried out. Sigsbee et al. (2004b) found that the characteristics of the waves were in agreement with most of the previous studies. In addition the Cluster observations were found to follow the log normal statistics predicted by stochastic growth theory (Cairns and Robinson, 1999); however, deviations from this prediction occurred at large wave amplitudes when electric fields measured at a wide range of distances to the boundary between the electron foreshock and solar wind were included. This finding generally agrees with the results of Bale et al. (1997) and Cairns and Robinson (1997). However, Sigsbee et al. (2004b) pointed out that the slope of the power law obtained from the Cluster data was steeper.

Sigsbee et al. (2004a,b) showed that the center of the probability distributions constructed for small bins of D_f, the distance from the spacecraft to the tangent field line in the x GSE direction, shifts to lower amplitudes as one goes deeper into the foreshock, as illustrated in Figure 2.2. The data plotted in this figure show that the power law tail on the distribution for all values of D_f may result from the sum of the log-normal distributions at different locations, which is a new result. Stochastic growth theory may thus still be correct in explaining the observed tail distribution of amplitudes.

All of the Cluster results thus far have dealt with case studies. Future work in this area will concentrate on statistical studies of various electron foreshock crossings by Cluster under varying solar wind conditions and IMF orientations.

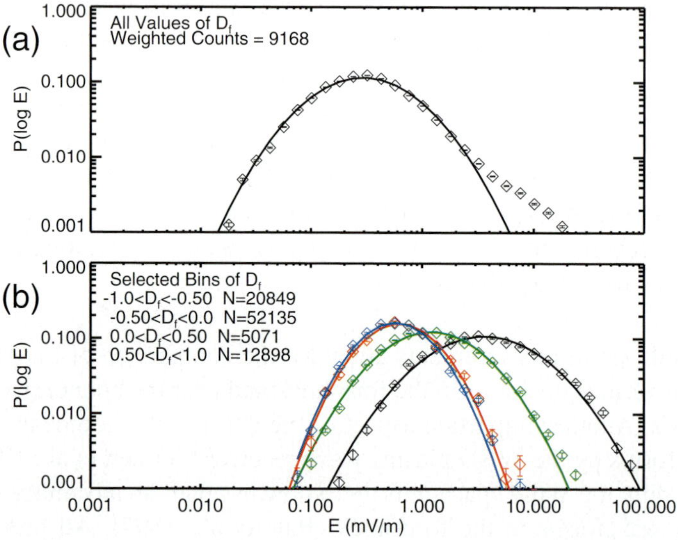

Figure 2.2. Probability distributions for the electric field amplitudes observed by the Cluster 3 WBD plasma wave receiver with fits to the Gaussian function predicted by stochastic growth theory (solid lines). (a) Probability distribution for all values of D_f (see text for description). (b) Probability distributions for selected ranges of D_f. From Sigsbee et al. (2004b).

2.2.2 Internal boundaries in the ion foreshock: Particle observations

Through the analysis of single and dual spacecraft foreshock observations, the basic morphology of the ion foreshock is reasonably well understood. Figure 2.3 shows a schematic picture of the ion foreshock for an IMF cone angle of 45°, here corresponding to a sunward pointing Parker spiral. The solar wind flows vertically from the top of the figure with velocity $\mathbf{v}_{sw}$. The tangent field line marks the point at which the solar wind becomes magnetically connected to the bow shock; behind this tangent point, ions can escape from the shock back into the upstream region. They are ejected with velocity $\mathbf{v}_{FAB}$, which here is shown to be field aligned. However, the backstreaming ions are also subject to $\mathbf{E} \times \mathbf{B}$ drift in the solar wind convection electric field, and therefore the upstream boundary of the ion foreshock is not aligned to the magnetic field. Note also that the upstream boundary of the ion foreshock does not intersect the tangent field line surface.

Field aligned distributions, which are not observed in conjunction with ULF waves, are typically observed at and near the leading edge of the ion foreshock, whereas intermediate and diffuse distributions, observed in the presence of ULF waves are observed deeper in the foreshock, upstream of the quasi-parallel shock (e.g., Russell and Hoppe, 1983). Consequently, there exists a second boundary within the ion foreshock confining the region of ULF wave activity. This boundary

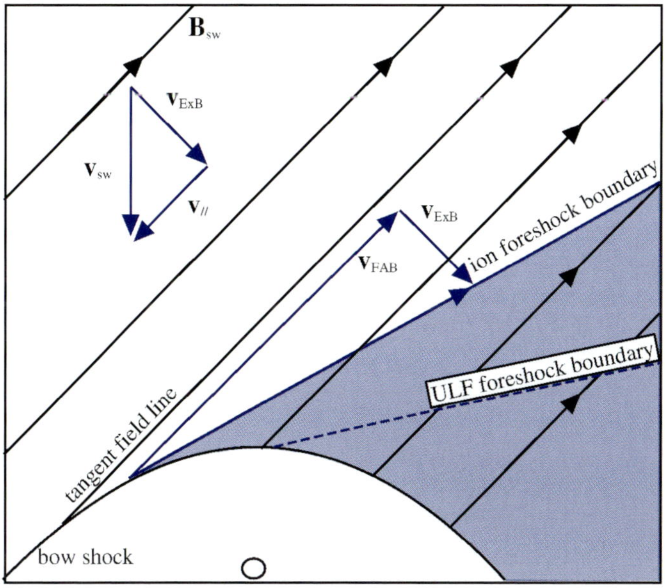

Figure 2.3. Schematic structure of the ion foreshock for a typical IMF configuration (cone angle = 45°). Note that the ion foreshock boundary is not field aligned, and exists behind the tangent field line. The ULF foreshock boundary indicates the upstream limit of the ULF wave field in the foreshock, as established experimentally. (Figure provided by J. P. Eastwood).

is the ULF foreshock boundary, or the ion foreshock wave boundary. Studies have shown that for IMF cone angles of $\geq 45°$, there is a well defined upstream boundary to the region of ULF wave activity, and that this boundary intersects the shock at $\theta_{Bn} \approx 50°$ (Le and Russell, 1992a). In Figure 2.3, the ULF foreshock boundary is shown as a dashed line.

Based on the analysis of ISEE data, it has been shown statistically that field-aligned beams are found in a layer of $\sim$ 0.4 R_E thickness at the leading edge of the ion foreshock, followed further downstream by a $\sim$3.5 R_E -wide layer of intermediate ions (Bonifazi and Moreno, 1981). Gyrophase bunched ions are usually found either inside the region containing intermediate distributions, or within a distinct thin layer between this region and the layer of field-aligned beams. This suggests the existence of a spatial boundary separating the field aligned beams from the intermediate/gyrophase bunched ions and also that field-aligned beams and gyrating/intermediate ions are not produced at the same region on the shock surface. Evidence for the existence of such a boundary has been reported by Meziane and d'Uston (1998), using data from the ISEE–1 unidirectional Berkeley-Toulouse particle detector. Meziane and d'Uston showed that for a given range of magnetic field

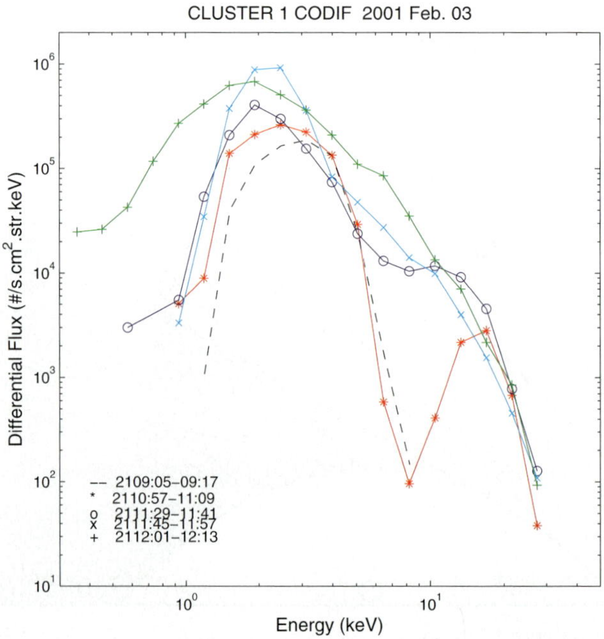

Figure 2.4. Energy-spectra of backstreaming ions, from Cluster 1 CIS/CODIF. The black dashed curve shows the initial proton energy-spectrum, taken at 21:09:05-09:17 UT on February 3, 2001, indicating the presence of a single component, corresponding to a field aligned beam. In the following spectra (black, red, blue, cyan, green), a secondary peak appears merging into a single warmer population by 21:12:01-12:13 UT. The final distribution is gyrating. (Adapted from Meziane et al., 2004b).

cone angle, flux enhancement onsets were spatially organised in the foreshock region. Because the field of view of this detector pointed in the -Z_{GSE} direction, the flux enhancement onsets were likely to be associated with heated backstreaming ions; i.e., intermediate, gyrating or possibly diffuse. For a magnetic field cone angle of 45 ° , this boundary was found to coincide with the ULF foreshock boundary previously reported previously by Greenstadt and Baum (1986). For magnetic field cone angles of 20-30 °, the orientation of the boundary was found to be significantly different from the ULF boundary.

Transitions between different regions of the foreshock are usually due to rotations of the interplanetary magnetic field. In general, these transitions are rapid and are consequently difficult to analyse. Recently, however, Meziane et al. (2004b) have presented Cluster observations of a slow transition from the field-aligned beam region to a region of gyrating ions, allowing the properties of this boundary to be studied in detail. The key observations are summarised in Figure 2.4, which shows the energy-spectra of backstreaming ions measured at successive times by Cluster 1 CIS/CODIF (Réme et al., 1997) during a 3 minute interval on February

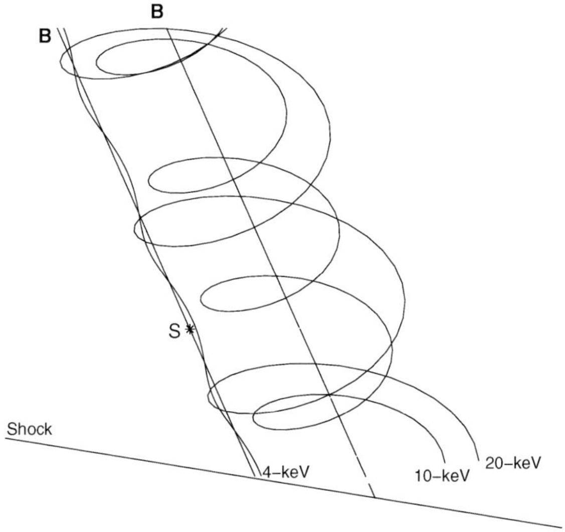

Figure 2.5. Interpretation of the energy spectra shown in Figure 2.4. Initially field aligned distributions are observed. As the spacecraft moves slowly relative to the boundary, it remote senses the most energetic gyrating ions first. The simultaneous observation is made possible by the fact that the spacecraft remain in the vicinity of the boundary for some time; the boundary itself is sharp, of the order of 1 gyroradius thick. (From Meziane et al., 2004b).

3, 2001. The initial proton energy-spectrum, taken at 21:09:05-09:17 UT (black dashed curve), shows the presence of a single component with an energy cut-off at $\sim$ 8 keV. In the following spectra, a secondary peak appears in the energy range 10-30 keV. The main beam gradually broadens while slightly shifting towards lower energy whilst the peak of the secondary beam shifts down to $\sim$ 10 keV. By 21:12:01-12:13 UT the spectrum is associated with is a single warmer population. The particle distribution functions (not shown here) indicate that the first main spectrum corresponds to a field-aligned beam with a speed of to 860 $\mathrm{km\,s^{-1}}$. The secondary beam and the final distribution are associated with gyrating ion distributions. During this time the angle between the IMF and the solar wind was $\sim$ 45°, and the spacecraft was connected to the bow shock at oblique geometries ($\theta_{Bn} \sim$ 50°). It should however be emphasised that the changes observed in the proton distributions are likely due to a very weak IMF rotation.

The onset of the gyrating ions is observed first at high energies. This has been interpreted as a remote sensing effect, illustrated in Figure 2.5. When the field-aligned beams are observed by the spacecraft (at point S), the gyrating ions located just downstream of the field-aligned region are remotely sensed. The high energy gyrating ions are detected first, because of their relatively large gyroradius, followed successively by lower energies. This simultaneous observation is made possible because the spacecraft remains in the vicinity of the boundary for some time. The boundary itself is sharp, of the order of 1 gyroradius.

The streaming speed of the field aligned distributions was found to be inconsistent with the predictions of current shock-related emission mechanisms. Furthermore, the detailed properties of the gyrating ions were also inconsistent with shock-related production mechanisms, suggesting that they may have been produced locally. However, Meziane et al. (2004b), reported that the boundary between the two types of backstreaming ions made an angle of 77° to the Sun-Earth line, in agreement with theoretical predictions and coincident with the expected location of the ULF foreshock boundary. It was noted that although the geometrical properties of this boundary between the two populations places strong constraints on the production mechanism, further work is required to fully account for the detailed properties of the backstreaming distributions.

2.2.3 Internal boundaries in the ion foreshock: Wave observations

In addition to studying the boundaries between different backstreaming particle distributions, Cluster can also be used to study the ULF foreshock boundary. As discussed in the previous section, this boundary is closely related to boundaries in the backstreaming ion distributions. Here we present two case studies into the onset of ULF foreshock waves, relating their appearance to changes in the orientation of the IMF.

Figure 2.6 shows an overview of the magnetic field observed by FGM (Balogh, 1997) on Cluster 1 between 12:00 UT on day 51 (February 20), and 06:00 UT on day 52 (February 21), 2001. At this time, the separation of the spacecraft was ~600 km. The data are shown at 4 s (spin) resolution, in GSE polar coordinates. At the start of the interval, the Cluster spacecraft were in the magnetosheath. Cluster 1 finally exited into the solar wind at 18:42 UT, where the magnetic field fluctuations were characteristic of unperturbed solar wind, and CIS energy-time spectrograms (not shown) did not indicate any back streaming ions. At 21:42 UT, the characteristics of the magnetic field clearly changed; Cluster 1 moved into the foreshock, where the fluctuations in the magnetic field significantly increased. At this time, the presence of back-streaming ions was recorded by CIS, confirming the transition into the foreshock. Cluster remained in the foreshock until just before midnight on day 52. The ULF wave activity included discrete wave packets and shocklets, characteristic of the diffuse ion foreshock (Greenstadt et al., 1995).

Figure 2.7 shows the ion plasma moments calculated on-board by Cluster 1 CIS/HIA, between 20:00 UT and 24:00 UT. The cadence of the onboard moments is 4 s, and HIA was operating in solar wind mode.

The transition into the foreshock ULF wave field results from the change in IMF orientation. This was investigated by tracing the observations to a model bow shock, and computing θ_{Bn} at the point of intersection as a function of time. The

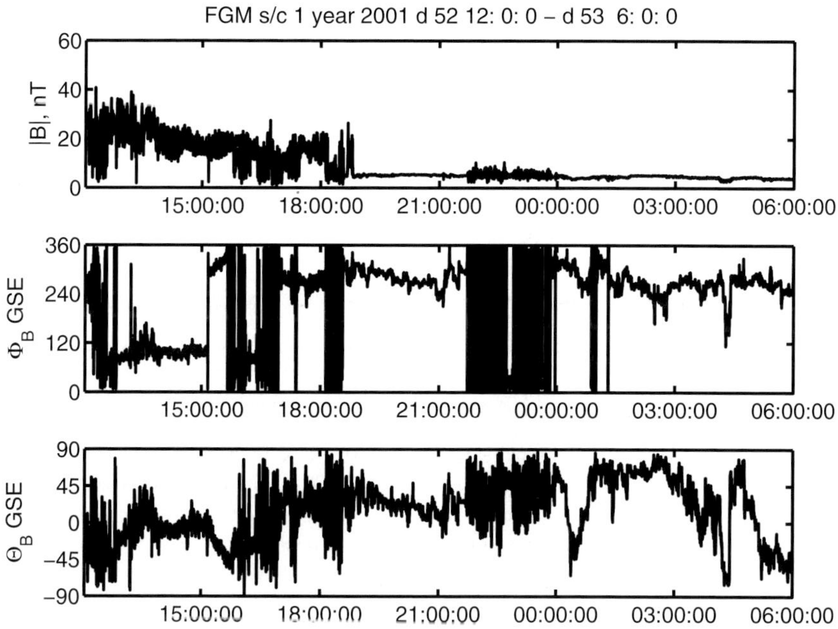

Figure 2.6. Magnetic field observed by Cluster 1 between 12:00 UT on February 20 and 06:00 UT on February 21, 2001. The spacecraft crossed from the magnetosheath into the unperturbed solar wind at 18:42UT. At 21:42 UT the spacecraft crossed into the foreshock, due to a change in the orientation of the interplanetary magnetic field (Figure provided by J. P. Eastwood).

Farris et al. model shock was used (Farris et al., 1991). If one assumes that Cluster is connected to the shock by straight magnetic field lines, the orientation of each magnetic field vector can be used to identify the intersection point on the model shock. Once this point has been identified, the angle between the magnetic field and the shock normal at the shock surface, θ_{Bn}, can be computed. The results of this calculation are shown in Figure 2.8. The transition into the foreshock and the subsequent increase in magnetic field activity occurred at 21:42 UT. The increased variation in θ_{Bn} after 21:42 UT reflects the input variation.

The increased wave activity is associated with the change in θ_{Bn} to quasi-parallel geometries associated with the IMF orientation becoming increasingly radial. Initially, the IMF is aligned to the Parker spiral and points sunwards. It then rotates towards the Earth-Sun line. The entry into the foreshock and the onset of wave activity is controlled by the magnetic field orientation. The transition into the foreshock is not accompanied by any dramatic changes: it would therefore appear that the solar wind field orientation is changing slowly enough for the foreshock to remain stable.

These Cluster findings are consistent with the previous statistical study of Le and Russell (1992a). They found that where $\theta_{Bn} \leq 50^\circ$, the foreshock wave field

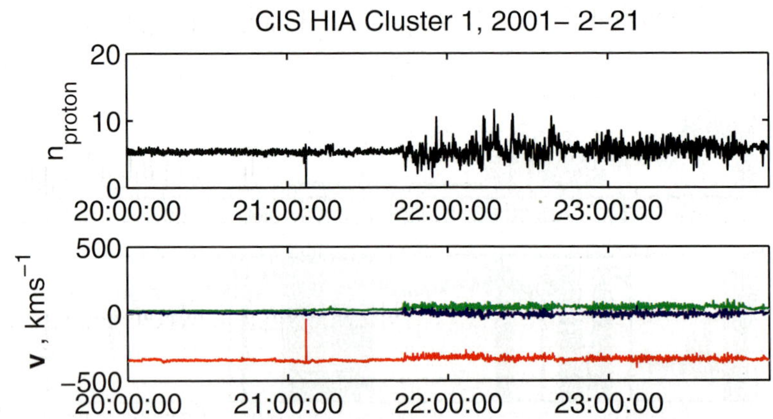

Figure 2.7. Plasma data recorded by CIS/HIA on Cluster 1 between 20:00UT and 24:00UT on February 21, 2001. The top panel shows the plasma density. The bottom panel shows the velocity, where the colours red, green and blue correspond to GSE V_x, V_y and V_z respectively. The HIA sensor, which does not discriminate between ion species, was operating in solar wind mode. Cluster 1 was upstream of the bow shock throughout this interval. At 21:42UT, the spacecraft crossed into the ion foreshock, and into the foreshock ULF wave field. (Figure provided by J. P. Eastwood).

is encountered. The foreshock ULF wave field boundary is controlled by changes in the orientation of the IMF. As the field changes direction, the θ_{Bn} of the magnetic field line connecting Cluster to the shock changes. There is no delay in the onset of the waves once $\theta_{Bn} \approx 50°$, suggesting that the foreshock retains a coherent structure and bodily moves as the orientation of the IMF changes. The foreshock ULF boundary, like the foreshock itself, responds quasi-statically to changes in IMF orientation.

At larger spacecraft separations, differences between the Cluster spacecraft become apparent. Figure 2.9 shows the magnetic field observed by Cluster 1 on January 16, 2003, where again a transition into and out of the foreshock was caused by a change in interplanetary magnetic field orientation. Initially, the spacecraft were located in the solar wind. We are interested in the burst of wave activity that occurred shortly after 04:00UT, but before the spacecraft crossed the bow shock at 05:54UT.

Figure 2.10 shows the GSE latitude, θ_B, of the magnetic field observed by each of the four Cluster spacecraft in the interval 04:05UT - 04:35UT. At this time, the solar wind magnetic field was largely confined to the ecliptic plane. The spacecraft encounter the wave activity in the order Cluster 1 - 2 - 3 - 4, but exit in order 4 - 3 - 2 - 1. This is more likely to be the signature of a boundary moving back and forth over the Cluster tetrahedron rather than a wave packet being convected in the solar wind. Referring to Figure 2.9, the dashed red line in the center panel indicates $\Phi_B = 315°$, i.e., the orientation of the Parker spiral. By 04:20, Φ_B had

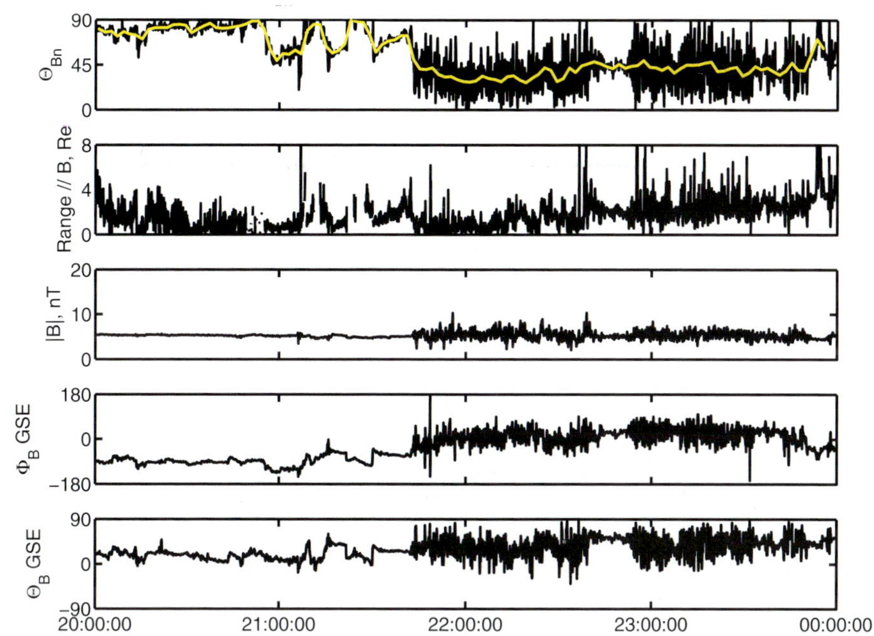

Figure 2.8. The association of foreshock wave activity with quasi-parallel shock geometries. The top panel shows θ_{Bn} for the magnetic field line that connects Cluster 1 to the shock as a function of time (yellow line shows 120 s averages of θ_{Bn}). The second panel shows the shock Cluster separation, along the magnetic field observed at Cluster. The bottom three panels show the magnetic field in GSE polar coordinates. The onset of wave activity (as seen in the bottom three panels) corresponds to a change in the IMF orientation, such that the Cluster spacecraft become connected to quasi-parallel shock geometries. This is consistent with previous statistical studies. (Figure provided by J. P. Eastwood).

increased to 335°, and waves were observed. Calculations suggest that the change in IMF orientation towards and then away from a more radial configuration caused the four spacecraft to briefly enter the foreshock and the foreshock ULF wave field. This example illustrates the potential of Cluster to resolve the motion of boundaries in the foreshock, and the possibility of studying the changes in plasma properties across such boundaries, a challenge that is well suited to the spacecraft separation strategy planned for the later stages of the Cluster mission.

2.3 Upstream Waves and Particles

It has been mentioned earlier that the main field signature of the ion foreshock is the presence of low frequency fluctuations. Identification of the properties of these waves, their nature, dispersion, diffusion and free energy source as well as their effect on the solar wind are therefore of prime interest. In this section we review Cluster observations and the Cluster-based analysis of these waves.

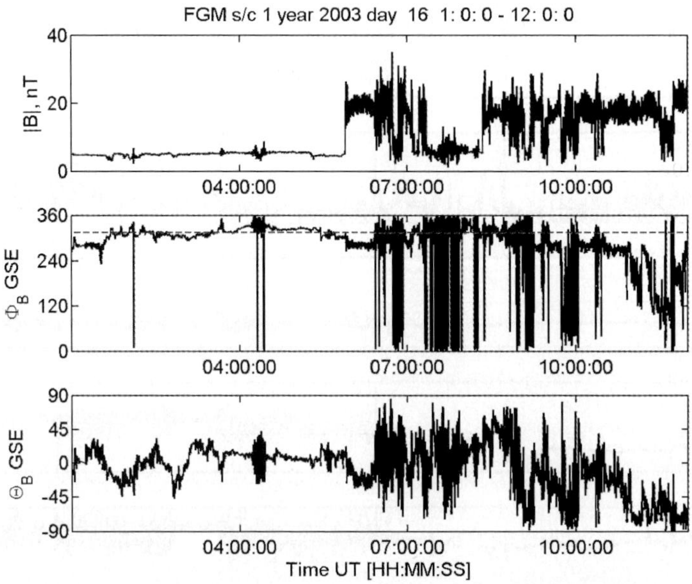

Figure 2.9. Magnetic field observed by Cluster 1 between 00:00 and 12:00UT, January 16, 2001. The data are expressed in GSE polar coordinates at spin resolution. The dashed red line in the middle panel represents the azimuth corresponding to a sunward-pointing Parker spiral. A short interval of foreshock wave activity was observed between 04:10UT and 04:30UT, when the azimuthal angle was at its largest. It is thought that the rotation of the magnetic field in the ecliptic plane caused the spacecraft to enter and exit the foreshock. (Figure provided by J. P. Eastwood).

2.3.1 Low-frequency waves

2.3.1.1 Monochromatic fast magnetosonic waves

A particular type of ULF waves are the so-called 30 s (period) waves, a term which corresponds to quasi-monochromatic ultra-low-frequency waves with characteristic periods of approximately 30 s in the magnetic field (e.g., Le and Russell, 1994). These waves are usually observed to be left handed in the spacecraft reference frame (Fairfield, 1969). Minimum variance analysis can be applied to single spacecraft observations and used to compute the direction of wave propagation, with a 180° ambiguity (e.g., Song and Russell, 1999), but it is not possible to identify the wave speed, the exact direction of propagation or the wavelength; for this, multi-point observations are required. Analyses of ISEE dual spacecraft data have shown such waves to in fact be intrinsically right handed, propagating against the solar wind flow (Hoppe et al., 1981; Hoppe and Russell, 1983). Their polarisation identifies them as kinetic fast magnetosonic waves (Krauss-Varban et al., 1994), generated via the ion-ion right-hand resonant beam instability (e.g., Brinca, 1991;

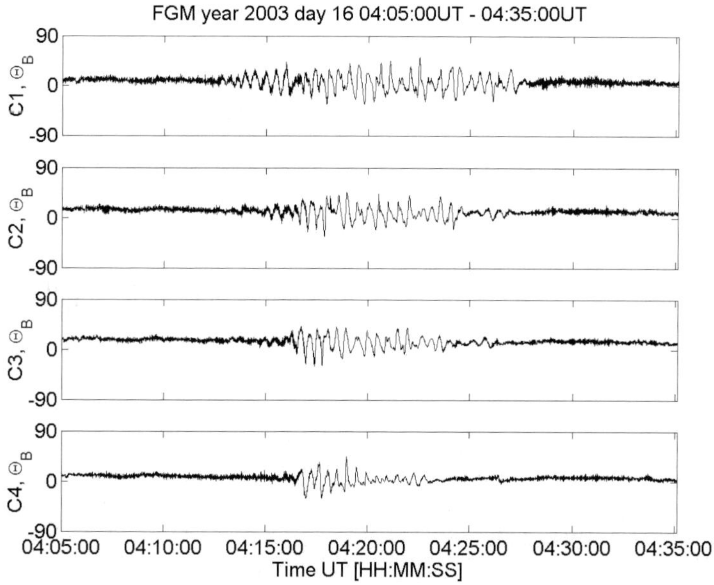

Figure 2.10. The GSE latitude, θ_B, of the magnetic field observed by each of the four Cluster spacecraft between 04:05UT and 04:35UT on January 16, 2001. The onset of wave activity is a function of spacecraft location. The signatures are nested, rather than convected; it is therefore likely that the spacecraft crossed into and out of the ULF wave field. Calculations (not shown) indicate that the rotation of the average IMF orientation is consistent with this interpretation. (Figure provided by J. P. Eastwood)

Gary, 1993) and transferring beam energy and momentum via the resonantly excited mode to the main plasma component.

Previous single and dual spacecraft analysis has been based on MVA of the magnetic field. Cluster can be used to test the performance of MVA and also provides the possibility of routinely determining the correlations between the plasma and field perturbations (Fazakerley et al., 1995). Figure 2.11 shows the spin resolution Cluster 1 FGM magnetic field between 00:00 UT and 15:00 UT on April 23, 2001, in GSE polar coordinates, as observed by Eastwood et al. (2002). The upper panel shows the magnetic field strength, the central panel shows Φ_B and the lower panel shows θ_B, the elevation of the magnetic field vector out of the ecliptic plane. From the start of this interval until 08:12 UT, Cluster was located in the solar wind. At this time, Cluster was moving Earthward from apogee in the solar wind towards a southern cusp entry. The spacecraft separation was $\sim$600 km.

Initially, the magnetic field was strongly southward, changing its orientation to lie in the ecliptic plane at 04:13 UT with the Cluster spacecraft becoming magnetically connected to the bow shock at this time. The magnetic field remained in this orientation until about 07:30 UT, when the field direction began to rotate such that

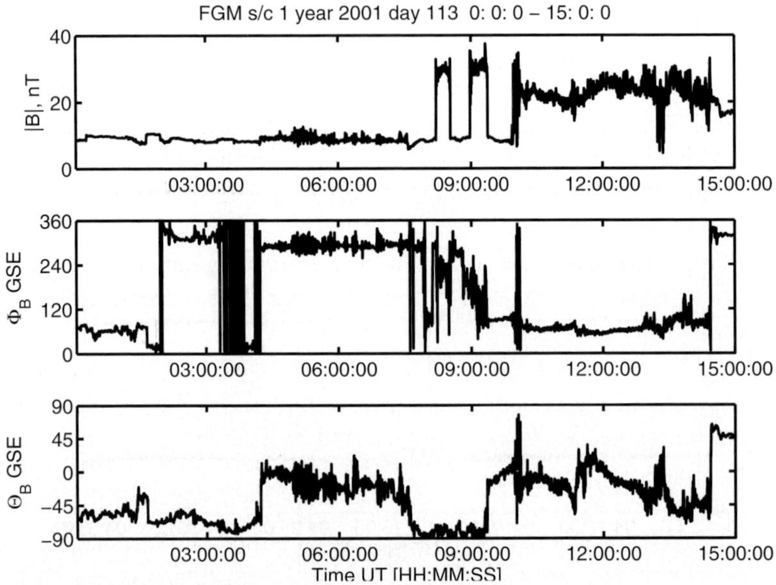

Figure 2.11. The spin resolution magnetic field observed by Cluster 1 on April 23 (day 113), 2001 between 00:00 UT and 15:00 UT (Figure provided by J. P. Eastwood).

it pointed southward again after 08:00 UT. From 08:12 to 10:00 UT, Cluster encountered the bow shock 5 times, the multiple encounters being due to the motion of the shock itself.

While Cluster was located in the foreshock, there was a prolonged period of wave activity starting just before 05:00UT and continuing, albeit intermittently, until the spacecraft became disconnected from the shock. Figure 2.12 shows the magnetic field measured by all four spacecraft in the interval 05:00-05:30 UT, when the spacecraft were located in the foreshock. 'Quasi-monochromatic' waves were observed at this time being largely coherent between the spacecraft. MVA showed that the waves were left-hand polarised in the spacecraft frame.

By assuming that the waves move at constant speed relative to the spacecraft, and also assuming that the waves are planar (their scale size is $\sim$1 R_E , see e.g., Le and Russell, 1990), an order of magnitude greater than the spacecraft separation), a timing analysis can be applied to the wave phase fronts (Schwartz, 1998). The time differences obtained from cross-correlation of the low-pass filtered time series (Bendat and Piersol, 1986) yield the wave phase velocity in the spacecraft frame, while the autocorrelation gives the wave frequency in the spacecraft frame, and consequently, the wavelength can be calculated. Finally, with the help of the known solar wind velocity the Doppler shift is obtained, and the phase speed and frequency in the solar wind rest frame can be determined.

Applying this technique to the interval 05:00-05:06 UT, the waves were found to have a speed of -220 km s^{-1} in the direction $\mathbf{n} = (0.67, -0.73, -0.13)$, which makes an angle of 25° against the magnetic field. The phase velocity in the solar wind rest frame was 65 km s^{-1}, compared to an Alfvén speed of 75 km s^{-1} and a sound speed of 30 km s^{-1}. The waves propagate sunwards in the solar wind rest frame and are intrinsically right handed, which identifies them as being on the fast magnetosonic branch of the kinetic dispersion relation (Krauss-Varban et al., 1994). The wavelength was found to be ~6000 km and the frequency 0.014 Hz, an order of magnitide below the local ion cyclotron frequency of 0.13 Hz. Hence this analysis is consistent with and confirms previous dual spacecraft analysis, and other multi-spacecraft analyses.

The multi-spacecraft capabilities of the Cluster mission are complemented by the existence of 3D plasma data at a resolution that allows the variations within a wave cycle to be resolved. This is demonstrated by Figure 2.13, which shows the correlation between the magnetic field strength (shown in black at 5 Hz resolution) and the density computed at 4 s resolution from the full 3D particle distribution on board the spacecraft (shown in red). The density and field perturbations are clearly correlated as already shown by ISEE observations (Paschmann et al., 1979). It is immediately apparent that these waves are fast magnetosonic, since the perturbations in the field strength are correlated with the density perturbations.

Figure 2.14 shows the results of a case study of MVA performance, reported by Eastwood et al. (2002). The interval 05:00-06:00 UT was filtered to remove high frequency noise and split into 30 2-min sections. Of these, 6 intervals were rejected on the basis of insufficient wave activity. For the other intervals, the minimum variance direction, averaged over the four spacecraft was calculated, as was the 4 spacecraft direction of propagation. The angle between the propagation direction and the average minimum variance direction (forced to point upstream) was then calculated. This angular difference is plotted as a function of intermediate/minimum eigenvalue (λ_2/λ_3) in Figure 2.14. In the case of linearly polarised signals, the variation is confined to one direction, identified as the maximum variance direction. The intermediate and minimum variance directions are degenerate, and consequently $\lambda_2/\lambda_3 = 1$. For circularly or elliptically polarised waves, $\lambda_2/\lambda_3 > 1$. If the ratio of eigenvalues is >10, it is clear that the two techniques are in agreement. If the ratio is <5, the estimates from the two techniques diverge. This is presumably due to the failure of MVA in the limit that the waves become more and more linearly polarised.

2.3.1.2 Kinetic Alfvén waves

Under typical foreshock conditions, the ion-beam instability generates kinetic fast magnetosonic waves (cf., e.g., Gary, 1993), which appear left handed in the spacecraft time series. However, if the beam is hot, or if the core distribution is anisotropic, the growth rate of the right-hand mode may be exceeded by the left-hand mode

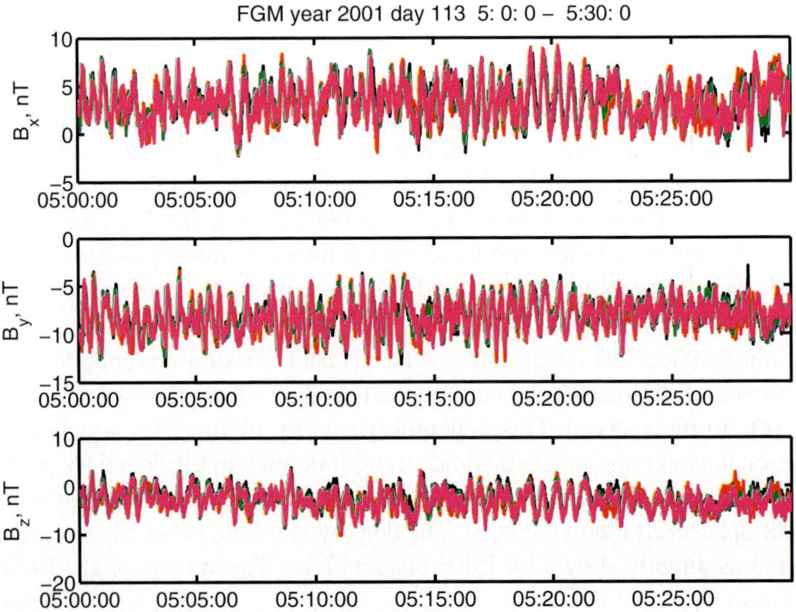

Figure 2.12. Magnetic field observed by the Cluster spacecraft between 05:00 and 05:30 UT on April 23, 2001. The time series are coloured black, red, green and magenta corresponding to spacecraft 1, 2, 3 and 4. Spacecraft 4 is plotted last: hence the magenta dominates because the time series are well correlated. The data are at 22 Hz resolution. (Figure provided by J. P. Eastwood).

growth rate, in which case the dominant unstable wave mode is the left-handed Alfvén/Ion cyclotron or 'kinetic Alfvén' wave.

Two possibilities arise for observation of left-hand polarised waves in the spacecraft time series. The first is that they are intrinsically right handed, and are propagating anti-sunwards, in the opposite direction to the beam that generated them. The solar wind Doppler shift would not cause a reversal in polarisation. The second is that they are intrinsically left handed, and propagate sunwards with the beam. The solar wind Doppler shift here causes the observed polarisation to be reversed.

In the first case, the ion beam excites such waves through the non-resonant firehose instability (Sentman et al., 1981). In the second case the left-hand resonant ion beam instability dominates (Gary, 1985). The firehose instability requires the beam to be fast and dense, whereas the left-hand resonant instability requires the beam to be hot. Before the launch of Cluster the exact nature of waves observed to be right-handed in the spacecraft frame was not clear. Single spacecraft studies comparing experimental wave transport ratios with those expected using kinetic theory suggested the existence of Alfvén waves (Blanco-Cano and Schwartz, 1997). Dual spacecraft studies were limited by geometry; reviews of foreshock physics before

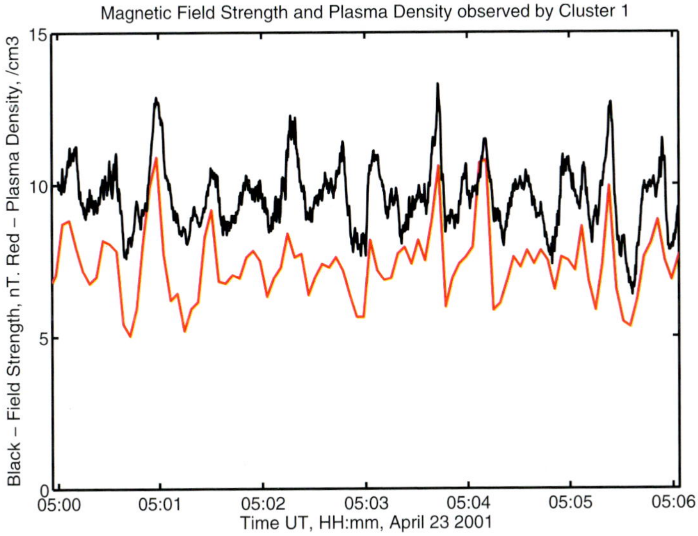

Figure 2.13. Correlation between magnetic field strength (black) and ion density as measured by HIA (red). Note that the cadence of the ion measurements is 4 s, whereas the magnetic field is shown at 5Hz resolution. The correlation between the two parameters indicates the existence of a fast mode wave according to MHD models. (After Eastwood et al., 2002).

the launch of Cluster noted that observations of right handed waves were not well understood (e.g., Le and Russell, 1994).

Figure 2.15 provides an overview of the data recorded by Cluster 1 FGM, as reported by Eastwood et al. (2003). The left three panels show the magnetic field recorded by the Flux Gate Magnetometer on Cluster 1 from 00:00-14:00 UT, at 4 s resolution. Initially, Cluster was in the solar wind, crossing extended bow shock structure just after 05:00 UT and entering the magnetosphere at approximately 09:15 UT. The four left panels show an interval of foreshock wave activity, between 04:00 UT and 04:10 UT. The top panel is the magnetic field strength, and the lower three panels are the components of the field in GSE coordinates. The data are shown at 22Hz resolution. The waves are exceptionally well defined, with very low turbulent noise, and have periods of approximately 10 s in the spacecraft frame.

Minimum variance analysis was applied to the interval 04:04:00-04:04:30 UT. The results are shown in Figure 2.16. The upper four panels show hodograms for each of the four spacecraft. In each case, the waves are right handed with respect to the magnetic field. The data are shown at 22 Hz resolution, and have not been smoothed or filtered. The waves are almost circularly polarised.

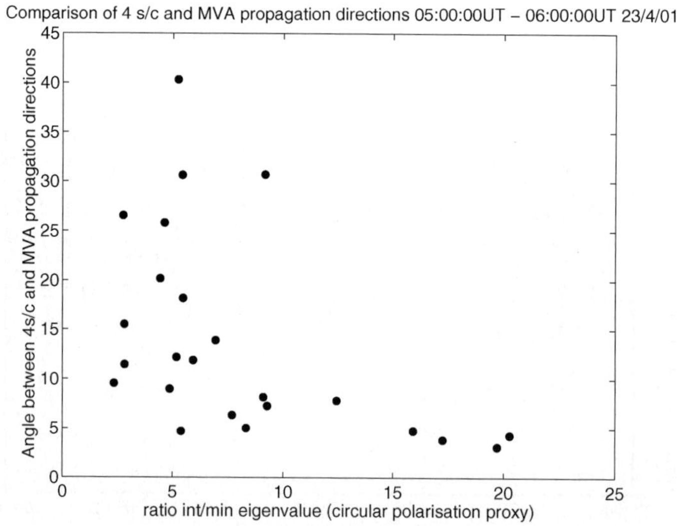

Figure 2.14. Performance of MVA: in the limit of linear wave polarisation, the MVA and 4 spacecraft estimates of wave propagation direction diverge. (From Eastwood et al., 2002).

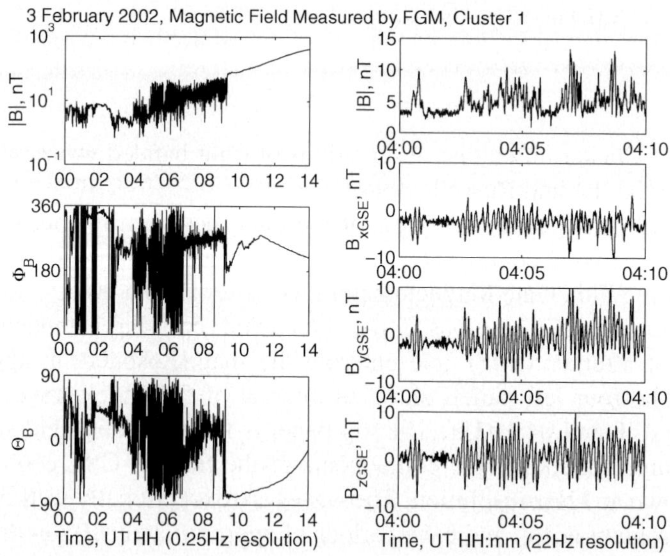

Figure 2.15. Overview of Cluster magnetic field observations on February 3, 2002. (From Eastwood et al., 2003).

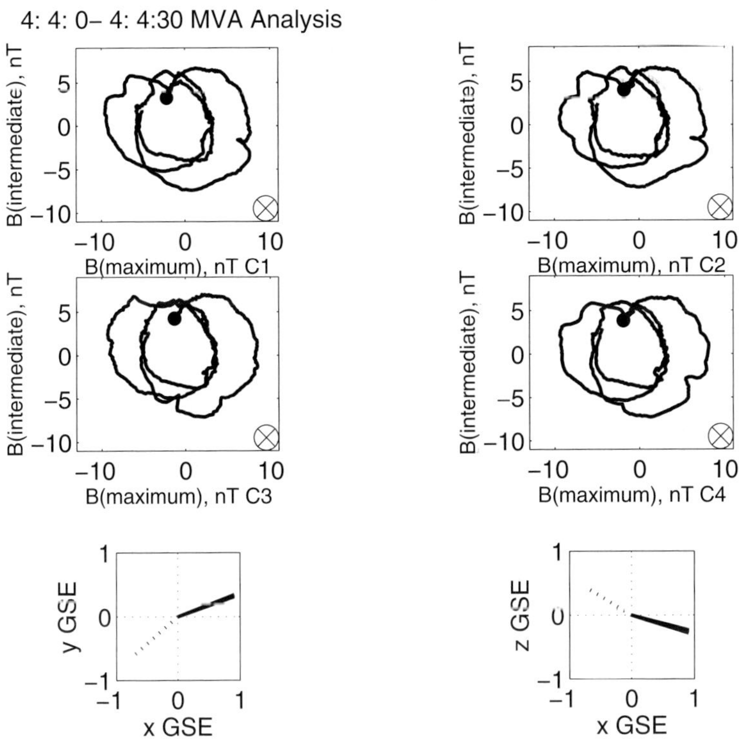

Figure 2.16. Results of MVA applied to Cluster magnetic field data for wave activity observed on February 3, 2002. The lower two panels show the projections of the field (dotted line) and minimum variance directions (solid line) into the $x-y$ and $x-z$ GSE planes. (From Eastwood et al., 2003).

Applying the same technique as before, Eastwood et al. (2003) found that in the solar wind rest frame the waves propagate away from the shock at speeds of the order of the local Alfvén speed. The k-filtering technique (Pinçon and Motschmann, 1998) was also applied to the data. The frequency at which the maximum power was carried was chosen, and the distribution of field energy density in k-space at this frequency was estimated. A single maximum was found in the distribution. Applying the solar wind Doppler shift, the phase speed in the plasma frame was again found to be of the order of the Alfvén speed, directed back into the upstream region.

Both techniques result in the same conclusion: The waves are intrinsically left handed attempting to propagate upstream but are blown anti-sunward, causing the observed polarisation to be reversed. These waves are therefore presumably generated by hot back-streaming distributions.

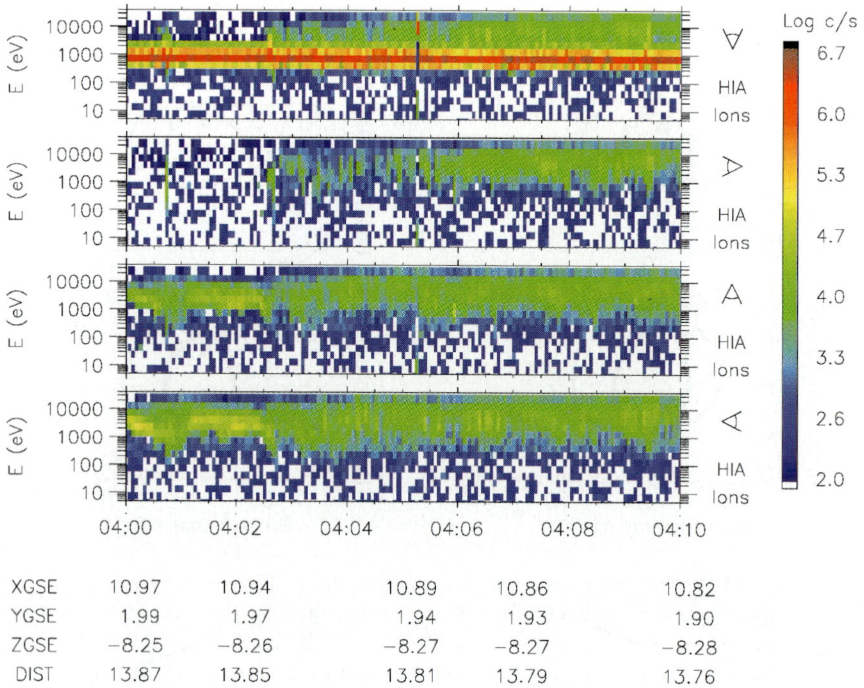

Figure 2.17. Ion energy spectrograms for the interval 04:00-04:10 UT on February 3, 2002. Counting from the top, the panels show the ion fluxes in the sunward, duskward, earthward and dawnward look directions, as obtained from the CIS/HIA sensor on Cluster 3. In the sunward direction, the solar wind is manifested as a continuous signal just below 1 keV. (Adapted from Eastwood et al., 2003).

Figure 2.17 shows energy spectrograms from the CIS on Cluster 3 for the interval 04:00-04:10 UT. After 04:02:30 UT, at the onset of the waves, there are significant fluxes of energetic ions in all look directions at energies extending to above 10 keV. Figure 2.18 shows the ion distribution from CIS/CODIF averaged over the interval 04:0304:10UT, where the hot backstreaming ('diffuse') ion distribution can clearly be seen (the backstreaming ions are to the left; the red region to the right corresponds to the unresolved solar wind). The ion distribution is qualitatively consistent with the wave properties. Before the onset of the waves, the ions are not diffuse; instead a dense backstreaming beam was observed. Further work quantifying this association is ongoing.

This result confirms the existence of waves in the foreshock that lie on the Alfvén/Ion-cyclotron branch of the kinetic dispersion relations. The waves are intrinsically left handed, but appear as right handed in the spacecraft time series. This result does not preclude the existence of waves driven by firehose instabil-

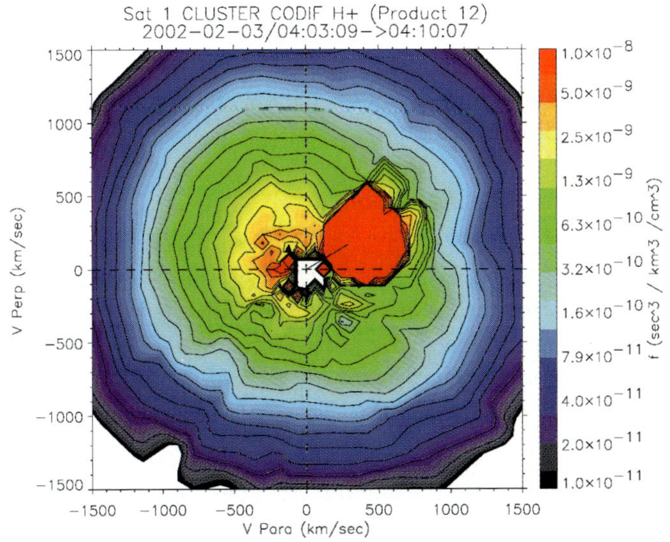

Figure 2.18. Ion distribution from CIS/CODIF on Cluster 1 on February 3, 2002, averaged over the interval 04:03 - 04:10UT, indicating the existence of a hot, diffuse, ion population. The back streaming ions fill the plot; the red region to the right corresponds to the unresolved solar wind. The existence of a hot ion distribution is necessary to produce the observed waves. (Figure provided by M. Scholer).

ities, which would have the same polarisation in the spacecraft time series. The waves were not specifically identified before the launch of Cluster at least not for their remarkably clear polarisation.

2.3.1.3 Oblique ULF waves

The left handed (in the spacecraft frame) quasi-monochromatic ULF foreshock waves with 30 s period are excited by back streaming ions through the right-hand resonant ion-ion beam instability. The linear growth rate of this instability is maximum for parallel propagation. This is in contrast to experimental observations, which have consistently shown that such waves propagate obliquely to the field. Eastwood et al. (2004) have recently shown that for large IMF cone angles, although the 30 s waves are observed to propagate obliquely, their propagation direction is confined to the plane defined by the x-GSE direction and the magnetic field.

This has been interpreted as a non-local effect. When the IMF cone angle is large, the waves are convected across energetic particle gradients by the solar wind. The upstream waves are excited by the reflected ions in the quasi-perpendicular region and then refracted as they enter the quasi-parallel region (Hada et al., 1987). As shown in 2-D simulations of a quasi-parallel shock by Krauss-Varban and

Omidi (1993) and Scholer et al. (1993), the waves excited far upstream by the diffuse ions are also refracted in the region of large diffuse ion gradients closer to the shock.

However, the term refraction should be used with care: firstly, the wavelength is comparable to the scale over which the superthermal particle distributions change, and secondly, the propagation of the waves through the inhomogeneous medium is altered by the convective plasma motion.

2.3.1.4 Dispersion relation of low frequency waves

One way to identify wave modes is by examining the dispersion relations directly. However, in the case of the foreshock, a correction must be made for the Doppler shift. The rest frame frequency is given by

$$\omega_{rest} = \omega_{sc} - \mathbf{k} \cdot \mathbf{V}_{\mathbf{sw}} \tag{2.1}$$

where ω_{rest}, ω_{sc}, $\mathbf{k}$ and $\mathbf{V}_{\mathbf{sw}}$ are the angular frequency in the plasma and spacecraft frames, wave vector, and solar wind velocity, respectively. The spacecraft velocity being of the order of a few km s^{-1} is neglected with respect to the solar wind speed.

Hoppe et al. (1981) and Hoppe and Russell (1983), based on ISEE 1 and ISEE 2 magnetic field data, determined plasma frame frequencies and wave numbers of magnetohydrodynamic waves in the foreshock. Dudok de Wit et al. (1995), Balikhin et al. (1997a,b) analysed AMPTE-UKS and AMPTE-IRM magnetic field data. Using MVA and projected wave numbers they derived approximate dispersion relations in the plasma frame assuming that the quasi-monochromatic wave assumption holds for the propagation direction.

Narita et al. (2003) used magnetic field and CIS/HIA plasma data from Cluster 3 for the time interval 1700-1730 UT on February 20, 2002 to study the dispersion relation of low frequency waves in the foreshock. They applied the wave telescope/k-filtering technique (Pinçon and Lefeuvre, 1991; Motschmann et al., 1996; Glassmeier et al., 2001) to determine the wave-vectors $\mathbf{k}$. After determination of the wave-vector corresponding to the largest wave power, the Doppler shift was calculated with the help of Eq. (2.1) using the fluid velocity moment. Applying principles of the optics of quasi-monochromatic waves (Fowler et al., 1967; Arthur et al., 1976; Born and Wolf, 1980), the ellipticity of polarisation in the frequency domain was found, being -1 for left-handed, 0 for linear, and 1 for right-handed waves.

Figure 2.19(a) displays the rest frame frequencies and their associated wave numbers with the sign of ellipticity overplotted, where wave number k is projected to the mean propagation direction $\theta_{kB} = 24°$ in the plasma rest frame. Right- and left-handed polarisations are represented by diamonds and plus signs, respectively. The frequency is normalised to the proton cyclotron frequency $\Omega_{cp} = 1.1$ Hz, and the wave number is normalised to $(V_A/\Omega_{cp})^{-1} = 10.3\,\mathrm{km}^{-1}$. The propagation direction is nearly aligned to the magnetic field direction at various frequencies.

The insert zooms into the dispersion relation at $-0.2 < \omega/\Omega_{cp} < 0.4$ for and $0.0 < kV_A/\Omega_{cp} < 0.3$.

Two different branches of the dispersion relation show up from this analysis, one being slightly curved and connected to $(\omega, k) = (0, 0)$, the other being a straight line connected to $(\omega, k) = (-\Omega_{cp}, 0)$ and intersecting the former branch. Both right and left handed polarisations are identified on both branches. The dispersion of these waves is consistent with the ion-ion interaction picture between the bow-shock reflected ion beam and the solar wind ion flow.

Figure 2.19(b) shows the theoretical cold plasma dispersion branches for a cold beam to plasma density ratio $n_b/n = 0.001$ (i.e., 0.1 %), $M_A = 5.6$ for the Alfvén Mach number of the beam, and quasi-parallel propagation ($\theta_{kB} = 24°$). The wave propagation angle θ_{kB} has been chosen to be the same as the mean propagation angle in the observation. Due to the anomalous Doppler effect, a right-handed wave exists at resonant frequency $\omega = kV_b - \Omega_p$ propagating along the ion beam. Four other wave modes turn up, all of them result from the single ion, cold plasma system: R^+, L^+, L^-, R^-, where '+' stands for a forward propagation with respect to the magnetic field and '-' for a backward propagation, and 'R' for a right-hand polarisation and 'L' for a left-hand polarisation. At and around a crossover frequency between 'R^+' and 'Res.', stationary nonlinear structures in a beam plasma system are found (Sauer et al., 2001; Sauer and Dubinin, 2003). They behave like solitons and so-called oscillitons.

Application to the results shown in Figure 2.19(a) yields that the major curved branch corresponds to the R^+ branch in Figure 2.19(b). The straight branch also corresponds to the resonant branch. However, not only right-handed but also left-handed polarisations are found in the observations, and some waves even deviate from the major branch around and above the proton cyclotron frequency. The curvature of the major branch is also slightly different from the R^+ branch. These disagreements must be resolved in the future by, for example, comparing different methods of polarisation analysis, testing various models of dispersion relation, including hot beams in a $\beta \simeq 1$, and investigating more events.

2.3.1.5 *Statistics of low frequency waves*

The Cluster spacecraft provide the unique opportunity to undertake a statistical study of LF wave structure and propagation in the foreshock based on FGM magnetic field and CIS/HIA plasma data.

Various distributions of the foreshock wave properties like frequencies, wave numbers, phase velocities, propagation directions, and polarisation have been investigated by Narita et al. (2004) who for the statistical study selected intervals of wavelengths more than ~ 200 km from the small spacecraft separation phase (February 3 - June 17, 2002) when the separation distance was ~ 100 km.

The wave telescope/k filtering technique was applied to 36 wave events selected in order to find the wave vectors $\mathbf{k}$ for wave frequencies < 0.5 Hz in the spacecraft

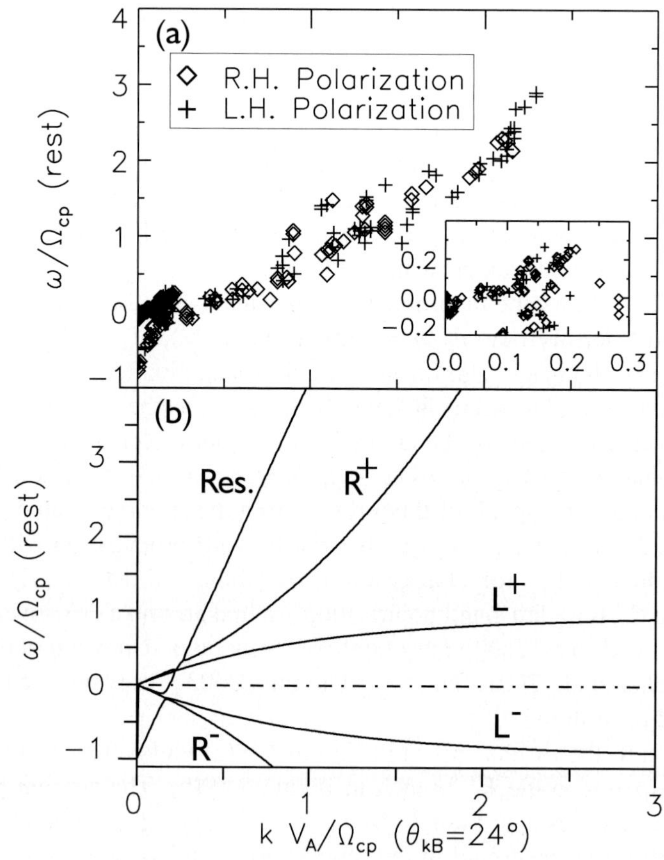

Figure 2.19. Dispersion relations in the rest frame derived from different sources: dispersion from the wave telescope and polarisation from the minimum variance analysis for (a) and computation using a multi-fluid plasma model for (b). Wave number k is projected to a mean propagation direction $\theta_{kB} = 24°$. Angular frequency ω and wave number k are normalised. (From Narita et al., 2003).

frame. Based on Cluster 3 CIS/HIA velocity moments this frequency was Doppler transformed into the plasma frame. Moreover, using quasi-monochromatic optical technique to determine the spectral density matrix (Bendat and Piersol, 1980; Born and Wolf, 1980) for the magnetic fluctuation field $\mathbf{B}(f,T)$ as function of frequency f and time record length T, the polarisation angle β was computed.

The results of this investigation are displayed in Figure 2.20 which shows the spatial distribution of phase velocities in the plasma rest frame projected into the xr plane in GSE ($r = \sqrt{y^2 + z^2}$). Small black filled circles give the locations, while the arrows indicate phase velocities normalised to the local Alfvén velocity V_A. The dashed curve is a nominal bow shock under quiet solar wind conditions.

Most of the waves propagate upstream and are more or less aligned with shock normal directions at various positions, near the x-axis to near shock flanks. Phase

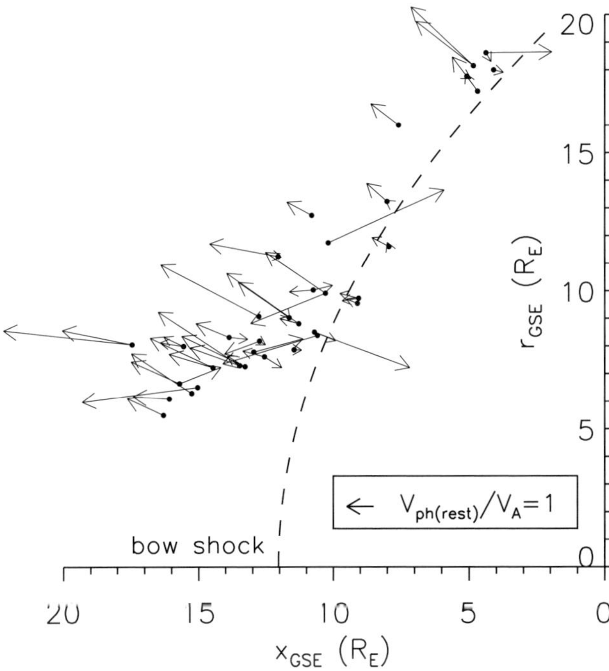

Figure 2.20. Spatial distribution of wave phase velocities in the plasma rest frame plotted together with the location of wave observations, projected into the xr plane in GSE ($r = \sqrt{y^2 + z^2}$). The phase velocities are normalised to the local Alfvén velocity. The dashed, curved line is a nominal bow shock. (From Narita et al., 2004).

velocities are found the order of the Alfvén speed. Some of the waves, however, propagate downstream in the plasma rest frame. Such waves tend to have phase velocities smaller than the Alfvén velocity, but a few cases near the shock show larger phase velocities.

Figure 2.21 shows distributions of frequencies, wave numbers, propagation angles, and ellipticities. The left panel in Figure 2.21 gives the distribution of normalised frequencies, $\omega_{rest}/\Omega_{cp}$, as function of the magnitudes of normalised wave numbers kV_A/Ω_{cp}, i.e., the average dispersion relation. A magnification of the region near $\omega/\Omega_{cp} \sim 0$ and $kV_A/\Omega_{cp} \sim 0.1$ has been embedded into the panel.

The dotted line represents Alfvén waves propagating along **B** with dispersion relation $\omega/k = V_A$. Most of the waves are found at $\omega/\Omega_{cp} \sim 0.1$ and $kV_A/\Omega_{cp} \sim 0.1$. The major wave population thus propagates at speeds below and up to the Alfvén velocity. Some waves are found up to $\omega/\Omega_{cp} \sim 5$ and $kV_A/\Omega_{cp} \sim 1.2$. These waves are scattered in dispersion, few of them following the Alfvén branch.

The middle panel in Figure 2.21 shows the distribution of frequencies as a function of θ_{kB}. Waves which propagate within 30° of the magnetic field line tend to be

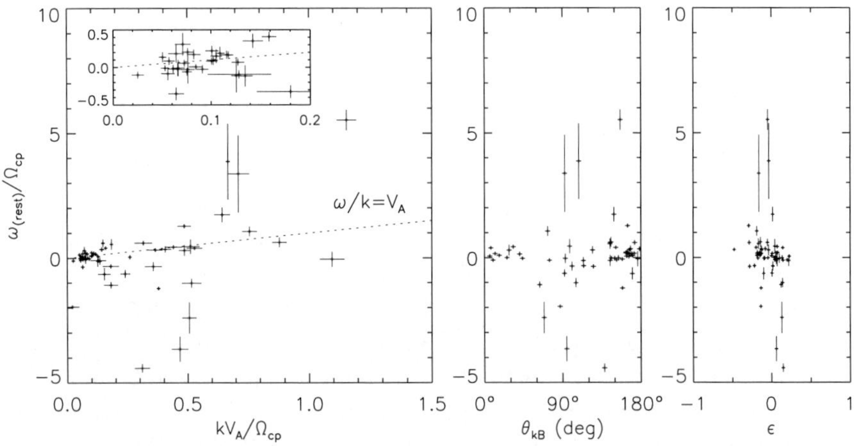

Figure 2.21. Distribution of rest frame frequencies, wave numbers, propagation angles, and ellipticity in the plasma rest frame including error bar estimates. The small plot embedded in the left panel magnifies the frequencies and the wave numbers near the origin. The dotted, straight line in the left panel is the dispersion relation for the extended linear Alfvén waves. (From Narita et al., 2004).

of lower frequency, with higher frequency waves only propagating perpendicular to the field.

The right-hand panel in Figure 2.21 shows the distribution of frequencies as a function of wave ellipticity. The average low frequency wave polarisation is left handed, with the higher frequency minor waves being left-handed when propagating upstream and right-handed for downstream propagation.

It is thus concluded that the foreshock in this frequency range is dominated by upstream propagation in agreement with other conclusions (Russell et al., 1971; Hoppe et al., 1981; Hoppe and Russell, 1983; Eastwood et al., 2002, 2003; Narita et al., 2003). In addition, waves at rest frame frequencies $\sim 0.1 \times \Omega_{cp}$ with wave numbers $\sim 0.1 \times \Omega_{cp}/V_A$ (wavelengths $\sim R_E$) have been identified. Some wave propagating nearly perpendicular to the magnetic field were also found. Their presence may be important in understanding the physical processes in this region.

2.3.2 Origin of gyrophase bunched ions

Backstreaming ion beams of several keV, collimated along interplanetary field lines, have been observed close to the ion foreshock boundary. Downstream of the field-aligned beam region, distributions characterised by a gyromotion around the magnetic field, i.e., a non-vanishing perpendicular bulk velocity with respect to the background magnetic field, have been reported. These gyrating ion distributions are nongyrotropic or nearly gyrotropic. Numerous studies concerning gyrating ions have been reported in earlier investigations mainly from ISEE 1 and 2 (Gosling et al., 1982; Thomsen et al., 1985; Fuselier et al., 1986; Fuselier et al., 1986a,b), AMPTE (Fazakerley et al., 1995) and WIND (Meziane et al., 1997; Meziane et al., 2001; Mazelle et al., 2000). Gyrating ions are often observed in association with ULF waves having substantial amplitude (Fuselier et al., 1986b). The waves are right-handed and propagate nearly along the ambient magnetic field (Thomsen et al., 1985). It is believed that the ULF waves are excited through a beam plasma instability (Gary et al., 1981) resulting from the propagation of field-aligned ions which precede them closer to the foreshock boundary.

Two mechanisms have been put forward to explain the origin of upstreaming gyrating ions. In one mechanism, a portion of the incoming solar wind is reflected in a specularly manner at the shock (Gosling et al., 1982; Gurgiolo et al., 1983). In the second mechanism, the waves produced through a beam-plasma instability can in turn trap the ions and cause the phase bunching of the distribution in what is called a beam disruption mechanism (Hoshino and Terasawa, 1985). Significant new results have been obtained from Cluster observations on this topic.

2.3.2.1 Bow shock specular reflection mechanism

Solar wind specular reflection at the quasi-parallel bow shock should produce non-gyrotropic ion distributions in the foreshock. When propagating upstream, the bunched ions undergo gyrophase mixing within a few Earth radii from their source on the shock leading to nearly gyrotropic distribution function (Gurgiolo et al., 1983). Gyrotropic distributions have not been observed in 3D measurements; however, 2D measurements suggest their presence within ~ 4 R_E from the shock (Fuselier et al., 1986b). This also indicates that shock-produced non-gyrotropic distributions are rarely observed beyond ~ 4 R_E . Very few observations of gyrophased bunched ions propagating away from the shock in the upstream region consistent with production by specular reflection have been reported in the literature. Gosling et al. (1982) were the first to report evidence of such specular reflection based on data from the FPE experiment obtained when ISEE was just upstream of the bow shock. Another event consistent with the specular reflection, observed at 4 R_E and in association with ULF wave activity, was also reported by Thomsen et al. (1985).

As shown by Gosling et al. (1982), only under quasi-parallel geometries are the guiding centers of the specularly reflected particles oriented upstream, thus allowing them to escape. In fact, because of kinematic considerations, $\theta_{Bn} < 39.9^\circ$ is the limiting shock geometry allowing ions to escape upstream (Schwartz et al., 1983). In presence of large amplitude ULF MHD waves near the shock front, always observed in association with gyrating ions, the angle between the shock normal and the magnetic field can vary substantially. This can cause the shock to switch back and forth from a quasi-parallel to quasi-perpendicular configuration (Greenstadt and Mellott, 1985) and is illustrated in Figure 2.22. As shown theoretically by Fuselier et al. (1986), this effect, which depends on the characteristics of the waves,

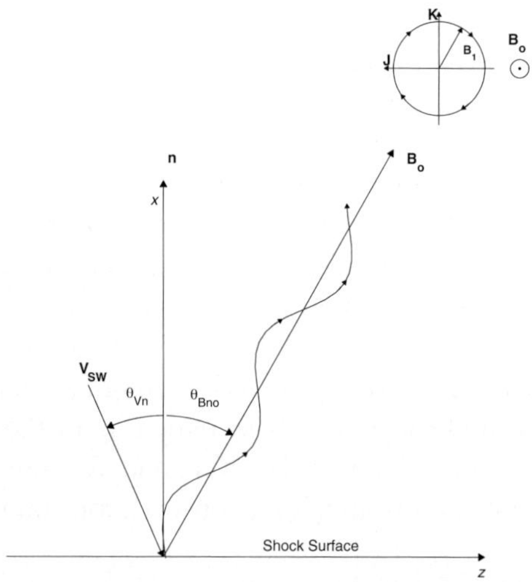

Figure 2.22. A sketch showing how θ_{Bn} may be modulated by the presence of an upstream wave. (From Meziane et al., 2004a).

inhibits the escape of specularly reflected ions. When the wave amplitude is high ($\delta|\mathbf{B}|/B_0 \approx 1$), this effect is relatively strong.

Meziane et al. (2004a) reported clear quantitative evidence for the specular reflection mechanism. They have studied an ion event (Figure 2.23) observed in association with quasi-monochromatic ULF waves which strongly modulate the ion fluxes by nearly two orders of magnitude for some energy ranges, the fluxes coming down close to the instrumental background level for the minima. The analysis of the three-dimensional angular distribution indicates that ions propagating roughly along the magnetic field direction are observed at the onset of the event. Later on, the angular distribution is gyrophase-bunched and the pitch-angle distribution is peaked at $\sim 150°$ in the solar wind frame. Analytic calculations show that the specular reflection of the solar wind proton population with a simple Maxwellian distribution should produced a reflected distribution peaked at a pitch-angle $\alpha_0 \sim \theta_{Bn}$. Since the measurement of particle pitch-angle and the derivation of θ_{Bn} are independent, they provide a good quantitative test of specular reflection. According to three statistical bow shock position models, the Cluster spacecraft were located at about 0.5 R_E from the shock with an averaged bow shock θ_{Bn} of about 30°. This

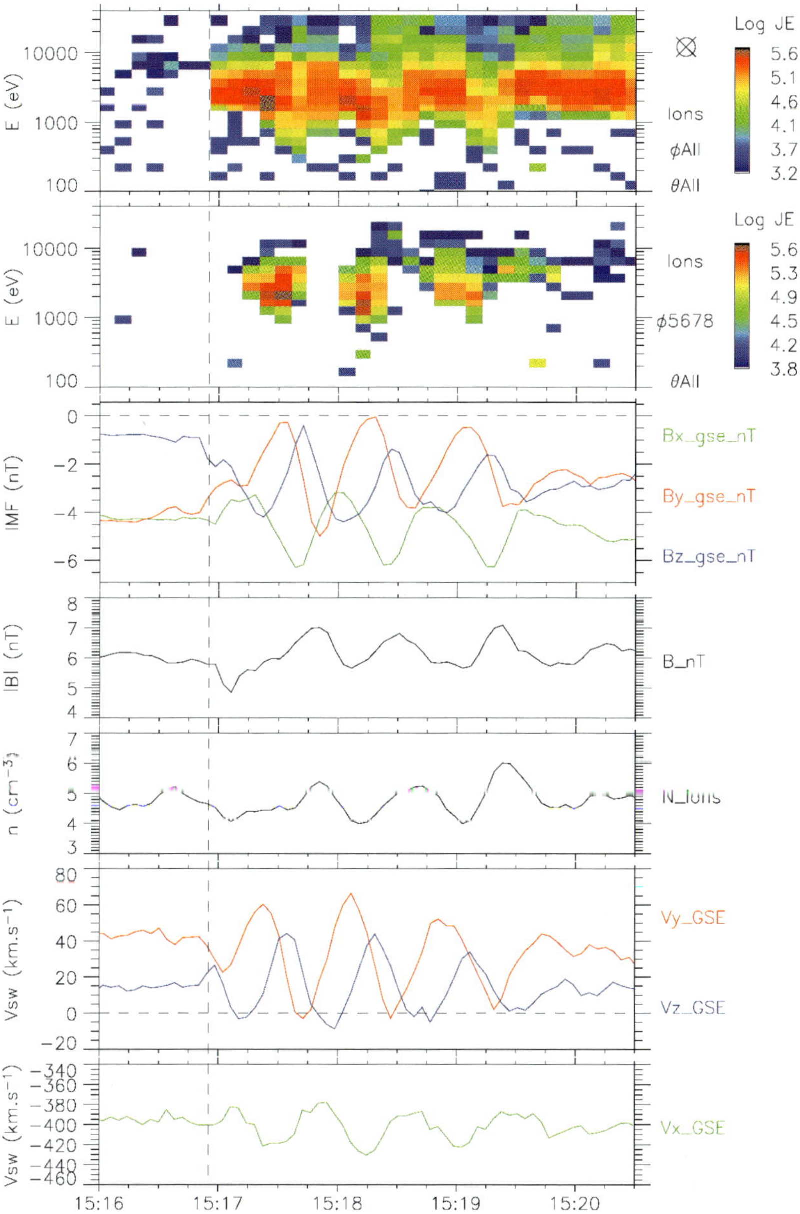

Figure 2.23. Ion and magnetic field observations from Cluster 1 between 15:16 - 15:20:30 UT on March 9, 2002. The top two panels show, respectively, energy spectrograms from CIS/HIA (large geometric factor section) integrated over all directions, and over the 145–235° azimuthal look directions; the third and fourth panels display the GSE components and magnitude of the IMF; and the last three panels present solar wind densities and GSE velocities from CIS/HIA (small geometric factor section). The dashed vertical line indicates the magnetic connection time. From Meziane et al. (2004a).

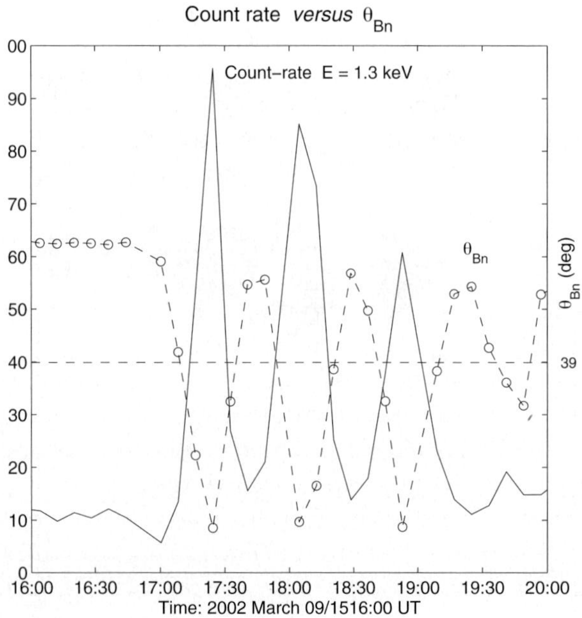

Figure 2.24. The changes of local shock θ_{Bn} angle (dashed line-open circles) and the particle count rates during the interval when the ULF waves are present. The dashed horizontal line corresponds to $\theta_{Bn} = 39.9°$. For clarity, the particle count rate values have been scaled by a factor 0.5. From Meziane et al. (2004a).

result is therefore fully consistent with the specular reflection production mechanism.

The analysis of the waves has shown that they are left-handed in the spacecraft frame and propagate approximately along the ambient magnetic field. Meziane et al. (2004a) have found that they are in cyclotron resonance with the field-aligned beam observed just upstream of the interval of gyrating ions. Using the measured properties of the waves and particles (Figure 2.23), Meziane et al. have explained the observed particle flux-modulation in term of changes in θ_{Bn} at the shock due to the low frequency waves. Figure 2.22 shows a sketch of the underlying geometrical model, which illustrates how the instantaneous θ_{Bn} deviates from the average θ_{Bn_0}, computed from the background field $\mathbf{B_0}$, when the the wave field is added. Significant ion fluxes were observed only when the instantaneous θ_{Bn} was less than 40°, consistent with specular reflection at the shock being the source of the ions.

2.3.2.2 Local production by wave pitch-angle trapping

Fuselier et al. (1986a) made a quantitative analysis of the particles and monochromatic waves from ISEE data, which strongly suggested that there was a coherent wave-particle interaction. They obtained a phase relationship between the gyrovelocity v_G and the transverse wave field B_t indicating that energy transfer occurred between the particles and the waves, and that gyrophase trapping by the wave was possible. Since field-aligned distributions propagate deeply into the foreshock, the local production of gyrating ions through this process ought to be able to be observed far from the shock. Meziane et al. (1997) reported the first observations from Wind spacecraft data of several gyrating ion distributions and their association with low frequency waves, at distances larger than 20 R_E from the shock. Clear indications of wave-particle interactions were observed. A more detailed study of the three-dimensional ion distributions with a large data set and the highest available time resolution (3s) has since shown that these observational features can be found up to more than 80 R_E from the shock (Meziane et al., 2001). Investigation of the non-linear wave trapping mechanism has shown that it can explain the properties of such gyrating ion distributions registered at large distances from the shock (Mazelle et al., 2000). It has been shown that the particles are not only bunched in gyrophase but also trapped in pitch-angle in velocity space around a value which is directly related to the amplitude of the wave self-consistently generated by the original field-aligned ion beam.

Mazelle et al. (2003) have investigated this local production mechanism to explain the existence of the well-defined gyrating ion distributions observed by the Cluster CIS instrument. The event shown in Figure 2.25 occurred during a long interval of foreshock wave activity. At 23:34:30 UT, energetic ions are revealed in the second energy spectrogram corresponding to measurements by the high-geometrical-factor side of the CIS/HIA instrument (the difference with the first panel showing the solar wind population is quite obvious). High fluxes are then continuously observed until 23:44 UT, followed by two small patches. These ions are mostly propagating sunward, as revealed from the analysis of their guiding center velocity, i.e., they are backstreaming ions. Before 23:34:30 UT, the IMF was nearly quasi-steady. Prominent large amplitude low frequency waves are observed after 23:35:45 UT both on the magnetic field and on the solar wind velocity. Figure 2.26 displays three-dimensional representations of the ion distribution functions registered by CIS/CODIF at 4s cadence. Nine consecutive distributions are shown for one energy channel ($\sim$8 keV) for which the observed backstreaming fluxes are maximum for a time interval inside the event displayed on Figure 2.25. Each frame in Figure 2.26 is a projection in gyrophase and pitch-angle with the **B**-direction located at the center. The three first snapshots indicate an ion beam propagating along the +**B** direction with a parallel velocity of 1100 km s^{-1}, but the third one also shows a second peak for a large pitch-angle of about 60°. Then after 23:35:45, the spacecraft has entered a gyrating ion region. Gyrating ions are identified by

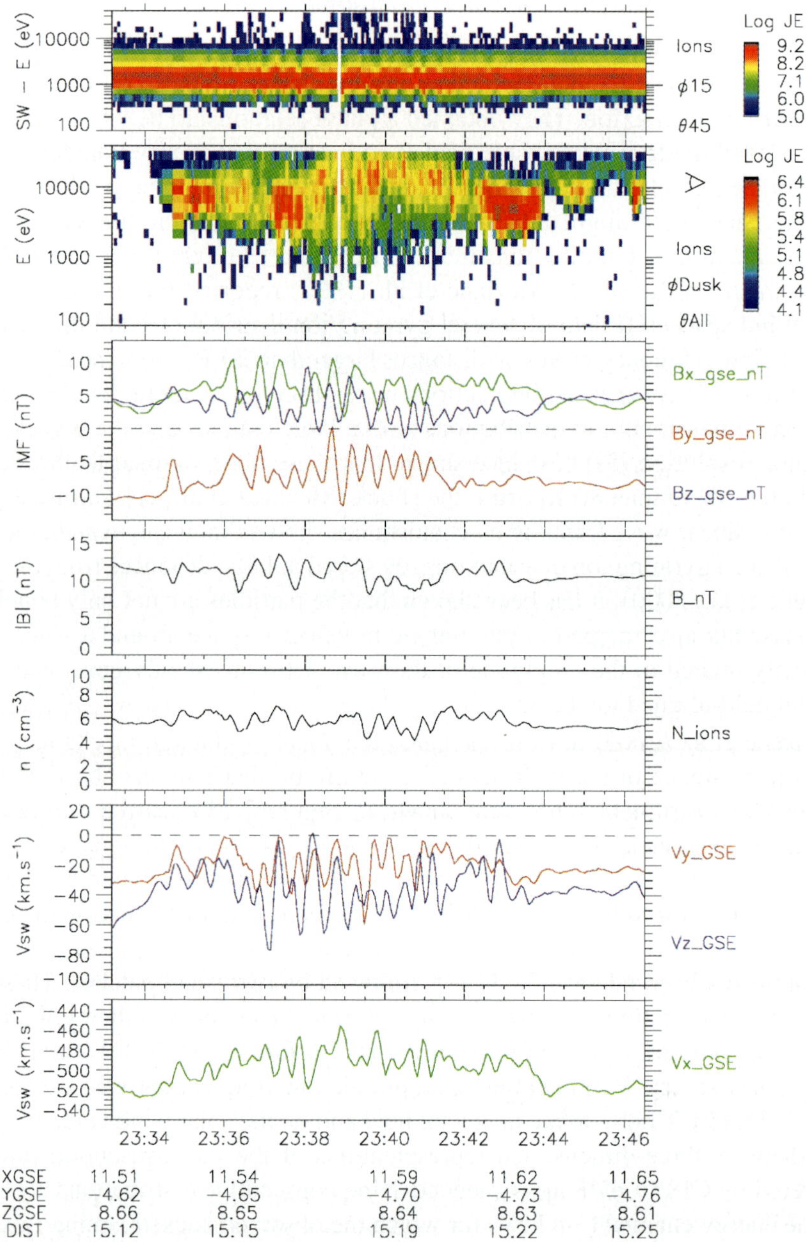

Figure 2.25. Observations from Cluster CIS and FGM on satellite 1 between 23:33-23:46 UT on April 7, 2001 : energy-time spectrograms of all ions from CIS/HIA for 'solar wind sectors' (sunward looking direction - upper panel) and 'dusk' solid angle (duskward looking direction - second panel), respectively; dc magnetic field components in GSE coordinates and its magnitude; ion density and bulk velocity in GSE coordinates derived from HIA measurements. From Mazelle et al. (2003).

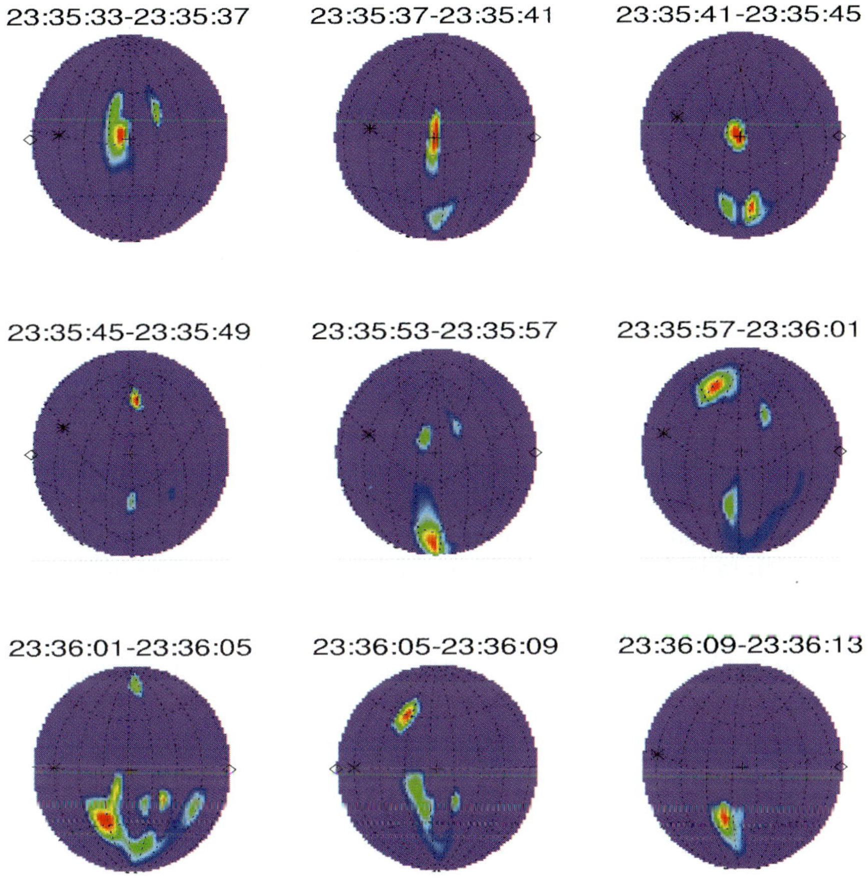

Figure 2.26. Sequence of consecutive three-dimensional 4-s display of the proton angular distributions registered by CIS/CODIF on Cluster 3 for an energy of $\sim$8 keV (flux maximum) on April 7, 2001. Each frame represents the normalised distribution function on a surface of constant energy in the solar wind frame of reference projected to display 4π-coverage. The $\mathbf{B}_0$ vector is located at the center of each plot and the '*' sign indicates the solar wind direction. For each frame, the maximum maximum value of the normalised phase space density is shown in red. From Mazelle et al. (2003).

their gyrophase-restricted distribution peaked off the magnetic field direction. The interplanetary magnetic field used to plot the distributions is averaged over the spin interval (4 s) while the local proton cyclotron period is 7 s (i.e., about two ion sampling intervals). The gyrating distributions show a clear rotation of their maximum phase density in the left-handed sense around the magnetic field with alternating values separated by about 180°. Such gyrating ion distributions are observed up to $\sim$23:44 UT.

Mazelle et al. (2003) have analysed the associated large amplitude low frequency waves using multi-spacecraft analysis techniques (e.g., Eastwood et al., 2002). The wave are right-hand mode waves ('30-s waves'). They have shown that these wave are in cyclotron resonance with the field-aligned beam observed just before the spacecraft entered the gyrating ion/ULF wave region. This is the first direct quantitative evidence so far of this cyclotron resonance from observations in the ion foreshock. Then, they have studied the possibility of resonantly driving these waves unstable from the electromagnetic ion/ion beam instability by field-aligned beam ions also observed in the same region. The results from the linear theory has led to a very good agreement with the observed wave mode.

The event reported is inconsistent with a specular reflection at the Earth's bow shock since the observed pitch-angle of the gyrating ions are much too large: it should be nearly equal to θ_{Bn}, which was $\sim 30°$ in this case. It was thus necessary to invoke a local production mechanism for these upstream distributions. The possibility of producing the observed gyrophase-bunched ion distributions from the disruption of the beam by the excited wave has led to a good quantitative agreement from nonlinear trapping theory which predicts that the pitch-angle of the final gyrating ion distribution is related to the wave amplitude (Mazelle et al., 2000). This result is very similar to those obtained from previous studies in the distant foreshock (up to 80 R_E) from Wind data with lower backstreaming ion densities and wave amplitude (Meziane et al., 1997; Meziane et al., 2001; Mazelle et al., 2000), which could mean that the present case study corresponds to the same mechanism observed by Cluster closer to the bow shock. Many other events consistent with this trapping mechanism have also been found while only one example of gyrophase-bunched ions produced by specular reflection has been identified by Cluster (Meziane et al., 2004a). However, better statistics are necessary to determine the percentage of gyrating events corresponding to either mechanism. It will be also interesting then to examine the location of such events and compare them with the forward boundary of the foreshock ULF waves.

2.4 Hot Flow Anomalies

2.4.1 Origin

Hot Flow Anomalies (HFAs) are transient perturbations of the bow shock and foreshock caused by its interaction with a tangential discontinuity (TD) embedded within the inflowing solar wind plasma. They were first identified using data from the ISEE and AMPTE missions (Schwartz et al., 1985; Thomsen et al., 1986; Paschmann et al., 1988). An explanation of their generation mechanism was developed as a result of both observations and modelling work (e.g., Burgess, 1989; Thomas et al., 1991).

Observational features of HFAs include (Paschmann et al., 1988; Schwartz et al., 1988; Thomsen et al., 1988; Schwartz et al., 2000): (1) hot central regions with isotropic ion and electron velocity distributions; (2) bulk velocities much slower than the ambient solar wind, and deflected in the same sense as the magnetosheath flow, but by a larger amount, often attaining even a sunward component; (3) noisy magnetic field in the central region, often with intervals of depressed and enhanced field magnitude; (4) durations that imply scale sizes of one or more R_E; (5) compressed edge regions leading and/or trailing the central region. The outer boundaries are consistent with weak shocks and the edge regions appear to be simply shocked solar wind, separated from the central region by a TD; (6) three-dimensional structure as inferred from the orientations of the leading and trailing shock normals, suggesting a bulge on the bow shock surface.

In order for a TD to generate an HFA, the discontinuity normal must be inclined at a large angle to the Earth-Sun line (e.g., Schwartz et al., 2000). The line along which the TD intersects the bow shock tracks slowly across the bow shock surface as the TD is convected anti-sunward in the solar wind flow. A fraction of the incident solar wind flow is reflected from the bow shock and those reflected ions in the vicinity of the TD experience a motional (convection) electric field arising from the convection of the interplanetary magnetic field by the solar wind.

The physics of the formation of a HFA can be briefly summarised as follows. For an HFA to form, the magnetic field direction on one or both sides of the TD must act to focus the reflected ions towards the TD plane. In addition the tracking velocity of the TD must be slow relative to the reflected ion velocity, so that many reflected ions are directed along the discontinuity plane. Lastly, the discontinuity must be thin relative to a reflected-ion gyroradius. Then a significant population of reflected ions can be channelled along the discontinuity plane. These ions interact with the solar wind beam, to generate waves, causing the two ion populations to scatter and form a single hot population. The over-pressure of this hot ion population flowing along the discontinuity place causes the expansion of a cavity centered on the discontinuity plane. Compression regions form at the edges of the cavity which can develop into shocks if the expansion rate is sufficiently rapid.

2.4.2 Cluster observations

Multi-spacecraft Cluster observations of HFAs allow the study of both the spatial structure and temporal evolution of HFAs in more detail than has been possible before. For example, when the spacecraft are close together, the orientation and motion of the compression regions at the edges of the HFA can be estimated, for comparison with the orientation of the underlying discontinuity orientation and the direction of the velocity perturbations arising from expansion of the cavity (Lucek et al., 2004). When the spacecraft tetrahedron is larger, $\sim$5000 km for example, evidence for time evolution of the HFA structure can be seen clearly.

Figure 2.27 shows data from the first of a series of HFAs observed by Cluster on April 2, 2002, when the spacecraft were at approximately (+9.9, -2.2, +8.11) R_E in GSE. Two discontinuities occurred during the interval in Figure (2.27). The first

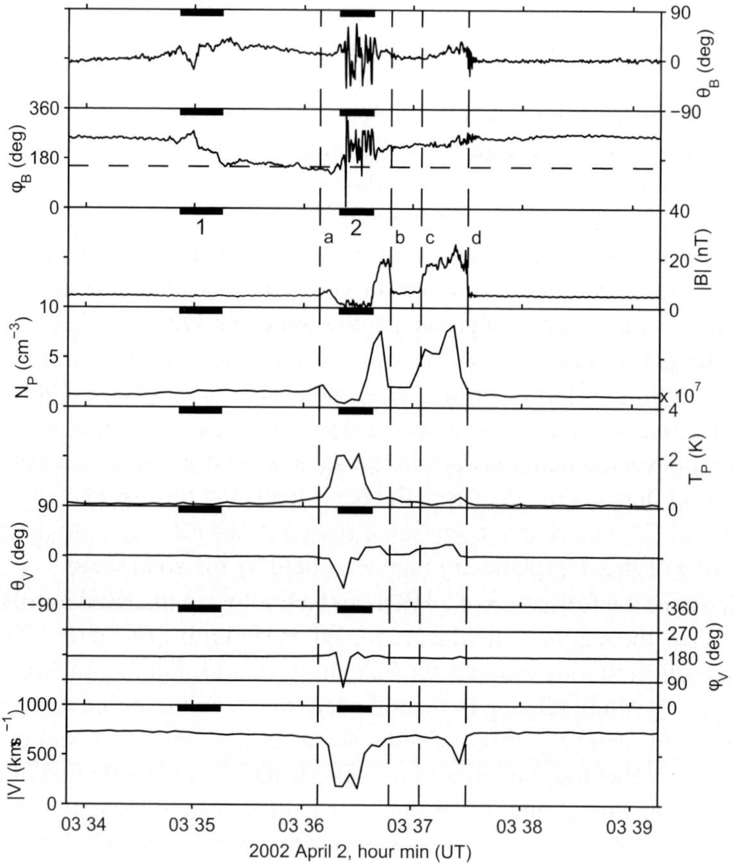

Figure 2.27. Magnetic field and plasma data from an HFA observed by Cluster 1. Panels show magnetic field elevation angle θ_B, longitude angle ϕ_B in GSE coordinates; magnetic field magnitude, $|\mathbf{B}|$; proton density, N_p; proton temperature, T_p; velocity vector elevation and longitude angles θ_v and ϕ_v; and velocity magnitude, $|\mathbf{V}|$. (From Lucek et al., 2004).

(labelled 1 in the figure) did not give rise to an HFA, since the magnetic field orientation was such that reflected ions would be directed away from the discontinuity plane. The second discontinuity (labelled 2 in the figure), visible by the change in magnetic field orientation across region 2, did generate an HFA. Calculation of the motional (convection) electric field for this case showed that it would act to focus reflected ions towards this discontinuity, consistent with an HFA being formed.

Cluster entered the HFA through a weak compression (a) and exited through a shock (b). Between these transitions Cluster observed a region of low magnetic field. Within this region, examination of the ion distributions showed the presence of two ion populations: the solar wind beam, and a beam of sunward flowing ions, consistent with ions reflected from the shock being focused along the discontinuity. The densities of the two populations were approximately equal. Such an ion distribution is unstable to wave generation which would tend to scatter the ions, forming a single distribution, and indeed the cavity already showed some evidence of wave activity, and HFAs later on the same orbit contained a single, hot plasma population. Since the two ion populations were distinct in this case, Lucek et al. (2004) suggest that this HFA was relatively young.

Using data from the four spacecraft Lucek et al. (2004) estimated the orientation of the underlying discontinuity, and consequently the velocity of the line of interaction between the discontinuity and the bow shock. In addition, the orientation and speed of the edges were estimated, for comparison with the observed plasma motion within the HFA core.

We briefly summarise the properties of this HFA. The orientation of the underlying discontinuity was estimated from the cross product of the magnetic field upstream and downstream of the HFA disturbance. The discontinuity normal was found to lie mainly in the GSE-Z direction, largely confined to the XZ-plane, $\mathbf{n}_{dis} = (0.17, -0.07, 0.98)$. Since Cluster was near noon, the sketch of this HFA can conveniently be projected onto the XZ-plane. The discontinuity first intersected the bow shock along a line northwards of Cluster, then tracked southwards at a speed of $\sim 110\,\mathrm{km\,s^{-1}}$. Assuming that the HFA did not develop significantly during the time it took to cross the spacecraft, which is supported by the similar magnetic profile seen at all four spacecraft, it can be considered that Cluster made a vertical cut through the HFA structure. It is then possible to transform from the time profile in Figure 2.27 to a spatial schematic in Figure 2.28. Projected onto the spacecraft track are the perturbations to the velocity vectors: these are the instantaneous differences between the local velocity vector and the solar wind velocity outside of the HFA. The filled grey circle indicates the satellite location at the time the HFA was observed. Also shown in Figure 2.28 is a cut through a model bow shock surface estimated for the time at which Cluster crossed into the solar wind, just a few minutes earlier. The satellite therefore appeared to be close to the bow shock at this time, as was also expected from the observation of two ion populations within the cavity.

The orientations of the edges of the HFA cavity were calculated using four spacecraft timing discontinuity analysis, together with the speed of motion along their normals. These orientations in the XZ-plane are indicated on Figure 2.28 at the time along the satellite track where they were observed. Cluster entered through a small compression region, labelled (a) in Figure 2.27, with a normal direction $\mathbf{n}_a = (0.46, 0.40, 0.79)$, travelling at only -15 $\mathrm{km\,s^{-1}}$ in the solar wind frame, i.e., anti-sunward. Cluster exited through a shock (b), with a normal $\mathbf{n}_b = (0.60, -0.26, 0.76)$, travelling sunward at 340 $\mathrm{km\,s^{-1}}$ in the solar wind frame, corresponding to a Mach number $M_{ms} \sim 5$. The vertical scale of the HFA (~ 3000 km) is given by its duration. However, the orientations of the magnetic compression and the shock are nearly parallel to the estimated orientation the underlying dis-

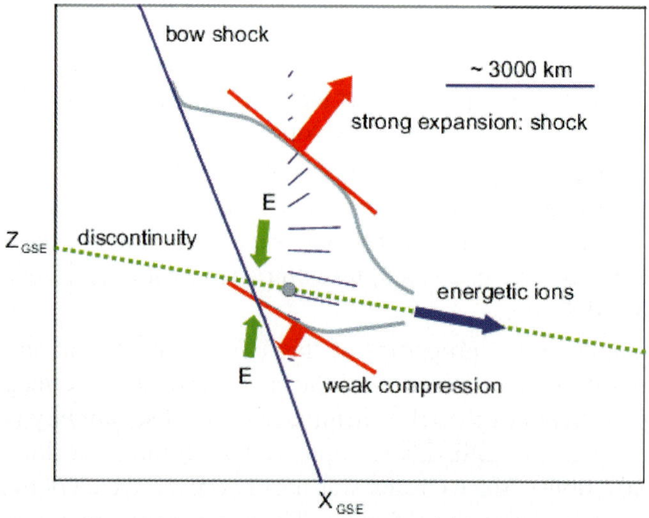

Figure 2.28. Schematic diagram of an HFA observed by Cluster, projected onto the $X - Z_{GSE}$ plane. The signatures seen by Cluster are interpreted as a bubble expanding under internal pressure. The form of the HFA is constrained by the following information: the spacecraft location (filled circle) relative to a model bow shock surface, the estimated discontinuity plane (heavy dashed line), the velocity perturbation vectors (faint lines) plotted assuming that the discontinuity tracks southwards at $110\,\mathrm{km\,s^{-1}}$, the orientations of the compression and shock at the edges of the event. (From Lucek et al., 2004).

continuity (within 12°), consistent with them being generated by expansion of the cavity perpendicular to the discontinuity plane. In addition, the perturbations to the plasma velocity projected onto the track of the tetrahedron through the cavity show flow along the discontinuity plane near the center of the cavity, and flow consistent with strong expansion nearer the edges. We therefore expect the extent of the HFA in the Y and X directions to be larger than its scale in Z .

The region of compressed, heated plasma between transitions (c) and (d) in Figure 2.27 was interpreted by Lucek et al. (2004) as a re-encounter with the compressed plasma at the edge of the HFA, which moved northwards, back over the spacecraft. Cluster did not see a signature of hot plasma embedded within the compressed region, suggesting that the spacecraft did not re-enter the HFA cavity, but the orientation and motion of the boundaries (c) and (d) was consistent with this interpretation. This re-encounter suggests either a change in the expansion rate of the cavity or a change in the motion of the whole structure. The former could arise from the mixing of the two ion populations, or the convection of the HFA closer to the nose of the bow shock where it is stronger. The latter could arise from nonplanarity of the underlying TD.

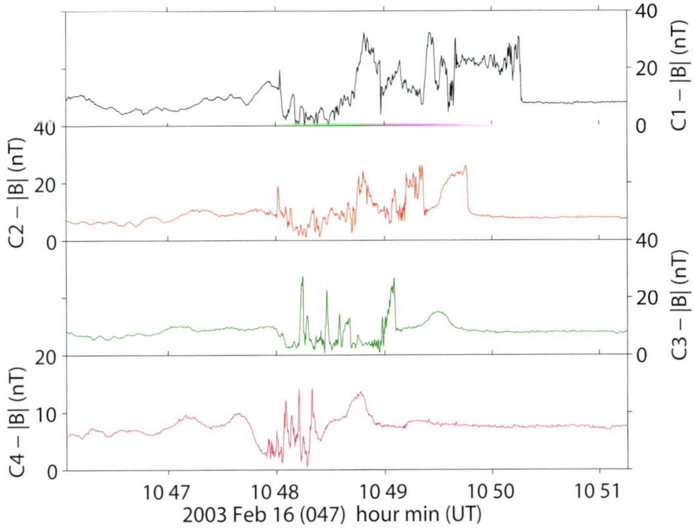

Figure 2.29. Magnetic field magnitude signature of an HFA observed by the four Cluster spacecraft when the tetrahedron scale was ~ 5000 km. (Figure provided by E. Lucek.)

Figure 2.29 shows the magnetic field magnitude through an HFA when the Cluster tetrahedron scale was of the order of 5000 km. The order in which the HFA crosses the four spacecraft is: 4, 3, 2, 1. On these separation scales the convection time is long enough for the HFA to develop as it crosses the tetrahedron. Cluster 4 observes a cavity bounded by compression regions, while later crossings show evidence for the development of a shock on the exit edge of the HFA.

2.5 Summary

Perhaps the most important consequence of the Cluster mission is that it has clearly demonstrated the benefits and therefore necessity of multi-satellite missions. A number of the results presented here, particularly involving the wave analysis, could only have been derived through a multi-point analysis. The following results have been derived from the Cluster data set:

1. The determination of the properties of the electron foreshock and electron Langmuir waves, using waveform capture techniques, in particular with regards to the applicability of stochastic growth theory.

2. Independent confirmation of previous dual spacecraft analysis demonstrating that the majority of foreshock ULF waves are indeed intrinsically fast magnetosonic in nature, and independent experimental investigation of single spacecraft techniques viz. minimum variance analysis.

3. The identification of circularly polarised, left handed (and therefore on the Alfvén/ion cyclotron branch of the kinetic dispersion relations), ULF waves in the foreshock, generated by hot backstreaming distributions.

4. Demonstration that the oblique propagation of foreshock ULF waves may in certain circumstances be confined to the plane defined by the magnetic field and the solar wind vector, when the magnetic field cone angle is large.

5. The experimental determination of the wave dispersion relation, independently confirming the typical wavelength and frequency of foreshock waves.

6. A demonstration that the upstream ULF wave field can modify θ_{Bn} at the shock, causing a modulation of the backstreaming ion beam.

7. A quantitative experimental test clearly demonstrating that quasi-monochromatic foreshock waves can be generated by field aligned beams which are subsequently trapped by the wave field, leading to the production of gyrophase bunched ion distributions.

8. The spatial structure and temporal evolution of HFAs, using multispacecraft analysis to quantify their size and motion.

The Cluster dataset has already proven to be a remarkably rich and extremely valuable resource for understanding the foreshock and its often-complex behaviour. Although a great deal is now known about the foreshock, many questions remain unanswered. Hidden within the Cluster dataset are the clues and the evidence that will make many of these problems tractable. Rather than simply outline a list of problems that the Cluster dataset can be applied to, we instead offer a list of themes into which current activities can be organised.

1. *Foreshock as a plasma physics laboratory.* This theme concerns the microphysics of the foreshock. Currently, there is a significant effort to understand in a quantitative manner the details of the non-linear ion beam interactions that occur in the foreshock. Examples include studying the deceleration of the solar wind in the foreshock, the manner in which beam instabilities saturate, and the conditions under which different instabilities dominate. In the electron foreshock, there is a great deal of work to be carried out comparing theoretical predictions with observations of Langmuir waves. These waves are often double-peaked near the plasma frequency, suggesting the existence of various instabilities that also lead to the generation of other waves, such as sound waves. In addition, it is not unusual to see such waves upshifted and downshifted from the plasma frequency in the foreshock. The physics of the instabilities that lead to all these phenomenon needs to be explored in greater depth by carrying out multi-point, multi-instrument studies using Cluster under varying solar wind conditions in addition to developing the associated theory.

2. *Multipoint, multiscale.* The spacecraft separation strategy plays a key role in enabling the types of investigation for which Cluster may be used. For example, the spacecraft separation in 2001 was of the order of 600km, which is ideal for multipoint wave analysis. In the later stages of the mission, the separation will increase to 10 000km or more, which will be ideal for investigating the morphology of the foreshock, for example investigating spatial structure in the ULF wave field, or the occurrence of gyrophase bunched ion distributions. As data accumulates, studies comparing observations of the same phenomena at different scales will become possible. Such studies are likely to provide valuable new insights into foreshock processes.

3. *Time dependence.* As stated in the introduction, many pictures of the shock/foreshock system assume the existence of a dynamical equilibrium, even though the foreshock can exhibit dramatic transient behaviour. Statistical studies need to be performed under varying solar wind conditions and IMF orientations to see how changing the system inputs affect the various foreshock boundaries. Other important investigations concern Hot Flow Anomalies and Foreshock Cavities, in particular using Cluster multi-point measurements to understand in a quantitative manner their onset, growth and decay and the rate at which particle energisation and thermalisation occur.

4. *The influence of the foreshock on the bow shock and magnetosphere.* In the introduction, the connection between the foreshock and the quasi-parallel shock was discussed. Further studies are required to fully understand the role of the foreshock in quasi-parallel shock dynamics, for example statistics concerning the modulation of θ_{Bn} at the shock by the foreshock wave field. The problem of SLAMS growth out of the foreshock wave field has also attracted significant interest. Additional investigations may usefully be made into the transmission of foreshock waves through the shock. Also, the foreshock interacts with the magnetopause and magnetosphere more directly, via hot flow anomalies and foreshock cavities. As the size of the Cluster dataset increases, it is likely that observations of these events will accumulate. By studying the different ways in which individual events impact on the magnetosphere, a better understanding of their importance will be established.

Acknowledgements

JPE holds a National Research Council Resident Research Associateship Award at NASA GSFC. JP acknowledges support from NASA under grants NAG5-9974 and NNG04GB98G. Cluster research in the UK and France is supported by PPARC and CNES, respectively.

References

Arthur, C., R. L. M. Pherron, and J. D. Means: 1976, 'A comparative study of three techniques for using the spectral matrix in wave analysis'. *Radio Sci.* **11**, 833.

Asbridge, J. R., S. J. Bame, and I. B. Strong: 1968, 'Outward flow of protons from the Earth's bow shock'. *J. Geophys. Res.* **73**, 5777–5782.

Bale, S. D., D. Burgess, P. J. Kellogg, K. Goetz, and S. J. Monson: 1997, 'On the amplitude of intense Langmuir waves in the terrestrial electron foreshock'. *J. Geophys. Res.* **102**, 11,281.

Balikhin, M., T. D. de Wit, H. S. C. K. Alleyne, L. J. C. Woolliscroft, S. N. Walker, V. Krasnosel'skikh, W. A. C. Mier-Jedrzejeowicz, and W. Baumjohann: 1997a, 'Experimental determination of the dispersion of waves observed upstream of a quasi-perpendicular shock'. *Geophys. Res. Lett.* **24**, 787–790.

Balikhin, M. A., L. J. C. Woolliscroft, H. S. C. Alleyne, M. Dunlop, and M. A. Gedalin: 1997b, 'Determination of the dispersion of low frequency waves downstream of a quasiperpendicular collisionless shock'. *Ann. Geophys.* **15**, 143–151.

Balogh, A., S. J. Schwartz, S. D. Bale, M. A. Balikhin, , D. Burgess, T. S. Horbury, , V. V. Krasnoselskikh, H. Kucharek, B. Lembège, E. A. Lucek, E. Möbius, M. Scholer, M. F. Thomsen, and S. N. Walker: 2005, 'Cluster at the Earth's Bow Shock: Introduction'. *Space Sci. Rev.* **this issue**.

Balogh, A. et al.: 1997, 'The Cluster magnetic field investigation'. *SpaceSci. Rev.* **79**, 65–91.

Bame, S. J., J. R. Asbridge, W. C. Feldman, J. T. Gosling, G. Paschmann, and N. Sckopke: 1980, 'Deceleration of the solar wind upstream from the Earth's bow shock and the origin of diffuse upstream ions'. *J. Geophys. Res.* **85**(6), 2981–2990.

Barnes, A.: 1970, 'Theory of generation of bow-shock-associated hydromagnetic waves in the upstream interplanetary medium'. *Cosmic Electrodyn.* **1**, 90–114.

Bendat, J. and A. G. Piersol: 1980, *Engineering applications of correlation and spectral analysis*, pp. 54–56. New York: John Wiley & Sons, Inc.

Bendat, J. S. and A. G. Piersol: 1986, *Random Data Analysis and Measurement Procedures*. John Wiley and Sons, 2 edition.

Blanco-Cano, X., G. Le, and C. T. Russell: 1999, 'Identification of foreshock waves with 3-s period'. *J. Geophys. Res.* **104**(3), 4643–4656.

Blanco-Cano, X. and S. J. Schwartz: 1997, 'Identification of low-frequency kinetic wave modes in the Earth's ion foreshock'. *Ann. Geophys.* **15**, 273–288.

Bonifazi, C. and G. Moreno: 1981, 'Reflected and diffuse ions backstreaming from the Earth's shock 2. Origin'. *J. Geophys. Res.* **86**(15), 4405–4414.

Born, M. and E. Wolf: 1980, *Principles of optics - 6th ed.*, pp. 503–504, 550. New York: Pergamon press.

Brinca, A. L.: 1991, 'Cometary Linear Instabilities: From Profusion to Perspective'. In: A. D. Johnstone (ed.): *Geophysical Monograph 61: Cometary Plasma Processes*. American Geophysical Union, pp. 211–221.

Burgess, D.: 1989a, 'Cyclic behavior at quasi-parallel collisionless shocks'. *Geophys. Res. Lett.* **16**, 345–348.

Burgess, D.: 1989b, 'On the effect of a tangential discontinuity on ions specularly reflected at an oblique shock'. *J. Geophys. Res.* **94**, 472–478.

Burgess, D.: 1995, 'Foreshock-shock interaction at collisionless quasi-parallel shocks'. *Adv. Space Res.* **15**(8/9), 159–169.

Burgess, D.: 1997, 'What do we really know about upstream waves?'. **20**, 673–682.

Burgess, D., E. A. Lucek, M. Scholer, S. D. Bale, M. A. Balikhin, A. Balogh, T. S. Horbury, V. V. Krasnoselskikh, H. Kucharek, B. Lembège, E. Möbius, S. J. Schwartz, M. F. Thomsen, , and S. N. Walker: 2005, 'Quasi-parallel Shock Structure and Processes'. *Space Sci. Rev.* **this issue**.

Cairns, I. H. and P. A. Robinson: 1997, 'First test of stochastic growth theory for Langmuir waves in Earth's foreshock'. *J. Geophys. Res.* **24**, 369.

Cairns, I. H. and P. A. Robinson: 1999, 'Strong evidence for stochastic growth of Langmuir-like waves in Earth's foreshock'. *Phys. Rev. Lett.* **82**, 3066.

Cairns, I. H., P. A. Robinson, R. R. Anderson, and R. J. Strangeway: 1997, 'Foreshock Langmuir waves for unusually constant solar wind conditions: data and implications for foreshock structure'. *J. Geophys. Res.* **102**, 24,249.

Drury, L. O.: 1983, 'An introduction to the theory of diffusive shock acceleration of energetic particles in tenuous plasmas'. *Rep. Prog. Phys.* **46**, 973–1027.

Dudok de Wit, T., V. V. Krasnosel'skikh, S. D. Bale, M. W. Dunlop, H. Lühr, S. J. Schwartz, and L. J. C. Woolliscroft: 1995, 'Determination of dispersion relations in quasi-stationary plasma turbulence using dual satellite data'. *Geophys. Res. Lett.* **22**, 2653–2656.

Eastwood, J. P., A. Balogh, M. W. Dunlop, T. S. Horbury, and I. Dandouras: 2002, 'Cluster observations of fast magnetosonic waves in the terrestrial foreshock'. *Geophys. Res. Lett.* **29**, 2046, doi: 10.1029/2002GL015582.

Eastwood, J. P., A. Balogh, and E. A. Lucek: 2003, 'On the existence of Alfvén waves in the terrestrial foreshock'. *Ann. Geophys.* **21**, 1457–1465.

Eastwood, J. P., A. Balogh, C. Mazelle, I. Dandouras, and H. Reme: 2004, 'Oblique propagation of 30s period fast magnetosonic foreshock waves: A Cluster case study'. *Geophys. Res. Lett.* **31**, L04804, doi:10.1029/2003GL018897.

Etcheto, J. and M. Faucheaux: 1984, 'Detailed study of electron plasma waves upstream of the Earth's bow shock'. *J. Geophys. Res.* **89**, 6631.

Fairfield, D.: 1969, 'Bow shock associated waves observed in the far upstream interplanetary medium'. *J. Geophys. Res.* **74**, 3541–3553.

Fairfield, D. H.: 1974, 'Whistler waves observed upstream from collisionless shocks'. *J. Geophys. Res.* **79**(10), 1368–1378.

Farris, M. H., S. M. Petrinec, and C. T. Russell: 1991, 'The thickness of the magnetosheath: Constraints on the polytropic index'. *Geophys. Res. Lett.* **18**(10), 1821–1824.

Fazakerley, A. N., A. J. Coates, and M. W. Dunlop: 1995, 'Observations of upstream ions, solar wind ions and electromagnetic waves in the Earth's foreshock'. *Adv. Space Res.* **15**(8/9), 103–106.

Feldman, W. C., J. R. Asbridge, S. J. Bame, and M. D. Montgomery: 1973, 'Solar wind heat transport in the vicinity of the Earth's bow shock'. *J.Geophys.Res.* **78**, 3697–3713.

Filbert, P. and P. J. Kellogg: 1979, 'Electrostatic noise at the plasma frequency beyond the Earth's bow shock'. *J. Geophys. Res.* **84**, 1369.

Fitzenreiter, R. J.: 1995, 'The electron foreshock'. *Adv. Space Res.* **15**(8/9), (8/9)9–(8/9)27.

Fitzenreiter, R. J., A. J. Klimas, and J. D. Scudder: 1984, 'Detection of bump-on-tail reduced electron velocity distributions at the electron foreshock boundary'. *Geophys. Res. Lett.* **11**, 496–499.

Fowler, R. A., B. J. Kotick, and R. D. Elliott: 1967, 'Polarization analysis of natural and artificially induced geomagnetic micropulsations'. *J. Geophys. Res.* **72**, 2871–2883.

Fredricks, R. W., F. L. Scarf, and L. A. Frank: 1971, 'Nonthermal electrons and high-frequency waves in the upstream solar wind'. *J. Geophys. Res.* **76**, 6691–6699.

Fuselier, S. A.: 1994, 'Suprathermal ions upstream and downstream from the Earth's bow shock'. In: M. J. Engebretson, K. Takahashi, and M. Scholer (eds.): *Solar Wind Sources of Magnetospheric Ultra-Low-Frequency Waves Geophysical Monograph 81*. American Geophysical Union, pp. 107–119.

Fuselier, S. A.: 1995, 'Ion distributions in the earth's foreshock upstream from the bow shock'. *Adv. Space Res.* **15**(8/9), (8/9)43–(8/9)52.

Fuselier, S. A., J. T. Gosling, and M. F. Thomsen: 1986, 'The motion of ions specularly reflected off a quasi-parallel shock in the presence of large-amplitude, monochromatic MHD waves'. *J. Geophys. Res.* **91**, 4163–4170.

Fuselier, S. A., D. A. Gurnett, and R. J. Fitzenreiter: 1985, 'The downshift of electron plasma oscillations in the electron foreshock region'. *J. Geophys. Res.* **90**, 3935.

Fuselier, S. A. and M. F. Thomsen: 1992, 'He^{2+} in field aligned beams: ISEE results'. *Geophys. Res. Lett.* **19**(5), 437–440.

Fuselier, S. A., M. F. Thomsen, S. P. Gary, S. J. Bame, C. T. Russell, and G. K. Parks: 1986a, 'The phase relationship between gyrophase-bunched ions and MHD-like waves'. *Geophys. Res. Lett.* **13**(1), 60–63.

Fuselier, S. A., M. F. Thomsen, J. T. Gosling, S. J. Bame, and C. T. Russell: 1986b, 'Gyrating and intermediate ion distributions upstream from the Earth's bow shock'. *J. Geophys. Res.* **91**(1), 91–99.

Gary, S.: 1993, *Theory of Space Plasma Microinstabilities*. Cambridge: Cambridge Atmos. Space Science Series.

Gary, S. P.: 1985, 'Electromagnetic ion beam instabilities: hot beams at interplanetary shocks'. *Astrophys. J.* **288**, 342–352.

Gary, S. P., J. T. Gosling, and D. W. Forslund: 1981, 'The electromagnetic ion beam instability upstream of the Earth's bow shock'. *J. Geophys. Res.* **86**(15), 6691–6696.

Glassmeier, K.-H., U. Motschmann, M. Dunlop, A. Balogh, M. H. Acuña, C. Carr, G. Musmann, K.-H. Fornaçon, K. Schweda, J. Vogt, E. Georgescu, and S. Buchert: 2001, 'Cluster as a wave telescope - first results from the fluxgate magnetometer'. *Ann. Geophys.* **19**, 1439–1447.

Gosling, J. T., J. R. Asbridge, S. J. Bame, G. Paschmann, and N. Sckopke: 1978, 'Observations of two distinct populations of bow shock ions in the upstream solar wind'. *Geophys. Res. Lett.* **5**, 957–960.

Gosling, J. T., M. F. Thomsen, S. J. Bame, W. C. Feldman, G. Paschmann, and N. Sckopke: 1982, 'Evidence for specularly reflected ions upstream from the quasi-parallel bow shock'. *Geophys. Res. Lett.* **9**, 1333–1336.

Gosling, J. T., M. F. Thomsen, S. J. Bame, and C. T. Russell: 1989, 'On the source of the diffuse, suprathermal ions observed in the vicinity of the Earth's bow shock'. *J. Geophys. Res.* **94**(4), 3555–3563.

Greenstadt, E. W.: 1985, 'Oblique, Parallel, and Quasi-Parallel Morphology of Collisionless Shocks'. In: B. T. Tsurutani and R. G. Stone (eds.): *Collisionless Shocks in the Heliosphere: Reviews of Current Research Geophysical Monograph 35*. American Geophysical Union, pp. 169–184.

Greenstadt, E. W. and L. W. Baum: 1986, 'Earth's compressional foreshock boundary revisited Observations by the ISEE 1 magnetometer'. *J. Geophys. Res.* **91**(10), 9001–9006.

Greenstadt, E. W., I. M. Green, G. T. Inouye, A. J. Hundhausen, S. J. B. I. B., and Strong: 1968, 'Correlated magnetic field and plasma observations of the Earth's bow shock'. *J. Geophys. Res.* **73**, 51–60.

Greenstadt, E. W., G. Le, and R. J. Strangeway: 1995, 'ULF waves in the foreshock'. *Adv. Space. Res.* **15**, 71–84.

Greenstadt, E. W. and M. M. Mellott: 1985, 'Variable field-to-normal angles in the shock foreshock boundary observed by ISEE 1 and 2'. *Geophys. Res. Lett.* **12**, 129–132.

Gurgiolo, C., G. K. Parks, and B. H. Mauk: 1983, 'Upstream gyrophase bunched ions: a mechanism for creation at the bow shock and the growth of velocity space structure through gyrophase mixing'. *J. Geophys. Res.* **88**(11), 9093–9100.

Gurnett, D. A., R. L. Huff, and D. L. Kirchner: 1997, 'The Wide-Band Plasma Wave Investigation'. *Space Sci. Rev.* **79**, 195–208.

Hada, T., C. F. Kennel, and T. Terasawa: 1987, 'Excitation of compressional waves and the formation of shocklets in the Earth's foreshock'. *J. Geophys. Res.* **92**(5), 4423–4435.

Hoppe, M. M. and C. T. Russell: 1983, 'Plasma rest frame frequencies and polarizations of the low-frequency upstream waves: ISEE 1 and 2 observations'. *J. Geophys. Res.* **88**, 2021–2028.

Hoppe, M. M., C. T. Russell, L. A. Frank, T. E. Eastman, and E. W. Greenstadt: 1981, 'Upstream hydromagnetic waves and their association with backstreaming ion populations: ISEE 1 and 2 observations'. *J. Geophys. Res.* **86**, 4471–4492.

Hoshino, M. and T. Terasawa: 1985, 'Numerical study of the upstream wave excitation mechanism 1. Nonlinear phase bunching of beam ions'. *J. Geophys. Res.* **90**(1), 57–64.

Ipavich, F. M., J. T. Gosling, and M. Scholer: 1984, 'Correlation between the He/H ratios in upstream particle events and in the solar wind'. *J. Geophys. Res.* **89**(3), 1501–1507.

Jones, F. C. and D. C. Ellison: 1991, 'The plasma physics of shock acceleration'. *Space Sci. Rev.* **58**(3–4), 259–346.

Kis, A., M. Scholer, B. Klecker, E. Möbius, E. Lucek, H. Réme, J. M. Bosqued, L. M. Kistler, and H. Kucharek: 2004, 'Multispacecraft observations of diffuse ions upstream of Earth's bow shock'. *Geophys. Res. Lett.* **31**, L20801, doi:10.1029/2004GL020759.

Krauss-Varban, D. and N. Omidi: 1993, 'Propagation characteristics of waves upstream and downstream of quasi-parallel shocks'. *Geophys. Res. Lett.* **20**, 1007–1010.

Krauss-Varban, D., N. Omidi, and K. B. Quest: 1994, 'Mode properties of low frequency waves: Kinetic Theory versus Hall-MHD'. *J. Geophys. Res.* **99**, 5987–6009.

Le, G. and C. T. Russell: 1990, 'A study of coherence length of ULF waves in the Earth's foreshock'. *J. Geophys. Res.* **95**, 10,703–706.

Le, G. and C. T. Russell: 1992a, 'A study of ULF wave foreshock morphology - I: ULF foreshock boundary'. *Planet. Space Sci.* **40**(9), 1203–1213.

Le, G. and C. T. Russell: 1992b, 'A study of ULF wave foreshock morphology - II: spatial variation of ULF waves'. *Planet. Space Sci.* **40**(9), 1215–1225.

Le, G. and C. T. Russell: 1994, 'The Morphology of ULF Waves in the Earth's Foreshock'. In: M. J. Engebretson, K. Takahashi, and M. Scholer (eds.): *Solar Wind Sources of Magnetospheric Ultra-Low-Frequency Waves*, Geophysical Monograph 81. American Geophysical Union, pp. 87–98.

Le, G., C. T. Russell, M. F. Thomsen, and J. T. Gosling: 1992, 'Observations of a new class of upstream waves with periods near 3 seconds'. *J. Geophys. Res.* **97**, 2917–2925.

Lin, R. P., C.-I. Meng, and K. A. Anderson: 1974, '30- to 100keV protons upstream from the Earth's bow shock'. *J.Geophys.Res.* **79**, 489–498.

Lucek, E. A., T. S. Horbury, A. Balogh, I. Dandouras, and H. Rème: 2004, 'Cluster observations of hot flow anomalies'. *J. Geophys. Res.* pp. A06207, doi:10.1029/2003JA010016.

Mazelle, C., D. Le Quéau, and K. Meziane: 2000, 'Nonlinear wave-particle interaction upstream from the Earth's bow shock'. *Nonlinear Proc. Geophys.* **7**, 185–190.

Mazelle, C., K. Meziane, D. Le Quéau, M. Wilber, J. P. Eastwood, H. Re, J. A. Sauvaud, J. M. Bosqued, I. Dandouras, M. McCarthy, L. M. Kistler, B. Kleckler, A. Korth, M. B. Bavassano-Cattaneo, R. Lundin, and A. Balogh: 2003, 'Production of gyrating ions from nonlinear wave-particle interaction upstream from the Earth's bow shock: a case study from Cluster-CIS'. *Planet. Space Sci.* **51**, 785–795.

Meziane, K. and C. d'Uston: 1998, 'A statistical study of the upstream intermediate ion boundary in the Earth's foreshock'. *Ann. Geophys.* **16**, 125–133.

Meziane, K., C. Mazelle, C. D'Uston, H. Rème, R. P. Lin, C. W. Carlson, D. Larson, J. P. McFadden, R. E. Ergun, K. A. Anderson, G. K. Parks, D. Berdichevsky, and R. P. Lepping: 1997, 'Wind observation of gyrating-like ion distributions and low frequency waves upstream from the Earth's bow shock'. *Adv. Space. Res.* **20**, 703–706.

Meziane, K., C. Mazelle, R. P. Lin, D. Le Quéau, D. E. Larson, G. K. Parks, and R. P. Lepping: 2001, 'Three-dimensional observations of gyrating ion distributions far upstream from the Earth's bow shock and their association with low-frequency waves'. *J. Geophys. Res.* **106**(4), 5731–5742.

Meziane, K., C. Mazelle, M. Wilber, D. Le Quéau, J. Eastwood, H. Rème, I. Dandouras, J. Sauvaud, J. Bosqued, G. Parks, L. Kistler, M. McCarthy, B. Klecker, A. Korth, M. Bavassano-Cattaneo, R. Lundin, and A. Balogh: 2004a, 'Bow shock specularly reflected ions in the presence of low-frequency electromagnetic waves: a case study'. *Ann. Geophys.* **22**, 2325–2335.

Meziane, K., M. Wilber, C. Mazelle, D. Le Quéau, H. Kucharek, E. A. Lucek, H. Rème, A. M. Hamza, J. A. Sauvaud, J. M. Bosqued, I. Dandouras, G. K. Parks, M. McCarthy, B. Klecker, A. Korth, M. B. Bavassano-Cattaneo, and R. N. Lundin: 2004b, 'Simultaneous observations of field-aligned beams and gyrating ions in the terrestrial foreshock'. *J. Geophys. Res.* **109**, A05107, doi:10.1029/2003JA010374.

Motschmann, U., T. I. Woodward, K.-H. Glassmeier, D. J. Southwood, and J. L. Pinçon: 1996, 'Wavelength and direction filtering by magnetic measurements at satellite arrays: Generalized minimum variance analysis'. *J. Geophys. Res.* **101**, 4961–4965.

Narita, Y., K.-H. Glassmeier, S. S. S., U. Motschmann, K. Sauer, I. Dandouras, K.-H. Fornaçon, E. Georgescu, and H. Rème: 2003, 'Dispersion analysis of ULF waves in the foreshock using Cluster data and the wave telescope technique'. *Geophys. Res. Lett.* **30**, 1710, doi:10.1029/2003GL017432.

Narita, Y., K.-H. Glassmeier, S. Schäfer, U. Motschmann, M. Fränz, I. Dandouras, K.-H. Fornaçon, E. Georgescu, A. Korth, H. Rème, and I. Richter: 2004, 'Alfvén waves in the foreshock propagating upstream in the plasma rest frame: Statistics from Cluster Observations'. *Ann. Geophys.* **22**, 2315–2323.

Ogilvie, K. W., T. von Rosenvinge, and A. C. Durney: 1977, 'International Sun-Earth Explorer: A Three-Spacecraft Program'. *Science* **198**(4313), 131–138.

Orlowski, D. S., C. T. Russell, D. Krauss-Varban, N. Omidi, and M. F. Thomsen: 1995, 'Damping and spectral formation of upstream whistlers'. *J. Geophys. Res.* **100**(9), 17117–17128.

Paschmann, G., G. Haerendel, N. Sckopke, E. Möbius, H. Lühr, and C. W. Carlson: 1988, 'Three-dimensional plasma structures with anomalous flow directions near the Earth's bow shock'. *J. Geophys. Res.* **93**, 11279–11294.

Paschmann, G., N. Sckopke, S. J. Bame, J. R. Asbridge, J. T. G. C. T. Russell, and E. W. Greenstadt: 1979, 'Association of low-frequency waves with suprathermal ions in the upstream solar wind'. *Geophys. Res. Lett.* **6**, 209–212.

Paschmann, G., N. Sckopke, I. Papamastorakis, J. R. Asbridge, S. J. Bame, and J. T. Gosling: 1981, 'Characteristics of reflected and diffuse ions upstream from the Earth's bow shock'. *J. Geophys. Res.* **86**, 4355–4364.

Pinçon, J. and F. Lefeuvre: 1991, 'Local characterization of homogeneous turbulence in a space plasma from simultaneous measurement of field components at several points in space'. *J. Geophys. Res.* **96**, 1789–1802.

Pinçon, J. and U. Motschmann: 1998, 'Multi-Spacecraft Filtering: General Framework'. In: G. Paschmann and P. Daly (eds.): *Analysis methods for multi-spacecraft data*, ISSI Sci. Rep. SR-001. Bern: ISSI, pp. 65–78.

Reiner, M. J., M. L. Kaiser, J. Fainberg, M. D. Desch, and R. G. Stone: 1996, '2fp radio emission from the vicinity of the Earth's foreshock: WIND observations'. *Geophys. Res. Lett.* **23**(10), 1247–1250.

Réme, H., J. M. Bosqued, J. A. Sauvaud, A. Cros, J. Dandouras, C. Aoustin, C. Martz, J. L. Médale, J. Rouzaud, E. Möbius, K. Crocker, M. Granoff, L. M. Kistler, D. Hovestadt, B. Klecker, G. Paschmann, M. Ertl, E. Künneth, C. W. Carlson, D. W. Curtis, R. P. Lin, J. P. McFadden, J. Croyle, V. Formisano, M. DiLellis, R. Bruno, M. B. Bavassano-Cattaneo, B. Baldetti, G. Chionchio, E. G. Shelley, A. G. Ghielmetti, W. Lennartson, A. Korth, H. Rosenbauer, I. Szemerey, R. Lundin, S. Olson, G. K. Parks, M. McCarth, and H. Balsiger: 1997, 'The CLUSTER Ion Spectrometry Experiment'. *Space Sci. Rev.* **79**, 303.

Robinson, P. A., I. H. Cairns, and D. A. Gurnett: 1993, 'Clumpy Langmuir waves in the type III radio sources: comparison of stochastic-growth theory with observations'. *Astrophys. J.* **407**, 790–800.

Russell, C. T.: 1994, 'Planetary Upstream Waves'. In: M. J. Engebretson, K. Takahashi, and M. Scholer (eds.): *Solar Wind Sources of Magnetospheric Ultra Low Frequency Waves*, Geophysical Monograph 81. American Geophysical Union, pp. 75–86.

Russell, C. T., D. D. Childers, and J. P. J. Coleman: 1971, 'OGO 5 observations of upstream waves in interplanetary medium: Discrete wave packets'. *J. Geophys. Res.* **76**, 845–861.

Russell, C. T. and M. M. Hoppe: 1983, 'Upstream waves and particles /Tutorial Lecture/'. *Space Sci. Rev.* **34**, 155–172.

Sarris, E. T., G. C. Anagnostopoulos, and S. M. Krimigis: 1987, 'Simultaneous measurements of energetic ion (50keV and above) and electron (220 keV and above) activity upstream of Earth's bow shock and inside the plasma sheet – Magnetospheric source for the November 3 and December 3, 1977 upstream events'. *J. Geophys. Res.* **92**(11), 12083–12096.

Sauer, K. and E. Dubinin: 2003, 'Oscillitons and gyrating ions in a beam-plasma system'. *Geophys. Res. Lett.* **30**, 2192, doi:10.1029/2003GL018266.

Sauer, K., E. Dubinin, and J. F. McKenzie: 2001, 'New type of soliton in bi-ion plasmas and possible implications'. *Geophys. Res. Lett.* **28**, 3589–3592.

Scarf, F. L., R. W. Fredricks, L. A. Frank, and M. Neugebauer: 1971, 'Nonthermal electrons and high-frequency waves in the upstream solar wind 1. Observations'. *J.Geophys.Res.* **76**, 5162–5171.

Scholer, M., M. Fujimoto, and H. Kucharek: 1993, 'Two-dimensional simulations of supercritical quasi-parallel shocks: upstream waves, downstream waves, and shock re-formation'. *J. Geophys. Res.* **98**, 18971–18984.

Schwartz, S.: 1998, 'Shock and discontinuity normals, Mach numbers, and related parameters'. In: G. Paschmann and P. W. Daly (eds.): *Analysis Methods for Multi-Spacecraft Data*, ISSI SR-001. ESA Publications Division, pp. 249–270.

Schwartz, S. J. and D. Burgess: 1991, 'Quasi-Parallel Shocks: A Patchwork of Three-Dimensional Structures'. *Geophys. Res. Lett.* **18**(3), 373–376.

Schwartz, S. J., C. P. Chaloner, P. J. Christiansen, D. S. Hall, A. D. Johnson, M. P. Gough, A. J. Norris, R. J. Rijnbeek, D. J. Southwood, and L. J. C. Woolliscroft: 1985, 'An active current sheet in the solar wind'. *Nature* **318**, 269–271.

Schwartz, S. J., R. L. Kessel, C. C. Brown, L. J. C. Woolliscroft, M. W. Dunlop, C. J. Farrugia, and D. S. Hall: 1988, 'Active current sheets near the Earth's bow shock'. *J. Geophys. Res.* **93**, 11295–11310.

Schwartz, S. J., N. Paschmann, N. Sckopke, T. M. Bauer, M. W. Dunlop, C. J. Farrugia, and D. S. Hall: 2000, 'Conditions for the formation of hot flow anomalies at the Earths bow shock'. *J. Geophys. Res.* **105**, 12639–12650.

Schwartz, S. J., M. F. Thomsen, and J. T. Gosling: 1983, 'Ions upstream of the Earth's bow shock - A theoretical comparison of alternative source populations'. *J. Geophys. Res.* **88**(17), 2039–2047.

Sentman, D., J. P. Edmiston, and L. A. Frank: 1981, 'Instabilities of low frequency, parallel propagating electromagnetic waves in the Earth's foreshock region'. *J. Geophys. Res.* **86**, 7487–7497.

Sentman, D. D., M. F. Thomsen, S. P. Gary, W. C. Feldman, and M. M. Hoppe: 1983, 'The oblique whistler instability in the Earth's foreshock'. *J. Geophys. Res.* **88**(3), 2048–2056.

Sibeck, D. G., T.-D. Phan, R. P. Lin, R. P. Lepping, and A. Szabo: 2002, 'Wind observations of foreshock cavities: A case study'. *J. Geophys. Res.* **107**(A10), 1271, doi:10.1029/2001JA007539.

Sigsbee, K., C. A. Kletzing, D. A. Gurnett, J. S. Pickett, A. Balogh, and E. Lucek: 2004a, 'Statistical behavior of foreshock Langmuir waves observed by the Cluster Wideband Data Plasma Wave Receiver'. *Ann. Geophys.* **22**, 2337–2344.

Sigsbee, K., C. A. Kletzing, D. A. Gurnett, J. S. Pickett, A. Balogh, and E. Lucek: 2004b, 'The dependence of Langmuir wave amplitudes on position in Earth's foreshock'. *Geophys. Res. Lett.* **31**, L07805, doi:10.1029/2004GL019413.

Song, P. and C. T. Russell: 1999, 'Time series data analyses in space plasmas'. *Space Sci. Rev.* **87**, 387–463.

Sonnerup, B. U. Ö. and M. Scheible: 1998, 'Minimum and Maximum Variance Analysis'. In: G. Paschmann and P. W. Daly (eds.): *Analysis Methods for Multi-Spacecraft Data*, ISSI SR-001. ESA Publications Division, pp. 185–220.

Stone, R. G. and B. T. Tsurutani (eds.): 1985, *Collisionless shocks in the heliosphere: A tutorial review*, Geophys. Monogr. Ser. vol 34. Washington, D.C.: American Geophysical Union.

Thomas, V. A., D. Winske, M. F. Thomsen, and T. G. Onsager: 1991, 'Hybrid simulation of the formation of a hot flow anomaly'. *J. Geophys. Res.* **96**, 11625–11632.

Thomsen, M. F., J. T. Gosling, S. J. Bame, K. B. Quest, and C. T. Russell: 1988, 'On the origin of hot diamagnetic cavities near the Earth's bow shock'. *J. Geophys. Res.* **93**(12), 11311–11325.

Thomsen, M. F., J. T. Gosling, S. J. Bame, and C. T. Rusell: 1985, 'Gyrating ions and large-amplitude monochromatic MHD waves upstream of the Earth's bow shock'. *J. Geophys. Res.* **90**(9), 267–273.

Thomsen, M. F., J. T. Gosling, S. A. Fuselier, S. J. Bame, and C. T. Russell: 1986, 'Hot, diamagnetic cavities upstream from the Earths bow shock '. *J. Geophys. Res.* **91**, 2961–2973.

Treumann, R. A. and M. Scholer: 2001, 'The magnetosphere as a plasma laboratory'. In: *The Century of Space Science*. Kluwer Academic, p. 1495.

Tsurutani, B. T. and R. G. Stone (eds.): 1985, *Collisionless shocks in the heliosphere: Reviews of current research*, Geophys. Monogr. Ser. vol 35. Washington, D.C.: American Geophysical Union.

Watanabe, Y. and T. Terasawa: 1984, 'On the excitation mechanism of the low-frequency upstream waves'. *J. Geophys. Res.* **89**(8), 6623–6630.

Wong, H. K. and M. L. Goldstein: 1987, 'Proton beam generation of whistler waves in the Earth's foreshock'. *J. Geophys. Res.* **92**(11), 12419–12424.

Chapter 3

The Magnetosheath

E. A. Lucek[1], D. Constantinescu[2], M. L. Goldstein[3], J. Pickett[4], J. L. Pinçon[5], F. Sahraoui[6], R. A. Treumann[7], and S. N. Walker[8]

3.1 Introduction

3.1.1 Average magnetosheath properties

The magnetosheath lies between the bow shock and the magnetopause and is formed mainly from decelerated and deflected solar wind, with a small contribution of plasma from the magnetosphere. The observed magnetosheath plasma parameters show both large scale spatial ordering, imposed by the shape of the magnetopause, and variability dependent on the solar wind input. Because the nature of the bow shock depends on the orientation of the interplanetary magnetic field with respect to the local bow shock normal (θ_{Bn}), the properties of the magnetosheath plasma just behind the bow shock depend also on whether the shock is quasi-perpendicular or quasi-parallel. In general, the magnetosheath tends to be in a more turbulent state

[1]Space and Atmospheric Physics, The Blackett Laboratory, Imperial College London, London, UK
[2]Institut für Geophysik und extraterrestrische Physik, Technische Universität, Braunschweig, Germany
[3]NASA Goddard Space Flight Center, Greenbelt, MD, USA
[4]Department of Physics and Astronomy, The University of Iowa, Iowa City, IA, USA
[5]LPCE/CNRS, Orléans, France
[6]CETP/IPSL, Vélizy, France
[7]Max-Planck-Institut für extraterrestrische Physik, Garching, Germany
[8]Automatic Control and Systems Engineering, University of Sheffield, Sheffield, UK

Space Science Reviews 118: 95–152, 2005.
DOI: 10.1007/s11214-005-3825-2

behind the spatially extended quasi-parallel bow shock than it is behind the quasi-perpendicular shock. Lastly the plasma properties of the magnetosheath depend on properties of the upstream solar wind, including density, velocity and β.

The average properties of the magnetosheath have been documented based on data from several missions including ISEE 1 and 2 (e.g., Kivelson and Russell, 1995), AMPTE-IRM (Phan et al., 1994; Hill et al., 1995; Li et al., 1995), and Wind (Phan et al., 1996, 1997). Magnetosheath plasma is characterised by the following: first, its average density and magnetic field strength are higher than in the upstream solar wind by a factor consistent on average with the Rankin-Hugoniot relation for the fast mode shock; second, the average flow direction deviates from the anti-solar direction such that the plasma flows around the blunt magnetosphere; third, the velocity downstream of the bow shock is lower than the local fast magnetosonic speed; fourth, the flow velocity increases again to supersonic speeds around the magnetopause flanks; fifth, the ion temperature of the sheath is higher than in the solar wind while the electron temperature does not increase very much over its upstream value, such that the ion to electron temperature ratio in the sheath is of order $6-7$; sixth, the plasma β shows large variations from the order of unity to values much greater than one; seventh, the magnetosheath plasma develops a pronounced temperature anisotropy ($T_\perp > T_\|$) behind the bow shock that increases toward the magnetopause and is more pronounced in the ions than in the electrons. As a consequence of this the magnetosheath seems to develop two regions of different turbulent behaviour: one behind the bow shock and the other closer to the magnetopause.

3.1.2 Sources of low frequency waves and turbulence

As in the foreshock, there are multiple sources of waves and turbulence in the magnetosheath, but the physics of the latter is more difficult to untangle since the low frequency magnetic field fluctuations can be of order of $\langle \delta B^2 \rangle / B_0^2 \sim \mathrm{O}(1)$, which is in the strong turbulence regime. Embedded in the magnetosheath plasma are fluctuations arising from intrinsic solar wind turbulence, investigated extensively in previous work (Goldstein et al., 1995; Horbury et al., 1995; Horbury and Balogh, 1997; Marsch and Tu, 1997; Matthaeus et al., 1990; Roberts and Goldstein, 1991; Tu and Marsch, 1995), that have been processed as the plasma passed through the bow shock. Other fluctuations come from the foreshock region, where they are generated by reflected particle components (e.g., Paschmann et al., 1980, 1981). These waves have group velocities slower than the undisturbed solar wind speed and are therefore convected with the solar wind toward the shock front and into the magnetosheath. Further magnetosheath fluctuations are generated at the bow shock itself (e.g., Sckopke et al., 1983, 1990; Brinca et al., 1990).

As the magnetosheath plasma convects from the bow shock to the magnetopause the pressure anisotropy increases, and the free energy in the anisotropy can drive plasma instabilities (Crooker and Siscoe, 1977). The sense of the anisotropy, $p_\perp / p_\| >$

1, is opposite to that which leads to the excitation of the Alfvénic firehose mode. Instead, in a bi-Maxwellian plasma such an anisotropy can drive two instabilities which generate waves with frequencies below the ion cyclotron frequency. The first of these, the ion cyclotron instability, dominates when the temperature anisotropy is high and the proton plasma $\beta \sim 1$ and generates transverse electromagnetic ion cyclotron waves through a resonant wave particle interaction (e.g., Schwartz et al., 1996). Ion cyclotron waves typically have phase velocities close to the Alfvén speed and propagate away from their source region. The second, the mirror instability, tends to dominate under conditions of moderate temperature anisotropy and high plasma β (e.g., Schwartz et al., 1996). It generates large amplitude, anti-correlated variations in the magnetic field magnitude and plasma density which are non-propagating in the plasma frame. These structures can act as magnetic bottles, trapping part of the particle distribution (e.g., Kivelson and Southwood, 1996). For example, 'lion roar' oscillations are thought to be generated by anisotropic electrons trapped inside mirror structures (Baumjohann et al., 1999). Since mirror modes structures can be of large amplitude, introducing excess energy into the spectrum over a finite frequency band, it has been suggested that they could lead to both a forward and inverse cascade of energy to larger and smaller wavenumbers (e.g., Pokhotelov et al., 2003).

3.1.3 Cluster achievements

As described in the previous two sections, the magnetosheath is inherently complex. It is rich in waves and turbulence, and its properties are strongly influenced by processes occurring at, and upstream of, the bow shock. Since the properties of the bow shock vary in both space and time, so does the magnetosheath. Untangling the physics here, to understand wave generation, properties of the turbulence, and the evolution of the plasma behind the bow shock is therefore a challenge.

Previous studies have been limited by having measurements from only single or dual spacecraft missions. There was insufficient information to be able to uniquely identify wave modes, and it was very difficult to separate signatures of time evolution from those of spatial variation. For the first time, using the four point Cluster data, significant progress has been made in both of these areas, each essential for understanding how the magnetosheath works, and already new discoveries have been made. In addition, the polar Cluster orbit, which crosses the magnetosheath at high latitudes near noon and at the equator far on the flanks, samples regions which have not, until now, been comprehensively studied using such a sophisticated payload.

In the following sections we present some of the new results discovered using data from the Cluster mission. The following is a brief outline of the chapter:

- In section 3.2 two methods are presented which allow the unique identification of low-frequency wave modes present in the plasma. The methods are shown to be complementary. One, k-filtering, allows the identification of multiple wave modes simultaneously supported by the plasma. It is found that, at frequencies below the ion gyro-frequency, these waves lie on MHD dispersion curves. The second, the phase difference method, allows the time variation of the dominant wave mode to be analysed on a time scale of only 15 seconds.

- In section 3.3 a study is presented of the low-frequency waves generated by particle temperature anisotropies downstream of the quasi-perpendicular bow shock, where they are for the first time unambiguously identified as Alfvén waves. These are then shown to evolve into field aligned current tubes, and it is suggested that this occurs under the action of the filamentation instability.

- Section 3.4 describes the derivation of a model of the 3D properties of local inhomogeneities introduced by the presence of mirror modes. This model is then compared with Cluster multi-point measurements, allowing an estimation of the mirror mode scale to be made.

- Section 3.5 describes the first statistical study of the occurrence and distribution of high frequency waves in the magnetosheath, including lion roars and a range of isolated electrostatic structures (IES). The observations suggest that magnetosheath IES might be being generated by a different mechanism, as yet unexplained, to that generating the IES seen in other regions.

- In Section 3.6 a type of dispersed ion signature found in the magnetosheath is described. The characteristics of these 'magnetosheath dispersed ion signatures' (MDSs) are used to develop a model which describes their generation by particle acceleration in a non-linear interaction between a solar wind discontinuity and the bow shock.

- Section 3.7 provides a summary and outlook.

3.2 Low Frequency Wave Mode Identification

3.2.1 Introduction

One major topic of study using Cluster data from the magnetosheath, has been the analysis of low frequency (LF) fluctuations, at frequencies below the ion gyrofrequency, where the new multi-spacecraft methods allow detailed analysis of the wave modes never previously possible with data from one or two spacecraft. The tetrahedron scale size relative to the wave vector is of particular importance in studying the magnetosheath wave field since many multi-spacecraft wave analysis methods require that the four satellites sample the same wave.

Many experimental studies of waves in the magnetosheath have focused on identifying linear modes at frequencies below the proton gyro-frequency, f_{ci}, (Anderson et al., 1994; Denton et al., 1994; Lacombe et al., 1995; Song et al., 1994;

Lucek et al., 1999). Before the Cluster mission, experimental analysis of the magnetosheath typically suffered from the spatio-temporal ambiguity that characterises single-satellite measurements. At times, analyses also lacked simultaneous field and plasma data. Under such conditions, wave mode identification was based on analysis of the relationship between the different components of the electromagnetic field and the plasma characteristics. Such methods were generally applied to 'wave-like' fluctuations, when the magnetic field power spectrum contained a clear peak, rather than to broadband fluctuations with power-law spectra. Typically these studies reported observations of Alfvénic fluctuations or ion cyclotron waves when the transverse magnetic field components were dominant, and mirror mode or slow waves when the fluctuations were compressional (e.g., Song and Russell, 1992). Moreover, the nature of the identified linear waves was found to depend on the depth in the magnetosheath, the geometry of the shock, β, the ion temperature anisotropy $A_i = T_\perp/T_\| - 1$ and the frequency range analysed. High β and small A_i (e.g., $\beta_{i\|} \approx 2$ and $A_i \approx 0.4$) were found to be favourable to mirror modes, whereas low β and large A_i (e.g., $\beta_{i\|} \approx 0.2$ and $A_i \approx 2$) favoured EMIC (Electromagnetic Ion Cyclotron) waves (Anderson et al., 1994).

With data from only a single satellite it appeared difficult to go beyond this set of well established results. Denton et al. (1995) described the difficulty found with some events in identifying any linear mode. They suggested that either multiple modes were present in the same frequency range, or that non linear effects made the linear analysis suspect at best.

In these previous studies the presence of a unique wave vector at each given frequency had to be assumed, with a direction determined, for instance using minimum variance analysis (MVA). This restriction to monochromatic plane waves (one **k** for one frequency), even when justified, can only provide information on the mode containing the bulk of the energy. No information is obtained about other weaker modes in the wave field, which, when present, can compromise further analysis, of the wave polarisation for example. No method which assumes the presence of a single wave vector can be applied when the wave field is more properly represented as a superposition of several plane waves (i.e., with several wave vectors) containing comparable energies. Lastly, an additional complication arises because the waves are sampled in the spacecraft frame as they are carried across the satellites by the flowing plasma, while the theoretical linear waves are derived in the plasma frame (Omidi et al., 1994; Song et al., 1994). In the case when the wave speed is much smaller than the plasma flow speed, then it can be assumed that the waves do not change significantly in the time they take to cross the spacecraft, i.e., Taylor's hypothesis is satisfied, and the time series can be converted into a spatial cut through the plasma by taking into account the Doppler effect of the plasma velocity on the waves. However, in the magnetosheath the characteristic wave speed can often be of the same order of magnitude as the flow velocity, and it is therefore not usually possible to assume that Taylor's hypothesis is satisfied.

Now, with four point Cluster data it becomes possible under many circumstances to remove this spatio-temporal ambiguity and explore the three-dimensional motion of the electromagnetic and particle structures. One of the first studies to analyse magnetosheath data using a multi spacecraft wave identification method was presented by Glassmeier et al. (2001). They used a technique called the 'wave telescope' (Motschmann et al., 1996, 1998), which is similar in formulation to the k-filtering method discussed later in this section. They chose an interval where a clear, quasi-monochromatic wave was observed in the magnetic field data. They were able to identify uniquely the **k**-vector of the wave, finding that it had a wavelength of $\simeq 12.5 \times 10^3$ km, and was propagating nearly parallel to the ambient magnetic field.

In the following sections one interval of magnetosheath data (February 18, 2002 at around 05:34 UT), containing turbulent ULF magnetic fluctuations commonly seen in the high β magnetosheath plasma, is analysed using two methods, which allow the **k**-vectors of magnetosheath waves to be calculated. The first study by Sahraoui et al. (2003) describes the application of the k-filtering technique to magnetosheath fluctuations. In this analysis the wave field was modelled as a linear superposition of monochromatic plane waves. The k-filtering method allows the identification of multiple wave vectors at a single frequency by the application of a series of filters applied to data from all four spacecraft (Pinçon and Motschmann, 1998; Pinçon and Lefeuvre, 1991). The derivation of the filters can be constrained by various physical criteria, and in this case the results were obtained from a set of filters which incorporated the condition that $\nabla \cdot \mathbf{B} = 0$. The second study, presented in Walker et al. (2004), describes the results of the phase difference method for the same interval and compares the two methods. The phase difference method (Balikhin and Gedalin, 1993) also represents the wave field as superposition of plane waves, but in addition it assumes the dominance of a single **k** vector at a given frequency. At each frequency the **k** vector is estimated from the phase delays in the Fourier components, calculated using data from each of the four spacecraft. Both studies reveal complex physics. Walker et al. (2004) concluded that there is good agreement between the two methods in their identification of the dominant mode in the data set. They also show, however, that at some frequencies there is significant power in modes other than the dominant one.

On February 18, 2002, Cluster was on an outbound trajectory, crossing the magnetopause into the magnetosheath at ~04:59 UT. The tetrahedron was located close to (5.6, 4.6, 8.4) R_E GSE, about 0.8 R_E outside the model magnetopause of Roelof and Sibeck (1993). The inter-spacecraft separation varied between 94 and 104 km. For both studies the shape of the tetrahedron is important and it has been shown that the two geometrical factors characterising the three-dimensional configuration of the spacecraft: elongation E and planarity P, have to take values ≤ 0.4 for the 3D configuration to be appropriate for application of the methods (Robert et al., 1998). For this interval both P and E were small: 0.04 and 0.08, respectively.

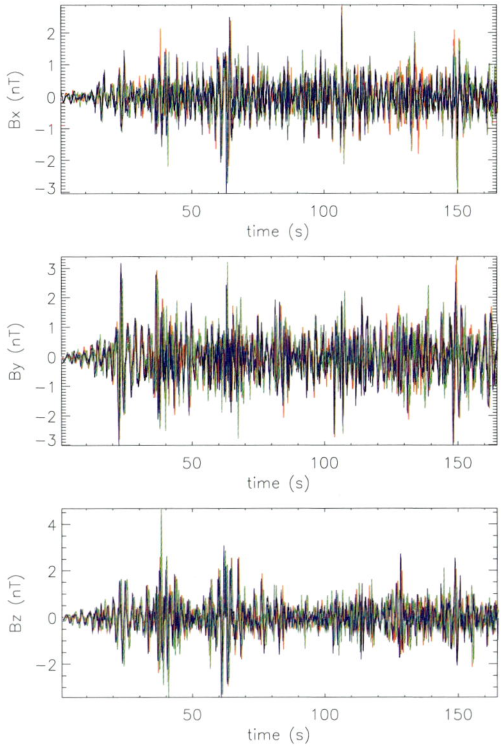

Figure 3.1. STAFF waveform data from February 18, 2002. The interval length is 164s, starting at 05:34:01.15 UT, and the data are high-pass filtered at the cut-off frequency $f_{cut-off} = 0.35$ Hz. From top to bottom the panels show the X, Y, and Z components of the magnetic field in the GSE frame. The coloured lines represent data from Cluster 1 (black), 2 (red), 3 (green), and 4 (blue). The waves in this interval were analysed using the k-filtering and phase difference methods. (From Sahraoui et al., 2003).

A comprehensive description of the magnetosheath plasma parameters for this interval derived from FGM, WHISPER and CIS can be found in Sahraoui et al. (2003). They used magnetic field data from both the STAFF search coil magnetometer (Cornilleau-Wehrlin et al., 1997) and FGM (Balogh et al., 2001) experiment, while Walker et al. (2004) used data from STAFF and from EFW (Gustafsson et al., 1997). The Cluster payload was operating in a telemetry mode, where STAFF and EFW were returning magnetic/electric field waveforms up to 10 Hz and sampled at 25 Hz. For both studies, the STAFF data were filtered using a high-pass filter with a cut-off frequency f_{co}=0.35 Hz. The purpose of this filter is to prevent any problems with the spacecraft spin, which might pollute the magnetic data at frequencies close to 0.25 Hz.

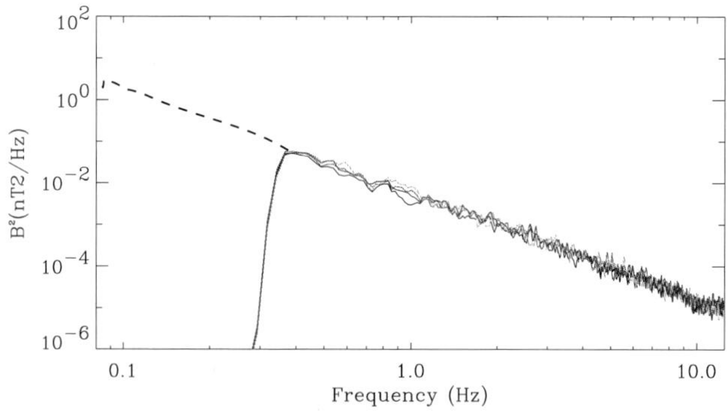

Figure 3.2. Power spectrum of the ULF magnetic fluctuations shown in Figure 3.1. This is close to a power-law $f^{-\alpha}$ with $\alpha \approx 2.2$. The coloured lines represent data from Cluster 1 (black), 2 (red), 3 (green), and 4 (blue). From Sahraoui et al. (2003).

3.2.2 Magnetic wave field energy distribution from *k*-filtering

The interval on February 18, 2002 selected for analysis by the *k*-filtering method was of length $\Delta T = 164\,s$, starting at 05:34:01.15 UT (see Figure 3.1). The modulus of the amplitude of the filtered magnetic fluctuations, normalised to the background magnetic field, was weaker than 15%, implying that the fluctuations in this interval can be interpreted using concepts from weak turbulence theory.

The power spectrum of the magnetic fluctuations in the frequency range 0.35 Hz to 12.5 Hz is shown in Figure 3.2. The spectral slope of the power-law fitted to this spectrum was ~ 2.2, and the spectrum was interpreted by the authors as being similar to those characterising the cascade of energy from large to small scales in turbulence theory. The spectra of the parallel and perpendicular components of the magnetic fluctuations (with respect to the background magnetic field) were compared to that of the sum of the component spectra, and it was found that both components looked similar.

Four frequencies were arbitrarily selected from the continuous spectrum between the low frequency cut off at $f_1 = 0.37$ Hz, and the maximum frequency which can be analysed accurately using the *k*-filtering technique, $\simeq 2$ Hz. The chosen four frequencies were: $f_1 = 0.37$ Hz, $f_2 = 0.49$ Hz, $f_3 = 0.61$ Hz, and $f_4 = 1.15$ Hz, and the *k*-filtering technique was applied. Here the results for $f = 0.61$ Hz are reviewed. Application of the *k*-filtering technique to the magnetic field data allows a comparison to be made between the distribution of the most significant portion of the magnetic field energy in the $(\omega, \mathbf{k})$ domain and the theoretical dispersion relation for propagating waves. For this purpose, it is convenient to display the data using magnetic field aligned (MFA) coordinates. In this coordinate system the z-axis is along the mean magnetic field $\mathbf{B}_o = B_o\mathbf{z}$, the x-axis is perpendicular to

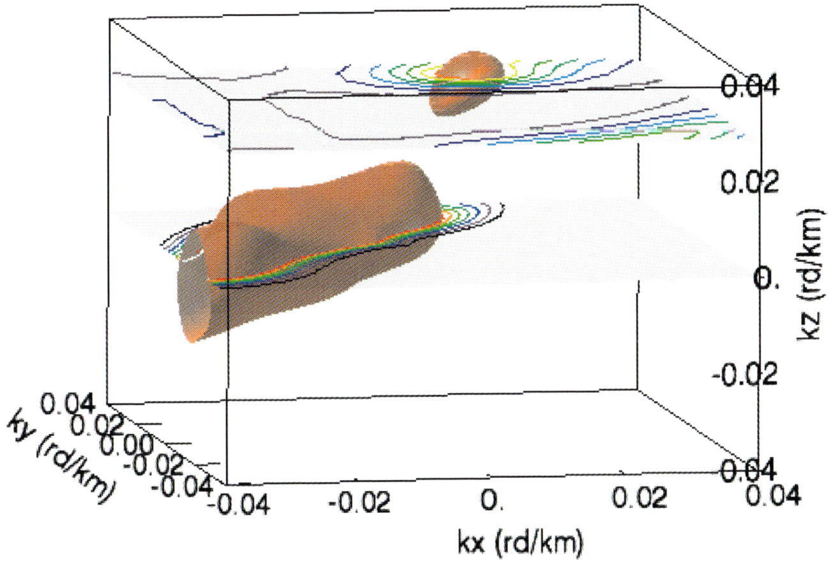

Figure 3.3. Three-dimensional display of the magnetic field energy distribution (MFED) in (k_x, k_y, k_z) space for $f = 0.61$ Hz showing the MFED isosurface corresponding to 33% of the MFED maximum value in the validity domain. Most of the energy is confined to two limited areas. Energy isocontours are drawn in the (k_x, k_y) plane for $k_z = -0.0007$ rad km^{-1} and $k_z = 0.0305$ rad km^{-1} with k_x and k_y values being (-0.04, 0.04) rad km^{-1} respectively. Figure 3.4 shows these two energy contribution plots. (Figure provided by J.-L. Pinçon.)

z-axis, in the plane containing the Sun-satellite line and the z-axis, and directed towards the Sun, and the y-axis completes the right-handed set.

A representation of the magnetic field energy distribution (MFED) at 0.61 Hz in the (k_x, k_y, k_z) domain is shown in Figure 3.3. The three-dimensional view was obtained by displaying the MFED isosurface corresponding to 33% of the maximum value. In this interval the most significant part of the field energy was confined to two distinct areas in (k_x, k_y): one plane at $k_z = -0.0007$ rad km^{-1} and the other at $k_z = 0.0305$ rad km^{-1}. Each (k_x, k_y) plane is restricted to the validity domain defined by $k_x, k_y \in [-0.04; 0.04]$ rad km^{-1}. The validity domain in the wave vector space is determined by the Cluster tetrahedron scale: to avoid aliasing, all the wavelengths have to be larger than the satellite separations, which are of the order of 100 km in the present case. The resolution in k-space along k_x, k_y, k_z was found to be sufficient to determine the MFED without loss of information about its three-dimensional shape.

Since the chosen frequency, $f = 0.61$ Hz, was of the same order as the local ion-gyro frequency, the MFED derived from k-filtering was compared to the lin-

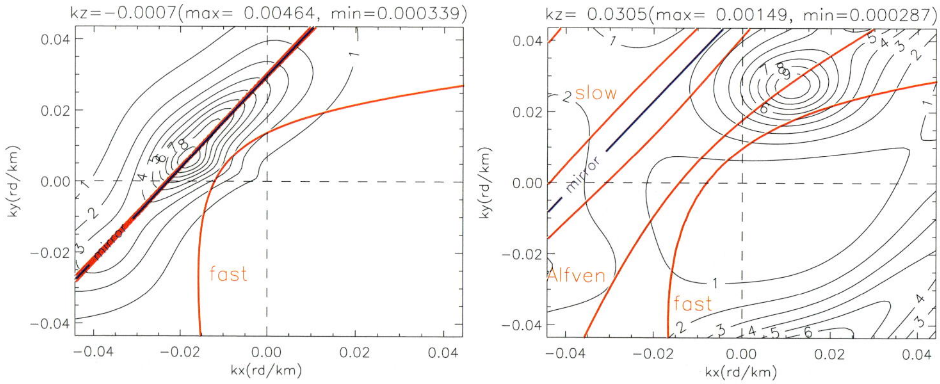

Figure 3.4. Two cuts through the magnetic wave field energy distribution (MFED), calculated for f = 0.61 Hz (as shown in Figure 3.3). The left panel is a cut at k_c = -0.0007 rad km^{-1} and the right panel at k_c = -0.0305 rad km^{-1}. Both panels show experimental magnetic energy (thin black lines) and the theoretical dispersion relations of the low frequency modes (coloured thick lines) as functions of k_x and k_y in the MFA frame. The blue line is the Doppler shift $\omega = \mathbf{k}.\mathbf{v}$. Two main peaks are identified: a mirror mode (left panel) and an Alfvén wave (right panel) having a frequency in the plasma frame close to the second gyroharmonic $f_{plasma} = 0.71\text{Hz} \sim 2f_{ci}$. (From Walker et al., 2004).

ear dispersion relations for the low frequency modes: mirror, Alfvén, fast and slow magnetosonic modes. These were derived from the WHAMP program (Rönnmark, 1983), using the plasma parameters applicable to this event. The mirror mode can be added by considering it as a non propagating mode $\omega_{mirror} = 0$ in the plasma frame. The MFED was computed in the MFA frame (which is at rest with respect to the satellite). Because the theoretical dispersion relations are obtained in the plasma frame, the influence of the Doppler effect was taken into account before the comparison was made. The relative velocity between the plasma and the spacecraft frames was derived from CIS data (Rème et al., 2001): $(V_x, V_y, V_z) \approx (-180, -130, -30)$ km s^{-1}in the MFA frame. The Doppler shift was estimated for each k-value and used to obtain the theoretical dispersion relations of the LF modes in MFA frame. The k-filtering results were then compared with the linear model of propagating waves.

Figure 3.4 shows a superposition of the theoretical LF dispersion relation (thick coloured lines) with isocontours of the MFED at the frequency $f = 0.61$ Hz. The results are presented for the two (k_x, k_y) planes corresponding to significant MFED maxima ($k_z = -0.0007$rad km^{-1} and $k_z = 0.0305$ rad km^{-1}) identified in Figure 3.4. The presence of both a mirror mode (left panel) and an Alfvén wave (right panel) is apparent. For a given frequency and k_z value, a mode is clearly identified when its theoretical curve in the (k_x, k_y) plane lies close to an observed magnetic energy peak. It should be noted that the superposition of the slow, Alfvén, and mirror dispersion curves in Figure 3.4 (left panel) reflects the degeneracy of these modes at low frequencies in the plasma frame (≈ 0 Hz), particularly for this quasi-perpendicular direction of propagation ($\approx 87^\circ$ with respect to the local magnetic field). Such coexistence of several waves for a given frequency is not exceptional. Using the selected STAFF data the same k-filtering analysis was performed for three more frequencies in the range 0.35 Hz to 1.4 Hz and for each case the k-filtering technique showed the presence of several modes at a single frequency.

3.2.2.1 *Discussion of the k-filtering analysis*

By using the k-filtering technique to four point magnetic field data, Sahraoui et al. (2003) determined the magnetic field energy distribution in (k_x, k_y, k_z) space for four different frequencies, of which the results for $f = 0.61$ Hz have been presented here. In each case the energy observed at a single frequency could be attributed to the superposition of more than one wave. The experimental results were compared with a model of low frequency waves. The distribution of maxima in the MFED, in the wave vector domain was found to be consistent with the dispersion relations of ULF wave modes. The mirror mode was confirmed to be the dominant mode in the high β plasma analysed in this case, Doppler shifted to non zero frequencies in the MFA frame, but Alfvén and slow modes were also identified. The imaginary parts of the theoretical solution obtained from the WHAMP model were non-zero, suggesting that weak non-linear interaction between low frequency modes might counteract the effects of linear kinetic damping. There are further arguments in support of the validity of a model of weak turbulence for studying the coupling between large and small scale magnetosheath fluctuations. Not only was the amplitude of the magnetic fluctuations less than 15% of the background magnetic field, as already mentioned earlier, but the observed magnetosheath magnetic energy was distributed over several eigenmodes close to the theoretical ULF mirror, Alfvén, and slow modes.

Similar results were obtained when the analysis was repeated for an interval 25 minutes later, when Cluster was still in the magnetosheath, but to construct a global view of the magnetic fluctuations in the magnetosheath, similar studies are needed of low beta magnetosheath plasma, for which it is expected that other waves than the mirror mode might be dominant.

The next section will present the results from a comparison between the wave vectors resulting from the application of the k-filtering and phase differencing methods.

3.2.3 Application of the phase difference method

Use of synchronous data from two closely separated satellites allows the determination of wave modes as well as the direction of wave propagation. This has been done by applying either spectral/coherency analysis methods together with time delay information to the components of the magnetic field (Gleaves and Southwood, 1991), magnetic and thermal pressures, and MVA (Song and Russell, 1992), or by computing the phase difference of the waves measured at each satellite (Balikhin and Gedalin, 1993; Balikhin et al., 2003; Dudok de Wit et al., 1995). The latter method yields the projection of the wave vector along the satellite separation direction. In certain circumstances the complete wave vector (both in magnitude and direction) can be constructed subsequently by employing MVA to find the wave propagation direction (Means, 1972) with respect to the vector of satellite separation. However, the use of MVA imposes limitations: it can only be applied to magnetic field wave data, and the wave in question should be monochromatic and elliptically polarised.

The previous section presented the application of the k-filtering method to an interval of magnetosheath data in order to derive the **k** vectors present in the wave field. In this section results are presented from analysis by Walker et al. (2004) of the same data set using the phase difference method to identify the main wave mode. A comparison of the results of the two methods is then made.

The k-filtering method was applied to 164 seconds of data from February 18, 2002, starting at 05:34:01.15 UT. Walker et al. (2004) analysed the same interval, both in its entirety and also by dividing it into sub-intervals in order to examine the time variation of the waves. As described in section 3.2.1 the phase difference method was applied to magnetic field data from STAFF, and to electric field measurements from EFW. The STAFF data were high pass filtered to remove spacecraft spin effects, and then converted from spacecraft coordinates to GSE. The EFW instrument measures the two components of the electric field in the spacecraft spin plane. The data were despun to remove most of the effects of spacecraft-spin. Since the GSE latitude of the spin vector of all spacecraft is $\approx$ -84 °, the satellite reference frame is very close to an inverted GSE frame. The spin axis geometry also implies that the electric field booms might traverse the spacecraft wake, causing interferences at frequencies twice or four times the spin frequency in the E_x component of the electric field. Due to operational reasons it was not possible to measure the electric field vector on spacecraft 1. Therefore although the phase difference method could not be used to determine the wave vector of oscillations observed in the electric field, data from the remaining satellites was used to check the projection of the wave vector onto the separation vectors of spacecraft 2, 3, and 4.

Figure 3.5 shows the waveform of the magnetic field measured by the STAFF search coil magnetometer during the period between 05:34:15 and 05:35:15 UT on February 18, 2002. This interval corresponds to approximately the first minute of the total period and was the portion of data used in the wave vector determination process using the phase difference method. This period was chosen because the nature of the waves remained relatively stable during this time, and the $\omega - k$ spectra, calculated from data within 15 s sub-intervals were fairly consistent from one interval to the next. During the rest of the period the $\omega - k$ spectra were less clean and more variable. This variability is addressed further in Section 3.2.4

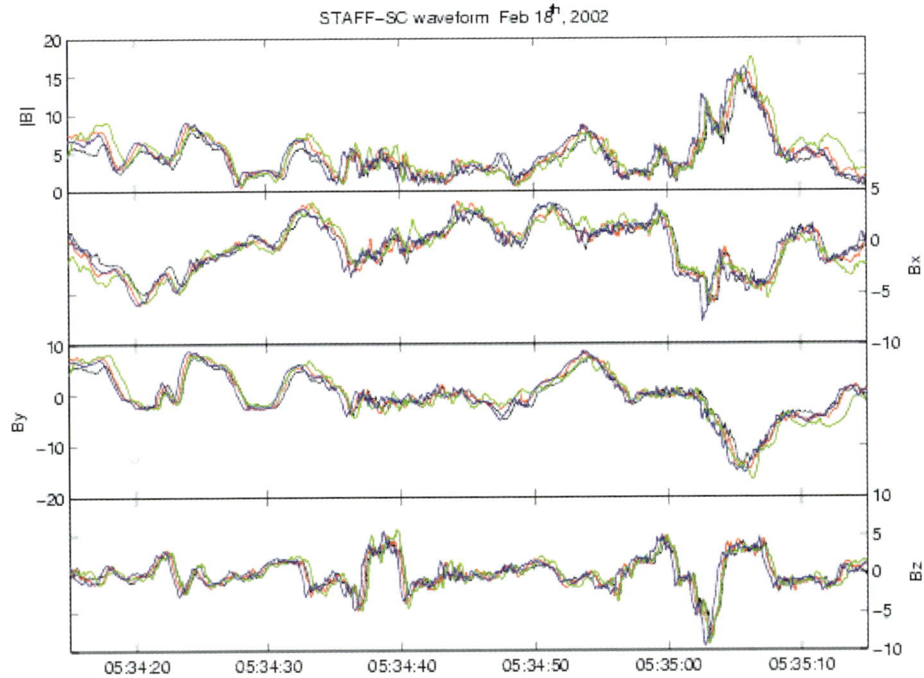

Figure 3.5. The magnetic field measured by the STAFF search coil between 05:34:15 and 05:35:15 UT on February 18, 2002. From top to bottom the panels show magnitude, and the X, Y, and Z components of the magnetic field in the GSE frame, all in nT. The coloured lines represent data from Cluster 1 (black), 2 (red), 3 (green), and 4 (blue). This interval corresponds approximately to the first minute of data shown in Figure 3.1. (From Walker et al., 2004).

The EFW waveforms are shown in Figure 3.6. The top panel gives the magnitude of the electric field measured in the spin plane of each satellite. The middle and bottom panels show the E_x and E_y spin plane components. The waveforms from satellites 3 and 4 are almost identical in nature. This is evident from the components of the electric field, in which fluctuations are observed first by Cluster 4 (blue) followed after a time delay of 0.4 seconds by Cluster 3. The waveforms show that the X component of the signal measured on satellites 3 and 4 was dominated by an oscillation with a period corresponding to half of the satellite spin period. The authors noted that the satellite spin signature was smaller in the Y component, and so analysis of this component was done for comparison with the results obtained from the STAFF search coil data.

Figure 3.7 shows the $\omega - k$ spectrum resulting from the analysis of satellite combinations (1,4), (2,4), and (3,4), calculated using data from STAFF for the time period 05:34:15-05:35:15 UT. A distinctive linear feature indicating an increase in

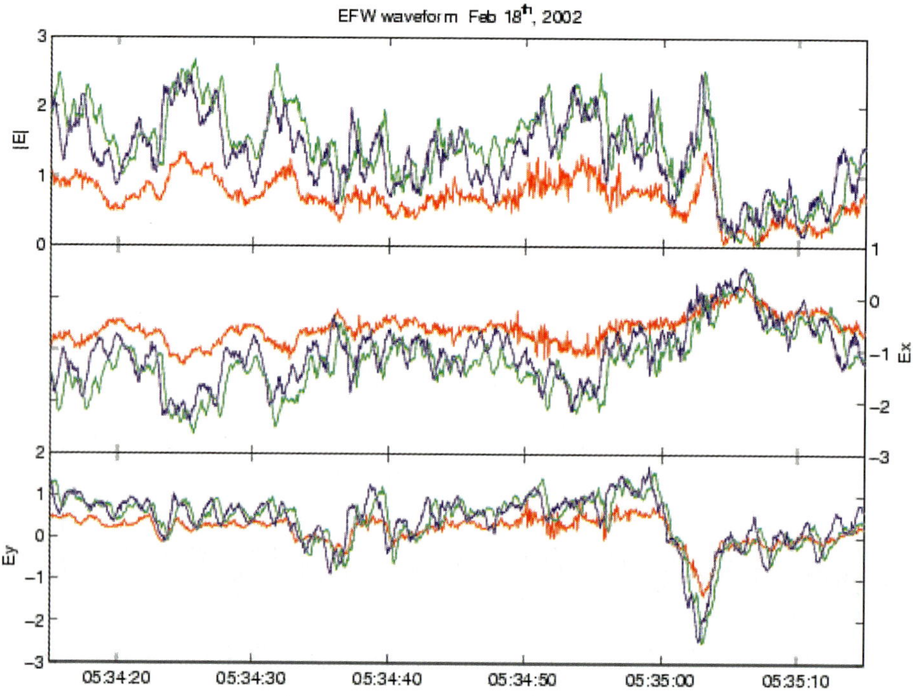

Figure 3.6. The electric field measured by the EFW instruments for the same interval as covered by Figure 3.5. From top to bottom the panels show the field magnitude, and the X and Y components in the spin plane. Units are in $\mathrm{mV\,m^{-1}}$. The coloured lines represent data from Cluster 2 (red), 3 (green), and 4 (blue). The X component seen on satellites 3 and 4 shows fluctuations at twice the spin frequency that are probably the result of one of the probe pairs passing through the wake of the satellites. (From Walker et al., 2004).

phase difference with frequency is visible in all three cases. The positive gradient of these features implies that the waves were propagating from the second satellite in each of the pairings towards the first (for example from satellite 4 towards 1). A comparison of the dispersions calculated for all satellite combinations showed that the waves crossed the satellites in the order 4, 2, 3, 1. This result is generally in agreement with the crossing order observed in the waveform data shown in Figure 3.5, in which satellite 4 (indicated by the blue line) was always the first to observe any fluctuation in the field.

Due to the ambiguity in the determination of the phase difference, the $\omega - k$ spectra are periodic with period 2π. Thus in order to compute the value of $\mathbf{k}$ correctly it is important to determine the correct dispersion branch. The similarity of the STAFF-SC waveforms during this period implied that all satellites saw the same wave with time differences smaller than the inverse frequency. Therefore the correct branch to examine in the dispersion plot is the central one, i.e., the branch that converges to the point $f = 0\,\text{Hz}$, $\Delta\psi/\mathbf{r} = 0$. This also implies that the wavelength of the waves being studied was much greater than the inter-spacecraft separation distance.

The white crosses in Figure 3.7 mark the peaks in the dispersion curve at frequencies of 0.37, 0.44, 0.62, and 1.1 Hz, corresponding approximately to those frequencies chosen by Sahraoui et al. (2003), as discussed in Section 3.2.2 The horizontal lines through the crosses represent the width of the peak in the dispersion at the frequencies studied. The **k** vector of the waves at each of these frequencies was determined and the results are shown in Table 3.1a. The typical error in the estimation of the projection of the wave vector along the satellite separation distance from the plots shown in Figure 3.7 is of the order of $\pm 1.7 \times 10^{-3}\,\text{km}^{-1}$. This translates to an error of $\simeq 10^\circ$ in the determination of the direction of **k**. The direction of the **k**-vectors differed considerably from the minimum variance direction computed using the same data period, the difference being $\approx 40^\circ$. This is understandable since mirror modes have a linear polarisation and so the similarity between the intermediate and minimum eigenvalues implies that the minimum variance direction is not well defined.

Once the wave vector direction had been determined it was possible to identify the propagation mode of the waves. Using the Doppler equation the plasma frame frequency f_{plasma} was determined, shown in Table 3.1a. At the lower frequencies (0.37 and 0.44 Hz) the plasma frame frequency was close to zero whilst at higher frequencies the phase velocity of the waves V_{phase} was determined to be of the order 200 km s^{-1}. The angle between the wave vector and the plasma bulk velocity was $\approx 25^\circ$ implying that the phase velocity of the wave was comparable with the projection of the plasma bulk velocity in the direction of the wave vector i.e., the wave was not propagating but was convected past the spacecraft in the plasma. Finally the angle of propagation with respect to the magnetic field was of the order of 85°. Walker et al. (2004) concluded from this evidence that the waves were likely to be mirror mode waves.

3.2.4 Comparison of *k*-filtering and phase difference results

For comparison with Table 3.1a, which shows the **k** vectors derived at four frequencies using the phase difference method, Table 3.1b shows the **k** vectors of the dominant wave mode identified using the *k*-filtering method, derived at nearly the same four satellite frame frequencies. As described in the previous section, the **k** vectors identified from the phase difference method were found to be propagating at an angle of 85° to the magnetic field, with plasma frame frequencies close to zero, consistent with the mirror mode. In addition, each of the dominant **k**-vectors found by the *k*-filtering method were approximately perpendicular to the background magnetic field ($\Theta_{Bk} \approx 87^\circ$). Once again, their corresponding frequencies in the plasma frame were almost zero and therefore these waves were also attributed

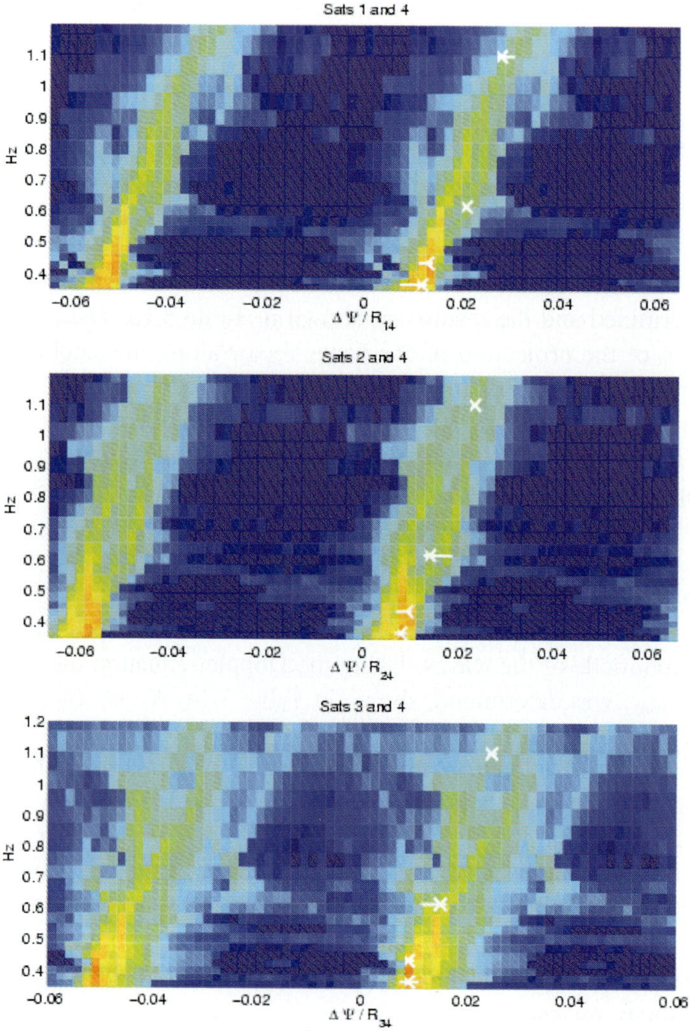

Figure 3.7. The $\omega - k$ joint spectrogram computed using STAFF search-coil data in the time period 05:34:15-05:35:15 UT for the satellite pairs (1,4), (2,4), and (3,4). The joint spectrogram is composed of a set of histograms, computed at a number of frequencies and stacked vertically. Each histogram shows the distribution of the observed phase differences between the two spacecraft, divided by the satellite separation distance, with the colour representing T the number of times a particular value was seen. The white crosses represent the phase differences ($\Delta\Psi/R_{xy}$) at the frequencies used in both the phase differencing and k-filtering analysis methods. (From Walker et al., 2004).

to the mirror mode. A comparison, therefore, of the results from the phase difference and k-filtering techniques shows that both methods identified the dominant fluctuation present in the data to be a mirror mode. The magnitudes of the **k** vectors

a) Wave vectors from phase difference method.

f_{sat} Hz	f_{plas} Hz	k_x rad km^{-1}	k_y rad km^{-1}	k_z rad km^{-1}	$\|\mathbf{k}\|$ rad km^{-1}
0.37	0.03	-0.009959	-0.003494	0.006239	0.0122
0.44	0.01	-0.009969	-0.00462	0.007797	0.01347
0.62	0.13	-0.016833	-0.006143	0.0119	0.02151
1.10	0.2	-0.0253	-0.01283	0.01264	0.03106

b) Wave vectors from **k**-filtering method.

f_{sat} Hz	k_x rad km^{-1}	k_y rad km^{-1}	k_z rad km^{-1}	$\|\mathbf{k}\|$ rad km^{-1}
0.37	-0.01097	-0.00236	0.00528	0.01241
0.44	-0.01241	-0.00279	0.00529	0.01378
0.61	-0.01671	-0.00404	0.00682	0.01849
1.12	-0.03065	-0.00941	0.01438	0.03514

Table 3.1. a) Wave vectors, **k**, computed for the four frequencies being investigated using the phase difference method. b) The computed characteristics of the most intense identified wave for the four studied frequencies using the k-filtering technique. (After Walker et al., 2004).

	Satellite pair (1,4)	(2,4)	(3,4)
$\mathbf{k}_{\mathrm{STAFF}}$	-0.01087	-0.00739	0.00837
$\mathbf{k}_{\mathrm{EFW}}$	-	-0.0074	0.0084
$\mathbf{k}_{k-filt}$	-0.00943	-0.00556	0.01055

Table 3.2. Wave vectors, **k**, projected along the satellite separation vectors. The frequency used was 0.37 Hz. The values were determined using the phase based method with both STAFF-SC ($\mathbf{k}_{\mathrm{STAFF}}$), and the EFW electric field ($\mathbf{k}_{\mathrm{EFW}}$), and by applying the k-filtering method to the STAFF-SC ($\mathbf{k}_{k-filt}$) data set. (From Walker et al., 2004).

were comparable: for example at a frequency of 0.37 Hz $\omega - k$ dispersion yields $|\mathbf{k}| = 0.0122$ rad km^{-1} versus the k-filtering results of $|\mathbf{k}| = 0.0124$ rad km^{-1}, while at a frequency of 0.44 Hz the magnitudes $|\mathbf{k}|$ are 0.0135 and 0.00138 rad km^{-1} for the phase difference and k-filtering methods, respectively. In the case of both frequencies the angles between the vectors was $\approx 8°$ which is less than the error in the phase difference technique.

The **k**-vectors derived from the phase difference method were also compared with results obtained using the same method applied to EFW electric field data.

However, as noted earlier, electric field data were not available from spacecraft 1 so it was not possible to calculate the full wave vector **k**. It was, however, possible to calculate the projection of the wave vector along the satellite separation directions and hence to determine the value of the **k** vector projected onto the satellite separation directions (2,4), and (3,4). Table 3.2 lists the projections of the wave vector along the satellite separation directions used in the analysis and shows that the **k** directions determined using the phase difference method with STAFF-SC and EFW data gave identical results. Thus the authors concluded that they could be fairly confident of the values of **k** they found. In a similar way, a further comparison between the phase difference and *k*-filtering methods, was made by comparing the projection of those two wave vectors onto the spacecraft separation vectors. This is also shown in Table 3.2, where Walker et al. (2004) note that, at low frequencies, the values are comparable.

Comparison of the **k**-vectors from two methods at higher frequencies in Table 3.1 shows that the difference increases for higher frequencies. This is because at higher frequencies the error in the phase difference method can be larger because of the presence of other modes of comparable energies, seen by the fact that the *k*-filter method identified other wave modes as well as the mirror mode. Table 3.3 shows the properties of the weaker waves identified by the *k*-filtering technique at each of the four frequencies studied. These modes correspond to slow, Alfvén and ion cyclotron modes. An example of such a coexistence between several waves for one given frequency was shown in Figure 3.4. Both an Alfvén wave (left panel) as well as a mirror mode (right panel) were found to be present in the data.

f_{sat} Hz	$\lvert\mathbf{k}\rvert$ rad km^{-1}	$\theta \equiv (\mathbf{k}, \mathbf{B}_0)$	f_{plasma} Hz	Mode	I
0.37	0.01914	$-30°$	-0.14$\sim 0.4 f_{ci}$	Slow	12%
0.44	0.02326	$-45°$	-0.29	Alfvén$\sim f_{ci}$	21%
0.61	0.04246	$44°$	0.71	Alfvén$\sim 2f_{ci}$	33%
1.12	0.04553	$-57°$	1.99	Cycl.$\sim 6f_{ci}$	98%

Table 3.3. Using the *k*-filtering technique other waves with lower intensities are identified for the four studied frequencies. Only the first secondary maxima are given here. In Column 'I' are given their intensities compared to that of the dominant wave for the corresponding frequency. f_{plasma} is the relative frequency in the plasma frame, $f_{ci} = 0.33$ Hz is the proton cyclotron frequency. (Reproduced from Walker et al., 2004).

The main result from this analysis was that up to a frequency ~ 0.7 Hz in the satellite frame, slow and Alfvén modes coexisted with the dominant mirror one. At higher frequencies, proton cyclotron harmonics were identified as having energies comparable to the mirror mode. The tendency of the mirror mode to lose its dominant character can be seen from column 'I' in Table 3.3 which shows the energy in

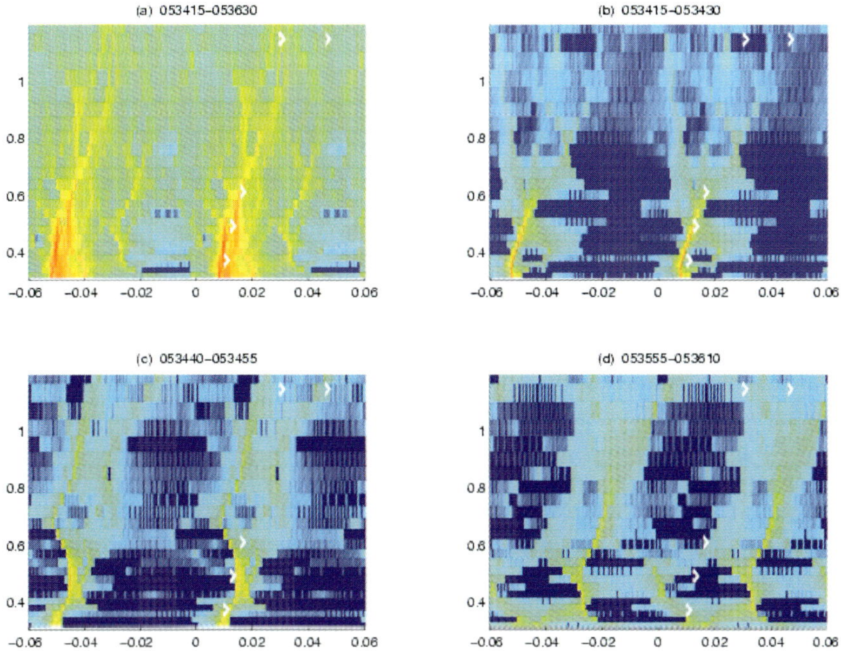

Figure 3.8. A set of snapshots of the $\omega - k$ dispersion. Panel (a) is calculated using data covering the full period 05:34:15-05:36:30 UT, in the same format as Figure 3.7. The other snapshots are computed using 15 s time intervals within the period under study. In all cases data from spacecraft 3 and 4 were used. Superimposed onto each plot, indicated by white symbols, are the mirror mode wave vectors determined from k-filtering, projected onto the satellite separation vectors. (From Walker et al., 2004).

the first secondary maximum as a percentage of the energy in the dominant mode. For $f = 1.12$ Hz the sixth cyclotron harmonic has comparable energy to that in the mirror mode.

The wave environment was highly variable on the time scale studied here. Figure 3.8 shows snapshots of the $\omega - k$ spectra, calculated using STAFF-SC data for the satellite pair (3,4), at different times within the time interval under investigation. Panel (a) shows the $\omega - k$ spectrum calculated for the whole period 05:34:00-05:36:44 UT. A number of different dispersion branches are visible. The main branch corresponds to the wave number range $0.01 < \Delta\psi/\mathbf{r}_{34} < 0.015$ at 0.35 Hz. At higher frequencies 0.4-0.5 Hz this branch exhibits a double peak in k-space and at the highest frequencies analysed it splits into two distinct branches. The projection of the mirror mode **k**-vectors determined from k-filtering are indicated with white symbols. For 0.35-0.8 Hz a second branch appears to the right of the main dispersion branch. It is of lower intensity and corresponds to a different mode.

The other three panels show results calculated from three intervals, each of 15 s duration, significantly shorter than the 1 minute interval used to generate the results shown in Figure 3.7. Panel (b) shows the $\omega - k$ spectrum calculated using data for a 15 s period starting at 05:34:15 UT. A well defined branch, corresponding to the mirror mode determined from k-filtering, appears below 0.7 Hz. The dispersion curves extend to higher frequencies but with lesser amplitudes. Panel (c) gives the $\omega - k$ spectrum for a 15 s snapshot starting near 05:34:40 UT. At low frequencies, the dispersion branch still resembles the mirror mode dispersion whilst at higher frequencies it splits into two sections, both of which were identified as mirror modes. Finally, panel (d) shows another 15 s snapshot towards the end of the investigated time interval, beginning at 05:35:55 UT. Now the mirror mode has disappeared completely and the branch lying to the right of the main branch identified in panel (a) becomes the dominant mode.

Walker et al. (2004) noted that by performing k-filtering on shorter time intervals it can be shown that the results change as a consequence of underestimating the energy contributed by various other modes, but that the determined wave vectors remain unaffected. They also describe the advantages and disadvantages of the two different methodologies used to identify wave modes. The phase difference method, implemented using a wavelet transform based on the Morlet wavelet, can be used to analyse a shorter period of data than the k-filtering technique, thus revealing time dependent behaviour of the plasma, while the Fourier transform in k-filtering enables the determination of the energy density in the individual wave modes, information not available from the Morlet wavelet transform used in the phase difference method. Also, the phase difference method works best when only one wave mode is present or when one wave mode dominates the wave environment, since multiple modes result in multiple dispersion curves rendering the wave vector directions unresolvable. Methods to resolve this problem are currently being pursued. Hence, the k-filtering technique can resolve the presence of multiple waves within the plasma, while the phase differencing technique can access the time dependence. Thus, in some sense, the two methods are complementary.

To summarise, two complementary methods were used to combine data from four spacecraft, during an interval when broadband waves were observed in the magnetosheath. The results demonstrate that during this time the plasma contained multiple wave vectors at each frequency studied. Furthermore, these wave modes were found to lie on MHD dispersion curves, allowing the identification of multiple MHD wave modes present simultaneously in the data. Thus, in this case, the broadband wave spectrum in the plasma was shown to arise from the superposition of linear MHD modes, leading to the conclusion that a weak turbulence approach is appropriate for understanding the plasma at this time.

3.3 Waves Behind the Quasi-Perpendicular Bow Shock

The structure of supercritical, quasi-perpendicular shocks is dependent on the motion of gyrating ions reflected from the shock ramp. The reflected ions pass downstream where they contribute to an ion temperature anisotropy such that $T_{i\perp} > T_{i\parallel}$. As described in Section 3.1.1, such an ion temperature anisotropy can drive either Alfvén ion cyclotron or mirror instabilities (e.g., Schwartz et al., 1996; Gary et al., 1994).

Alexandrova et al. (2004) presented an analysis of waves and small scale filamentary structures observed downstream of a quasi-perpendicular, super-critical bow shock. They found that the plasma conditions immediately downstream of the shock ramp were appropriate for the generation of electromagnetic Alfvén ion cyclotron waves (AIC) waves. Using data from the four spacecraft, the authors were able to demonstrate for the first time that the two quasi-monochromatic waves observed downstream of the shock had properties consistent with generation by proton and alpha cyclotron waves. Further downstream the AIC waves ceased abruptly. At the same time the temperature anisotropy decreased and signatures consistent with cylindrical current tubes aligned with the magnetic field were found. The authors concluded that the current tubes evolved downstream of the shock from the Alfvén waves under the action of the filamentation instability (Laveder et al., 2002) occurring where the total plasma $\beta > 1$.

3.3.1 Observations

Alexandrova et al. (2004) examined several crossings from March 31, 2001 previously studied by Maksimovic et al. (2003), presenting in detail the results from one crossing which occurred at 18:02:17 UT. Four-spacecraft timing discontinuity analysis was used to estimate a normal for the shock surface, **n**, which was then refined using the constraint that the observed direction of maximum magnetic field variance **b** should be orthogonal to the bow shock normal: $\mathbf{n} = -\mathbf{b} \times (\mathbf{b} \times \mathbf{n})$. This led to a small correction in the shock normal. The unit vectors **b** and **n** were then used to define the shock normal co-ordinate system. When the average upstream and downstream magnetic fields were estimated, their components in the shock normal direction were almost the same, supporting the shock normal estimation.

Data from the STAFF search coil experiment (Cornilleau-Wehrlin et al., 1997, 2003), high pass filtered to remove the spacecraft spin signal at 0.25 Hz, were combined with four second averaged magnetic field data from FGM (Balogh et al., 2001) using a Haar wavelet transform, in order to obtain a 'mixed' signal covering both low frequency variations and high frequency waves. This procedure is described in detail in Alexandrova et al. (2004). Using data from ACE (for the solar wind plasma parameters) and the instruments on Cluster, characteristic upstream and downstream plasma parameters were calculated, including average magnetic field (B), proton and alpha gyrofrequencies (f_{cp}, $f_{c\alpha}$), plasma flow velocity (V),

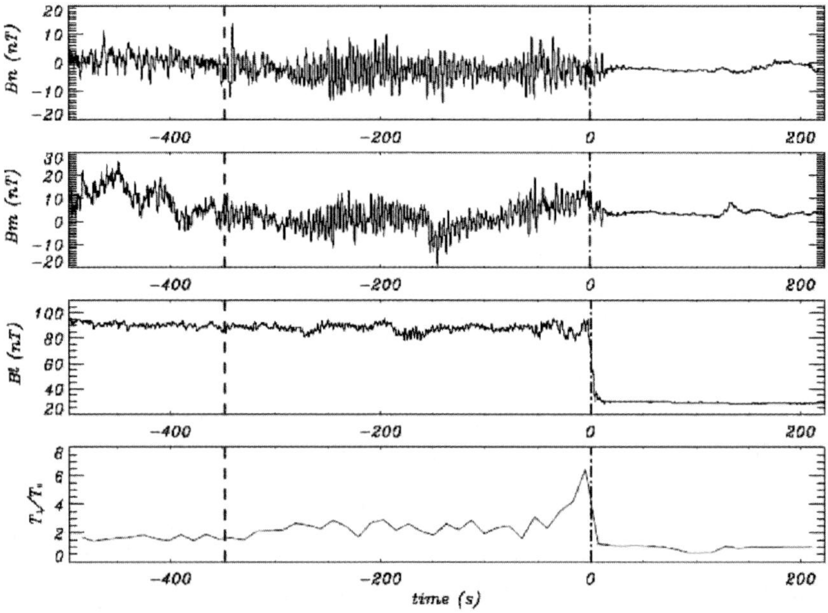

Figure 3.9. Combined magnetic profile from STAFF and FGM field data, from Cluster 1, between 17:54 and 18:06 UT on March 31, 2001 in shock normal co-ordinates: panel 1, B_n, the magnetic field component along the shock normal; panel 2, B_m, the magnetic field component in the direction perpendicular to the maximum variance and the shock normal directions; panel 3, B_l, the magnetic field component along the shock maximum variance direction; panel 4, the proton temperature anisotropy. The origin for time is taken at the shock crossing, at 18:02:17 UT (dash-dotted line), so that negative times correspond to data from the magnetosheath. The vertical dashed line, at 17:56:30 UT, marks the transition between two regions: the first contains signatures of current tubes and the second contains well developed, quasi-monochromatic, transverse wave activity. (From Alexandrova et al., 2004).

proton and electron temperatures (T_p, T_e), proton, alpha and electron densities (N_p, N_e, N_α), Alfvén velocity (V_A), ion inertial length (c/ω_{pi}), and ion and electron plasma beta (β_p, β_e). The shock was found to have a $\theta_{Bn} \sim 82°$, and a Mach number $M_A \sim 4.4$, and was therefore both quasi-perpendicular and supercritical. In addition, the parallel electron and proton temperatures ($T_{e\|}$, $T_{p\|}$) and corresponding anisotropies ($T_{e\perp}/T_{e\|}$, $T_{p\perp}/T_{p\|}$) were estimated.

3.3.2 Alfvén ion cyclotron waves

Figure 3.9 shows the 'mixed' magnetic field data, obtained by combining STAFF and FGM data, for the shock and the downstream region. Immediately downstream of the shock, well developed transverse wave activity was observed for ~ 300 seconds. An analysis of the observed waves was made in the context of expectations from linear wave theory: using the background magnetosheath plasma parameters, and assuming bi-Maxwellian particle distributions for electrons, protons and alpha particles, a linear model, WHAMP (Rönnmark, 1983), was used to calculate the most unstable modes. These were found to be Alfvén waves propagating parallel to the background magnetosheath magnetic field. Positive growths rates in two frequency bands were obtained: 0.24 f_{cp} and 0.51 f_{cp}, corresponding to wave modes excited by the alpha and proton temperature anisotropies respectively. The power in the lower frequency wave was found to be greater than that in the higher frequency. The authors represented these wave modes in terms of the normalised wave vector, kr_p, parallel to the magnetic field, where r_p is the proton Larmor radius based on the proton parallel thermal velocity $r_p = (2k_B T_{p\|})$. Described in this way the two unstable wave modes were centred on $k_1 r_p = 0.19$ and $k_2 r_p = 0.45$.

A power spectrum of the fluctuations observed by Cluster showed two peaks separated by approximately a factor of 2, at 0.29 and 0.57 Hz. The minimum variance direction during the wave activity was found to be within 10° of the average magnetosheath magnetic field, suggesting that the waves were nearly field aligned. The waves were also left-hand, nearly circularly polarised with respect to the background magnetic field. The wave vectors of the two waves were estimated by applying the four-spacecraft phase difference method to magnetic field data during the wave activity, using data from Cluster 1 as the reference set and solving

$$[\phi_1(f) - \phi_i(f)] = (\mathbf{k} \cdot \delta \mathbf{r}_{1i}) + 2\pi n_{1i}, \qquad i = 2,3,4 \tag{3.1}$$

The 2π ambiguity was removed by constraining the solution such that (1) the solution was parallel to the minimum variance direction, (2) the angle between the wave vector and the background magnetic field was small, as expected from linear theory, (3) the sum of the differences in Equation 3.1 was minimised and (4) the value of n_{1i} in Equation 3.1 was the smallest value satisfying the constraints listed above. The following results were obtained for the two waves: $\mathbf{k}_1 = 14 \times 10^{-3}(0.27, -0.26, -0.93)$ km s^{-1}for the wave at $f_1 = 0.29$ Hz and $\mathbf{k}_2 = 25 \times 10^{-3}(-0.07, -0.23, -0.97)$km s^{-1}for the wave at $f_2 = 0.57$ Hz. The waves had unit vectors, expressed in shock normal co-ordinates, of $\mathbf{e}_1 = (0.08, 0.12, 0, 99)$ and $\mathbf{e}_2 = (-0.25, 0.18, 0, 95)$, which were approximately parallel (within $\sim 10°$ and $\sim 15°$ respectively) to the background magnetic field in the magnetosheath.

The properties of the observed transverse waves were compared with the expectations from linear Vlasov theory. The observed waves had plasma rest frame frequencies of $f_{0,1} = 0.43 f_{cp}$ and $f_{0,2} = 0.58 f_{cp}$, with normalised k vectors of $k_{1\|} r_p = 0.31$ and $k_{2\|} r_p = 0.55$. Therefore, although there was qualitative agreement between the observed and predicted waves, there was not exact correspondence. It was suggested that a better quantitative agreement might be achieved by having better defined background plasma parameters to use as input to the linear model. The authors concluded however that the observed wave frequencies, polarisation, dispersion characteristics and correspondence with linear theory were sufficient to positively identify the waves as Alfvén ion cyclotron waves, excited by proton and alpha anisotropies.

It was noted that, although the AIC waves were periodic, they were not sinusoidal. Instead they had steepened wave fronts, which the authors suggested was indicative of the presence of relatively strong non-linearities. A systematic analysis of the regions of steepened magnetic field profile was made. Regions of steep local gradient were defined by the energy density in the smallest scale of a Haar wavelet transform exceeding the average energy density at this scale by a threshold factor, ν. The identification of these regions was found to be independent of the value of ν, as long as it exceeded $3-5$. Analysis of 2 seconds of magnetic field data centred on each region of local steepening showed that their magnetic signatures were consistent with those of current sheets, lying approximately parallel to the AIC wave fronts, and the authors suggested that they were produced by non-linear steepening of the waves.

3.3.3 Filamentary current tubes

Well developed AIC waves were observed for about 350 seconds in the region downstream of the outbound quasi-perpendicular shock crossing. The plasma populated by these waves had a proton $\beta_p < 1$ and $T_{p\perp}/T_{p\|} > 1$ and the onset of the waves occurred simultaneously with an abrupt change in $T_{p\perp}/T_{p\|}$. Prior to the interval of transverse wave activity, i.e., further downstream of the shock, magnetic perturbations were identified which were interpreted as three-dimensional current tubes aligned with the background magnetosheath magnetic field. Transformation of the magnetic field data into local minimum variance co-ordinates showed a systematic variation of all three components, with the minimum variance direction approximately aligned with the background magnetic field.

The Cluster tetrahedron scale at this time was ~600 km, and two spacecraft: Cluster 1 and Cluster 4, had a separation vector inclined at only $\sim 20^\circ$ to the magnetic field. Comparison of data from this spacecraft pair revealed a good correlation between the magnetic field signatures for two structures, confirming that the structures were field aligned. The authors concluded on this basis that the structures had a field parallel dimension in excess of 620 km. Using the time delay between the structures being seen at the spacecraft, together with the separation of the two satellites in the plane perpendicular to the average magnetic field, allowed the authors to estimate the speed of the structures perpendicular to **B**. They estimated a velocity of $\sim$ 140 km/s in this plane, with the uncertainty of 10° in the orientation of the structures with respect to **B** giving a range for the velocity of $70-230$ km. The upper limit of this range was close to the background plasma flow velocity.

The observed magnetic perturbation was compared with a simple 2-D model of a localised current tube. In the model the current tube was assumed to have a cylindrical shape, to be aligned with the magnetic field (which was in the **z** direction), and to have properties which were independent of distance along the tube axis. The model was characterised by two functions: the perturbation of the magnetic field component parallel to the tube axis ($\delta B_z(x,y)$), and the variation of the z component of the vector potential ($\mathbf{A} = A\mathbf{e_z}$), both as functions of position in the (x,y) plane perpendicular the tube axis:

$$\delta B_z = \Delta B_{\parallel} \exp^{(-R^2/R_0^2)} \quad (3.2)$$

$$A = \left(\frac{\Delta B_{\perp}}{\Delta B_{\parallel}}\right) R_0 \delta B_z \quad (3.3)$$

where $\Delta B_{\perp}$ and $\Delta B_{\parallel}$ characterise the amplitude of the parallel and perpendicular magnetic field perturbations generated by the current tube, R_0 is the radius of the current tube and R is the distance from the tube axis. By assuming that the spacecraft crossed the tube in the (x,y) plane, the best fit between the model and the observations was obtained with $R_0 = 60$ km, $\Delta B_{\parallel} \sim 15$ nT, $\Delta B_{\perp}/\Delta B_{\parallel} \sim 2$ and $R = 30$ km. Figure 3.10 shows a comparison of the data and the model fit for one current tube signature observed by Cluster 1 at 17:55:16 UT. There is good agreement between the magnetic field components transverse to the inferred current tube axis ($\delta B'_x, \delta B'_y$) with some deviation for the axis aligned component ($\delta B'_z$), which has the smallest amplitude variation of the three.

Similar structures were also observed by Cluster 2 during this interval, but the inter-spacecraft separation between Cluster 2 and spacecrafts 1 and 4 were $\delta R_{\perp 12} = 650$km $\delta R_{\perp 24} = 720$km: in each case much larger than the estimated current tube diameter and so the authors concluded that they could not be certain that Cluster 2 was observing the same current structure. Magnetic field data measured by Cluster 3 during this time contained similar signatures, but there was a data gap at the time that this particular current tube might have been observed.

3.3.4 Conclusions

Alexandrova et al. (2004) analysed magnetic perturbations downstream of a supercritical quasi-perpendicular shock observed on March 31 2001 when the Cluster tetrahedron scale was ~ 600 km. Figure 3.11 illustrates the processes which they suggested occurred downstream of the shock in this case. The shock generated a region of high temperature anisotropy immediately downstream, which was unstable to the excitation of proton and alpha AIC waves. Using data from the four spacecraft, they were able to make the first accurate identification of AIC waves in this region by comparison of the wave properties with those expected from linear theory (Rönnmark, 1983). Good qualitative agreement was found, although there were some differences, perhaps arising from uncertainties in the background plasma parameters. The authors proposed that these waves then collapsed into field aligned current filaments, through the action of the filamentation instability (Laveder et al., 2002). They used dual spacecraft observations of the same structure to demonstrate that the current tubes had an extent along the magnetic field of more than 600 km. They then fitted the observations to a simple two-dimensional model of a field

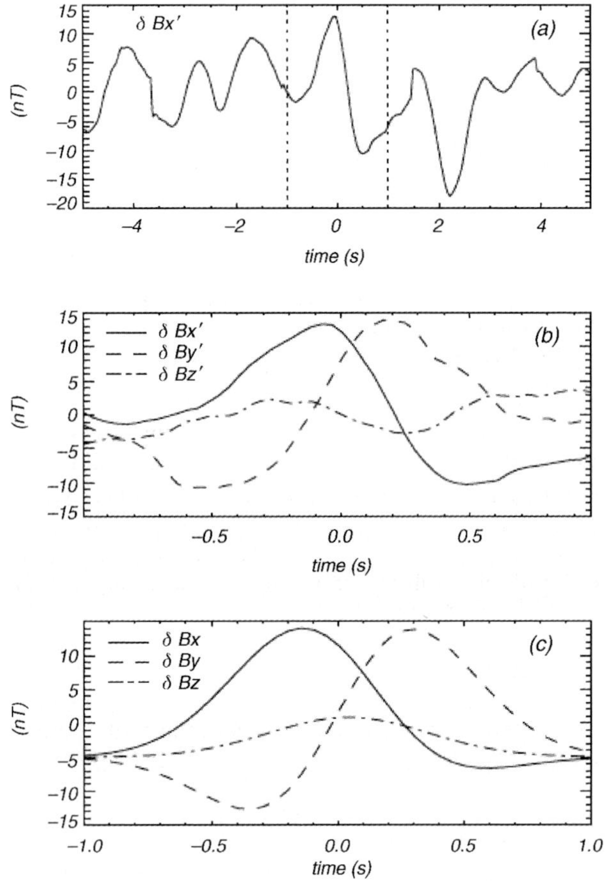

Figure 3.10. Magnetic field fluctuations plotted as a function of time, taking 17:55:16 UT as the time origin, in co-ordinates defined by MVA applied to the interval indicated by the vertical dashed lines in panel (a). In this co-ordinate system x' is in the maximum variance direction, and y' and z' in the intermediate and minimum variance directions respectively. Panel (a) shows 10 seconds of data of the B_x' magnetic field component, (b) shows the three components for a single current tube, for the time interval between the two dotted vertical lines in panel (a), and (c) shows the model data, where z is along the tube axis and x and y are in the plane perpendicular to the axis. There is good agreement for the components in the plane of the current tube perpendicular to its axis. (From Alexandrova et al., 2004).

aligned current tube, and estimated that the filaments had a radius of ~ 60 km: of the order of the local ion inertial length ($c/\omega_{pi} = 35$ km). The onset of the region populated by current filaments was associated with a decrease in the temperature anisotropy, which led the authors to suggest that the filamentation instability might play an important role in the relaxation of the plasma downstream of the shock.

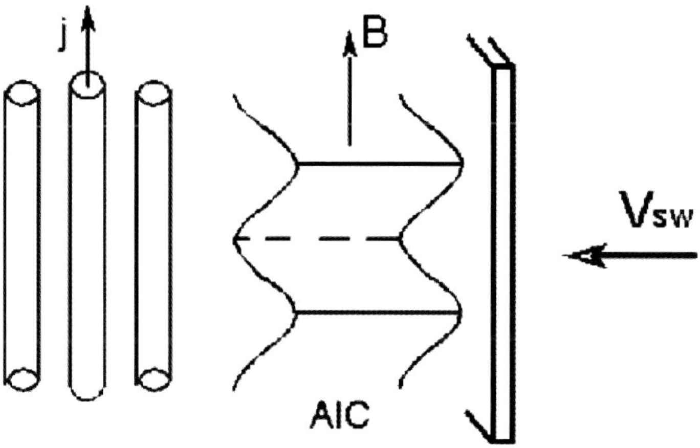

Figure 3.11. Schematic presentation of the generation and filamentation of AIC waves downstream of the perpendicular bow shock. The solar wind flows from the right into the shock (shown as a vertical plane). AIC waves are generated as a result of the strong temperature anisotropy downstream of the shock. The AIC waves there give way to current tubes aligned with the magnetic field (shown on the left of the figure) via the filamentation instability. (From Alexandrova et al., 2004).

However, detailed comparisons of this non-linear instability with theoretical models, including kinetic effects, will form part of future studies.

3.4 Mirror Mode Theory and Modelling

Mirror modes, as described in Section 3.2.1, frequently occur in the magnetosheath under conditions of enhanced ion temperature anisotropy $(T_\perp > T_{||})$ and high $\beta_{\perp i} \sim 2$. They are non-propagating magnetic bottle structures, characterised by large amplitude variations in the magnetic field magnitude, $\Delta B/B \sim 1$, anti-correlated with variations in the plasma number density introducing inhomogeneity into the plasma. Anticorrelation between magnetic field and density perturbations is not unique to mirror modes. It is also typical of slow mode waves, and can lead to large amplitude soliton chains (as has recently been shown by Stasiewicz, 2004), in which case no anisotropy is required. Such structures may appear in the solar wind. In the magnetosheath, mirror modes are dominant because the presence of the large anisotropy and high plasma temperature, damps slow mode waves. It has been suggested that such large amplitude variations might be a significant source of turbulent energy (Pokhotelov et al., 2003).

Previous studies of mirror modes have been limited to the use of single and dual spacecraft. However, four spacecraft Cluster data, available for a variety of tetrahedron scales, allow analysis of mirror structure three-dimensional shape and size that have not previously been possible.

3.4.1 Brief review of mirror mode theory

The existence of mirror modes was predicted by Rudakov and Sagdeev (1959) and Chandrasekhar et al. (1958) from anisotropic plasma fluid theory followed by kinetic approaches by Tajiri (1967), Hasegawa (1969), and Pokhotelov and Pilipenko (1976).

The starting point in a quasi-hydrodynamic approach is the pressure equilibrium condition for an anisotropic plasma

$$\delta p_\perp + \frac{B\,\delta B_\parallel}{\mu_0} = -\frac{k_\parallel^2}{k_\perp^2}\left[1+\frac{1}{2}(\beta_\perp - \beta_\parallel)\right]\frac{B\,\delta B_\parallel}{\mu_0}. \tag{3.4}$$

Here $\delta p_\perp$ is the variation of the perpendicular plasma pressure, $B = |\boldsymbol{B}|$ is the magnitude of the ambient magnetic field $\boldsymbol{B}$, $\delta B_\parallel$ is the compressional magnetic field perturbation, $k_\perp$ and $k_\parallel$ are the components of the wave vector $\boldsymbol{k} = (k_\perp, k_\parallel)$ perpendicular and parallel to the ambient field, respectively, and μ_0 is the free space permeability. The ratio of kinetic to magnetic energy density is given by $\beta = nk_BT/(B^2/2\mu_0)$, with indices $\perp, \parallel$ indicating perpendicular or parallel pressures; n is the plasma number density. The perturbed quantities in Equation (3.4) are assumed to vary in both time and space as $\sim \exp(-i\omega t + i\boldsymbol{k}\cdot\boldsymbol{r})$, where $\omega \ll \omega_{ci}$ is the wave frequency.

The variation in the perpendicular plasma pressure is obtained from the perturbed particle distribution (Pokhotelov et al., 2001)

$$\delta F_j = -\frac{\mu\,\delta B_\parallel}{B}\frac{\partial F_j}{\partial \mu} + q_j\phi\frac{\partial F_j}{\partial W} - \frac{\omega(q_j\phi + \mu\delta B_\parallel)}{\omega - k_\parallel v_\parallel}\frac{\partial F_j}{\partial W} \tag{3.5}$$

where $F_j(W,\mu)$ is the particle distribution function which depends on the energy, W, and magnetic moment, μ, of the jth species of mass m_j, charge q_j and parallel speed $v_\parallel = \sigma[2(W-\mu B)/m_j]^{\frac{1}{2}}$ ($\sigma \pm 1$ and indicates the direction of $v_\parallel$). Here, ϕ is the scalar potential, with the wave electric field given by $E_\parallel = -ik_\parallel\phi + i\omega A_\parallel$, $A_\parallel$ being the parallel vector potential.

The ordinary ion-mirror mode is only one of the possibly unstable solutions of the dispersion relation (Pokhotelov et al., 2003) resulting from pressure balance and Maxwell's equations $\mathsf{D}(\omega,\mathbf{k})\cdot\Psi = 0$ for the wave field vector Ψ, whose components are

$$\Psi_\parallel = A_\parallel - \frac{k}{\omega}\phi, \quad \Psi_A = -\frac{k_\perp}{\omega}\phi, \quad \Psi_M = \frac{(\mathbf{k}\times\mathbf{A})_\parallel}{k_\perp} - \frac{k_y\kappa_B}{k_\perp\omega}\phi \tag{3.6}$$

The last term includes the background magnetic field inhomogeneity $\kappa_B = |\nabla \ln B|$ which yields the drift frequency $\omega_D = (k_y v_\perp^2/2\omega_{ci})\kappa_B$. The simplest case, neglecting the drift frequency contributions and assuming a two component Maxwellian plasma with cold electrons, yields the ordinary ion-mirror mode which becomes unstable when the pressure anisotropy $A \equiv p_\perp / p_\parallel$, satisfies

$$A - 1 > \beta_\perp^{-1} \tag{3.7}$$

$\beta_\perp = 2\mu_0 n k_B T_{i\perp}/B^2$ is the perpendicular ion β, and the electrons do not play any role in the instability. The growth rate of this mode (Hasegawa, 1969) is proportional to the ratio $(k_\parallel/k_\perp)^2$ where $(k_\parallel/k_\perp)^2 \ll 1$. The **k**-vector is thus nearly perpendicular to the magnetic field and the mode has a small growth rate. However, because it is practically non-propagating and is therefore convected with the plasma flow, it has plenty of time to grow and so can reach large amplitudes which ultimately cannot be described by simple linear theory. In the limit of $T_e \rightarrow 0$, theory predicts that the cold electrons will wipe out any parallel electric field and therefore that $k_\parallel$ should be zero and the mode cannot exist. However, a small but finite temperature of the electrons will allow for the mode to exist in slightly oblique direction (Pantellini and Schwartz, 1995; Pokhotelov et al., 2001, 2003, 2004).

The ordinary ion-mirror mode grows fastest Pokhotelov et al. (2004) at perpendicular wavelengths comparable to the ion gyroradius, $k_\perp \rho_i \sim 1$. The above threshold for the short wavelength mirror mode is higher by a factor of 2 than in the very long wavelength case $k_\perp \rho_i \ll 1$. Thus, depending on the anisotropy, the fastest growing waves will be those which have a wavelength just long enough for the anisotropy to exceed the instability threshold. The inclination of the mode with respect to the magnetic field implies that the bottles are no longer symmetric around the field direction. Field aligned currents should flow within the structure, generating a non-coplanar magnetic field component, which twists the mirror mode magnetic field around the bottle.

The mirror mode is never observed in the state of linear small magnetic field compressions. Magnetic field compression ratios of 30-80% are observed, deep in the nonlinear regime. Since the mode is non-oscillatory, it is unsurprising that a quasilinear approach (Treumann and Baumjohann, 1997) does not explain the observations. That particle trapping occurs has been suggested by Kivelson and Southwood (1996). Such trapping is inferred from lion roar excitation (Treumann et al., 2000) and observation within mirror modes (Baumjohann et al., 1999), as well as by direct electron observation (Chisham et al., 1998). It has been recognised recently (Treumann et al., 2004) that in the nonlinear marginally stable state the mirror modes should evolve into three-dimensional cylindrical structures with zero parallel wave number extended along the ambient magnetic field. Any remaining inclination with respect to the field then indicates that the mode is still in evolution.

3.4.2 Nonlinear static bottle model of mirror modes

Constantinescu (2002) used the marginal mirror equilibrium condition to consider the stationary equilibrium state of a mirror bottle. Pressure equilibrium in the plasma reference frame is written

$$\nabla\left(p_{\perp}+\frac{B^2}{2\mu_0}\right)+\nabla\left[\left(p_{\parallel}-p_{\perp}-\frac{B^2}{\mu_0}\right)\frac{\mathsf{B}}{B^2}\right]=0 \tag{3.8}$$

where μ_0 is susceptibility, B magnetic field strength, B is a tensor with elements $(\mathsf{B})_{ij}=B_iB_j$, and $p_{\parallel}$ and $p_{\perp}$ are plasma pressure components. The temperature anisotropy in a bi-Maxwellian plasma is

$$A(\mathbf{r})=\frac{T_{\perp}(\mathbf{r})}{T_{\parallel}(\mathbf{r})}=\left[1-\left(1-\frac{1}{A_0}\right)\frac{B_0}{B(\mathbf{r})}\right]^{-1} \tag{3.9}$$

where $A(\mathbf{r})$ and $B(\mathbf{r})$ are the final anisotropy and magnetic field, and A_0 and B_0 are unperturbed anisotropy and magnetic field. Equation (3.9) holds for 2 (Lee et al., 1987) and 3 (Constantinescu, 2002) dimensions.

In cylindrical symmetry: $\rho, z, \partial/\partial\varphi=0$, one has $\mathbf{B}(\rho,z)=(B_0+\delta B_z(\rho,z))\,\mathbf{e}_z+\delta B_\rho(\rho,z)\mathbf{e}_\rho$, leading to a set of Bessel differential equations

$$\rho^2\frac{d^2}{d\rho^2}\delta B_\rho^n+\rho\frac{d}{d\rho}\delta B_\rho^n+\left[\left(\frac{n\alpha\rho}{L}\right)^2-1\right]\delta B_\rho^n=0 \tag{3.10}$$

where α is a dimensionless parameter:

$$\alpha=\pi\sqrt{\frac{\frac{1}{2}\left(1-\frac{1}{A_0}\right)+\frac{1}{\beta_{0\perp}}}{A_0-1-\frac{1}{\beta_{0\perp}}}} \tag{3.11}$$

and $\beta_{0\perp}$ is the plasma parameter, i.e., the ratio between the orthogonal plasma pressure, $p_{0\perp}$ and the magnetic pressure, $B_0^2/2\mu_0$.

The solution of Equation (3.10)

$$\{\delta B_\rho^n(\rho),\delta B_z^n(\rho)\}=\{\frac{i\pi}{\alpha}C_nJ_1\left(\frac{n\alpha\rho}{L}\right),C_nJ_0\left(\frac{n\alpha\rho}{L}\right)\} \tag{3.12}$$

holds for $\alpha^2>0$, and for physically realistic solutions $C_{-n}=C_n^*$. In terms of the initial anisotropy and plasma β this is equivalent to:

$$A_0>1+\frac{1}{\beta_{0\perp}}\qquad\text{or}\qquad A_0<\frac{\beta_{0\perp}}{\beta_{0\perp}+2} \tag{3.13}$$

The first inequality in Equation (3.13) is the mirror instability condition, and the second the firehose condition (cf, e.g., Baumjohann and Treumann, 1996).

Figure 3.12 shows the onion layer like structure of the magnetic field of a mirror mode bubble. For values of ρ for which $J_1(\rho)=0$, the field lines become straight

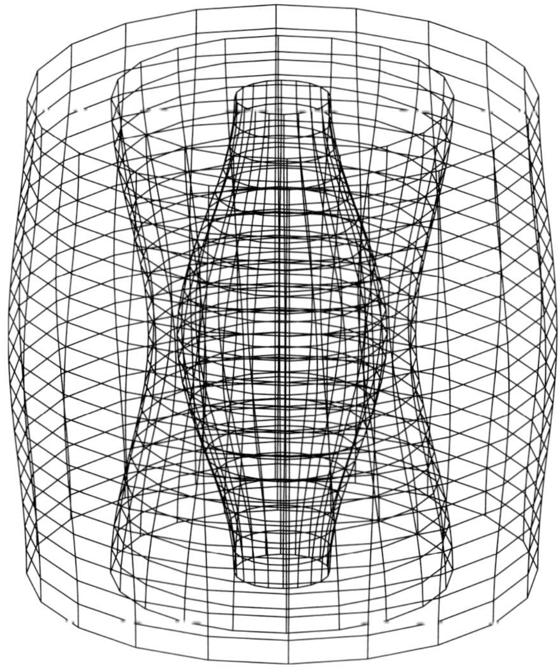

Figure 3.12. Magnetic field surfaces of the first Fourier component of the magnetic mirror perturbation. The main structure, closest to the axis, has a typical bottle shape. Moving radially away from the axis one encounters a series of nested structures which have a similar symmetry to the central one. The surfaces shown in the figure represent surfaces of constant magnetic field, with the field increasing radially outward from the axis. The ambient field direction in this model is parallel to the axis. (From Constantinescu, 2002).

lines on the surface of the cylinder, defining the border between two layers of opposite curvature. The position of the first border, which defines the main structure, is given by the ratio of radius to length of the central bubble

$$\alpha R/L = 3.832 \tag{3.14}$$

and α thus determines the elongation of the bubble.

3.4.3 Fitting model to data

Fitting the model to measured mirror magnetic field allows for the determination of the dimensions of a mirror bubble in equilibrium and the determination of the bubble structure (Equation 3.14). For two reasons the measured mirror field signature is unlikely to contain a significant contribution from more distant layers. Firstly, the amplitude of the Bessel functions decays rapidly with increasing radial distance; and secondly the magnetic field at larger distances from the core of the structure is likely to be affected by the interaction of neighbouring mirror modes. Application

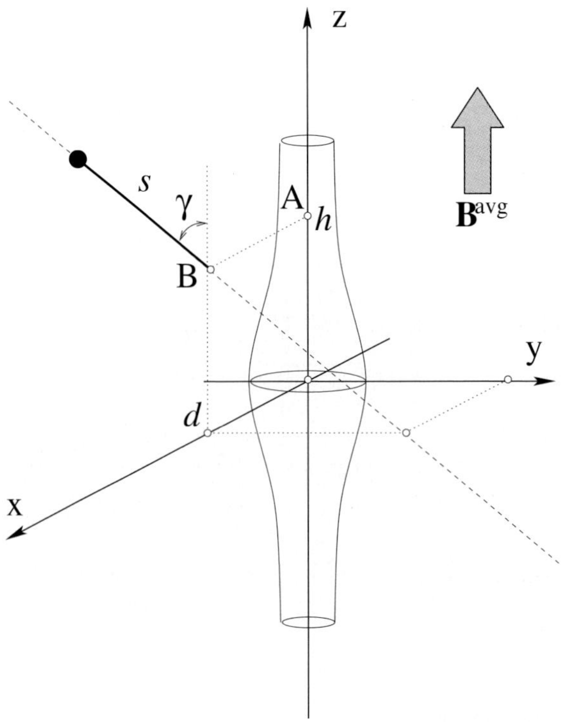

Figure 3.13. The normal coordinate system. The dashed line is the spacecraft path which intersects the (x-z)-plane of the MMS system in the point B(d,0,h) and is parallel with the (y,z)-plane. The angle between the trajectory and the z-axis is γ and the distance between the spacecraft and the point B is s. (From Constantinescu et al., 2003).

of the method requires the introduction of a normal coordinate system (h, d, γ, s) (see Figure 3.13) in which the spacecraft track is parallel to the (y, z)-plane. In these coordinates the spacecraft trajectory can be defined by three parameters: the angle $\gamma \in [0, \pi/2]$ between spacecraft path and the axis of the mirror mode structure (MMS); the distance $d \in [0, \infty)$ between spacecraft path and the MMS axis; and the distance $h \in [-L, L]$ between the centre of the MMS and point A in Figure 3.13. The position of the spacecraft relative to the MMS is then specified by one additional coordinate: $s \in (-\infty, +\infty)$, which is the distance between the point B in Figure 3.13 and the spacecraft.

The parameters used for the fitting are: trajectory normal coordinates (h, d, γ), initial spacecraft position on its path (s_0), length of the MMS (L), the unperturbed magnetic field intensity (B_0) and α. Components up to $n \leq 4$ in the Fourier expansion of the fields have been included. Higher orders merely reproduced irrelevant 'small scale' fluctuations.

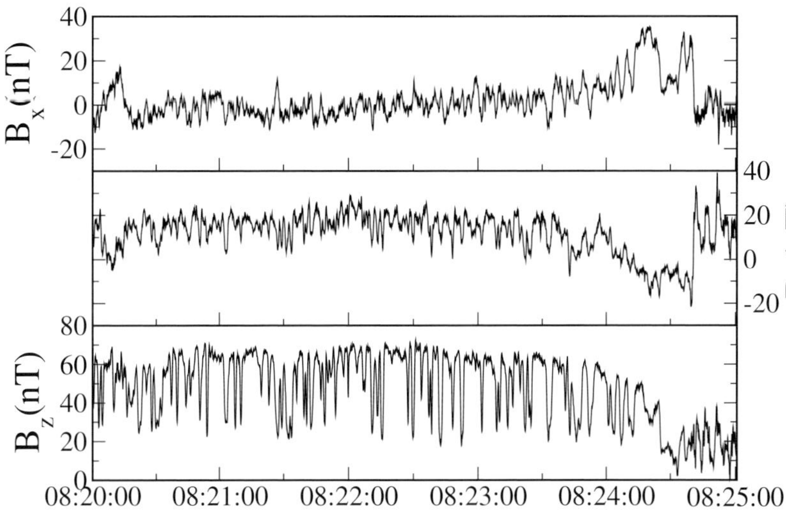

Figure 3.14. Magnetic field data in GSE coordinates from C1 on November 10, 2000 between 08:20 and 08:25 UT. Panels show the magnetic field components at 22 vectors/sec. (Figure provided by D. Constantinescu).

The fit to the data from a single spacecraft is sensitive to the initial choice for the parameters yielding a non-unique solution. With the availability of Cluster data, this ambiguity was avoided (Constantinescu et al., 2003) by simultaneously fitting data from $1 < n \leq 4$ spacecraft. After having identified an MMS from one 'reference spacecraft', and knowing the normal coordinates of the others the model magnetic field for the remaining $4 - n$ 'witness' spacecraft can be determined and the correlations constructed between the measured magnetic field and the model magnetic field.

3.4.4 Case study

In the way described above, Constantinescu et al. (2003) analysed an interval of strong mirror wave activity (Lucek et al., 2001) from November 10, 2000 08:20-08:25 UT (Figure 3.14). Cluster was in the dusk side magnetosheath with a typical spacecraft separation of 1000 km, moving at GSE velocities about 1 km s^{-1}. Using correlation analysis, corroborated by Wind observations, Lucek et al. (2001) concluded that the plasma flow velocity was of 815 km s^{-1} in the direction C1 - C3.

Magnetic field data at a resolution of 22 vectors per second were analysed using a fitting window with a width of 200 data points, corresponding to about 9 s or 7000 km. Depending on the values of L and α (Equation 3.14) this corresponded to several layers of the MMS. Assuming that $L = 5000$ km the spacecraft would have passed through 1-2 layers for $\alpha = 5$, and through 3-4 layers for $\alpha = 10$. In order to identify subintervals where the fit procedure was stable with respect to small changes in the data selected for analysis a sliding window technique was applied and the variability in the resulting parameters examined.

The fit produced a set of parameters for the data subinterval $[i, i+200]$. Those cases where the minimisation converged and the parameters had physically reasonable values were used as start parameters for the next data subinterval $[i+n, i+n+200]$. Otherwise the default values were used. A value of $n = 2$ (99% overlapping intervals) was normally used.

The subinterval 08:20:00 - 08:20:10 UT is shown in detail. Here agreement between the model and the witness spacecraft was reasonably good. The reference spacecraft C1 and spacecraft C2 were chosen to participate in the fit. C3 and C4 were chosen as witness spacecraft. C1 and C2 had a large separation in the direction orthogonal to the average magnetic field, and were expected to sample different MMS layers. The velocity vector was aligned with the C1 - C3 separation vector, as shown in the investigation of mirror modes by Lucek et al. (2001), and so the magnetic field measured by C3 was very similar to the magnetic field measured by C1. As a consequence the magnetic field measured by C3 did not contribute any additional information to the fit. A comparison between the measured and model magnetic field intensity for each spacecraft produced the following cross correlation coefficients: $C_1 = 0.81$, $C_2 = 0.83$, $C_3 = 0.78$, $C_4 = 0.64$. The radius of the main structure was found to be $R = 2061$ km. It is interesting to note that this is larger than what had previously been determined from Cluster observations of mirror modes in the magnetosheath. Lucek et al. (2001) in their investigation of the same day found from correlation analysis that, although variable, the observed mirror structures showed differences on scales of 750-1000 km at spacecraft separated along the magnetic field, while the mirror mode radii appeared to be less than 600 km. Their extent along the flow, however, was of the order of 1500-3000 km. Possibly the correlational method favours the smaller scale local structures while the model assumption provides larger and more global properties.

Figure 3.15 shows the measured and model magnetic field intensities. The vertical continuous lines mark the intersection with the XZ plane, i.e., the time when the distance (d) between spacecraft and the axis of the MMS was minimum. The vertical dotted lines mark the boundaries between different layers. Multiple minima result from passing through different regions of the same magnetic mirror structure. This can also be seen from Figure 3.16, which shows the way in which the spacecraft passes through the layers of the structure.

Figure 3.17 shows the reconstructed magnetic field in the XZ plane. This figure reveals that C1, C3 and C4 probably passed through a central uniform field region of the structure yielding plateaus in Figure 3.15. C2 was much closer to the axis measuring large changes in the magnetic field inside the central structure.

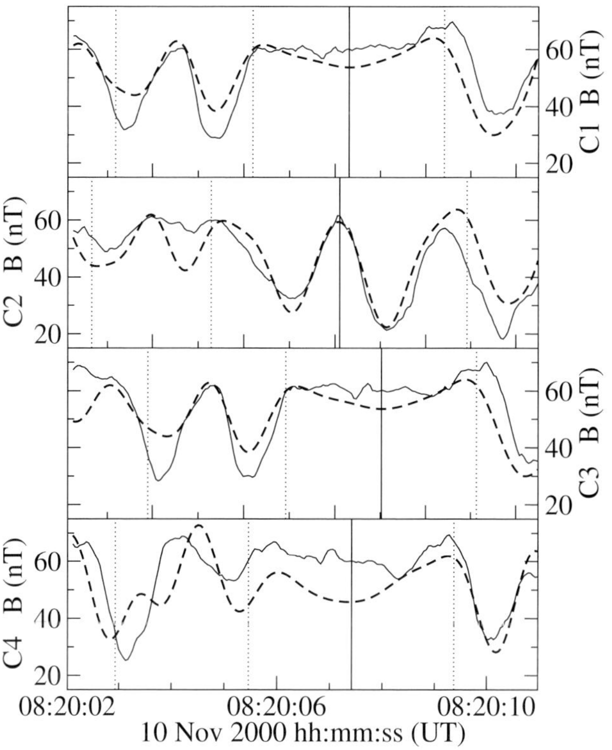

Figure 3.15. Measured (solid lines) and model (dashed lines) magnetic field intensity for the four Cluster spacecraft. C1 and C2 were participating in the fit, C3 and C4 were witness spacecraft. The vertical solid lines represent the intersection with XZ plane (i.e., $s = 0$) and the vertical dotted lines separates different layers of the MMS. (From Constantinescu et al., 2003).

The quality of the parameters can be improved by performing new fits, using data from all four spacecraft and the previous parameters as start parameters. For this particular event, however, Constantinescu et al. (2003) found that no improvement could be achieved. Of course, these values should taken with caution. The precision of the fit to the measurements though being robust contains a number of systematic errors. First, there are the assumptions of the model: it is two-dimensional, assumes stationarity and stability in a non-linear sense, and does not include Finite-Larmour-Radius effects or any kinetic effects. Each of these assumptions has to be proved. Moreover, the model is applied here to only one single mirror structure. The agreement is reasonable, though the Figures 3.15 and 3.16 show that there remain large uncertainties in the fit and the form of the structure relative to the paths of the spacecraft. Hence a conservative conclusion is that the model is still too rough to unambiguously reproduce mirror structures. In future it will be improved in different directions, in particular by applying it to a larger statistical ensemble.

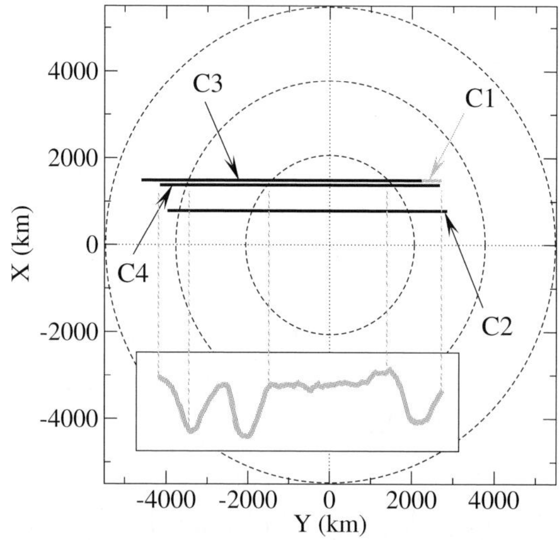

Figure 3.16. The projections of the spacecraft trajectories in the (x-y) plane(straight lines) and the magnetic field measured by C1 (gray curve). The circles represent the boundaries between different layers of a mirror mode structure. (From Constantinescu et al., 2003).

Horbury et al. (2004) recently performed a statistical survey of anti-correlated magnetic field and density structures observed in the magnetosheath. Their study estimated the motion of 39 isolate magnetic dips and 12 peaks, spread over four days, using the application of four-spacecraft timing analysis. Their results were not therefore dependent on the use of a particular mirror mode model. The important finding is that within timing errors the most probable (average) velocity is very close to zero indicating that the structures are convected with the flow. The scattering of the velocity was ~ 21 km s^{-1} much less than the local wave speeds ($<5\%$ of the local Alfvén speed and $<15\%$ of the local slow mode velocity). Hence, observations do not support the view that propagating slow mode waves or solitons are responsible for the formation of such structures in the magnetosheath. Neither are they standing structures, i.e., propagating against the flow. The directions of the normals derived for the mirror modes from timing analysis were consistent with the structures having a cylindrical cross section, inclined by a small angle ($< 20°$) to the background magnetic field. These results, therefore, suggest a mirror mode geometry more similar to the idealised model presented in this section, rather than a soliton model, where the structures would be expected to be sheet-like.

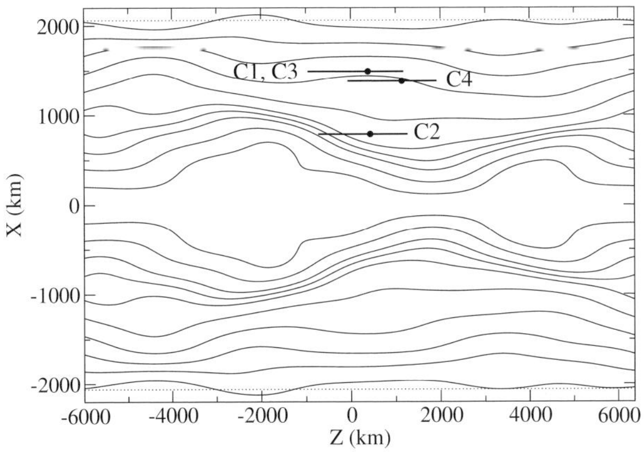

Figure 3.17. The magnetic field lines derived from the model of a mirror mode structure. The straight lines are the projections of the spacecraft trajectories on the XZ plane. The dots are the points where the spacecraft paths intersect the XZ plane. (From Constantinescu et al., 2003).

3.5 Observations of High-Frequency Waves

3.5.1 Electromagnetic waves: Lion roars

Lion roars are intense, short-duration, narrow-band packets of whistler mode waves observed in Earth's magnetosheath, first reported by Smith et al. (1967) using data from OGO 1. The average mean frequency of these waves in the magnetosheath is ~ 100 Hz (0.25-0.5 f_{ce}) with typical amplitudes of 0.1 nT and burst durations of 1 - 2 s (Smith and Tsurutani, 1976). When Lion roars are excited inside mirror modes, their frequency is much lower because of the strongly depressed magnetic field. A more detailed study of lion roars found that these emissions were most often observed in the inner region of the sunward magnetosheath, and the distribution of their intensity at 200 Hz was highest near the subsolar magnetopause (Rodriguez, 1985). Tsurutani et al. (1982) found that lion roars often coincide with minima of the magnetic field strength and maxima of plasma density, indicative of mirror mode structures. Zhang et al. (1998), using large amounts of data collected by the Geotail spacecraft, argued that only 30% of the lion roars were associated with mirror modes. This was probably due to a selection effect of the magnetometer aboard Geotail as discussed by Baumjohann et al. (1999) who, using the very sensitive magnetometer on Equator-S, found that low frequency lion roars are nicely associated with mirror modes. They consist of very narrow band emissions of a few tens of Hz frequency following the local electron cyclotron frequency and indicating the presence of trapped electrons in mirror modes. Lion roars are most probably generated by unstable anisotropic electron distributions, when the perpendicular electron temperature is larger than the parallel temperature. These are the low parallel energy electrons. In the mirror mode structures, the number of resonant electrons is larger for a given anisotropy than outside in the undisturbed magnetosheath (Smith and Tsurutani, 1976).

Early studies of lion roars in the magnetosheath found that they typically propagate with wave vectors at angles less than 30° relative to the magnetic field (Smith and Tsurutani, 1976). Baumjohann et al. (1999) found that the wave vectors were very close to parallel to the magnetic field, taking into account measurements of nearly monochromatic magnetic wave packets detected by the Equator-S spacecraft in the minima of the magnetic field. Some of the lion roars contained in the Geotail study (Zhang et al., 1998) were found to have wave vectors near 90°, suggesting they were downstream propagating whistler waves from the bow shock. Zhang et al. (1998) also found that although the majority of lion roars that were detected by Geotail were propagating in one single direction, one class of lion roars were found to be propagating in two directions simultaneously, suggesting the local plasma as the source.

The multipoint measurements of Cluster are ideally suited for determining the location of the source region and characteristics of the magnetosheath lion roars. Maksimovic et al. (2001), using data obtained in the duskside magnetosheath from the Cluster STAFF Spectrum Analyser (Cornilleau-Wehrlin et al., 1997), found that close to the magnetopause, lion roars in deep magnetic troughs are observed to propagate simultaneously in two directions, both parallel and anti-parallel to the magnetic field, as shown in Figure 3.18. Panels (a) and (b) of this figure show that the lion roars (solid line) are found to be more circular and more right-handed than the other whistler waves (dotted line). Panel (c) shows two peaks for the wave vector of the lion roars. After taking into account Doppler effects and the plasma velocity, these results imply that some lion roars on this date were propagating both upstream and downstream. This suggests that the Cluster spacecraft were in the source region of the lion roars, consistent with the results of Zhang et al. (1998). Far from the magnetopause, the waves were found to propagate in only one direction, roughly anti-parallel to the magnetic field. In addition Maksimovic et al. (2001) found that the lion roars were propagating at angles of 30° to 50° from the local magnetic field direction, which is inconsistent with whistler mode waves which would normally be Landau damped in a bi-Maxwellian plasma. This differs from many of the earlier works which found much smaller angles, and may be due to the sampling and bandwidth characteristics of the STAFF-SA instrument. It is also possible that this angular difference has a physical explanation that will become apparent through future statistical studies of lion roars. One should, however, take into account that the time resolution of the Cluster measurements is conciderably lower than that of the Equator-S measurements, and because of this, the direction finding is less accurate for Cluster measurements than for Equator-

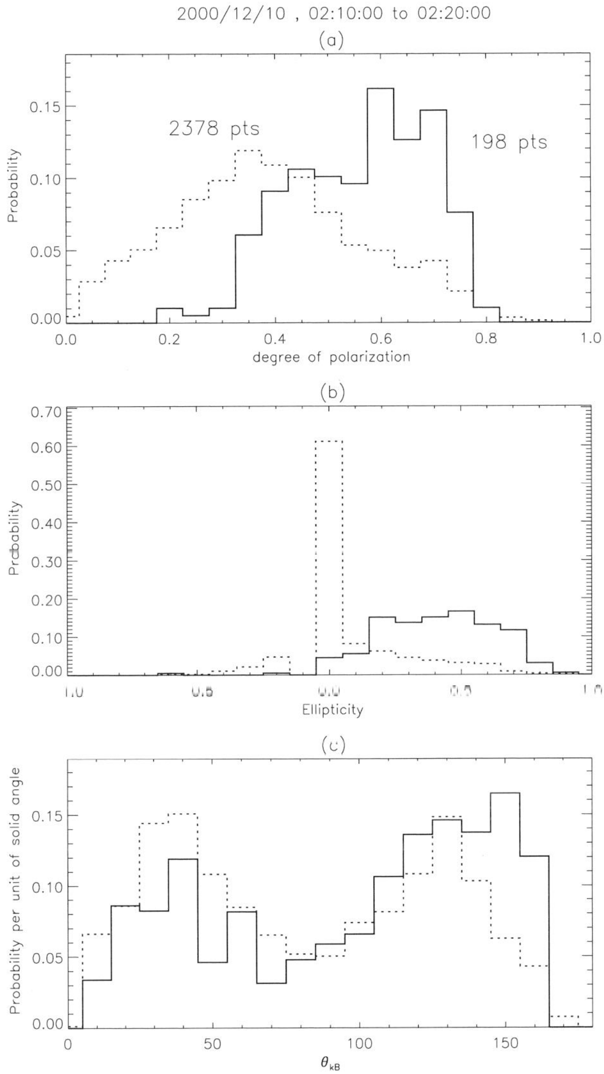

Figure 3.18. Histograms showing various characteristics of the whistler waves observed on December 10, 2000 near the magnetopause by the STAFF-SA instrument on the Cluster spacecraft: (a) the degree P of polarisation, with a value of 1 indicating the three components are fully coherent and the wave field is fully polarised, (b) the ellipticity (+1 for circular right-hand polarisation), and (c) the angle between the wave vector and B_0, for the lion roars (solid line) and for other whistlers (dotted line). (From Maksimovic et al., 2001).

S, as discussed by Baumjohann et al. (1999) with respect to the measurements of Zhang et al. (1998) for lion roars related to mirror modes. The more sensitive magnetometer aboard Equator-S which allowed for extraordinarily high time resolution clearly identified narrow band low frequency lion roars to propagate very close to parallel or antiparallel to the local magnetic field thus suggesting that the lion roars observed in relation to mirror modes originate from various trapped electron components in different places inside mirror mode bubbles.

The work of Maksimovic et al. (2001) was an introductory study of lion roars using the Cluster fleet. The primary contribution that Cluster can make to the knowledge already gained on the characteristics of lion roars is performing a statistical study of the direction of propagation using the multi-spacecraft measurements of STAFF-SA obtained in various regions of the magnetosheath as well as outside and inside mirror modes. These measurements will be supplemented with information obtained from various Cluster instruments, such as FGM, to determine the fraction of lion roars observed within mirror mode structures, or PEACE to determine whether an electron temperature anisotropy exists for those lion roars observed to be propagating in two directions. In addition, the very high time resolution waveform data of the Wideband (WBD) plasma wave receiver (Gurnett et al., 1997) should help in this regard. Since WBD has a time resolution of 35 μs and time of measurement accuracies to 10 μs on all 4 spacecraft, cross-spacecraft correlations of the lion roar waveforms can be performed, yielding information on the propagation of these lion roar wave packets from the source region. A similar correlation analysis has been successfully carried out for the chorus emission region in the inner magnetosphere (Santolik and Gurnett, 2003).

3.5.2 Electrostatic waves: Broadband structures

All of the early plasma wave measurements made by spectrum analyzers showed the prevalence of Broadband Electrostatic Noise (BEN) in many regions of Earth. First discovered by Scarf et al. (1974) and Gurnett et al. (1976) in the distant tail of Earth, BEN was characterized as being bursty and consisting of broadband spectral features usually extending from the lowest frequencies measured up to as high as the plasma frequency. The intensity of BEN was found to decrease with increasing frequency. Rodriguez (1979) provided the first comprehensive survey of BEN in the magnetosheath. He found that the magnetosheath electrostatic turbulence was almost continually present throughout the magnetosheath with broadband (20 Hz to 70 kHz) rms field intensities typically 0.01 - 1.0 mV m^{-1}. He also found the turbulence to consist of two or three components: a high frequency component ($\geq$ 30 kHz) peaking at the electron plasma frequency f_{pe}, a low frequency component with a broad intensity maximum below the nominal ion plasma frequency f_{pi}, and a less well defined intermediate component in the range $f_{pi} < f < f_{pe}$.

Dubouloz et al. (1993) carried out a theoretical investigation that showed that electron acoustic solitons passing by a satellite would generate spectra that could explain the high frequency part of BEN, above the electron plasma frequency, that had been observed in the dayside auroral zone by the Viking satellite. This was followed in 1994 by the findings of Matsumoto et al. (1994) that solitary waves of a few milliseconds in duration were responsible for the high frequency part of BEN observed by the Geotail satellite in the plasma sheet boundary layer. These electrostatic solitary waves appeared in the form of bipolar pulses (one positive electric field peak followed by one negative peak, or vice versa) in the time series data obtained by the Geotail Plasma Wave Instrument. An electron two-stream instability that produced nonlinear Bernstein-Greene-Kruskal (BGK) type isolated potentials was proposed by Matsumoto et al. (1994) as the generation mechanism for the electrostatic solitary waves. The Fourier spectrum of the solitary waves represented by the bipolar pulses was thus the reason for the broad band signature since a single pulse contains all frequencies up to a frequency determined by the time duration of the pulse. Subsequently similar solitary waves were found to be responsible for the BEN observed in several other regions of Earth, c.f. Franz et al. (1998) and Ergun et al. (1998) using Polar and FAST satellite data, respectively. Interferometry data obtained on these satellites allowed for the identification of these solitary waves as coherent potential structures, either electron or ion phase-space holes determined by the direction of propagation of the solitary waves, the hemisphere in which they were detected, and the initial direction (positive or negative electric field) of the pulses. A statistical survey of electron solitary waves observed by the Polar satellite at 2 to 9 R_E was carried out by Cattell et al. (2003) with the following findings: 1) the mean solitary wave duration was about 2 ms; 2) the waves have velocities from $\sim 1000\,\mathrm{km\,s^{-1}}$ to $> 2500\,\mathrm{km\,s^{-1}}$; 3) the observed scale sizes (parallel to the magnetic field) are on the order of 1-10 λ_D, with $e\phi/kT_e$ from ~ 0.01 to $O(1)$.

Cluster observations of solitary waves in the magnetosheath were first reported by Pickett et al. (2003). Using Cluster WBD data (Gurnett et al., 1997), the bipolar pulse solitary waves were found to have time durations of $\sim$25-100 μs in the dayside magnetosheath near the bow shock. These solitary waves were found to be consistent with electron phase space holes. They were detected when the magnetic field was contained primarily in the spin plane, indicating that they propagate along the magnetic field. It was not possible for Pickett et al. (2003) to determine the velocity of the structures since the Cluster WBD instrument makes a one axis measurement, that being the average potential between the two electric field spheres. It was also not possible to correlate individual solitary pulses across different Cluster satellites, due to the 1/8 duty cycle of the WBD instrument when using this wideband 77 kHz filter mode and/or the solitary waves evolving (growing/damping) over the distance from one spacecraft to the next. Limited success of correlating solitary waves across satellites has been attained thus far only for the tripolar type solitary wave (discussed below) observed in the auroral region at 4.5 - 6.5 R_E, well above the auroral acceleration region (Pickett et al., 2004b).

By using spectrograms of the waveform data obtained from two of the Cluster spacecraft separated by over 750 km in the magnetosheath, Pickett et al. (2003) found that the overall profile of the broad-band noise associated with the solitary waves was remarkably similar in terms of onset, frequency extent, intensity and termination on both spacecraft. This similarity implies that the generation region of the solitary waves observed in the magnetosheath near the bow shock is very large. The generation region may be located at or near the bow shock, or it may be local in the magnetosheath but related to processes occurring at the bow shock. Figure 3.19 shows an example of the broadband structures observed during a Cluster magnetosheath pass at high magnetic latitude on 05 April 2004 when the spacecraft were separated by as little as 150 km (spacecraft 1 and 2) and by as much as 500 km (spacecraft 1 and 4). The top three panels contain data from each of Cluster spacecraft 1, 2 and 4 showing the broadband structures spanning the range from the lowest frequency measured, 1 kHz, up to as great as 60 kHz. Intense waves around 2 - 3 kHz are also observed throughout the interval. Interference from the emission of electron beams from the EDI experiment shows up as horizontal lines in the Cluster 1 data. The waveforms used to create the spectrograms for each of the spacecraft for a small time period around 13:37:00.33 UT are shown in the bottom three panels. Each panel contains 4 ms of data and except for parts of the bottom two panels, the data shown are from slightly different time periods based on the offset time below each panel from the start time shown at the top of the three panels. Solitary waves of a few tenths of mV m^{-1} peak-to-peak and time durations of around 90 μs are seen to dominate the waveform data leading to the broad bands seen in the spectrograms. Evidence of the 2 - 3 kHz sinusoidal type wave is also seen in the Cluster 1 data. Just as for the case with larger separations reported in Pickett et al. (2003), the overall profiles of the broad-band structures associated with the solitary waves are very similar on all spacecraft. Even so, the solitary waves as seen in the bottom two panels of the figure during the parts that overlap in time do not show a correlation.

Because the Cluster WBD plasma wave receiver is particularly suited to making measurements of solitary waves over a large range of time scales (a few tens of microseconds to several milliseconds) and over a wide range of amplitudes due to its automatic gain control feature, data from this instrument were used (Pickett et al., 2004a) in carrying out a survey of solitary waves observed throughout the Cluster orbit, including the magnetosheath. These solitary waves have been referred to as isolated electrostatic structures (IES) because they are isolated pulses in the electric field waveform data and previous studies have found them to be potential structures propagating primarily along magnetic field lines. Pickett et al. (2004a) found that throughout the Cluster orbit, two dominant types of IES are observed, the bipolar pulse type already discussed, and the tripolar pulse, consisting of one positive and

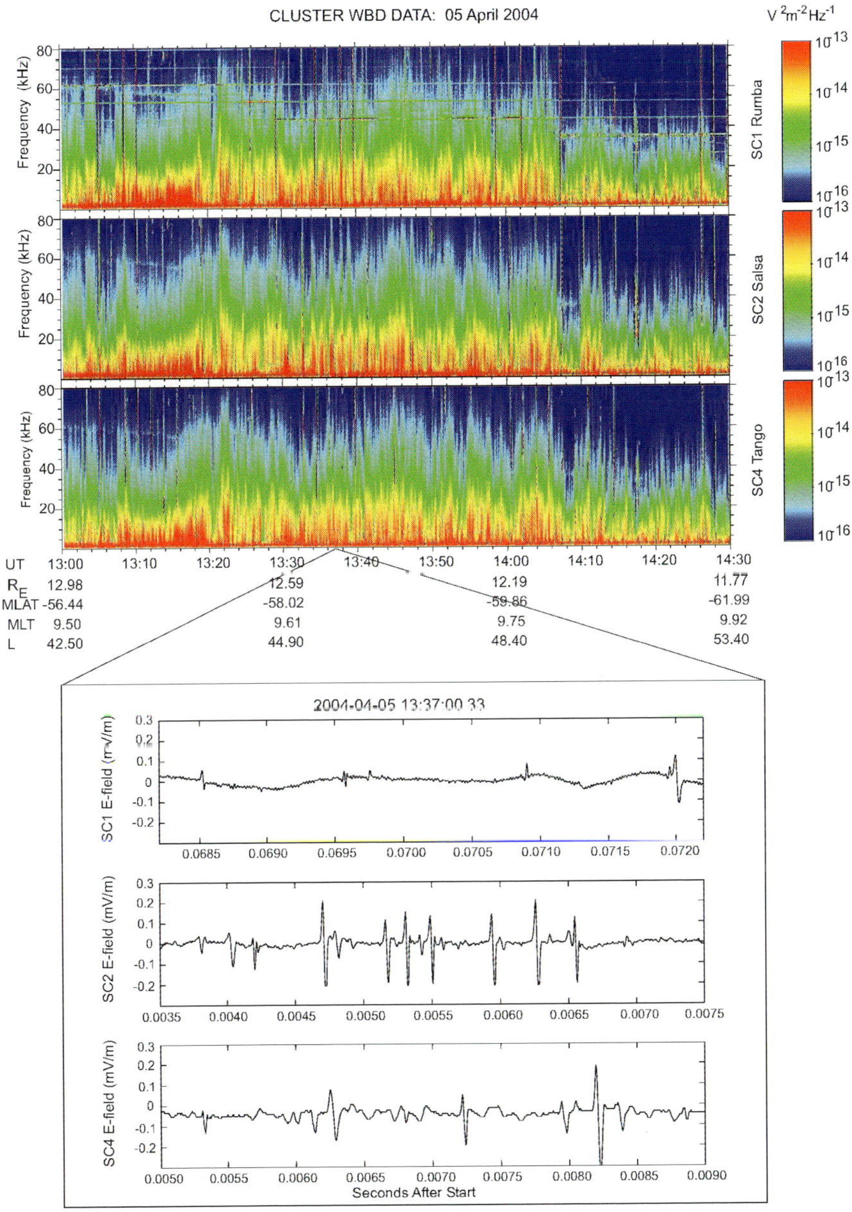

Figure 3.19. Cluster WBD observations of the waves in the magnetosheath. The top three panels contain the data in spectral form produced from high time resolution waveforms obtained on Cluster 1, 2 and 4, respectively, and the bottom three panels show very short time period snapshots of the waveforms from each of these three spacecraft. (From Pickett et al., 2003).

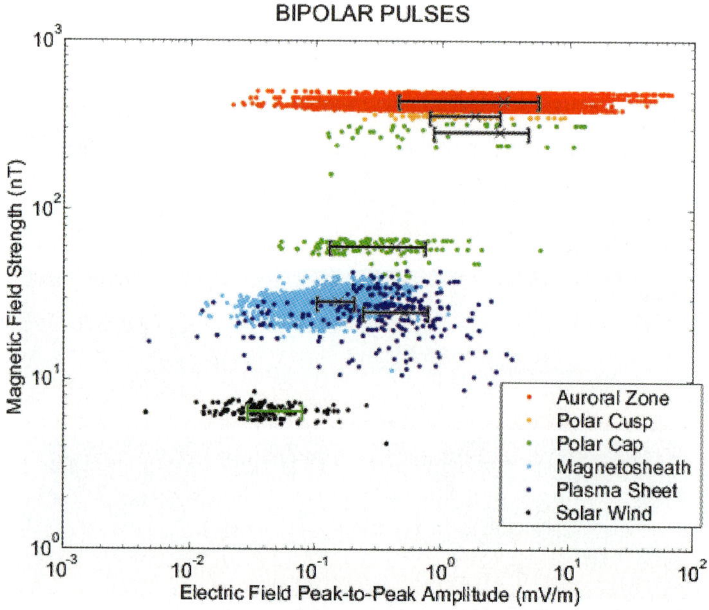

Figure 3.20. Survey of the bipolar pulses observed by Cluster WBD over a two-year period. This plot shows the electric field amplitude vs. magnetic field strength, with the over plotted bracketed lines with the embedded 'x' within each regional grouping representing the standard deviation and mean of that group, respectively. (From Pickett et al., 2004a).

two negative peaks, or vice versa, thought to be weak double layers (Mangeney et al., 1999; Pickett et al., 2004b). In carrying out the survey, Pickett et al. (2004a) found that there is a broad range of electric field amplitudes at any specific magnetic field strength as shown in Figure 3.20 for the bipolar pulses. More surprising was their finding that there is a general trend for the electric field amplitudes of the structures to increase as the strength of the magnetic field in which the measurements are made increases, as shown in Figure 3.20 for the bipolar pulses. A similar trend exists for the tripolar pulses (see Figure 4a of Pickett et al., 2004a). The magnetosheath IES were thus found to have some of the smallest amplitudes observed throughout Cluster's orbit with a mean value of about 0.2 mV m^{-1} since the local magnetic field strength is relatively small there (few tens of nT). Only the IES observed in the solar wind had smaller amplitudes.

An explanation for the relationship between the solitary wave electric field amplitude and magnetic field strength is not yet fully developed. However, Pickett et al. (2004a) considered the possibility that the solitary waves are BGK mode (Bernstein et al., 1957). This being the case, the observed data trend of increasing electric field amplitudes with increasing magnetic field strength could be rooted in the stability requirements of the BGK mode in finite magnetic fields. Chen et al.

(2004) have found a set of inequalities that describe the stability of BGK mode solitary waves. These conditions point to a trend that in a much weaker magnetic field, either the potential amplitude would decrease or the size would increase in order for the structures to be stable, which results in smaller electric field amplitudes.

One other important finding came out of the survey by Pickett et al. (2004a). They found that when the time durations of the bipolar and tripolar IES were plotted with respect to the magnetic field strength, there was no similar trend as with the electric field amplitudes (see Figures 3b and 4b of Pickett et al., 2004a). However, plotting the data in this fashion clearly pointed out that the magnetosheath IES, both bipolar and tripolar, are clearly of much shorter time duration than the structures observed in all the other regions sampled by Cluster, including those in the solar wind. Although Pickett et al. (2004a) speculated that this may imply that the magnetosheath solitary waves are being generated by a different mechanism (perhaps through turbulence rather than a beam instability) than the IES observed in all other regions by Cluster, clearly more work, both experimental and modelling, is required in order to solve this mystery. In particular, the Cluster electron and ion data, together with other wave data, will be evaluated with respect to the observance of solitary waves. These data will then be used as inputs to models seeking to determine the generation mechanism of the solitary waves observed in the magnetosheath. Modelling has been very effective in shedding light on the generation of the solitary waves observed in the auroral acceleration region (for a short review, see Pickett et al., 2003).

3.6 Observations of Dispersed Ion Signatures

Observations made using data from previous missions have shown that Hot Flow Anomalies (HFAs), discussed in some detail in Eastwood et al. (2005), can generate a signature in the magnetosheath (Paschmann et al., 1988; Schwartz et al., 1988; Safránková et al., 2000), and it was found later that they can even affect the magnetopause sufficiently for a ground response to occur (Sibeck et al., 1998, 1999). Recently, however, it has been proposed that in addition to HFAs generating hot, turbulent plasma in the magnetosheath, a different, subtle, interaction between ions reflected from the bow shock and an interplanetary discontinuity can generate dispersed bursts of anomalously high energy particles in the magnetosheath (Louarn et al., 2003). These have been called 'magnetosheath dispersed signatures', or MDSs. Louarn et al. (2003) reported energy-time dispersed ion signatures in the magnetosheath observed when Cluster was located over 12 R_E from Earth. They presented CIS and FGM data from three events, noting that more than 20 events occurred during a 2-month period between February and March 2001.

MDSs are characterised by a dispersed ion signature, with the most energetic ions being observed first, with significant fluxes at energies > 5 keV, followed by a decrease in the ion energy over approximately 1-2 min. CIS observations show ions appearing first in the highest energy range of the instrument (30 keV), but RAPID data indicate that the maximum MDS ion energy extends well above the energy range measured by CIS, up to energies of 30-60 keV. The ion signature of an MDS is not just restricted in energy. The pitch angle distribution is very anisotropic, with the ions systematically organised in a clear ring, with the central axis of the ring close to the magnetic field direction. The 3-D ion distribution, therefore, forms a hollow cone, which has its axis close to the local magnetic field direction. MDS ion distributions can also be non-gyrotropic. Although the ion energy decreases with time during each MDS, the characteristics of the cone do not appear to change systematically throughout the event. The characteristics of the thermal plasma are not affected by the MDS and the absence of Oxygen ions suggest that the spacecraft are not magnetically connected to the magnetosphere during these events.

MDSs are distinct from the magnetosheath signatures of HFAs. The latter are typically characterised by heated plasma, associated with a region of disturbed magnetic field and significant plasma velocity perturbations. The HFA signatures are centered on a change in magnetic field direction and show depressed magnetic field magnitude, flanked by enhancements in $|\mathbf{B}|$ and N_p. The ion flux associated with HFAs at, for example 20 keV can be up to two orders lower than the flux at that energy within an MDS. MDSs do, however, appear to be associated with HFAs signatures in the magnetosheath, and several were observed to occur just before HFA signatures.

Louarn et al. (2003) proposed that the acceleration process generating MDSs is located at the shock. However, the profile of the time-dispersed ion signature cannot be explained just by invoking a time of flight effect from a remote, stationary, source. A model was proposed by which the MDSs ion spectra profiles could be explained by a combination of effects. The first effect arises from the motion of the injection site of the particles along the bow shock. The injection site is assumed to be the region where a discontinuity interacts with the bow shock, and this moves across the bow shock as the discontinuity is convected anti-sunwards by the solar wind. Once they enter the magnetosheath, the motion of the highest-energy ions is dominated by the time of flight effect, so the highest-energy ions arrive first. At the other end of the energy range, motion of lower energy particles is dominated by anti-sunward convection of the plasma, and these particles show almost no time-of-flight dispersion. The combination of these effects is demonstrated by the use of a 2D model to give rise to a particle dispersion profile consistent with MDSs (Figure 3.21).

The organisation of the particles in a ring distribution places some limits on the acceleration mechanism. In addition, since particles are accelerated to >30 keV, into the RAPID energy range, this implies that the acceleration process must be efficient, and occur in only a few gyro-periods. Shock drift acceleration (e.g., Armstrong et al., 1985), which operates at quasi-perpendicular shocks, accelerates particles in only a few gyro-periods, but the energy gain is limited to only a few times the initial particle energy, which is lower than the energy gains observed in MDSs.

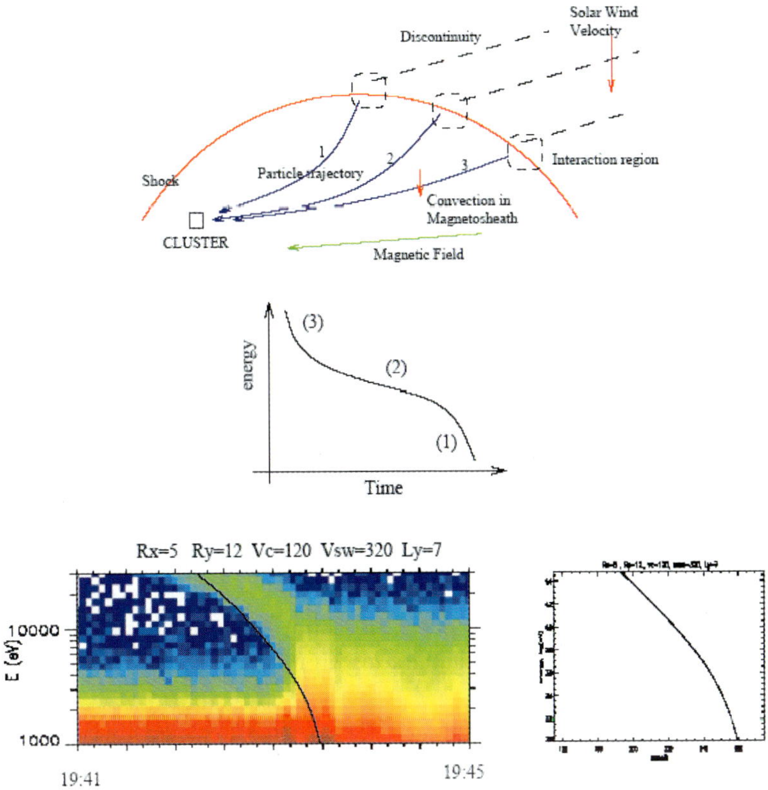

Figure 3.21. A schematic of the interaction between a discontinuity and the bow shock, generating the dispersive ion signature of a magnetosheath dispersed structure. From top to bottom: (1) global configuration of the interaction and example of trajectories (2) typical dispersion calculated from the model and, (3) fit of the observations. (From Louarn et al., 2003).

Fermi-acceleration (Bell, 1978), which operates when particles scatter from irregularities with converging velocities, can lead to greater energy gain, but operates more slowly than shock drift acceleration. In order to explain the observations an acceleration process was proposed which combines aspects of both mechanisms.

The authors estimate that the interaction time between the particles and the shock, during which acceleration occurs, is approximately 100 s. The model is as follows: A fraction of incident solar wind ions are reflected from the shock. Some of these particles are then turned by their interaction with the moving discontinuity, gaining energy in the process, and then re-encounter the shock. A fraction of these particles are also reflected, leading to a further possible interaction with the discontinuity. In this way, multiple bounces between the converging shock face and discontinuity plane lead to energy gains into the 20-30 keV range, as observed at an MDS. A fraction of the particles escapes multiple interactions with the shock, because they exit the discontinuity at more than 2 Larmor radii from the shock, and hence do not re-cross the shock face.

The proposed process predicts an asymmetry in the source of the particles since the mechanism only operates if the gyratory motion carries the ions further from the shock. Such an asymmetry is observed in MDSs. Also the characteristic pitch angle selection of the energised particles can be explained in terms of the interaction of the particles with the quasi-perpendicular bow shock. Lastly, in this scenario, when the discontinuity interacts with the shock, particles which bounce between the two form the MDS. A further population of particles would form an HFA, and thus give rise to hot turbulent plasma in the magnetosheath which is commonly associated with HFA formation. The two types of magnetosheath signature would reach the observer almost simultaneously, thus explaining why an MDS is often observed as a precursor to the magnetosheath signature of an HFA.

These dispersed structures do not lead to significant pressure disturbances in the magnetosheath, and so they therefore probably play only a minor role in solar wind/magnetosphere coupling. However, a greatly enhanced efficiency of particle acceleration, in time, flux and energy, is achieved through the interaction of ions reflected from a quasi-perpendicular bow shock with the moving interplanetary discontinuity. This demonstrates the potential importance of shock/shock or discontinuity/shock interactions in generating significant fluxes of accelerated particles.

3.7 Summary and Outlook

Cluster has already contributed substantially to our understanding of the physics of the magnetosheath. Though it is not the first time that more than one spacecraft has been used to study the spatial structure of space plasmas, it is the first time that structures in the magnetosheath can be resolved in three dimensions and in the space-time domain, by combining data from the four Cluster spacecraft. In particular the ability to calculate the wave vectors of magnetosheath waves, and to examine the distribution of energy between different modes simultaneously, revolutionises the study of many magnetosheath phenomena, as illustrated by the results presented in this chapter. In addition, more sensitive and sophisticated instrumentation on the Cluster spacecraft, measuring phenomena occurring over a wide range of spatial scales, has opened further new opportunities for research. The richness of the Cluster data set has also allowed new analysis of such fundamental topics as particle acceleration and plasma instabilities. Finally, this chapter demonstrates that Cluster offers us the best opportunity yet to study the highly complex plasma phenomena in this fascinating region.

The main achievements described in this chapter are the following:

1. The unambiguous identification of wave modes present in a broad band magnetosheath fluctuation spectrum. Complementary methods allow the identifi-

cation of the dominant wave mode, together with time variations in the wave mode power. At low frequencies the mirror mode was found to dominate but results at higher frequencies showed that the plasma supported multiple waves simultaneously, and that their wave vectors lay on MHD dispersion curves.

2. The first accurate identification of Alfvén waves generated by proton and alpha particle temperature anisotropies immediately downstream of the quasi-perpendicular shock. In addition, these waves then appeared to evolve into thin current tubes, aligned with the magnetic field. The tubes have a radius of only 100 km - the order of the local ion inertial length, but a length exceeding 600 km. It is proposed that these current tubes are generated from the Alfvén waves, through the action of the filamentation instability.

3. In the highly turbulent magnetosheath plasma the mirror mode is probably the lowest frequency energy source in the turbulence. With Cluster it is possible to make the first estimate of the spatial scale of mirror modes in directions other than along the magnetosheath flow. Results show that they are nearly aligned with the magnetic field and that they have an extent exceeding 1000 km in this direction.

4. Cluster data can also be used to estimate the motion of mirror modes, which are predicted, from theory, to be stationary in the plasma frame. A statistical investigation of a set of mirror structures showed that in general mirror modes are close to stationary in the plasma frame. They are convected by the magnetosheath plasma flow and in this study their remnant plasma frame velocity was found to be much less than the slow mode speed, indicating that they were neither slow modes nor solitons. These observations are also consistent with the mirror modes having a cylindrical cross section. In contrast, analysis of a different time interval as part of a different study, suggests that, under some circumstances, slow magnetosonic solitons can exist in the magnetosheath.

5. Fitting a nonlinear stationary model to an observed mirror structure allows the determination of its scale in different directions. Mirror structure sizes vary between 600-2000 km in diameter and several thousand km in length along the ambient field. Nonlinear theory suggests that they should not be inclined to the magnetic field, and the observation of a small inclination to the magnetic field indicates that they are driven, and are still evolving.

6. In the higher frequency range Cluster observations of lion roars have been analysed, yielding controversial results on their origin. Their propagation seems to be too oblique for whistler mode waves, thereby posing an important problem for future research.

7. Waveforms of broadband electrostatic noise in the magnetosheath between 1 - 60 kHz indicate that most of this noise results from large numbers of phase

space holes; Cluster is the first spacecraft to detect such holes in the magnetosheath. From the similarity of the waveforms and spectra of broadband electrostatic noise on spacecraft separated by several hundred kilometres, close to the bow shock, it is concluded that the noise source region was extended, and that the generation region was probably close to the bow shock.

8. Cluster has detected two types of noise leading to Isolated Electrostatic Structures in the magnetosheath: bipolar and tripolar structures, the latter belonging to weak double layers, showing that the magnetosheath contains small-scale regions of stationary, localised electrostatic fields.

9. Discovery of energy dispersion in ion fluxes originating from the bow shock suggests that in some cases, when an interplanetary discontinuity interacts with the bow shock, energetic ions can bounce between the shock and the discontinuity, and in this way become energised by the Fermi mechanism. A proportion of the hot ions enter the magnetosheath, giving a dispersed burst of energised particles with maximum energies exceeding 30 keV.

Some of these findings are unprecedented and many were possible only with the multi-spacecraft capabilities of Cluster. In the light of these results, in which direction should the Cluster investigation of the magnetosheath go in the near future? It is clear that the multi-spacecraft aspect of Cluster will be essential to future magnetosheath research, even though the instrumentation provides excellent single spacecraft science as well. Many of the studies performed so far will only reach maturity with more extensive analysis of magnetosheath phenomena under the wide range of boundary conditions provided by the solar wind and the magnetosphere. There are many ideas to be pursued:

1. One most tantalizing question is the understanding of the turbulence in the magnetosheath plasma. Cluster provides a unique opportunity to study the spatio-temporal properties of this turbulence, its spatial variation and its wave vector spectrum, at least in the volume of **k**-space made accessible by the range of spacecraft separation scales. The k-filtering technique has provided a tool for investigating the dispersion relation of multiple waves present simultaneously in the plasma. Its application demonstrates that multiple wave modes are present in the plasma, and that the relative contributions of the different modes varies with frequency. A complete description of the fluctuation spectrum at all frequencies and all wave vectors is the purpose of active studies. As these are fundamental to all turbulence theory, this is a key topic for study with Cluster. A further opportunity for Cluster is to examine the time evolution of wave modes, for example using the phase difference method which can be applied under circumstances when one wave mode dominates the spectrum.

2. The lowest frequency turbulent mode is the mirror wave. Its complete investigation requires a more precise determination of its scales, growth rates and

saturation mechanisms than available today. This problem can be addressed partially by Cluster. It is of particular interest to decide whether mirror modes are in their final nonlinear state or still evolving; whether they propagate in the low frequency slow mode; what their cause is: either they are unstably excited by the mirror instability, as is believed, or result from an inverse cascade in magnetised plasma turbulence. This requires investigation of the turbulent modes and their interaction, again requiring the spatial properties of the turbulence to be resolved. Particle measurements must also be considered in order to decide what the role the ions and electrons play respectively in the evolution of turbulence in general, and in the development of mirror modes in particular.

3. An interesting question concerning mirror modes is whether, when they are close to the magnetopause, they can trigger reconnection by lowering the magnetic field. However, the observation that mirror modes tend to occur as large amplitude holes near the magnetopause may, on the other hand, suggest that they are themselves the result of the magnetopause reconnection process. Cluster could settle this problem by investigating the properties and propagation characteristics of mirror modes near the magnetopause, and their relation to reconnection.

4. The waves generated by the bow shock form another component of the low frequency magnetosheath fluctuation spectrum. Use of the four spacecraft data together, at a range of separation scales, will allow the waves generated by the bow shock and their subsequent evolution to be analysed in detail, and comprehensive comparisons to be made with predictions from wave and instability theory.

5. An important consideration in the high frequency electro-static range is the nature of phase space holes throughout the magnetosheath. What is their excitation mechanism? Are they related to the bow shock? Are weak double layers formed locally in the magnetosheath? Do they contribute to electron or ion heating? Do they generate transport coefficients which affect the large-scale turbulence or, close to the magnetopause, even reconnection? By combining knowledge of their propagation and orientation with particle observations, Cluster can address some of these questions.

6. Other facets of high frequency electrostartic fluctuations observed by the Cluster wave instrument package in the magnetosheath are yet to be interpreted. For example, intriguing sporadic bursts of a few kHz width, and of short duration (of the order of 100 ms), which are observed at frequencies shifted by a few kHz above the local plasma frequency.

7. The unexpected observation of dispersed ion bursts in the magnetosheath might be one signature of a potentially important mechanism operating at the shock.

They seem to arise from the bouncing of particles between the shock and an oblique discontinuity, as the discontinuity tracks across the shock carried by the solar wind flow. This might be the injection process by which particles are energised, to form a seed population for further Fermi acceleration, a process for which scientists have been searching for many years.

Acknowledgements

Cluster research in the UK and France is supported by PPARC and CNES, respectively. MLG and JP acknowledge support from NASA.

References

Alexandrova, O., A. Mangeney, M. Maksimovic, C. Lacombe, N. Cornilleau-Wehrlin, E. A. Lucek, P. M. E. Décréau, J. M. Bosqued, P. Travnicek, and A. N. Fazakerley: 2004, 'Cluster observations of finite amplitude Alfvén waves and small-scale magnetic filaments downstream of a quasi-perpendicular shock'. *J. Geophys. Res.* **109**, A05207, doi: 10.1029/2003JA010056.

Anderson, B., S. A. Fuselier, S. P. Gary, and R. E. Denton: 1994, 'Magnetic spectral signatures in the Earth's magnetosheath and plasma depletion layer'. *J. Geophys. Res.* **99**, 5877–5891.

Armstrong, T. P., M. E. Pesses, and R. B. Decker: 1985, 'Shock drift acceleration'. In: B. T. Tsurutani and R. G. Stone (eds.): *Collisionless shocks in the heliosphere: reviews of current research*, AGU Monograph Series 35. Washington, D.C.: Amer. Geophys. Union, pp. 271–285.

Balikhin, M. A. and M. E. Gedalin: 1993, 'Comparative analysis of different methods for distinguishing temporal and spatial variations'. In: *Proc. of START Conf., Aussois, France*, Vol. ESA WPP 047. pp. 183–187.

Balikhin, M. A., O. A. Pokhotelov, S. N. Walker, E. Amata, M. Andre, M. Dunlop, and H. S. K. Alleyne: 2003, 'Minimum variance free wave identification: Application to Cluster electric field data in the magnetosheath'. *Geophys. Res. Lett.* **30**(10), 1508, doi:10.1029/2003GL016918.

Balogh, A., C. M. Carr, M. H. Acuña, M. W. Dunlop, T. J. Beek, P. Brown, K.-H. Fornaçon, E. Georgescu, K.-H. Glassmeier, J. Harris, G. Musmann, T. Oddy, and K. Schwingenschuh: 2001, 'The Cluster Magnetic Field Investigation: overview of in-flight performance and initial results'. *Ann. Geophys.* **19**, 1207–1217.

Baumjohann, W. and R. A. Treumann: 1996, *Basic space plasma physics*. London: Imperial College Press.

Baumjohann, W., R. A. Treumann, E. Georgescu, G. Haerendel, K.-H. Fornaçon, and U. Auster: 1999, 'Waveform and packet structure of lion roars'. *Ann. Geophys.* **17**, 1528–1534.

Bell, A. R.: 1978, 'The acceleration of cosmic rays in shock fronts'. *Mon. Not. R. Astron. Soc.* **182**, 147–156 and 443–455.

Bernstein, I. B., J. M. Greene, and M. D. Kruskal: 1957, 'Exact nonlinear plasma oscillations'. *Phys. Rev.* **108**, 546–550.

Brinca, A. L., N. Sckopke, and G. Paschmann: 1990, 'Wave excitation downstream of the low-beta, quasi-perpendicular bow shock'. *J. Geophys. Res.* **95**, 6331–6335.

Cattell, C., C. Neiman, J. Dombeck, J. Crumley, J. Wygant, C. A. Kletzing, W. K. Peterson, F. S. Mozer, and M. Andre: 2003, 'Large amplitude solitary waves in and near Earth's magnetosphere, magnetopause and bow shock: Polar and Cluster observations'. *Nonlinear Proc. Geophys.* **10**, 13–26.

Chandrasekhar, S., A. N. Kaufman, and K. M. Watson: 1958, 'The stability of the pinch'. *Proc. R. Soc. London, Ser. A* **245**, 435–455.

Chen, L.-J., D. J. Thouless, and J.-M. Tang: 2004, 'Bernstein-Greene-Kruskal solitary waves in three-dimensional magnetized plasma'. *Phys. Rev. E* **69**, 055401–1–055401–4.

Chisham, G., D. Burgess, S. J. Schwartz, and M. W. Dunlop: 1998, 'Observations of electron distributions in magnetosheath mirror mode waves'. *J. Geophys. Res.* **103**, 26765–26774.

Constantinescu, O. D.: 2002, 'Self-consistent model for mirror structures'. *J. Atmos. Terr. Phys.* **64**, 645–649.

Constantinescu, O. D., K.-H. Glassmeier, R. Treumann, and K.-H. Fornaçon: 2003, 'Magnetic mirror structures observed by Cluster in the magnetosheath'. *Geophys. Res. Lett.* **30**, 1802, doi:10.1029/2003GL017313.

Cornilleau-Wehrlin, N., G. Chanteur, S. Perraut, L. Rezeau, P. Robert, A. Roux, C. Villedary, P. Canu, M. Maksimovic, Y. Conchy, D. Lacombe, F. Lefeuvre, M. Parrot, J. Pinçon, P. Décréau, C. Harvey, P. Louarn, O. Santolík, H. Alleyne, M. Roth, T. Chust, O. Contel, and The Staff Investigator Team: 2003, ' First results obtained by the Cluster STAFF experiment '. *Ann. Geophys.* **21**, 437–456.

Cornilleau-Wehrlin, N., P. Chauveau, S. Louis, A. Meyer, J. M. Nappa, S. Perraut, R. L., P. Robert, A. Roux, C. De Villedary, Y. De Conchy, L. Friel, C. C. Harvey, D. Hubert, C. Lacombe, R. Manning, F. Wouters, F. Lefeuvre, M. Parrot, J. L. Pincon, B. Poirier, W. Kofman, P. Louarn, and The Staff Investigator Team: 1997, 'The Cluster Spatio-Temporal Analysis of Field Fluctuations (STAFF) Experiment'. *Space Sci. Rev.*. **79**, 107–136.

Crooker, N. U. and G. L. Siscoe: 1977, 'A mechanism for pressure anisotropy and mirror instability in the dayside magnetosheath'. *J. Geophys. Res.* **82**, 185–186.

Denton, R. E., S. P. Gary, B. Anderson, S. A. Fuselier, and M. K. Hudson: 1994, 'Low frequency magnetic spectra in the magnetosheath and plasma depletion layer'. *J. Geophys. Res.* **99**, 5893–5901.

Denton, R. E., S. P. Gary, X. Li, B. J. Anderson, J. W. Labelle, and M. Lessard: 1995, 'Low-frequency fluctuations in the magnetosheath near the magnetopause'. *J. Geophys. Res.* **100**, 5665–5679.

Dubouloz, N., R. A. Treumann, R. Pottelette, and M. Malingre: 1993, 'Turbulence generated by a gas of electron acoustic solitons'. *J. Geophys. Res.* **98**(A10), 17415–17422, 10.1029/93JA01611.

Dudok de Wit, T., V. V. Krasnosel'skikh, S. D. Bale, M. W. Dunlop, H. Lühr, S. J. Schwartz, and L. J. C. Woolliscroft: 1995, 'Determination of dispersion relations in quasi-stationary plasma turbulence using dual satellite data'. *Geophys. Res. Lett.* **22**, 2653–2656.

Eastwood, J., E. A. Lucek, C. Mazelle, K. Meziane, Y. Narita, J. Pickett, R. Treumann, and P. Décréau: 2005, 'The Foreshock'. *Space Sci. Rev.* **this issue**.

Ergun, R. E., C. W. Carlson, J. P. McFadden, F. S. Mozer, G. T. Delory, W. Peria, C. Chaston, M. Temerin, I. Roth, I. Muschietti, R. Elphic, R. Strangeway, R. Pfaff, C. A. Cattell, D. Klumpar, E. Shelley, W. Peterson, E. Moebius, and L. Kistler: 1998, 'FAST satellite observations of large amplitude solitary structures'. *Geophys Res. Lett.* **25**, 2041–2044.

Franz, J. A., P. M. Kintner, and J. S. Pickett: 1998, 'POLAR observations of coherent electric field structures'. *Geophys. Res. Lett.* **25**, 1277–1280.

Gary, S. P., P. D. Convery, R. E. Denton, S. A. Fuselier, and B. J. Anderson: 1994, 'Proton and helium cyclotron anisotropy instability thresholds in the magnetosheath'. *J. Geophys. Res.* **99**, 5915–5921.

Glassmeier, K.-H., U. Motschmann, M. Dunlop, A. Balogh, M. H. Acuña, C. Carr, G. Musmann, K.-H. Fornaçon, K. Schweda, J. Vogt, E. Georgescu, and S. Buchert: 2001, 'Cluster as a wave telescope - first results from the fluxgate magnetometer'. *Ann. Geophys.* **19**, 1439–1447.

Gleaves, D. G. and D. J. Southwood: 1991, 'Magnetohydrodynamic fluctuations in the earth's magnetosheath at 1500 LT - ISEE 1 and ISEE 2'. *J. Geophys. Res.* **96**, 129–142.

Goldstein, B. E., E. J. Smith, A. Balogh, T. S. Horbury, M. L. Goldstein, and D. A. Roberts: 1995, 'Properties of magnetohydrodynamic turbulence in the solar wind as observed by Ulysses at high heliographic latitudes'. *Geophys. Res. Lett.* **22**, 3393–3396.

Gurnett, D. A., L. A. Frank, and R. P. Lepping: 1976, 'Plasma waves in the distant magnetotail'. *J. Geophys. Res.* **81**, 6059–6071.

Gurnett, D. A., R. L. Huff, and D. L. Kirchner: 1997, 'The Wide-Band Plasma Wave Investigation'. *Space Sci. Rev.* **79**, 195–208.

Gustafsson, G., R. Bostrom, B. Holback, G. Holmgren, A. Lundgren, K. Stasiewicz, L. Ahlen, F. S. Mozer, D. Pankow, P. Harvey, P. Berg, R. Ulrich, A. Pedersen, R. Schmidt, A. Butler, A. W. C. Fransen, D. Klinge, C.-G. Falthammar, P.-A. Lindqvist, S. Christenson, J. Holtet, B. Lybekk, T. A. Sten, P. Tanskanen, K. Lappalainen, and J. Wygant: 1997, 'The Electric Field and Wave Experiment for the Cluster mission'. *Space Sci. Rev.* **79**, 137–156.

Hasegawa, A.: 1969, 'Drift mirror instability in the magnetosphere'. *Phys. Fluids* **12**, 2642–2650.

Hill, P., G. Paschmann, R. A. Treumann, W. Baumjohann, N. Sckopke, and H. Lühr: 1995, 'Plasma and magnetic field behavior across the magnetosheath near local noon'. *J. Geophys. Res.* **100**, 9575–9584.

Horbury, T. S. and A. Balogh: 1997, 'Structure function measurements of the intermittent MHD turbulent cascade'. *Nonlinear Proc. Geophys.* **4**, 185–199.

Horbury, T. S., A. Balogh, R. J. Forsyth, and E. J. Smith: 1995, 'Observation of evolving turbulence in the polar solar wind'. *Geophys. Res. Lett.* **22**, 3401–3404.

Horbury, T. S., E. A. Lucek, A. Balogh, I. Dandouras, and H. Rème: 2004, 'Motion and orientation of magnetic field dips and peaks in the terrestrial magnetosheath'. *J. Geophys. Res.* **109**(A18), doi:10.1026/2003JA0101237.

Kivelson, M. G. and C. T. Russell: 1995, *Introduction to space physics*. Cambridge ; New York : Cambridge University Press, 1995.

Kivelson, M. G. and D. J. Southwood: 1996, 'Mirror instability. II - The mechanism of nonlinearsaturation'. *J. Geophys. Res.* **101**, 17365–17371.

Lacombe, C., G. Belmont, D. Hubert, C. C. Harvey, A. Mangeney, C. T. Russell, J. T. Gosling, and S. A. Fuselier: 1995, 'Density and magnetic field fluctuations observed by ISEE 1-2 in the quiet magnetosheath'. *Ann. Geophys.* **13**, 343–357.

Laveder, D., T. Passot, and P. L. Sulem: 2002, 'Transverse dynamics of dispersive Alfvén waves: Direct numerical evidence of filamentation'. *Phys. Plasmas* **9**, 293–304.

Lee, L. C., C. S. Wu, and C. P. Price: 1987, 'On the generation of magnetosheath lion roars'. *J. Geophys. Res.* **92**, 2343–2348.

Li, X., H. R. Lewis, J. Labelle, T.-D. Phan, and R. A. Treumann: 1995, 'Characteristics of the ion pressure tensor in the Earth's magnetosheath'. *Geophys. Res. Lett.* **22**, 667–670.

Louarn, P., E. Budnik, J. A. Sauvaud, G. Parks, K. Meziane, J. M. Bosqued, I. Dandouras, H. Rème, U. Mall, P. Daly, M. Dunlop, A. Balogh, L. M. Kistler, and E. Amata: 2003, 'Observations of energy-time dispersed ion signatures in the magnetosheath by CLUSTER: possible signatures of transient acceleration processes at the shock'. *Ann. Geophys.* **21**, 1483–1495.

Lucek, E. A., M. W. Dunlop, A. Balogh, P. Cargill, W. Baumjohann, E. Georgescu, G. Haerendel, and K. H. Fornaçon: 1999, 'Identification of magnetosheath mirror modes in Equator-S magnetic field data'. *Ann. Geophys.* **17**, 1560–1573.

Lucek, E. A., M. W. Dunlop, T. S. Horbury, A. Balogh, P. Brown, P. Cargill, C. Carr, K. H. Fornacon, E. Georgescu, and T. Oddy: 2001, 'Cluster magnetic field observations in the magnetosheath: four-point measurements of mirror structures'. *Ann. Geophys.* **19**, 1421–1428.

Maksimovic, M., S. D. Bale, T. S. Horbury, and M. Andre: 2003, 'Bow shock motions observed with Cluster '. *Geophys. Res. Lett.* **30**, 1393, doi:10.1029/2002GL016761.

Maksimovic, M., C. C. Harvey, O. Santolik, C. Lacombe, Y. de Conchy, D. Hubert, P. F., N. Cornilleau-Werhlin, I. Dandouras, E. A. Lucek, and A. Balogh: 2001, 'Polarisation and propagation of Lion Roars in the dusk side Magnetosheath'. *Ann. Geophys.* **19**, 1429–1438.

Mangeney, C., C. Salem, C. Lacombe, J.-L. Bougeret, C. Perche, R. Manning, P.-J. Kellogg, K. Goetz, J. Monson, and J.-M. Bosqued: 1999, 'WIND observations of coherent electrostatic waves in the solar wind'. *Ann. Geophys.* **17**, 307–320.

Marsch, E. and C.-Y. Tu: 1997, 'Intermittency, non-Gaussian statistics and fractal scaling of MHD fluctuations in the solar wind'. *Nonlin. Proc. Geophys.* **4**, 101–124.

Matsumoto, H., H. Kojima, T. Miyatake, Y. Omura, M. Okada, I. Nagano, and M. Tsutsui: 1994, 'Electrostatic solitary waves (ESW) in the magnetotail: BEN wave forms observed by GEOTAIL'. *Geophys. Res. Lett.* **21**, 2915–2918.

Matthaeus, W. H., M. L. Goldstein, and D. A. Roberts: 1990, 'Evidence for the presence of quasi-two-dimensional nearly incompressible fluctuations in the solar wind'. *J. Geophys. Res.* **91**, 20673–20683.

Means, J. D.: 1972, 'Use of the three-dimensional covariance matrix in analyzing the polarization properties of plane waves'. *J. Geophys. Res.* **77**, 5551–5559.

Motschmann, U., K.-H. Glassmeier, and J. L. Pinçon: 1998, 'Plasma mode recognition'. In: G. Paschmann and P. W. Daly (eds.): *Analysis Methods for Multi-Spacecraft Data*, ISSI SR-001. ESA Publications Division, pp. 79–89.

Motschmann, U., T. I. Woodward, K.-H. Glassmeier, D. J. Southwood, and J. L. Pinçon: 1996, 'Wavelength and direction filtering by magnetic measurements at satellite arrays: Generalized minimum variance analysis'. *J. Geophys. Res.* **101**, 4961–4965.

Omidi, N., A. O'Farrell, and D. Krauss Vurban: 1994, 'Sources of magnetosheath waves and turbulence'. *Adv Space Res.* **14**, 745–754.

Pantellini, F. G. E. and S. J. Schwartz: 1995, 'Electron temperature effects in the linear proton mirrorinstability'. *J. Geophys. Res.* **100**, 3539–3549.

Paschmann, G., G. Haerendel, N. Sckopke, E. Möbius, H. Lühr, and C. W. Carlson: 1988, 'Three-dimensional plasma structures with anomalous flow directions near the Earth's bow shock'. *J. Geophys. Res.* **93**, 11279–11294.

Paschmann, G., N. Sckopke, J. R. Asbridge, S. J. Bame, and J. T. Gosling: 1980, 'Energization of solar wind ions by reflection from the earth's bow shock'. *J. Geophys. Res.* **85**, 4689–4693.

Paschmann, G., N. Sckopke, I. Papamastorakis, J. R. Asbridge, S. J. Bame, and J. T. Gosling: 1981, 'Characteristics of reflected and diffuse ions upstream from the Earth's bow shock'. *J. Geophys. Res.* **86**, 4355–4364.

Phan, T. D., D. Larson, J. McFadden, R. P. Lin, C. Carlson, M. Moyer, K. I. Paularena, M. McCarthy, G. K. Parks, H. Rème, T. R. Sanderson, and R. P. Lepping: 1997, 'Low-latitude dusk flank magnetosheath, magnetopause, and boundary layer for low magnetic shear: Wind observations'. *J. Geophys. Res.* **102**, 19883–19896.

Phan, T. D., D. E. Larson, R. P. Lin, J. P. McFadden, K. A. Anderson, C. W. Carlson, R. E. Ergun, S. M. Ashford, M. P. McCarthy, G. K. Parks, H. . Reme, J. M. Bosqued, C. D'Uston, K.-P. Wenzel, T. R. Sanderson, and A. Szabo: 1996, 'The subsolar magnetosheath and magnetopause for high solar wind ram pressure: WIND observations'. *Geophys. Res. Lett.* **23**, 1279.

Phan, T.-D., G. Paschmann, W. Baumjohann, N. Sckopke, and H. Lühr: 1994, 'The magnetosheath region adjacent to the dayside magnetopause: AMPTE/IRM observations'. *J. Geophys. Res.* **99**, 121–141.

Pickett, J. S., L.-J. Chen, S. W. Kahler, O. Santolik, D. A. Gurnett, B. T. Tsurutani, and A. Balogh: 2004a, 'Isolated electrostatic structures observed through the Cluster orbit: Relationship to magnetic field strength'. *Ann. Geophys.* **22**, 2515–2523.

Pickett, J. S., S. W. Kahler, L.-J. Chen, R. L. Huff, O. Santolik, Y. Khotyaintsev, P. M. E. Decreau, D. Winningham, R. Frahm, M. L. Goldstein, G. S. Lakhina, B. T. Tsurutani, B. Lavraud, D. A. Gurnett, A. Andre, A. Fazakerley, A. Balogh, and H. Reme: 2004b, 'Solitary waves observed in the auroral zone: the Cluster multi-spacecraft perspective'. *Nonlinear Proc. Geophys.* **1**, 1–14.

Pickett, J. S., J. D. Menietti, D. A. Gurnett, B. Tsurutani, P. M. Kintner, E. Klatt, and A. Balogh: 2003, 'Solitary potential structures observed in the magnetosheath by the Cluster spacecraft'. *Nonlinear Proc. Geophys.* **11**, 183–196.

Pinçon, J. and F. Lefeuvre: 1991, 'Local characterization of homogeneous turbulence in a space plasma from simultaneous measurement of field components at several points in space'. *J. Geophys. Res.* **96**, 1789–1802.

Pinçon, J. and U. Motschmann: 1998, 'Multi-Spacecraft Filtering: General Framework'. In: G. Paschmann and P. Daly (eds.): *Analysis methods for multi-spacecraft data*, ISSI Sci. Rep. SR-001. Bern: ISSI, pp. 65–78.

Pokhotelov, O. A., M. A. Balikhin, R. A. Treumann, and V. P. Pavlenko: 2001, 'Drift mirror instability revisited: 1. Cold electron temperature limit'. *J. Geophys. Res.* **106**, 8455–8464.

Pokhotelov, O. A. and V. A. Pilipenko: 1976, 'Contribution to the theory of the drift-mirror instability of the magnetospheric plasma'. *Geomagnetizm i Aeronomiia* **16**, 504–510.

Pokhotelov, O. A., R. Z. Sagdeev, M. A. Balikhin, and R. A. Treumann: 2004, 'The mirror instability at finite ion-Larmor radius wavelengths'. *J. Geophys. Res.* **109**, A09213, doi:10.1029/2004JA010568.

Pokhotelov, O. A., I. Sandberg, R. Z. Sagdeev, R. A. Treumann, O. G. Onishchenko, M. A. Balikhin, and V. P. Pavlenko: 2003, 'Slow drift mirror modes in finite electron-temperature plasma: Hydrodynamic and kinetic drift mirror instabilities'. *J. Geophys. Res.* **108**, 1098, doi:10.1029/2002JA009651.

Rème, H., C. Aoustin, J. M. Bosqued, I. Dandouras, B. Lavraud, J. A. Sauvaud, A. Barthe, J. Bouyssou, T. Camus, O. Coeur-Joly, A. Cros, J. Cuvilo, F. Ducay, Y. Garbarowitz, J. L. Medale, E. Penou, H. Perrier, D. Romefort, J. Rouzaud, C. Vallat, D. Alcaydé, C. Jacquey, C. Mazelle, C. Uston, E. Möbius, L. M. Kistler, K. Crocker, M. Granoff, C. Mouikis, M. Popecki, M. Vosbury, B. Klecker, D. Hovestadt, H. Kucharek, E. Kuenneth, G. Paschmann, M. Scholer, N. Sckopke, E. Seidenschwang, C. W. Carlson, D. W. Curtis, C. Ingraham, R. P. Lin, J. P. McFadden, G. K. Parks, T. Phan, V. Formisano, E. Amata, M. B. Bavassano-Cattaneo, P. Baldetti, R. Bruno, G. Chionchio, A. D. Lellis, M. F. Marcucci, G. Pallocchia, A. Korth, P. W. Daly, B. Graeve, H. Rosenbauer, V. Vasyliunas, M. McCarthy, M. Wilber, L. Eliasson, R. Lundin, S. Olsen, E. G. Shelley, S. Fuselier, A. G. Ghielmetti, W. Lennartsson, C. P. Escoubet, H. Balsiger, R. Friedel, J.-B. Cao, R. A. Kovrazhkin, I. Papamastorakis, R. Pellat, J. Scudder, and B. Sonnerup: 2001, 'First multispacecraft ion measurements in and near the Earth's magnetosphere with the identical Cluster Ion Spectrometry (CIS) experiment'. *Ann. Geophys.* **19**, 1303–1354.

Robert, P., A. Roux, C. C. Harvey, M. W. Dunlop, P. W. Daly, and K.-H. Glassmeier: 1998, 'Tetrahedron geometric factors'. In: G. Paschmann and P. W. Daly (eds.): *Analysis Methods for Multi-Spacecraft Data*, ISSI SR-001. ESA Publications Division, pp. 332–348.

Roberts, D. A. and M. L. Goldstein: 1991, 'Turbulence and waves in the solar wind'. *Rev. Geophys., Suppl.* **29**, 932–943.

Rodriguez, P.: 1979, 'Magnetosheath electrostatic turbulence'. *J. Geophys. Res.* **84**, 917–930.

Rodriguez, P.: 1985, 'Magnetosheath whistler turbulence'. *J. Geophys. Res.* **90**, 6337–6342.

Roelof, E. C. and D. C. Sibeck: 1993, 'Magnetopause shape as a bivariate function of interplanetary magnetic field b_z and solar wind dynamic pressure'. *J. Geophys. Res.* **98**, 21421–21450.

Rönnmark, K.: 1983, 'Computation of the dielectric tensor of a Maxwellian plasma'. *Plasma Phys.* **25**, 699–701.

Rudakov, L. I. and R. Z. Sagdeev: 1959, 'A quasi-hydrodynamic description of a rarefied plasma in a magnetic field'. *Plasma Physics and the problem of controlled thermonuclear reactions* **3**, 321–335.

Safránková, J., L. Prech, Z. Nemecek, D. G. Sibeck, and T. Mukai: 2000, 'Magnetosheath response to the interplanetary magnetic field tangential discontinuity'. *J. Geophys. Res.* **105**, 25113–25121.

Sahraoui, F., J. L. Pinçon, G. Belmont, L. Rezeau, N. Cornilleau-Wehrlin, P. Robrt, L. Mellul, J. M. Bosqued, A. Balogh, P. Canu, and G. Chanteur: 2003, 'ULF wave identification in the magnetosheath: The k-filtering technique applied to Cluster II data'. *J. Geophys. Res.* **108**, 1335, doi:10.1029/2002JA009587.

Santolik, O. and D. A. Gurnett: 2003, 'Transverse dimensions of chorus in the source region'. *Geophys. Res. Lett.* **30**, 1031, doi:10.1029/2002GL016178.

Scarf, F. L., L. A. Frank, K. L. Ackerson, and R. P. Lepping: 1974, 'Plasma wave turbulence at distant crossings of the plasma sheet boundaries and the neutral sheet'. *Geophys. Res. Lett.* **1**, 189–192.

Schwartz, S. J., D. Burgess, and J. J. Moses: 1996, 'Low frequency waves in the Earth's magnetosheath: Present status'. *Ann. Geophys.* **14**, 1134–1150.

Schwartz, S. J., R. L. Kessel, C. C. Brown, L. J. C. Woolliscroft, M. W. Dunlop, C. J. Farrugia, and D. S. Hall: 1988, 'Active current sheets near the Earth's bow shock'. *J. Geophys. Res.* **93**, 11295–11310.

Sckopke, N., G. Paschmann, S. J. Bame, J. T. Gosling, and C. T. Russell: 1983, 'Evolution of ion distributions across the nearly perpendicular bow shock - Specularly and non-specularly reflected-gyrating ions'. *J. Geophys. Res.* **88**, 6121–6136.

Sckopke, N., G. Paschmann, A. L. Brinca, C. W. Carlson, and H. Lühr: 1990, 'Ion thermalization in quasi-perpendicular shocks involving reflected ions'. *J. Geophys. Res.* **95**, 6337–6352.

Sibeck, D. G., N. L. Borodkova, S. J. Schwartz, C. J. Owen, R. Kessel, S. Kokubun, R. P. Lepping, R. Lin, K. Liou, H. Lühr, R. W. McEntire, C.-I. Meng, T. Mukai, Z. Nemecek, G. Parks, T. D. Phan, S. A. Romanov, J. Safrankova, J.-A. Sauvaud, H. J. Singer, S. I. Solovyev, A. Szabo, K. Takahashi, D. J. Williams, K. Yumoto, and G. N. Zastenker: 1999, 'Comprehensive study of the magnetospheric response to a hot flow anomaly'. *J. Geophys. Res.* **104**, 4577–4593.

Sibeck, D. G., N. L. Borodkova, G. N. Zastenker, S. A. Romanov, and J.-A. Sauvaud: 1998, 'Gross deformation of the dayside magnetopause'. *Geophys. Res. Lett.* **25**, 453–456.

Smith, E. J., R. E. Holzer, M. G. McLeon, and C. T. Russell: 1967, 'Magnetic noise in the magnetosheath in the frequency range 3-300 Hz'. *J. Geophys. Res.* **72**, 4803–4813.

Smith, E. J. and B. T. Tsurutani: 1976, 'Magnetosheath lion roars'. *J. Geophys. Res.* **81**, 2261–2266.

Song, P., C. T. Russell, and S. P. Gary: 1994, 'Identification of low-frequency fluctuations in the terrestrial magnetosheath'. *J. Geophys. Res.* **99**, 6011–6025.

Song, P., C. T. Russell, and M. F. Thomsen: 1992, 'Waves in the inner magnetosheath: a case study'. *Geophys. Res. Lett.* **19**, 2191–2194.

Stasiewicz, K.: 2004, 'Theory and observations of slow-mode solitons in space plasmas'. *Physical Review Letters* **93**(12), 125004.

Tajiri, M.: 1967, 'Propagation of hydromagnetic waves in collisionless plasma II, Kinetic approach'. *J. Phys. Soc. Jpn.* **22**, 1482–1498.

Treumann, R. A. and W. Baumjohann: 1997, *Advanced space plasma physics*. London : Imperial College Press, 1997.

Treumann, R. A., E. Georgescu, and W. Baumjohann: 2000, 'Lion roar trapping in mirror modes'. *Geophys. Res. Lett.* **27**, 1843–1846.

Treumann, R. A., C. H. Jaroschek, O. D. Constantinescu, R. Nakamura, O. A. Pokhotelov, and E. Georgescu: 2004, 'The strange physics of low frequency mirror mode turbulence in the high temperature plasma of the magnetosheath'. *Nonlin. Proc. Geophys.* **11**, 647–657.

Tsurutani, B. T., E. J. Smith, R. R. Anderson, K. W. Ogilvie, J. D. Scudder, D. N. Baker, and S. J. Bame: 1982, 'Lion roars and nonoscillatory drift mirror waves in the magnetosheth'. *J. Geophys. Res.* **87**, 6060–6072.

Tu, C.-Y. and E. Marsch: 1995, 'MHD structures, waves and turbulence in the solar wind: Obsevations and theories'. *Space Sci. Rev.* **73**, 1–210.

Walker, S. N., F. Sahraoui, M. A. Balikhin, G. Belmont, J. L. Pinçon, L. Rezeau, H. Alleyne, N. Cornilleau-Wehrlin, and M. André: 2004, 'A Comparison of Wave Mode identification Techniques'. *Ann. Geophys.* **22**, 3021–3032.

Zhang, Y., H. Matsumoto, and H. Kojima: 1998, 'Lion roars in the magnetosheath: the Geotail observations'. *J. Geophys. Res.* **103**, 4615–4626.

II

THE BOW SHOCK

Chapter 4

Cluster at the Bow Shock: Introduction

A. Balogh[1], S. J. Schwartz[2 3], S. D. Bale[4], M. A. Balikhin[5], D. Burgess[2], T. S. Horbury[1], V. V. Krasnoselskikh[6], H. Kucharek[7], B. Lembège[8], E. A. Lucek[1], E. Möbius[7 9], M. Scholer[10], M. F. Thomsen[11], and S. N. Walker[5]

The terrestrial bow shock is formed in the solar wind when the supersonic plasma emitted from the Sun encounters the Earth's magnetic field. The dipole magnetic field of the Earth acts, in the first approximation, as an impenetrable barrier to the solar wind which therefore has to slow down and flow around the obstacle. In this process, the magnetopause is formed, separating the magnetic field inside from the solar wind that flows around it. Ahead of the magnetopause, the bow shock forms a surface across which the solar wind plasma is heated and slowed down from supersonic to subsonic speeds. The Earth's bow shock is the best known and most studied example of a collisionless plasma shock and has been the subject of extensive observational and theoretical investigations since the start of the space age (see, e.g., Fairfield, 1976; Tsurutani and Stone, 1985; Burgess, 1995; Russell, 1995, and references therein). Collisionless plasmas make up a large fraction of

[1]Space and Atmospheric Physics, The Blackett Laboratory, Imperial College London, London, UK
[2]Astronomy Unit, Queen Mary, University of London, London, UK
[3]Now at Space and Atmospheric Physics, The Blackett Laboratory, Imperial College London, London, UK
[4]Department of Physics and Space Sciences Laboratory, University of California, Berkeley, CA, USA
[5]Automatic Control and Systems Engineering, University of Sheffield, Sheffield, UK
[6]LPCE/CNRS, Orléans, France
[7]Space Science Center, Institute for the Study of Earth, Oceans, and Space, University of New Hampshire, Durham, New Hampshire, USA
[8]CETP/IPSL, Velizy, France
[9]Also Department of Physics, University of New Hampshire, Durham, New Hampshire, USA
[10]Max-Planck-Institut für extraterrestrische Physik, Garching, Germany
[11]Los Alamos National Laboratory, Los Alamos, NM, USA

Space Science Reviews 118: 155–160, 2005.
DOI: 10.1007/s11214-005-3826-1

the astrophysical world. Shocks are believed to play critical roles in flow dynamics and heating under a wide variety of circumstances as well as providing prime acceleration environments for cosmic rays.

The overall, schematic view of the average location and shape of the bow shock is shown in Figure 4.1 (see, e.g., Formisano, 1979; Peredo et al., 1995). A typical distance from the Earth to the subsolar point of the bow shock is $\sim 14R_e$, but the location of the bow shock is highly variable, dependent on the speed and density of the solar wind. In general terms, the large scale geometry of the bow shock depends on the solar wind pressure. As for all collisionless plasma shock waves, the nature of the shock transition from supersonic to subsonic flow depends, primarily, on two parameters. One is the Mach number of the shock wave, the ratio of upstream velocity to the characteristic wave speed (e.g., Alfvén or magnetosonic); for the terrestrial bow shock this is usually in the range from ~ 3 up to 10. The second is the angle θ_{Bn} between the upstream magnetic field direction and the normal direction to the shock surface. The physics within the transition is also influenced by the upstream plasma β (the ratio of the thermal pressure to magnetic pressure).

The manner in which the dominant plasma heating and dissipation that occur at the shock transition depends on θ_{Bn}. Figure 4.1 shows the average direction of the interplanetary magnetic field lines upstream of the bow shock. Across the surface of the bow shock, θ_{Bn} ranges from close to 90° to close to 0° (or from quasi-perpendicular to quasi-parallel). In the quasi-perpendicular case, the shock transition tends to be abrupt in time (in the frame of the solar wind) and spatially well-defined, although the detailed physics within the shock layer involves multi-scale, time-dependent phenomena. In the quasi-parallel case, the transition occurs over an extended region which contains inhomogeneous and transient field and shock-related particle features.

The average direction of the interplanetary magnetic field shown in Figure 4.1 is in fact a gross oversimplification of the conditions actually observed upstream of the bow shock. Both the direction and strength of the IMF, together with other plasma parameters, are highly variable on a range of spatial scales relative to the dimensions of the bow shock. Accordingly, the geometry (quasi-parallel vs. quasi-perpendicular) and Mach number of individual relatively rapid shock transitions are controlled by the prevailing solar wind conditions.

Studies of the bow shock using single- and dual-spacecraft observations are too numerous to quote here individually (see, e.g., Thomsen, 1988). Equally, many theoretical and numerical modelling investigations have addressed the different aspects of the bow shock formation, structure, parametric dependence and dynamics (selected recent advances can be found in Lembège et al., 2004, and references therein).

Such single or two-point studies have intrinsic limitations. The bow shock is in constant motion as it moves in and out in response to changes in the solar wind ram pressure at speeds of ~ 5 to over 100 km s^{-1} (Lepidi et al., 1996). Its struc-

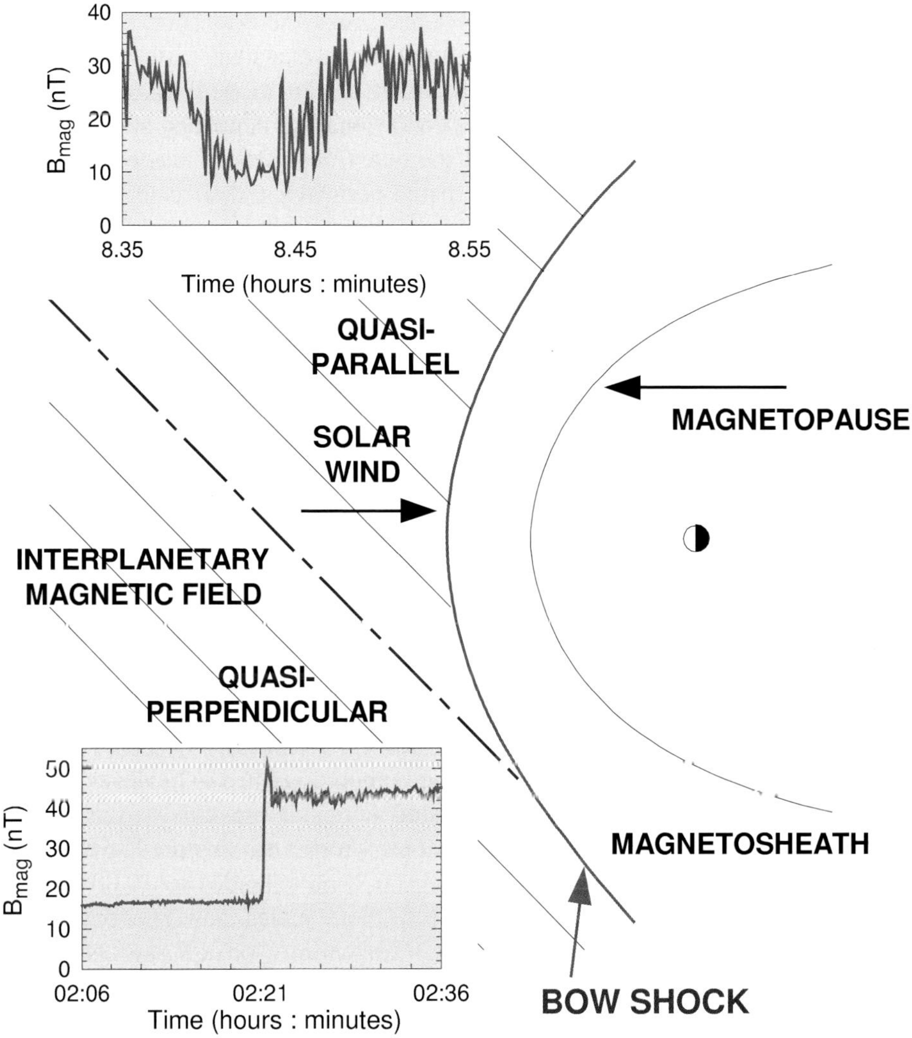

Figure 4.1. Sketch of the Earth's bow shock ahead of the magnetosheath and magnetosphere. The angle between the direction of the interplanetary magnetic field lines and the normal to the shock surface ranges (for the average direction of the IMF shown here) from quasi-parallel on the dawn side to quasi-perpendicular on the dusk side of the bow shock. The scales of the shock transition and dissipation regions are significantly different for the quasi-perpendicular and quasi-parallel cases as illustrated by the insets showing the evolution of the magnetic field magnitude measured by Cluster across two shock transitions. Figure provided by A. Balogh.

ture is also variable, mainly in response to local changes in θ_{Bn} and to magnetic and plasma structures swept into the shock by the solar wind. Relatively small directional changes in the IMF, as it is swept against the bow shock surface, may alter the physical parameters that would be observed at a small distance away from the actual crossing. Single- and dual spacecraft observations have clarified many properties of the bow shock under most conditions and described many details of its phenomenology. However, many of the quantities that are needed to describe the bow shock processes are related to spatial derivatives, such as the geometry of the shock surface, associated wave fields and the shock's velocity. The determination of such quantities using single spacecraft measurements requires supporting assumptions that may be approximately suitable or even questionable. Statistical studies of many observations have, however, alleviated some of the inherent shortcomings of single spacecraft measurements (Peredo et al., 1995).

The four-spacecraft measurements of Cluster have been able to contribute to many of the topics that are related to the physics of the bow shock. First, by making the first detailed, three-dimensional studies of individual shock crossings, the phenomenology and physical processes within and in the vicinity of the bow shock, under specific conditions, could be clarified. Second, through the ability to make unambiguous determinations of the vector quantities associated with the shock it has been possible to underpin and re-examine the statistical studies of shock motion, and both local and overall shock orientation.

The Cluster data set, now extending over four years from late 2000, has proved to be very rich. Single- or even dual-spacecraft observations and studies have not proved to be a fully adequate preparation for the complexity observed at all the separation scales near the bow shock. New methodology had to be developed (e.g. Paschmann and Daly (eds.), 1998) and tested alongside previously used methods, and new ideas confronted by the observations. Cluster has explored spatial scales from $\sim$ 100 km to 5000 km and this range will be extended to 10,000 km and beyond before the end of the mission. At all these scales, new phenomena were observed at shock crossings with potential implications for new aspects of shock physics to be studied in the future.

In a number of topics, the contribution by Cluster is already significant. These topics are extensively discussed in the following chapters.

- **Definitive studies of absolute shock scales: variations with shock parameters**
  Cluster studies have measured the width of the ramp at the quasi-perpendicular bow shock over a range of upstream parameters (Mach number, etc.). The width is a critical indicator of the internal shock processes which in turn govern the partition of energy amongst the incident particle populations.

- **Temporal/spatial variability: motion and internal dynamics**
  Cluster determination of the speed of the bow shock has shown that variations

in the upstream parameters have an immediate and direct impact on the location and gross motion of the shock. However, Cluster electric and magnetic field observations have also highlighted considerable variability in the shock structure and profile even over relatively small scales associated with particle kinetic behaviour.

- **Proof that ion beams manage to emerge from particles reflected at quasi-perpendicular shocks**
  Simultaneous Cluster ion observations at several locations have provided unambiguous evidence that field aligned beams found upstream of the quasi-perpendicular bow shock emerge out of the reflected and partially scattered population at the shock itself rather than originating deeper in the magnetosheath.

- **Surprisingly small-scale structure within large (discrete?) entities at quasi-parallel shocks**
  Cluster measurements of large-amplitude magnetic structures which are believed to be an integral part of collisionless shocks under quasi-parallel conditions have revealed the surprising result that they appear quite different even at scales 10% of their overall size. Moreover, these differences are not the same in the electric and magnetic components. Thus the previously-believed monoliths are in fact quite filamentary.

- **Determination of spatial gradients of diffuse energetic ions, and hence a definitive measure of the scattering mean free path**
  Spatial gradients of diffuse ions in the foreshock have provided a direct measure of the scattering mean free path in the self-consistent local turbulence. Thus the first order Fermi acceleration of particles in these regions can be quantified.

Acknowledgements

The contributions of HK and EM were partially supported under NASA Grants NAG5-10131 and NAG5-11804. PPARC (UK) support for this work includes fellowships (TSH and EL) and research grants (DB, SJS, SNW).

References

Burgess, D.: 1995, 'Collisionless Shocks'. In: *Introduction to Space Physics, M. G. Kivelson and C. T. Russell (eds).* Cambridge: Cambridge University Press, Chapt. 5, pp. 129–163.

Fairfield, D. H.: 1976, 'A summary of observations of the Earth's bow shock'. In: *Physics of Solar Planetary Environments*. pp. 511–525.

Formisano, V.: 1979, 'Orientation and shape of the Earth's bow shock in three dimensions'. *Planet. Space Sci.* **27**, 1151–1161.

Lembège, B., J. Giacalone, M. Scholer, T. Hada, M. Hoshino, V. Krasnoselskikh, H. Kucharek, P. Savoini, and T. Terasawa: 2004, 'Selected problems in collisionless-shock physics'. *Space Science Reviews* **110**, 161–226.

Lepidi, S., U. Villante, A. J. Lazarus, A. Szabo, and K. Paularena: 1996, 'Observations of bow shock motion during times of variable solar wind conditions'. *J. Geophys. Res.* **101**(.10), 11107–11124.

Paschmann, G. and P. W. Daly (eds.): 1998, *Analysis methods for multi-spacecraft data*, ISSI Sci. Rep. SR-001. Bern: ISSI.

Peredo, M., J. A. Slavin, E. Mazur, and S. A. Curtis: 1995, 'Three-dimensional position and shape of the bow shock and their variation with Alfvenic, sonic and magnetosonic Mach numbers and interplanetary magnetic field orientation'. *J. Geophys. Res.* **100**(9), 7907–7916.

Russell, C. T. (ed.): 1995, 'Physics of collisionless shocks: Proceedings of the Symposium of COSPAR Scientific Commission D', Vol. D2.1. Pergamon Press.

Thomsen, M. F.: 1988, 'Multi-spacecraft observations of collisionless shocks'. *Advances in Space Research* **8**, 157–166.

Tsurutani, B. T. and R. G. Stone: 1985, 'Collisionless shocks in the heliosphere: Reviews of current research'. *Washington DC American Geophysical Union Geophysical Monograph Series* **35**.

Chapter 5

Quasi-perpendicular Shock Structure and Processes

S. D. Bale[1], M. A. Balikhin[2], T. S. Horbury[3], V. V. Krasnoselskikh[4], H. Kucharek[5], E. Möbius[5 6], S. N. Walker[2], A. Balogh[3], D. Burgess[7], B. Lembège[8], E. A. Lucek[3], M. Scholer[9], S. J. Schwartz[7 10], and M. F. Thomsen[11]

5.1 Introduction

In the two decades prior to the launch of Cluster, collisionless shocks at which the magnetic field in the unshocked plasma is nearly perpendicular to the shock normal ('quasi-perpendicular shocks') received considerable attention. This is due, in part, to their relatively clean, laminar appearance in the time series data. The tendency of the magnetic field to bind particles together owing to their (perpendicular) gyromotion gives rise to this appearance, which facilitated deeper studies into the collisionless processes responsible for the overall thermalization of the principle plasma populations as well as the acceleration of an energetic non-thermal component. Despite the considerable effort, key questions remained unanswered or re-

[1]Department of Physics and Space Sciences Laboratory, University of California, Berkeley, CA, USA
[2]Automatic Control and Systems Engineering, University of Sheffield, Sheffield, UK
[3]Space and Atmospheric Physics, The Blackett Laboratory, Imperial College London, London, UK
[4]LPCE/CNRS, Orléans, France
[5]Space Science Center, Institute for the Study of Earth, Oceans, and Space, University of New Hampshire, Durham, New Hampshire, USA
[6]Also Department of Physics, University of New Hampshire, Durham, New Hampshire, USA
[7]Astronomy Unit, Queen Mary, University of London, London, UK
[8]CETP/IPSL, Velizy, France
[9]Max-Planck-Institut für extraterrestrische Physik, Garching, Germany
[10]Now at Space and Atmospheric Physics, The Blackett Laboratory, Imperial College London, London, UK
[11]Los Alamos National Laboratory, Los Alamos, NM, USA

Space Science Reviews 118: 161–203, 2005.
DOI: 10.1007/s11214-005-3827-0

mained open to interpretation. Single, and at best dual, spacecraft studies were unable to place quantitative limits on the important spatial scales, nor assess the role of non-stationary aspects in the overall shock transition.

By taking advantage of the sharp, quasi-perpendicular shock transitions, Cluster investigations have been able to address the shock orientation and motion via now-standard four spacecraft techniques. As a consequence, Cluster has been able to probe the internal shock scales (and hence physics). Additionally, the multi-spacecraft strategy has enabled definitive studies of where energetic particles do, and don't, come from. This Chapter summarises many of these achievements.

5.2 Structure and Thermalization

5.2.1 Bow shock orientation and global structure

Knowledge of the basic parameters of a shock, such as Mach number and angle θ_{Bn} between the shock normal and (unshocked) magnetic field, is essential for a quantitative analysis of shock dynamics. However, such parameters are often difficult to determine in practice since they require both accurate measurement of plasma and field values around the shock, as well as estimates of characteristics of the shock itself. Most obvious among the latter are the shock orientation and speed. These parameters are difficult to estimate with a single spacecraft, although various techniques such as coplanarity (e.g., Schwartz, 1998) can be used to estimate the shock orientation. New methods of determining shock orientation and speed are therefore of interest.

The measurement of the same shock transition by the four Cluster spacecraft in close succession allows us to estimate the orientation and velocity of the structure in several ways which have not previously been possible. Each of these methods requires assumptions to be made about the properties of the shock, and each has advantages and disadvantages in different situations. A number of such methods have been applied to Cluster crossings of the quasi-perpendicular bow shock, as we report below.

5.2.1.1 Comparison of methods to determine shock orientation

Horbury et al. (2002) used four spacecraft timings of magnetic field data to estimate the orientation of the quasi-perpendicular bow shock at a number of crossings and compare these results with magnetic coplanarity estimates as well as orientations predicted by parametric models of the large scale bow shock shape.

If we assume the shock to be planar on the scale of the spacecraft separations, and to be travelling at a constant speed as it passes over the four spacecraft, then the times at which it crosses them can be used to estimate the normal of the shock plane and the speed at which it is travelling along that normal (see Dunlop and Woodward, 1998, for more details). These assumptions will not always be satisfied. Non-planarity of the shock can be caused by large scale curvature or by rippling and waves around the shock. Horbury et al. (2002) considered shock crossings in late 2000 and early 2001, when the Cluster spacecraft were typically around 600 km apart, much smaller than the scale of the bow shock curvature around the magnetopause. They also only considered fairly sharp, clean crossings without extensive wave activity, which restricted them to quasi-perpendicular shocks. They rejected any shock with significantly different magnetic field profiles at different spacecraft, or any nearby changes in upstream conditions. For example, acceleration of the shock (discussed in detail in section 5.2.4.4) can often be identified as a different width of the profile at different spacecraft. Horbury et al. (2002) identified such cases, but did not analyse them.

For a reliable estimate of the shock orientation using timings, the error on the relative times at which the shock crossed the four spacecraft should be small. In practice, the finite shock width, combined with the presence of waves around the shock, can result in uncertainty in the shock time of around a second. The need to minimise this uncertainty for the timing method makes relatively sharp quasi-perpendicular crossings much easier to analyse than more oblique, structured shocks. When the Cluster spacecraft are around 100 km apart, the shock can travel between them in one or two seconds, leading to a large fractional error in the relative timings. If the spacecraft are a few thousand km apart, the shock has often changed speed or profile in the tens of seconds it takes to travel between them. In early 2001, however, the spacecraft were typically around 600 km apart at apogee in the solar wind, corresponding to shock crossings around ten seconds apart between spacecraft, resulting in small fractional errors in relative timings. However, with the assumptions used in this method (constant motion, planarity, accurate timings) it is important to test its accuracy before using it routinely.

Horbury et al. (2002) considered 48 quasi-perpendicular shock crossings in 2001, and estimated the speed and orientation of each, using inter-spacecraft timings (see Figure 5.1). They compared the timing-based orientation estimates with those from two models of the bow shock shape (Peredo et al., 1995; Formisano, 1979). They found that the timing-based estimates of the shock orientation agreed very well with the models, with nearly 80% being less than 10° apart (Figure 5.2, left panel). This result implies that both the timing method and the models are usually good estimators of the shock orientation, to around 10° accuracy. Indeed, the accuracy of the timing-based estimates can be seen qualitatively from the consistency of the normals shown in Figure 5.1. The consequences of the stability of the bow shock surface for its large scale structure are discussed in section 5.2.1.2.

Horbury et al. (2002) also compared timing-based bow shock normal estimates with magnetic field coplanarity, and found large discrepancies (Figure 5.2, middle panel). Coplanarity is known to be a poor estimator of the shock orientation for nearly perpendicular shocks (those where $\theta_{Bn} \approx 90°$; see, e.g Lepidi et al., 1997). This is a consequence of the up and downstream magnetic field vectors being nearly parallel for nearly perpendicular shocks. The large uncertainty in coplanarity nor-

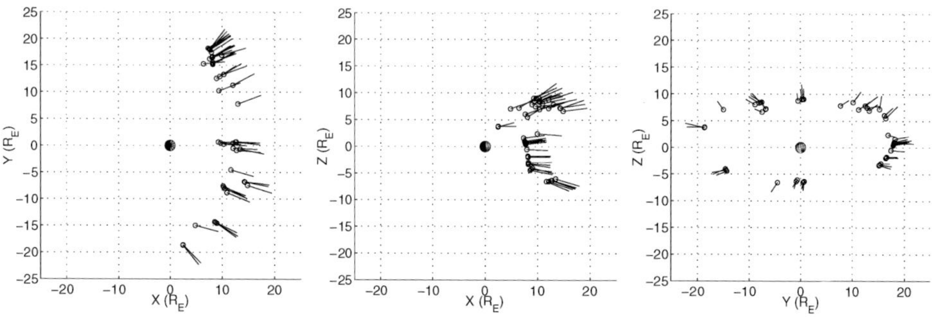

Figure 5.1. Bow shock normals (short lines) deduced from four spacecraft timings, plotted extending from their measured locations (circles). The three panels show the projection of the positions and normals onto the X-Y, X-Z and Y-Z GSE planes. Note that shocks near the nose are not sampled due to the polar Cluster orbit. From Horbury et al. (2002).

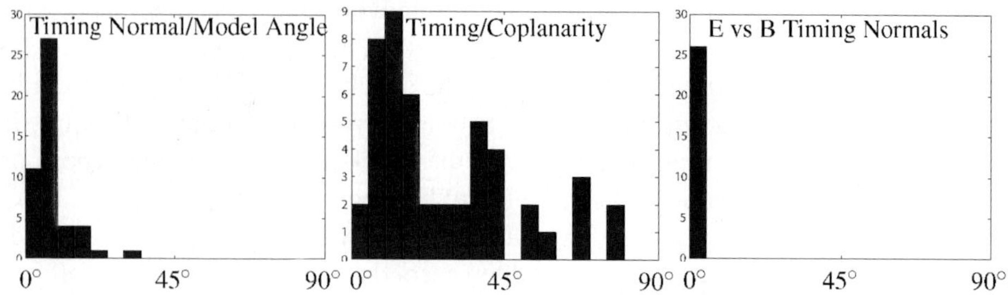

Figure 5.2. Histograms of angular difference between different estimates of quasi-perpendicular bow shock normals. Left: comparison of Cluster four-spacecraft timing normals with normals from a parameterised bow shock model (Peredo et al., 1995). The agreement is good, implying both are typically reliable estimators of the normal. Middle: comparison of timing normals with magnetic field coplanarity. Right: comparison of timing-based normal estimates using magnetic field and spacecraft potential to calculate timings. Left and middle panels from Horbury et al. (2002). Right panel provided by T. S. Horbury and S. D. Bale.

mals for $\theta_{Bn} \approx 90^\circ$, is apparent in Figure 5.3, which shows the deviation of coplanarity vectors from timing-based normal estimates for the shocks considered by Horbury et al. (2002): when $\theta_{Bn} \approx 90^\circ$, the scatter is very large. However, Horbury et al. found that deviations were still large (on average $22^\circ \pm 4^\circ$) for shocks with $\theta_{Bn} < 90^\circ$. This implies that coplanarity estimates of shock orientation can have significant errors even for moderate θ_{Bn} and they must therefore be treated with caution when using single spacecraft data. This is an example of how Cluster multi-spacecraft analysis can help us to interpret other, single spacecraft, data sets.

Horbury et al. (2002) used magnetic field profiles to estimate the shock crossing time at the Cluster spacecraft. However, other parameters can be used: for example, Maksimovic et al. (2003) and Bale et al. (2003) used the spacecraft potential, a

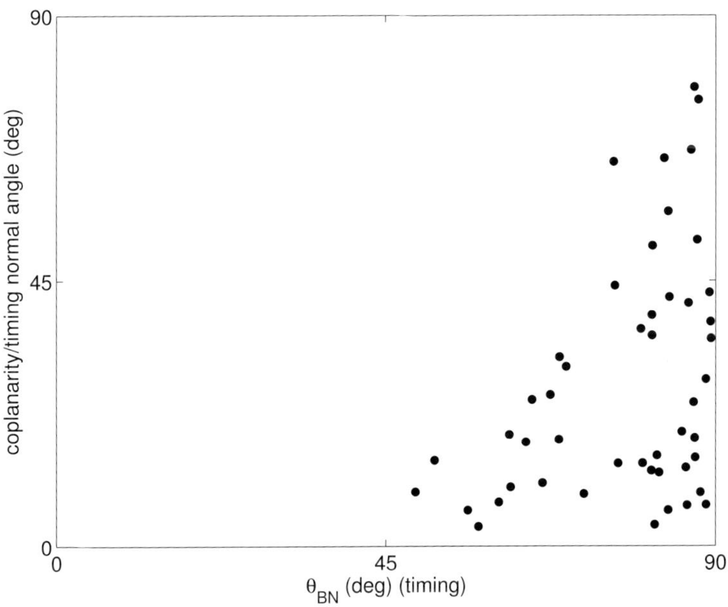

Figure 5.3. Angle between coplanarity and four spacecraft timing estimates of bow shock normals, as a function of θ_{Bn}. Coplanarity is an unreliable estimator of bow shock orientation for $\theta_{Bn} \gtrsim 70°$ but is still only accurate to around 20° for $\theta_{Bn} \lesssim 70°$. From Horbury et al. (2002).

proxy of the local plasma density. It is therefore of interest to compare four spacecraft timing estimates of the shock orientation using these different estimators. We have compared the orientations deduced from EFW (spacecraft potential) and FGM (magnetic field) timings for 26 quasi-perpendicular shocks in 2001 that could be identified cleanly in both EFW and FGM data at all four spacecraft, a subset of those used by Horbury et al. (2002). The agreement between the resulting shock normals is remarkable (Figure 5.2, right panel), with the mean angular deviation being 1.8° and the largest deviation being only 3.9°. The mean absolute difference in the deduced velocity was 2 km/s. While this comparison cannot tell us about the reliability of some of the assumptions (such as constant motion and planarity) of the timing method for quasi-perpendicular shocks, it confirms that it is not sensitive to the physical parameter used.

5.2.1.2 Large scale structure of the bow shock

As discussed above in section 5.2.1.1, Horbury et al. (2002) found close agreement between bow shock normal estimates based on four spacecraft timings and those of bow shock models such as that of Peredo et al. (1995). This implies that the bow shock stays close to the parabolic shape of the Peredo et al. model, at least under steady solar wind conditions. It also places an upper limit on the size of any large scale 'ripples' on this surface: if they were present to a significant degree, the local normals deduced from the Cluster measurements would not agree with the normal derived from the model normal.

Bow shock models have been derived in the past by estimating the parameters of a conic section from thousands of single spacecraft shock crossing locations, parameterised by upstream conditions such as ram pressure and magnetic field direction. It is therefore remarkable that these models of the large scale shape agree so closely with the local normal estimates from Cluster found by Horbury et al. (2002).

5.2.2 Large- and meso-scale shock structure

5.2.2.1 Dissipation at quasi-perpendicular shocks

Fast-mode collisionless shocks grow from magnetosonic waves when the incoming flow exceeds the fast magnetosonic speed; the wave steepens and eventually becomes a standing shock in the plasma. To stand as a steady-state shock, the plasma must dissipate small-scale structure to slow the steepening and prevent the shock from overturning; furthermore, the Rankine-Hugoniot (shock jump) relations tell us that the shock must convert incoming flow energy to electron and ion heating and magnetic field energy downstream. Early theories of energy dissipation at shocks sought a single mechanism to directly provide both small-scale dissipation *and* plasma heating (viz., Papadopoulos, 1985). Ion and electron heating mechanisms appear to be mostly unrelated.

At quasi-perpendicular collisionless shocks above a critical Mach number (Kennel et al., 1985), a significant fraction of incident ions are reflected within the shock transition. They gyrate in the upstream region, where the magnetic field is slightly increased to form a 'foot' before returning to the steep ramp region. Having gained energy due to the solar wind $\mathbf{v} \times \mathbf{B}$ electric field, they traverse the ramp and become temporarily trapped in the adjacent overshoot region (Paschmann et al., 1982; Sckopke et al., 1983; Sckopke et al., 1990). The ions ultimately convect further downstream, leading to a magnetic undershoot and series of decreasing oscillations accompanied by ion mixing and thermalisation.

The spatial scales over which the shock dissipates energy, and slows the incoming flow, are thought to be related to the nature of the dissipation mechanism itself. Hence, knowing these scales and their dependence on macroscopic plasma parameters is tantamount to knowing the dissipation physics at the shock. Quasi-perpendicular ($Q_\perp$) shocks have been traditionally targeted for dissipation scale studies, including aspects such as:

1. the role/interpretation of competing dissipation mechanisms within more classical frameworks (anomalous resistivity, viscosity, Hall physics)

2. differing scales for the transition of different bulk parameters (magnetic field, density, velocity)

3. Ohm's law, including contributions from electron inertia and departures from isotropy

4. the role of stationary (DC) fields in the dissipation processes (electron kinetics, ion reflection)

5. the role of non-stationary fields in scattering and shaping the particle distributions at, and downstream of, the main shock transition.

6. the competition between dissipation and dispersion in effecting and limiting the steepening of the shock profile.

5.2.2.2 *Shock ramp scales*

The shock ramp is the region of steepest spatial gradients. The steepening is limited and balanced by dispersion and/or dissipation. The nature of the dissipation differs according to the strength of the shock, i.e. low or high value of the Mach number (Alfvénic M_A or magnetosonic M_{ms}). Resistive dissipation alone is enough at low Mach number, while an additional dissipation (e.g., viscosity) is required at high Mach number. Low and high M_A (or M_{ms}) correspond to subcritical and supercritical Mach regime defined below and above a certain threshold (Tsurutani and Stone, 1985). This balance will define the width of the shock front and in particular the ramp width.

Theoretical and kinetic simulations (Leroy et al., 1982) suggested, together with previous observations, that the magnetic ramp occurs on either an ion inertial scale (c/ω_{pi}) or the gyro-radius of an ion moving at the upstream flow velocity in the downstream magnetic field. While the plasma density tends to follow the magnetic field (Scudder et al., 1986), the electric field shows fine scale features discussed in more detail in Section 5.2.3.

As a multi-spacecraft mission, Cluster was designed precisely to measure spatial scales in the magnetosphere. Typical Cluster spacecraft separations are 100-1000 km which correspond to crossing times of 1-100 s for boundary (shock) speeds of 10-100 km s^{-1}. Hence, sample speeds of 1-10 samples per second are sufficient to sample the shock transition and find a spatial transformation by the techniques discussed above and elsewhere (e.g., Paschmann and Daly (eds.), 1998).

Bale et al. (2003) used the Cluster EFW spacecraft potential as a proxy for electron density to study the ramp transition scale at approximately 100 Q⊥ bow shock crossings. A shock speed (and normal) was found using the timing technique and then each shock profile was fitted with a hyperbolic tangent function $n(x) = n_0 + n_1 \tanh(x/\chi)$. Figure 5.4 shows an example fit at a $M_{ms} \approx 3.5$, $\theta_{Bn} \approx 81°$ shock.

A characteristic scale size for the shock ramp was then given to be $L = n/|dn/dx|$ evaluated at the middle of the ramp and this was expressed in terms of the fit coef-

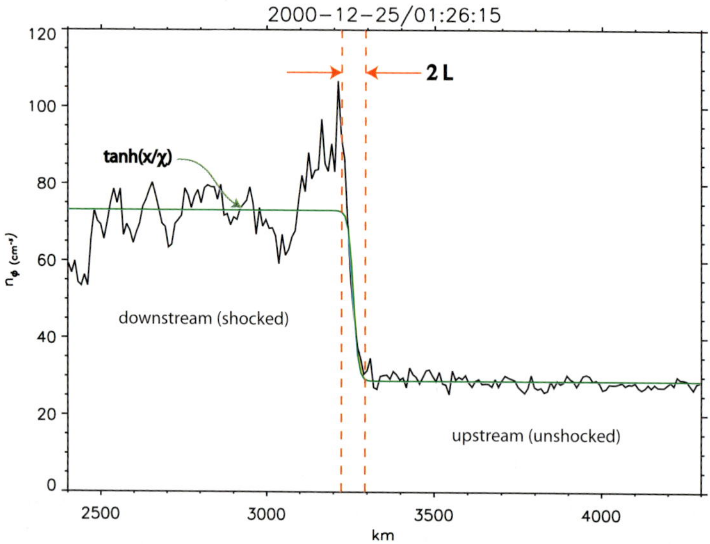

Figure 5.4. Density transition from downstream (shocked) to upstream (unshocked) states for a Mach $M_{ms} \approx 3.5$, $\theta_{Bn} \approx 81°$ shock. The green line is the hyperbolic tangent fit; red vertical lines show the density transition scale. From Bale et al. (2003).

ficients $L = n_0/n_1\ \chi$. Bale et al. (2003) then showed that statistically the measured ramp scale size was proportional to $v_{sh}/\Omega_{ci,2}$, the gyroradius of trapped ions, over a large range of Mach numbers. When compared with the ion inertial scale, $L/c/\omega_{pi}$ is seen to increase monotonically. This is the expected behaviour if the true shock ramp scale is like $v_{sh}/\Omega_{ci,2}$, since $v_{sh}/\Omega_{ci,2}/c/\omega_{pi} \propto M_A$. Figure 5.5 shows these trends.

Similar, supporting results (Horbury, 2004, unpublished) have been obtained using magnetic field data. Taken together, these scalings strongly suggest that the density, magnetic field, and velocity transition scales of the quasi-perpendicular shock are proportional to the gyroradius of the trapped ion population. At low Mach numbers, the two scales $v_{sh}/\Omega_{ci,2}$ and c/ω_{pi} are of similar magnitude and some ambiguity remains. However, the implication of this result is that dissipation at $Q_\perp$ shocks is related to the motion of the trapped ions. In a fluid sense, this corresponds to a viscosity term in Ohm's law associated with gradients in the ion pressure tensor as discussed above.

5.2.2.3 Overshoot/Undershoot structure

It is well known that supercritical shocks exhibit overshoot and undershoot behaviour of the magnetic field just downstream of the shock (Heppner et al., 1967; Russell and Greenstadt, 1979). Since this structure is only observed at supercriti-

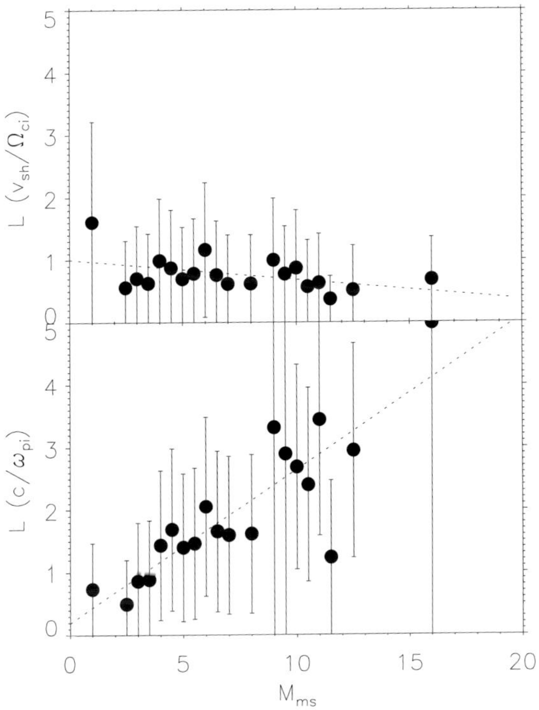

Figure 5.5. Relationship between scale size and magnetosonic Mach number. $L/(v_{sh,n}/\Omega_{ci,2})$ (upper panel) is approximately constant over a large range of Mach number, while the ion inertial scaling (lower panel) increases with Mach number. From Bale et al. (2003).

cal shocks, it was suggested (Morse, 1976) and confirmed by computer simulation (Leroy et al., 1982) that the overshoot structure is associated with a reflected and heated ion population. It is also known that overshoot phenomena play a role in ion acceleration (Giacalone et al., 1991) and electron heating (Gedalin and Griv, 1999). ISEE-1 and -2 measured magnetic overshoot thicknesses using two-point timing to obtain shock speed (Livesey et al., 1982) and found that the observed thickness was ordered by the downstream ion gyroradius. However, Cluster has made the first density measurements of well-defined overshoot-undershoot structure at $Q_\perp$ shocks (Saxena et al., 2004). Using Cluster EFW spacecraft potential as a density proxy, 56 $Q_\perp$ shocks have been analysed using techniques similar to those of Bale et al. (2003).

Figure 5.6 shows typical overshoot/undershoot structure at the same $M_{ms} \approx 3.5$, $Q_\perp$ shock as Figure 5.4. Subtracting the fitted hyperbolic tangent (top panel) leaves a clear 'chirp' signature associated with the overshoot/undershoot (middle panel).

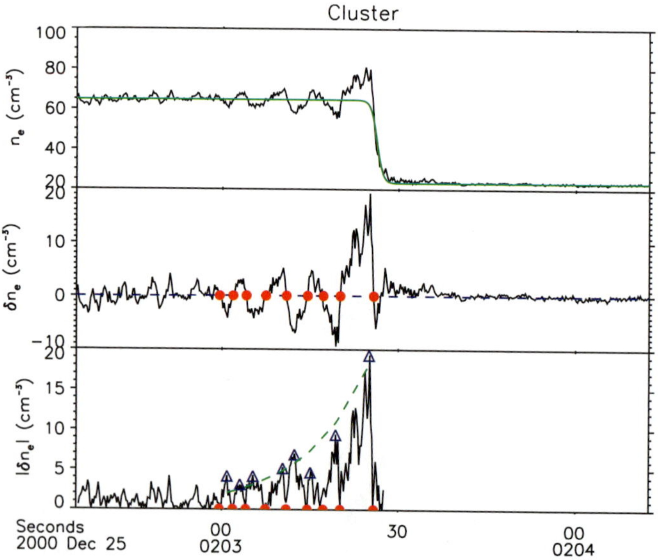

Figure 5.6. Overshoot/undershoot structure at the shock of Figure 5.4 The hyperbolic tangent trend is removed to show a 'chirp' (middle panel); red dots show the location of zero-crossings that are used to measure the overshoot wavelength. The maximum density perturbation between zero-crossings decays spatially (lower panel); blue diamonds show the maxima which are fitted to an exponential to retrieve a decay scale. From Saxena et al. (2004).

Then a zero-crossing algorithm is applied to the chirp (red dots, middle panel) to produce an estimated wavelength for the shock overshoot. Finally, the overshoot amplitude is seen to decay systematically (bottom panel). An exponential function is fitted to the maxima of $|\delta n|$ between each pair of zero-crossings (blue diamonds) to estimate a decay scale λ. *Both* the overshoot/undershoot wavelength and decay scale are found to be organised by the gyroradius of trapped ions, $v_{sh}/\Omega_{ci,2}$, (rather than the ion inertial length). The measured wavelength is consistent with ISEE magnetic observations, while the measurement of an overshoot exponential decay scale is a new result for Cluster.

5.2.3 Fine-scale features in the electric field

Within a collisionless shock front, energy transfer is achieved through the interaction between electric/magnetic waves and particles rather than the normal collisional processes that occur within common hydrodynamic shocks. The spatial scales over which these particles and fields can interact is important when trying to ascertain the energy transfer processes that may occur within the shock front. The determination of magnetic field structure and the spatial scales over which the field varies in the foot, ramp, and overshoot/undershoot regions has been intensively studied since shocks were first observed in the 1960's.

Typically, the foot width is of the order of $0.68V_{sw}/\Omega_{ci}$ where V_{sw} is the solar wind velocity and Ω_{ci} is the upstream ion gyrofrequency (Sckopke et al., 1983; Livesey et al., 1984). The ramp scale has been estimated to be less than an ion inertial length (e.g., see Balikhin et al., 1995, and references therein) with reports of one or two shocks whose ramp scale was of the order $0.1c/\omega_{pi}$ (Newbury and Russell, 1996; Walker et al., 1999). Figure 5.5 shows, however, that at larger Mach numbers the shock ramp is typically larger than an ion inertial length.

Reports of observations of the electric field, on the other hand, are very sparse. This was probably due to the lack of high quality, high time resolution measurements. Based on initial results from ISEE, Heppner et al. (1978) reported that short duration spike-like features were occasionally observed in the electric field as the satellite crossed the shock front. Further investigations of subcritical oblique shocks by Wygant et al. (1987) showed spike-like features with amplitudes up to 100 mV m^{-1} and a strong component along the shock normal. The observations were, however, not good enough to determine the free energy source, mode or scale size of these structures. They speculated that these waves may be either lower-hybrid or possibly Doppler shifted ion-acoustic waves. Based on spin averaged electric field measurements from ISEE, Formisano (1982) reported that the increase in the electric field observed at quasi-perpendicular shocks began just upstream of the magnetic ramp and lasted longer than the ramp crossing itself.

One key aspect to determine is the spatial scale over which changes in the electric field occur and its relation to the scale size over which changes in the magnetic field occur. Several differing points of view have been published. The first (Eselevich et al., 1971; Balikhin et al., 1993; Formisano and Torbert, 1982; Formisano, 1982, 1985; Balikhin et al., 2002; Krasnosel'skikh, 1985; Leroy et al., 1982; Liewer et al., 1991; Scholer et al., 2003) is that the spatial scales of the potential and magnetic field in the ramp region are similar whilst Scudder (1995) proposed the potential scale length is larger than that of the magnetic scale length. Others have suggested that the potential varies predominantly within iso-magnetic jumps, i.e., on a smaller scale than the magnetic field. In laboratory plasmas, such a short scale of the cross-shock electrostatic potential ('isomagnetic jump') was observed by Eselevich (1982). This isomagnetic jump is often attributed to the ion sound subshock (see the review by Kennel et al., 1985).

Using data generated from numerical simulations, Lembège et al. (1999) analysed simultaneous measurements of the scale size of both the magnetic ramp region and the region in which the change in potential was observed. Their results showed that the scale lengths were of the same order. This view is also supported by the simulations of Scholer et al. (2003). The latter authors show that the main potential drop can occur over several ion scales in the foot region, while the steepened

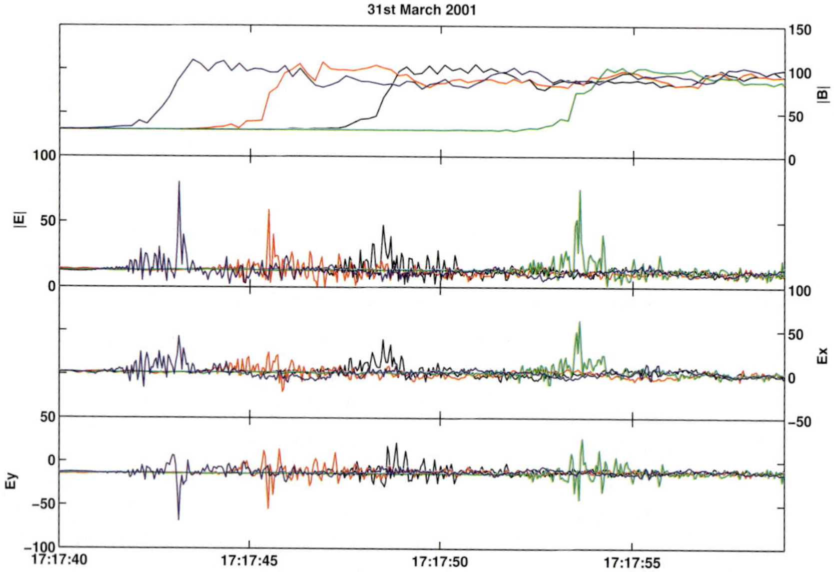

Figure 5.7. Overview of the shock crossing on March 31, 2001 at 1718 UT. The top panel shows the magnitude of the magnetic field measured by FGM. The second panel shows the magnitude of the electric field measured in the satellites' spin plane. The lower two panels show the spin plane components E_x and E_y. The four traces correspond to Cluster 1 (black), 2 (red), 3 (green) and 4 (blue). From Walker et al. (2004).

magnetic ramp region also contributes a significant fraction of the change in total potential over much smaller scales, typically 5-10 electron intertial lengths.

In this section we present the results of an investigation by Walker et al. (2004) into the spatial size of high amplitude, short lived spike like features observed in the electric field during a number of encounters with the quasi-perpendicular terrestrial bow shock. All electric field data presented were collected by the Cluster EFW instruments (Gustafsson et al., 1997), were sampled at 25 Hz, and have an upper cutoff frequency of 10 Hz.

As an example of the type of phenomena being investigated, Figure 5.7 shows an overview of the FGM magnetic field and EFW electric field for the shock encountered on March 31, 2001 at around 17:40 UT. The shocks observed on this particular day are not typical due to the passage of a CME that erupted from the sun a couple of days before. As a result, the upstream parameters were as follows: $|\mathbf{B}| \approx 30$ nT, $\theta_{Bn} \approx 87^\circ$, particle density $n \approx 19\,\mathrm{cm}^{-3}$, and an Alfvén Mach number (M_A)$\approx$ 3.6. The normal for this shock (based upon FGM crossing times) is $n_B = (0.94, -0.17, 0.293)$ (in the GSE frame), and the shock velocity was determined to be $48.92\,\mathrm{km\,s}^{-1}$.

The top panel shows the magnitude of the magnetic field measured by the four Cluster spacecraft. The data show that the satellites crossed the shock at 17:17:43.5 (Cluster 4), 17:17:45.5 (Cluster 2), 17:17:48.5 (Cluster 1), and 17:17:53.5 (Cluster 3). and that the foot, ramp and overshoot regions are clearly identifiable. The second panel shows the magnitude of the electric field measured in the spin planes of the satellites ($|E| = \sqrt{E_x^2 + E_y^2}$). The disturbances in the electric field begin in the magnetic foot region and continue through the overshoot/undershoot region. For this particular shock the field increase is around 6-7 mV m^{-1} when compared with the solar wind in both the foot and overshoot regions. However, in the region of the magnetic ramp the electric field exhibits increases of between 30 and 65 mV m^{-1} lasting a few hundred milliseconds. It is also obvious that the peaks appear in pairs. Inspection of the lower two panels (electric field spin plane components) shows that these peaks correspond to a rotation in field away from the projection of the normal onto the spin plane. By using the timings of the first peak observed in the electric field it is possible to compute a normal direction and velocity. This normal lies within a degree of that determined by FGM, and produces a similar value for the velocity. Thus it appears that these large amplitude, short duration features in the shock front appear to be some form of structured layer within the shock front.

Similar large amplitude, short duration features in the electric field have been observed at a number of shock crossings although their amplitudes are typically 10-20 mV m^{-1} above that observed in the solar wind. Their spatial scale and its variation with respect to the upstream shock parameters have been investigated. Figure 5.8 shows the distribution of scale sizes measured in electron inertial lengths (c/ω_{pe}). This measurement is unaffected by the fact that only two of the three electric field components are measured. It shows that the majority of scale sizes are of the order 1-5 c/ω_{pe}, a value much smaller than that of the magnetic ramp scale reported in Section 5.2.2.2. This measurement also shows that substantial contributions ($\approx 50\%$) to the overall cross shock potential occur within small regions of the shock.

The relationship between the scale size of the electric field spikes and the Alfvén Mach number is shown in Figure 5.9. It would appear that the scale size has an upper limit that increases as the Mach number decreases.

Figure 5.10 shows a scatter plot of the scale size of the electric field enhancements as a function of θ_{Bn}. The range of scale sizes appears to decrease as $\theta_{Bn} \rightarrow 90^\circ$. Karpman (1964) proposed that for shock with $\theta_{Bn} \approx 90^\circ$ the scale lengths are of the order of an electron inertial length. Indeed, although the errors in the determination of θ_{Bn} are probably $\sim 5^\circ$, the most nearly perpendicular Cluster shocks show scale lengths on the order of $2c/\omega_{pe}$.

Analysis of the field increase ($\Delta E = E_{spike} - E_{upstream}$) observed during these events show that their amplitudes vary in the range 4-70 mV m^{-1} above that measured in the solar wind upstream of the shock front. A scatter plot of the variation

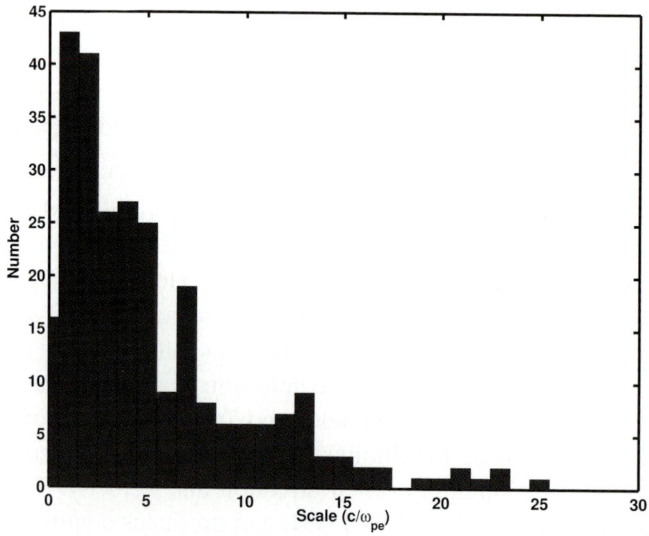

Figure 5.8. A histogram of the scale sizes for the spike-like enhancements observed during a number of crossings of the quasi-perpendicular bow shock. The typical width of a few electron inertial lengths (c/ω_{pe}) is much less than that of the magnetic field and density ramps. From Walker et al. (2004).

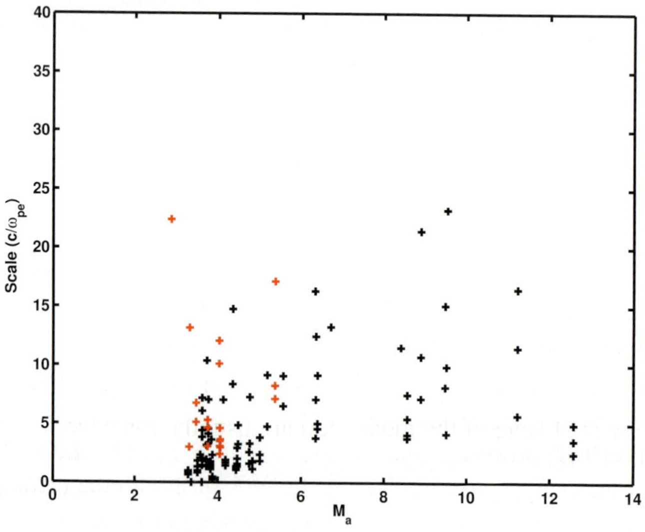

Figure 5.9. The dependence of electric field scale size on upstream Mach number. The red crosses are used to highlight the data for the shocks that occurred on March 31, 2001 under unusual conditions (Balikhin et al., 2002). From Walker et al. (2004).

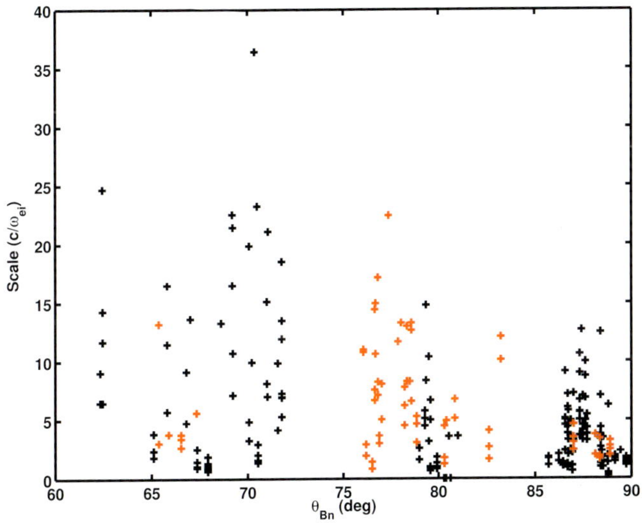

Figure 5.10. The dependence of electric field scale size on θ_{Bn}. From Walker et al. (2004).

of the peak field increase as a function of Mach number is shown in Figure 5.11. It can be seen that for those shocks whose Mach number $M_A > 5$ the value of ΔE is fairly constant and typically $\Delta E < 15$ mV m^{-1}. In contrast, for Mach numbers in the range $3 < M_A < 5$ the electric field amplitudes show a much larger spread, covering a range 5 - 60 mV m^{-1}. The shocks marked by red crosses correspond to those observed on March 31, 2001, a subset of shocks in a particularly interesting range of Mach numbers discussed in detail in Balikhin et al. (2002).

The relationship between ΔE and θ_{Bn} is shown in Figure 5.12. It clearly shows that as $\theta_{Bn} \rightarrow 90^\circ$ the range of the observed amplitudes of the electric field spikes increases.

5.2.4 Shock variability and non-stationarity

5.2.4.1 Introduction

Shock waves in plasmas as well as in gases and other media are nonlinear waves that cause changes of state of the media and are usually considered to be stationary in some reference frame. However, even early work in the subject (Morse et al., 1972) revealed evidence of non-stationarity in laboratory experiments. They revealed that in the fast magnetosonic mode Mach number range $M_{ms} \simeq 4$–8 the shock wave oscillates with a frequency comparable to the upstream ion gyrofrequency. The extent to which the physics of collisionless shocks involves intrinsically non-stationary processes has remained an open question. A compounding problem is the non-steady propagation speed of a shock which leads, for example

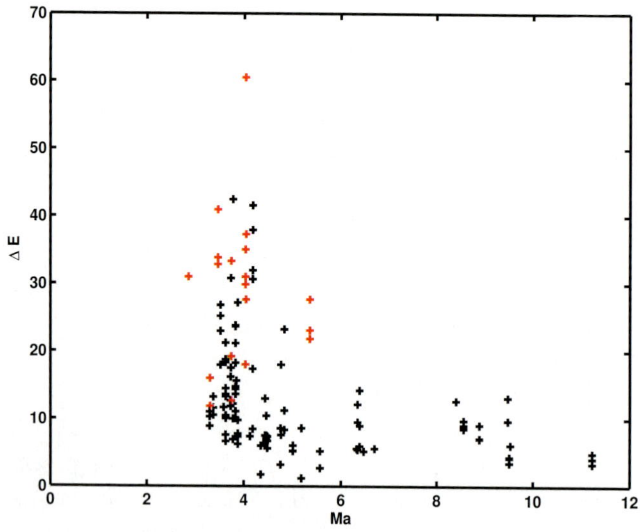

Figure 5.11. Scatter plot showing the relationship between the amplitude of the electric field spikes as a function of Mach number. From Walker et al. (2004).

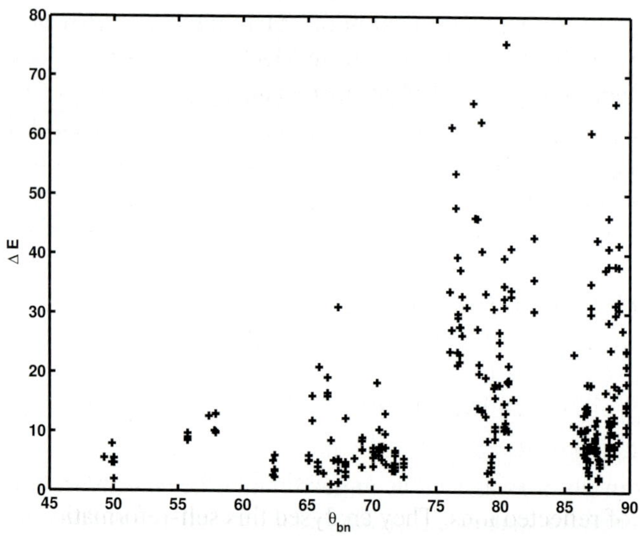

Figure 5.12. The relationship between the amplitude ΔE of the electric field spikes and θ_{Bn}. Shocks closer to 90° show a higher range of amplitudes. From Walker et al. (2004).

in the case of the Earth's bow shock, to relative motion of the bow shock with respect to the Earth in response to changes in solar wind conditions.

In the early 1980s, in the response both to new observations of the Earth's bow shock and computational capabilities, collisionless shock physics matured rapidly (e.g., Kennel et al., 1985, and accompanying papers). Indications of non-stationarity were found in low frequency oscillations of the ion flux at the bow shock (Vaisberg et al., 1984, 1986a,b) and at the bow shock of Uranus (Bagenal et al., 1987). Kinetic hybrid simulations (Leroy et al., 1981, 1982) for parameters typical at the Earth's bow shock ($M_A = 8$ and $\beta_e = \beta_i = 0.6$, where M_A is the Alfvén Mach number, $\beta_{e,i}$ is the ratio of the thermal and magnetic pressures, and 'e' and 'i' refer to electrons and ions respectively) showed that the shock structure varies with time. For example, the maximum value of the magnetic field exhibits temporal variations with a characteristic time of the order of the ion gyroperiod, the magnitude of these variations being about 20%.

Quest (1985) modelled high Mach number perpendicular shocks ($M_A = 22$, $\beta = 0.1$). In the absence of electron resistivity the ion reflection process is periodic, alternating between periods of 100% ion reflection and 100% ion transmission. As a result, a periodic shock front reformation was observed rather than a stationary structure. Quest (1986) extended these preliminary simulations to perform a systematic study of high Mach number perpendicular shocks. For $\beta = 0.1$ he revealed that the previously found (Leroy et al., 1982) tendency of a shock to become increasingly time-dependent as M_A increases was also observed for $M_A \geq 10$ and resulted in cyclical wave breaking for $M_A \geq 20$. In addition, for $\beta = 1$ and $M_A \geq 10$ a non-trivial dependence of the shock front structure on the resistivity was found.

Krasnosel'skikh (1985) and Galeev et al. (1988c) proposed models that attribute the shock front instability to the domination of nonlinear effects over dispersion and dissipation. Non-stationary whistler wave trains, which had been previously suggested (Galeev et al., 1988c,b,a), were reported in observations of the Earth bow shock onboard Intershock-Prognoz-10 and AMPTE UK spacecraft (Krasnosel'skikh et al., 1991). Recently, theoretical work and 1D full particle simulations have been used to analyse this mechanism in detail. Its application to obliquely propagating shocks has revealed a critical Mach number above which these non-stationarity processes operate (Krasnosselskikh et al., 2002).

Lembège and Dawson (1987b) have shown that the non-stationarity of the shock front can be due to the cyclic self-reformation of the shock front, and have recovered fluctuation levels of 20% in the magnetic field at the overshoot amplitude and in the density of reflected ions. They analysed this self-reformation in detail for an exactly perpendicular low-beta non-resistive shock in 1D full particle simulations and showed that this reformation persists even for a moderate (still supercritical) Mach number ($M_A = 2 - 4$). This non-stationary process persists over an angular range below $\theta_{Bn} = 90°$ as long as the density of reflected ions is high enough to feed the reformation (Lembège and Dawson, 1987a). Lembège and Savoini (1992) confirmed the previous results with the help of 2D full particle simulations and showed that reformation continues to occur even when finite resistivity effects due to cross field current instabilities are included self-consistently. In addition, the shock front appears to be rippled rather than uniform for both perpendicular and oblique planar shocks. Moreover, the reformation is expected for relatively low ion β_i (i.e. relatively cold upstream plasma) and/or high Mach number shocks, but disappears as β_i reaches relatively high values as shown by both 1D hybrid (Hellinger et al., 2002) and 1D full particle simulations (Scholer et al., 2003; Hada et al., 2003).

The problem of shock front stationarity described above gives several indications about possible manifestations of these effects in observations. Most of the results indicate that the characteristic timescale of the shock front variations is of the order of one ion gyroperiod or less, related to either the physics of the whistler mode expected to dominate the overall transition and/or the overturning due to non-steady ion reflection. This time period is comparable to and often shorter than that required to obtain full ion and electron distributions by Cluster. As noted above, shock motion can complicate this matter. Indeed, most of what we know about the position and shape of the bow shock is based on statistical studies together with modelling (e.g., Peredo et al., 1995, and references therein); the detailed response of the bow shock position to fluctuations in the upstream solar wind conditions has not been practical prior to the multi-spacecraft approach of Cluster.

In the following sections we provide an overview of the key Cluster results in this area. Evidence for intrinsic non-stationarity comes by studying variations of the shock profile, and by inferences on the variability of the ion reflection process(es). Cluster has also addressed directly the motion of the bow shock.

5.2.4.2 Shock profile variability

The near-simultaneous measurement of the shock profile by four spacecraft allows us to study spatial and temporal variability in ways that have not previously been possible. Different physical parameters such as the density, electric field and magnetic field would be expected to vary in different ways. The variability of one of these, the magnetic field, through the quasi-perpendicular shock was considered briefly by Horbury et al. (2001).

By considering the magnetic field profile through a nearly perpendicular supercritical shock ($\theta_{Bn} \approx 86^\circ$, plasma $\beta \approx 0.1$, Alfvén Mach number $M_A \approx 4.8$), Horbury et al. (2001) could identify structures which were stationary (i.e., phase standing) relative to the main shock ramp, and others that were not. The shock is shown in the top panel of Figure 5.13: the four profiles look superficially very similar. When the profiles are synchronised at the time of the crossing (Figure 5.13, bottom panel), some other features become visible. In particular, the shock overshoot, undershoot and subsequent oscillations in the magnetic field magnitude are fairly well synchronised between the four spacecraft, implying that this field magnitude structure does not vary significantly over the spacecraft separation (around 600 km) or the time differences between the shock passages of the different spacecraft (up to 30s).

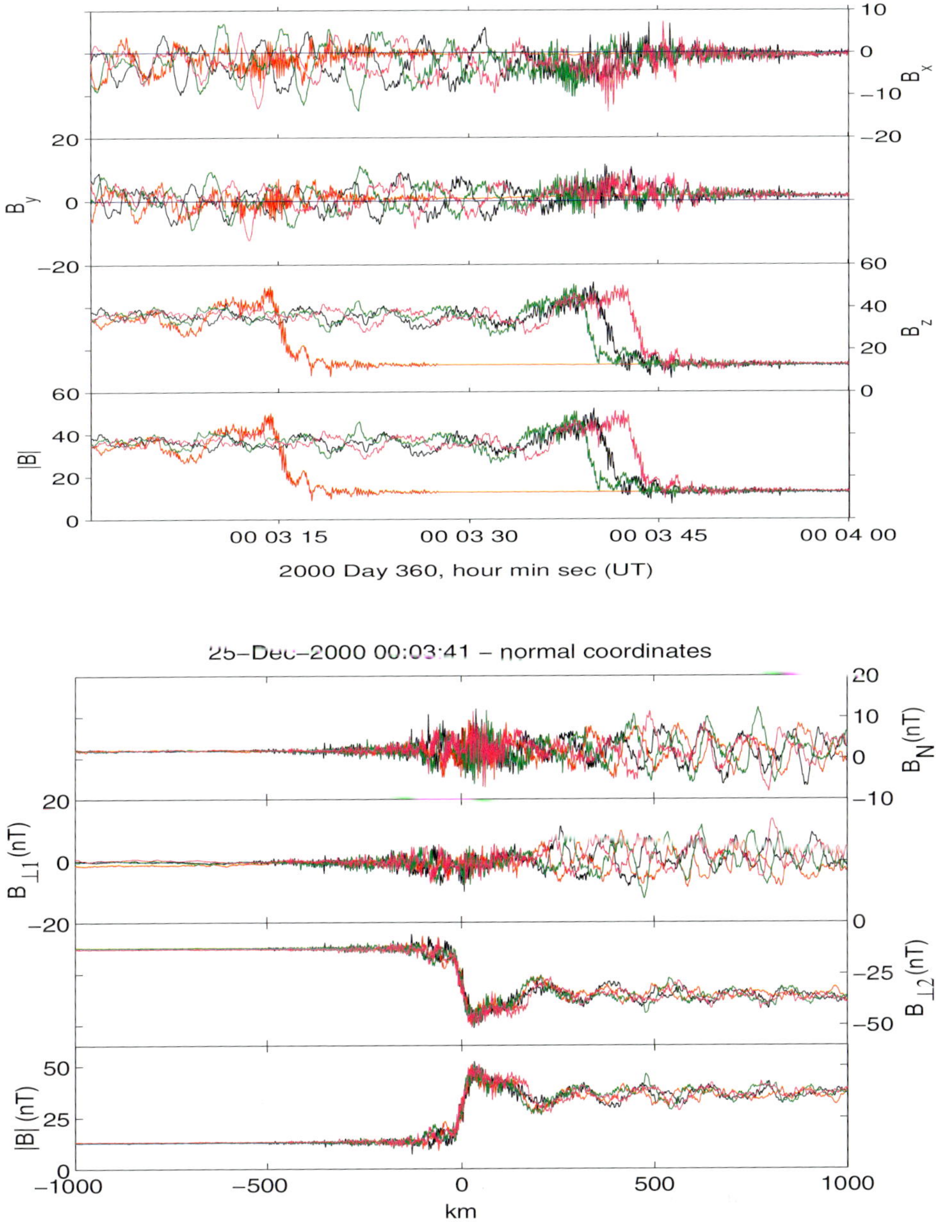

Figure 5.13. A supercritical quasi-perpendicular shock encountered by the Cluster spacecraft. Top: magnetic field data in GSE coordinates. Data from all four spacecraft are shown in different colours. Bottom: the same data, synchronised at the shock transition and transformed into a spatial scale and a shock-aligned coordinate system. The downstream oscillations in the field magnitude are well synchronised between the spacecraft, while the downstream field-perpendicular waves are not, implying that the latter are not phase standing with respect to the shock front. From Horbury et al. (2001).

In contrast to the shock profile in the magnetic field magnitude, the downstream large amplitude waves (with a polarisation and frequency consistent with ion cyclotron waves generated by non-gyrotropic ion distributions) varied significantly between spacecraft, confirming that these waves were not stationary with respect to the shock transition - again, this is clear from the synchronised profiles in Figure 5.13. In addition, it was not possible to identify the same waves at different spacecraft, implying that the scale sizes of these waves along the shock front were not larger than the spacecraft separations of around 600 km - their wavelength along the shock front was around 100 km. Analysis of shocks when the spacecraft are closer together will make it possible to identify the same wave at different spacecraft, and unambiguously determine the wavelength, propagation direction and speed of the downstream waves. There were also high frequency waves around the shock foot and ramp, which were consistent with whistlers.

Unlike the shock shown in Figure 5.13, which exhibits a rather steady background profile, Horbury et al. (2001) also considered a shock with a varying profile (Figure 5.14). The parameters of this shock were $\theta_{Bn} \approx 89°$ and $M_A \approx 5.6$. Here, significant differences are visible in the magnetic field profile measured by different spacecraft - for example, at spacecraft 3 (shown in green in Figure 5.14) the magnetic field magnitude actually decreases from its upstream value before increasing up the ramp, which is not the case at any of the other three spacecraft.

To study this variability in more detail, Horbury et al. (2001) synchronised the four shock profiles (in the same way as the right panel of Figure 5.13), calculated the average profile, and the deviations from the average at each of the four spacecraft; both are shown in Figure 5.15. The average profile exhibits a foot - an increase in the field magnitude before the main ramp - despite this not being clear in any of the four individual profiles in Figure 5.14. This highlights the difficulties of analysing shocks with single spacecraft: it is extremely difficult to determine which features are variable, and which are steady. With four profiles, however, we can begin to distinguish these effects.

The variability of the foot profile is clear in the right panel of Figure 5.15, which shows the deviation of the shock measured at each spacecraft from the average shown in the left panel. Deviations are apparent in the shock foot, but they are not present in the ramp. This implies that the fluctuations in the foot do not propagate into the ramp, and therefore cannot be a source for downstream waves. Horbury et al. (2001) noted that these may be signatures of unsteady reformation (Lembège and Savoini, 1992). The polarisation of these fluctuations – left-handed with respect to the magnetic field in the spacecraft frame - is consistent with whistler, but not Alfvén, waves. The large differences in the foot profile at the different spacecraft, even though the measurements are less than 10 s and 1000 km apart, show the small spatial or temporal scales of these fluctuations.

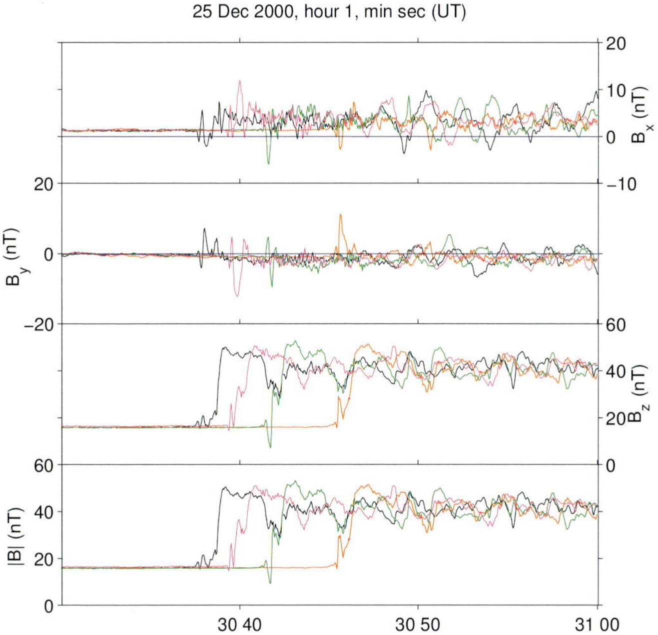

Figure 5.14. A quasi-perpendicular shock with a non-steady magnetic field profile. Clear differences between the profiles at different spacecraft are visible at the beginning of the magnetic field increase of the shock. From Horbury et al. (2001).

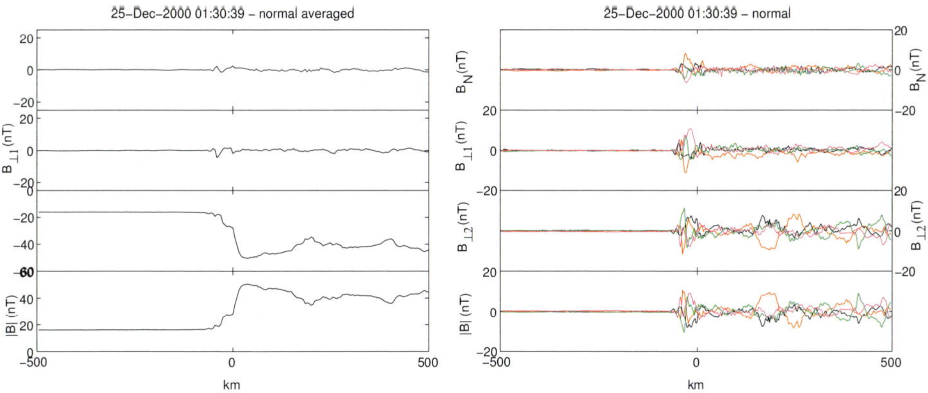

Figure 5.15. Left: Magnetic field profile of the shock shown in Figure 5.14, averaged over the four Cluster spacecraft in shock normal coordinates. A foot is clearly visible. Right: deviations of the magnetic field profile at each spacecraft from the average shown in the left panel. Variations are apparent in the foot, but not the ramp, demonstrating that the foot profile is variable, but that these fluctuations do not propagate into the shock ramp. From Horbury et al. (2001).

5.2.4.3 Variability of particle distribution functions

The bow shock is nearly always supercritical, so that the required heating cannot be accomplished purely resistively. The primary ion heating mechanism involves the reflection of a fraction of the incident ion population, which then gyrate around the magnetic field and return to the shock, thereby spreading the ion population in velocity space (Paschmann et al., 1982; Sckopke et al., 1983). While Cluster observations (Möbius et al., 2001; Kucharek et al., 2004) show that the fluxes of field-aligned beams upstream of the shock can be quite steady, they nonetheless reveal periodic variations by a factor ~ 3 (see Section 5.3.1.3). Closer to the shock the flux of reflected ions is modulated in response to changes in the local shock geometry (Meziane et al., 2004). This topic is addressed more completely in Eastwood et al. (2005, this issue). Identifying the signatures of particle variations associated with intrinsic shock front non-stationarity is more challenging since the spin period of the spacecraft, which limits the time resolution of the particle populations, is comparable to the relevant ion timescales.

5.2.4.4 Bow shock motion

In addition to variability of the shock front structure, the shock itself is in constant motion. It is important to be able to distinguish shock front variability from acceleration and motion in spacecraft data. Shock motion itself is also a topic of interest: changes in upstream conditions can alter the equilibrium position of the shock, but it is apparent that the shock moves even when the incoming solar wind is very steady. The shock motion, especially its acceleration, provides insight into the dynamics of its role in slowing the incident solar wind, and hence on the aspects of the shock which affect the global solar wind-magnetospheric interaction. Four spacecraft Cluster data can help us to measure this acceleration with far more confidence than estimates based on single spacecraft measurements.

Some crossings of the bow shock by Cluster show significantly different profiles at the four spacecraft. While these differences can reflect genuine temporal variability in the shock, for example due to reformation (see Section 5.2.4.1) or a change in upstream conditions, they can also be the result of changes in the speed of the shock. Shock crossings consistent with such acceleration - with profiles at some spacecraft significantly compressed or expanded in time compared to others, but otherwise similar - are relatively common in the Cluster data set. As ever, the spacecraft separation is important: at the smallest Cluster separations (around 100 km), where the times between crossings are at most a few seconds, evidence of acceleration is less common than at 5000 km separations, where there is more time between crossings for the shock to change speed.

Dunlop et al. (2002) used the 'discontinuity analyzer' (Dunlop and Woodward, 1998) to study the acceleration of the quasi-perpendicular bow shock. The discontinuity analyzer procedure requires an independent method to determine the shock orientation. Then, by determining the relative times at which pairs of spacecraft encountered the shock, the average speed of the shock between the two points can be estimated. Finally, by considering how this speed changes between successive pairs of spacecraft, the acceleration of the shock can be estimated.

Dunlop et al. (2002) used both coplanarity and minimum variance estimates of bow shock orientations, and found that they were consistent and close to model normals, and therefore were likely to be reliable estimators of the shock normal. They considered quasi-perpendicular bow shock crossings at ≈600 km spacecraft separations and showed that there was evidence of considerable deceleration of the shock transition - in one case, from $147\,\mathrm{km\,s^{-1}}$ to $27\,\mathrm{km\,s^{-1}}$ in around 10 s. This deceleration was usually smooth from one spacecraft to another, although sometimes it was more variable.

Maksimovic et al. (2003) used Cluster measurements of eleven quasi-perpendicular bow shocks on March 31, 2001 to measure the speed and location of the shock, and therefore estimate its large scale motion over more than two hours. They used simple four spacecraft timings, assuming planarity and constant motion during each shock crossing. On the basis of the shock location and orientation and a parabolic model of the shock surface, they could estimate the distance to the shock at the nose, i.e., the sub-solar distance. Using the measured shock speed, they then estimated the instantaneous speed of this sub-solar point, and interpolated it for times between the measured crossings using a cubic polynomial (Figure 5.16, panel a). For one pair of shock crossings, only one of the four spacecraft encountered the shock, so Maksimovic et al. (2003) assumed that it reversed its motion at that location and time.

A number of models have been developed that relate the bow shock sub-solar distance to upstream conditions such as the solar wind ram pressure. Many of these models are based on gas dynamic simulations by Spreiter et al. (1966). Maksimovic et al. (2003) used a combination of the Farris and Russell (1994) and Sibeck et al. (1991) models, with a dependence of the sub-solar distance a_s (in R_E) on solar wind ram pressure P (in nPa) and Alfvén Mach number M_A that scaled as

$$a_s = 12.2\left(\frac{2}{P}\right)^{1/6}\left[1+1.1\frac{(\gamma-1)M_A^2+2}{(\gamma+1)(M_A^2-1)}\right] \tag{5.1}$$

Maksimovic et al. (2003) used ACE data for upstream solar wind and magnetic field conditions, estimated the sub-solar distance and compared it with their estimates based on the Cluster crossings (Figure 5.16, panel b). The agreement was reasonably good, both in absolute position at various times, and in the amplitude of the variations in sub-solar distance (around 6 R_E from the Cluster observations, and 4-5 R_E from the model), implying that the model is valid not only in a statistical sense but also on shorter timescales as the shock responds to changing upstream conditions. Maksimovic et al. (2003) pointed out that their study could be refined further, allowing for effects such as changes in solar wind magnetic field, relative

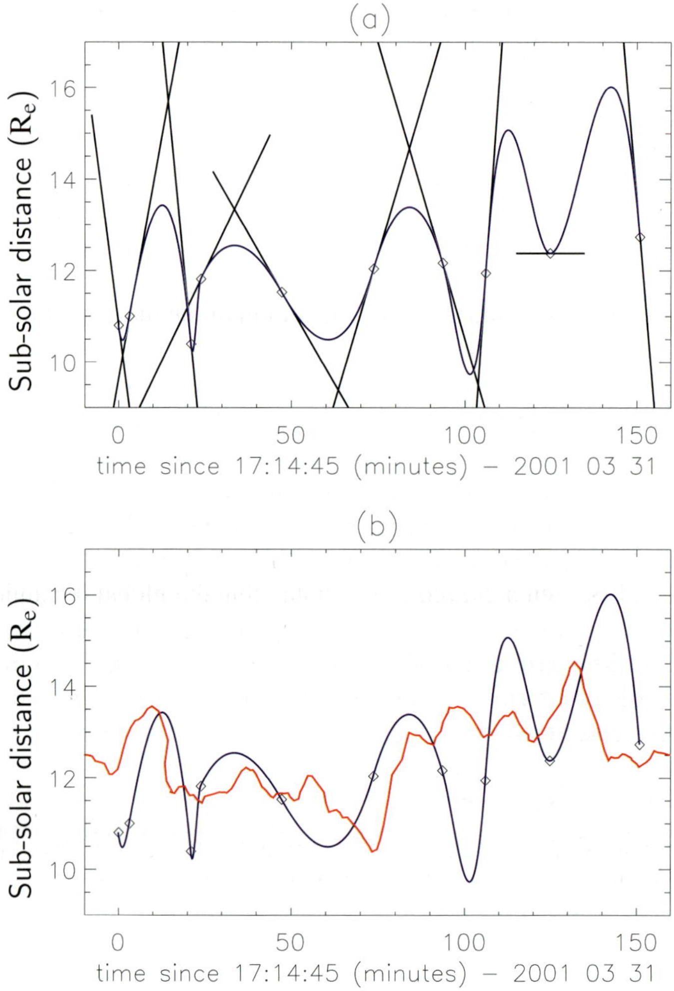

Figure 5.16. Motion of the bow shock sub-solar distance deduced from the measured location and speed of the shock at Cluster. Panel a: Sub-solar distance deduced from individual crossings (diamonds), and extrapolated from the estimated shock speed (black lines). The blue line shows a cubic interpolation through these points, allowing for the measured speeds. Panel b: Deduced sub-solar position in blue, as panel a, compared with the position calculated from a gas dynamic model using upstream solar wind parameters as input (red line). From Maksimovic et al. (2003).

propagation delays between ACE and the bow shock, and local measurement of shock acceleration from the Cluster formation.

5.3 On the formation and origin of field-aligned ion beams

5.3.1 Ion distributions at the quasi-perpendicular bow shock

A very prominent feature at the Earth's bow shock is the presence of backstreaming ions (Asbridge et al., 1968; Lin et al., 1974). The properties and morphology of these ion populations have been intensively studied over the past 30 years using in situ spacecraft observations. With the International Sun Earth Explorer (ISEE) spacecraft substantial progress was made with detailed observations in the foreshock region. This work resulted in the discovery of the different types of upstream distributions depending on the orientation of the magnetic field at the shock (Gosling et al., 1978; Bame et al., 1980; Greenstadt et al., 1980; Paschmann et al., 1981). A very collimated ion beam is found upstream of the quasi-perpendicular shock, and reflected gyrating ions are seen within one gyro-radius of the shock front.

Over the past 15 years the combination of computer simulations and detailed spacecraft observations has improved our knowledge of the processes at the bow shock as a prime example for collisionless shocks considerably. Although significant progress has been achieved in understanding the global dynamics of the ion distributions in the foreshock region, the underlying production mechanisms are still largely unexplained, and models may sometimes even be contradictory to observations. The origin and the basic production mechanism of field-aligned ion beams is such an example.

5.3.1.1 Characteristics of field-aligned beams

ISEE observations have provided well-documented characteristics of field-aligned ion beams, with most of them observed at $\theta_{Bn} < 70^o$. Studies by Paschmann et al. (1981) and Bonifazi and Moreno (1981) suggested that the characteristics of these beams are largely independent of θ_{Bn}. Typically they show an average flow speed which is on the order of twice the solar wind speed, but sometimes it can be significantly higher. Their density reaches up to about 1% of the solar wind, but it can be much smaller. Paschmann et al. (1981) found that these beams consistently exhibit a temperature anisotropy ($T_\perp/T_{||} \approx$ 4-9), where $T_\perp$ and $T_{||}$ are the temperatures perpendicular and parallel to the magnetic field. The perpendicular temperature is usually much larger than the solar wind temperature, which is in the keV range. Beams at larger shock normal angles ($\theta_{Bn} > 70°$) have been observed reaching much higher energies. Their bulk speed was higher by a factor of 5 (or even larger) than the solar wind bulk flow. Usually, these beams show a consistently higher temperature anisotropy.

Although previous observations have documented the characteristics of the upstream ion distributions rather well, they have not been sufficient to determine the source and production mechanism of the ion beams. Multi-spacecraft missions, such as Cluster, are bound to improve our understanding substantially, as they provide simultaneous observations at different locations and thus unravel temporal and spatial variations, even in 3 dimensions. Simultaneous observations upstream and downstream and/or in the shock ramp allow decisive tests on proposed sources.

5.3.1.2 Characteristics of reflected gyrating ions

In contrast to field-aligned beams, under quasi-perpendicular shock geometries gyrating ions only occur immediately in front of the shock and do not really escape upstream. The maximum of this distribution is seen at a specific pitch-angle that is determined by the orientation of the shock normal relative to the IMF and the solar wind direction. From detailed observations with ISEE-1 and -2 a firm picture of the evolution of these ions has developed.

A small percentage of the incoming protons is specularly reflected. During their gyration in the upstream magnetic field they gain energy by their motion parallel to the convection electric field of the solar wind. They subsequently encounter the shock and are transmitted downstream, where they constitute a gyro-phase bunched distribution (Sckopke et al., 1983; Sckopke et al., 1990). This gyro-phase bunched distribution develops into a ring in velocity space and, through interaction with the directly transmitted solar wind protons, into a pancake-like distribution. The large $T_{\perp}$ vs. $T_{||}$ anisotropy excites Alfvén ion cyclotron and/or mirror waves which, by pitch angle scattering, further downstream lead to the final hot magnetosheath distribution. In other words, specular reflection, gyration, wave excitation, and isotropization of the ions is essential for the formation of the hot magnetosheath flow and provides the mechanism for dissipation of bulk energy at the quasi-perpendicular shock.

In the past this reflected gyrating ion distribution has been distinguished from the field-aligned beam distribution (e.g., Gosling and Robson, 1985; Thomsen, 1985). While the specular reflection of the gyrating ring distribution has been explained in a straight-forward manner as a reflection by the shock potential, for specific phases of the incoming ions and assisted by gyro-motion in the compressed IMF downstream of the shock, the generation of the beam was not so readily understood.

5.3.1.3 Cluster observations of ion distributions upstream of the quasi-perpendicular bow shock

Recent Cluster results show that the ion beam distribution is closely related to the gyrating ion distribution formed by specularly reflected ions (Möbius et al., 2001). In their work, CODIF and HIA data of the CIS instrument were used to study the time evolution of the velocity distribution of reflected ions during a time period of repetitive shock crossings of the Cluster spacecraft. There is clear evidence that the gyrating and beam distributions are intimately connected. In fact, the beam distribution that escapes from the shock along the magnetic field lines emerges from the low pitch-angle wing of the specularly reflected ion distribution in the shock ramp under flux conservation.

The key results from this study are illustrated in Fig. 5.17. Shown are full sky maps of the H^+ distributions at 2.2 – 3.5 keV for four time periods on the left and the magnetic field strength on all 4 spacecraft as well as ion fluxes for the gyrating ions from Cluster 4 and the beam from Cluster 1, 3, and 4 as a function of time on the right. The ramp location is indicated in the figure for Cluster 1 as are the times corresponding to the displayed angular distributions.

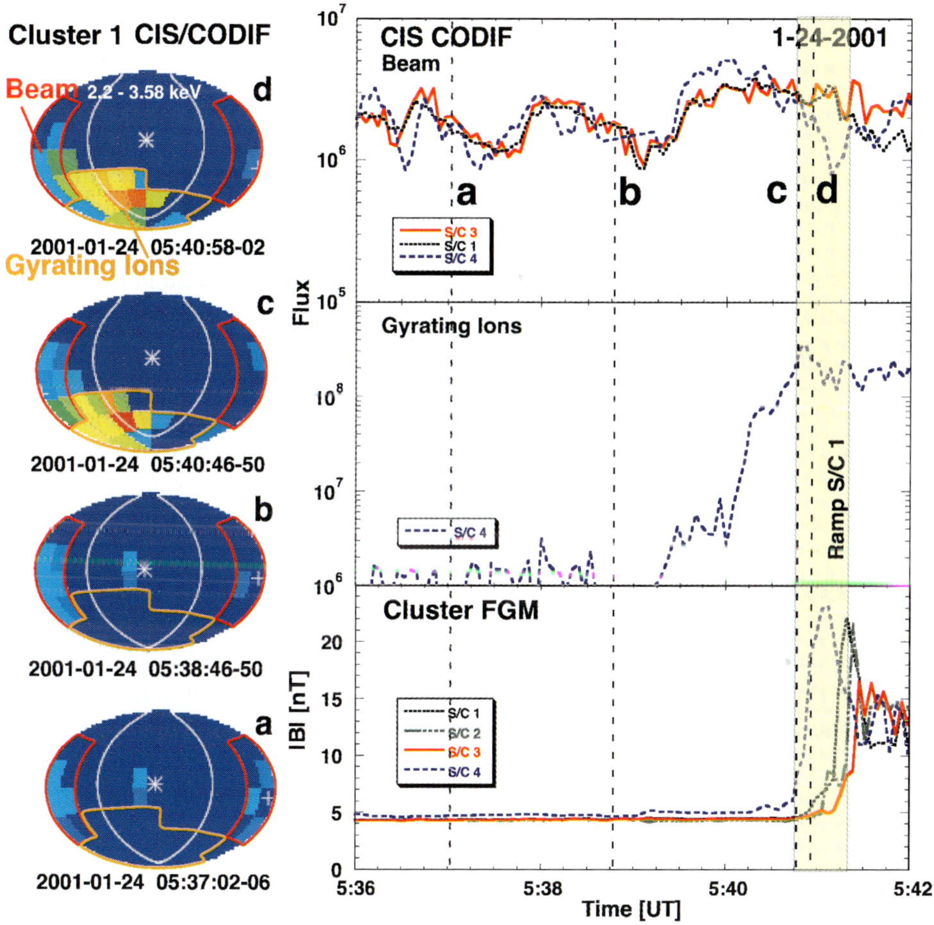

Figure 5.17. Left column: Angular distributions as seen on Cluster 1 in the shock ramp (d), at the ramp edge (c), and upstream of the shock (b, a). The angular regions of the beam (red) and ring (orange) are indicated; a '$+(*)$' denotes velocities in the direction $-(+)\mathbf{B}$. Right column: Integrated H^+ flux in the phase space portions that represent the beam from Cluster 1, 3 and 4 (upper panel), reflected ring from Cluster 4 (centre panel), and magnetic field strength from all four spacecraft (lower panel). They are shown from approximately 5 minutes before the bow shock encounter through the shock ramp. From Möbius et al. (2001).

From downstream of the shock through the shock ramp, the flux of the gyrating ion distribution (centre right panel in Fig. 5.17) remains high and falls off quickly by more than two orders of magnitude with distance from the shock. The remaining fluxes seen further upstream are on the level of one or a few counts and thus at the detection threshold for Cluster 4.

The total flux in the beam distribution is about one order of magnitude lower than that in the gyrating ions in the shock ramp, but it remains approximately constant (with quasi-regular fluctuations by a factor of three) with distance from the shock. The beam appears to originate in the combined reflected ion distribution in the shock ramp. It emerges from the edge of this wide pitch-angle distribution and then escapes more or less along the field lines under flux conservation. Therefore, the flux in the beam must depend on the angular width of the reflected distribution and on the original pitch-angle after ideal specular reflection in the shock potential. When relating the efficiency for beam generation to the shock parameters in comparison with models, these factors have to be taken into account in addition to the overall reflection efficiency for solar wind ions.

The results demonstrate quite clearly the value of two substantially different geometric factors in CODIF and of multi-spacecraft observations. While the high sensitivity side of CODIF is saturated by the ring distribution, which represents a large fraction of the solar wind flux, the low sensitivity side of CODIF is starved for counts in the beam distribution for the one-spin resolution observations used here. Only the split geometric factor allows for a simultaneous quantitative study of both, beam and gyrating ions. Furthermore, the high sensitivity side of CODIF, with a much higher count rate, shows remarkable structure and time variation in the beam. The beam flux varies substantially on a minute time scale, simultaneously on all spacecraft, while their distance from the shock differs, as indicated by the consecutive crossing of the shock: a clear indication for an intrinsic time variation of the beam. In addition, in view (a), i.e. substantially further upstream of the shock, the field aligned beam is more gyrotropic than about two minutes later. Closer to the shock the beam distribution apparently still has a memory of the original gyro-phase, with which it was injected at the shock. With increasing distance this memory effect is reduced because ions with different parallel velocities mix in gyrophase.

5.3.2 Source and generation of field-aligned beams

While the specular reflection of the gyrating ring distribution has been explained in a straight-forward manner as a reflection in the shock potential (Paschmann et al., 1980), the generation of the beam was not so readily understood. Although the kinematics and energetics of the beams can be derived correctly in terms of a perfect reflection of the incoming ions under energy conservation along the upstream IMF in the de Hofmann-Teller (dHT) frame (Sonnerup, 1969; Paschmann et al., 1980), the microphysics of their generation at the shock is still under debate. Considerable progress has been made through the use of numerical simulations, commencing with hybrid simulations by Leroy and Winske (1983) and Burgess (1987) which explored the acceleration at the shock front itself. Field-aligned beam generation and so-called shock drift acceleration invoked at interplanetary and astrophysical shocks are, in fact, closely related phenomena (Burgess, 1987).

How can these ions escape upstream? The escape of the particles depends on their guiding centre motion, which has to carry them away from the shock along the field line against the convection of the magnetic field with the solar wind bulk flow. The escape speed increases enormously if θ_{Bn} approaches 90°. Any process that can provide ions with high enough velocity parallel to the magnetic field so that their guiding centre velocity along the shock normal is larger than the convection speed will create an upstream ion beam. In principle beam ions can originate from downstream or from the shock ramp. However, as we shall see below, the physical processes responsible for the beam are still unknown. Additionally, although the case study of the preceding section suggests that the beam was drawn directly from the specularly reflected component, this has not yet been universally established.

5.3.2.1 Direct reflection under conservation of μ

A number of models to produce field-aligned ion beams have been proposed. Following the suggestion by Schwartz et al. (1983) we will discuss them within a similar representation in velocity space in the dHT-frame. This frame of reference simplifies the discussion as the flow of incoming ions is field-aligned and the motional electric field is zero. Figure 5.18 shows two possible scenarios of direct ion reflection at the bow shock. Sonnerup (1969) demonstrated that solar wind protons could easily be energised to form a rather energetic ion beam if the bow shock managed to turn the incoming ions around in such a way that they left the shock reasonably well field-aligned after reflection (left panel in Figure 5.18). He assumed that the particle energy was preserved in the dHT frame and the motion remained field-aligned after reflection, but he did not specify a reflection process. In observations with ISEE, Paschmann et al. (1980) actually found that the peak energy of ion beams as a function of the magnetic field orientation relative to the solar wind and to the shock normal agreed well with the prediction of this model.

This scenario is also referred to as 'adiabatic reflection' because of the apparent conservation of the magnetic moment μ. However, in observational studies (Paschmann et al., 1982) and numerical simulations (Leroy et al., 1981, 1982), μ is far from constant during ion reflection at the quasi-perpendicular bow shock. In their observations Paschmann et al. (1982) studied ions that were clearly specularly reflected by the shock potential. Such ions start to gyrate and then are swept downstream with the IMF, but do not show a field-aligned beam. This situation is shown in the right hand panel of Figure 5.18. The energy is conserved in the dHT frame in both scenarios, as indicated by the vectors for the reflected ion velocity ending on the same circle around the origin of the dHT frame. As a physical ion reflection

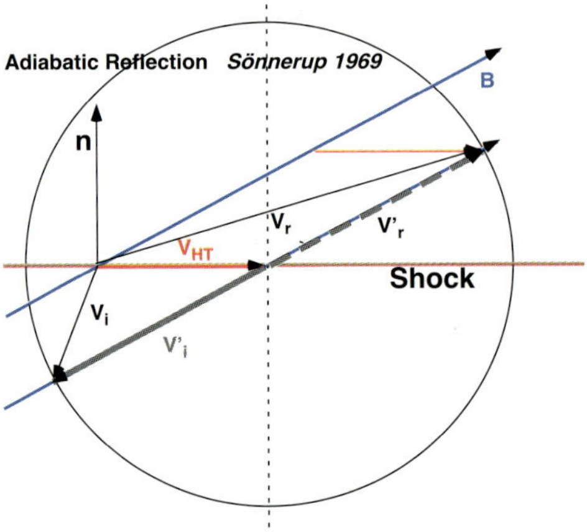

Figure 5.18. Kinematics of reflected ion beams and specularly reflected ions in velocity space at the quasi-perpendicular shock, shown in the de-Hoffmann-Teller (dHT) frame, i.e. translated along the shock with the dHT velocity V_{HT} so that there is no motional electric field. In this frame the incoming ions (V_i) arrive parallel to **B** (shown for upstream conditions) with V_i'. Ions that conserve energy in the dHT frame fall along the circle in velocity space. Left: Reflection along **B** under conservation of magnetic moment μ leads to escape along **B** with V_r' in the dHT frame and with V_r in the observer's frame (Sonnerup, 1969). Right: Specular reflection at the shock transforms part of the velocity into gyromotion (V_{rg}' perpendicular to **B**), and the velocity component parallel to **B**, $V_{r||}'$, of the reflected ion may point downstream or upstream, depending on θ_{Bn}. Figure adapted from Thomsen et al. (1983b) and other sources by E. Möbius.

mechanism that would conserve μ, Terasawa (1979) suggested the magnetic mirror force, but it cannot reflect solar wind ions at the shock because of their usually narrow pitch angle distribution. Apparently none of the known physical reflection mechanisms would cooperate toward the favourable energy condition found by Sonnerup (1969). Therefore, alternative generation mechanisms for beams were explored.

5.3.2.2 Leakage of heated magnetosheath ions

An alternate source for field-aligned beams could be leakage of magnetosheath ions that have been heated downstream of the shock. In an idealised model Edmiston et al. (1982) proposed that plasma is heated and thermalised in a thin layer at the shock front.

They calculated how ions from a hot Maxwellian distribution in this layer can return upstream. Magnetosheath ions, which arrive with zero velocity at the maximum of the shock potential, taken in the dHT frame, could be accelerated by falling through the potential drop at the shock ramp and escape parallel to **B**. However, in this model ions can only outrun the shock parallel to **B** and escape upstream for θ_{Bn} between 40° and 55°, yielding fractional densities up to at most 1%. For large θ_{Bn}

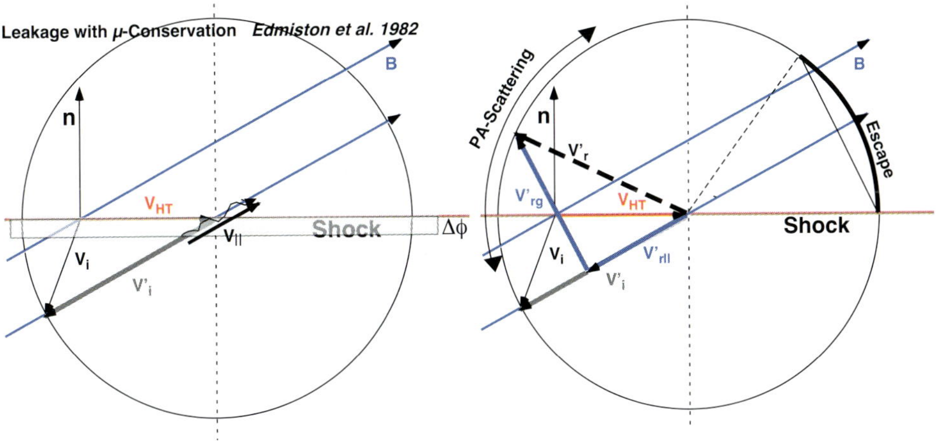

Figure 5.19. Kinematics of leakage and specularly reflected ions with subsequent pitch angle scattering in a similar representation as in Figure 5.18. Left: Leakage along **B** under conservation of magnetic momentum μ facilitated by the cross-shock potential $\Delta\phi$ (after Edmiston et al., 1982). Right: Specular reflection of incoming ions at the shock (after Paschmann et al., 1982) and subsequent pitch angle scattering with energy conservation in the dHT frame. Ions that get a large parallel velocity component away from the shock (dark section on the circle) can escape and form a beam-like distribution. Figure adapted from Thomsen et al. (1983b) and other sources by E. Möbius.

the parallel speed of the heated ions in the dHT frame is by far too small to allow escape. Because only the most energetic tail of the heated distribution downstream of the shock will even reach the top of the potential well at the shock with a speed ≥ 0, the peak of the emerging ion beam stems from those ions, which arrive with $v = 0$ and are ejected with a final speed equivalent to the shock potential (Schwartz et al., 1983). Ion distributions that are consistent with this picture, i.e. low energy ion beams at shocks with moderate values for θ_{Bn}, have been observed by Thomsen et al. (1983a).

Realising that the thickness of the layer with the shock potential is small compared with the ion gyro radii and that the ions most likely will not be magnetised during their transition of the shock, Schwartz et al. (1983) proposed a modified version of this model. They suggested that magnetosheath particles are accelerated by the shock potential mainly along the shock normal and that its component parallel to the magnetic field constitutes the resulting guiding centre motion back upstream. This poses an even more stringent constraint on the angles between shock normal and **B**, for which ions can escape upstream along the magnetic field, than the model by Edmiston et al. (1982).

Contrary to the earlier very local leakage scenarios, Tanaka et al. (1983) proposed a more self-consistent non-local model based on observations by Paschmann et al. (1982), simulations by Leroy et al. (1981), and the work by Edmiston et al. (1982). As solar wind encounters the quasi-perpendicular section of the bow shock, part of the incoming solar wind distribution is specularly reflected and creates a gyrating ion distribution that is swept downstream. Its significant temperature anisotropy, due mainly to its high perpendicular temperature, is the source of free energy for electromagnetic ion cyclotron (EMIC) waves downstream of the shock. As a result, efficient pitch angle scattering will produce particles with a high enough velocity parallel to the magnetic field so that they can escape upstream, although test particle calculations suggest that ions have difficulty trying to leak from the magnetosheath (Burgess and Luhmann, 1986). Tanaka et al. (1983) pointed out that this model is consistent with a large fraction of the beams observed by Paschmann et al. (1980), but fails to explain the most energetic ion beams. This could indicate that direct reflection may also contribute to the production of ion beams.

In Figure 5.20 ISEE-2 observations of ion beams are compiled in comparison with the expectations according to the theoretical considerations mentioned above (Schwartz and Burgess, 1984). Shown is the bulk velocity of the beams normalised to the solar wind velocity parallel to the magnetic field in the dHT frame as a function of the shock normal angle θ_{Bn}. As already mentioned above part of the field aligned beams observed in the Earth's foreshock region with ISEE-2 appear to be consistent with the leakage hypothesis, while another fraction appears to support the relation found for adiabatic reflection. Almost all of them fall into the space between these relations as boundaries. However, it should be noted that the majority of the beams achieve velocities much higher than supported by leakage from downstream and that the adiabatic reflection hypothesis does not contain a physical reflection mechanism.

The recent observation with Cluster that the beam distribution and the specularly reflected ions are intimately connected and that the beam appears to emerge from the wing of the combined distribution (Möbius et al., 2001) provides important evidence of the processes responsible for the beam. Early work (Burgess and Schwartz, 1984) showed how pure dc fields at the shock could lead to some reflected ions suffering multiple encounters with the shock, as confirmed in later self-consistent simulations (Leroy and Winske, 1983; Burgess, 1987). Such particles, when viewed in the dHT frame, must emerge on the constant energy circle in Figure 5.19, and will escape upstream if after a final encounter they have sufficient parallel guiding centre velocity to prevent their return to the shock surface. Certainly, particles finding themselves in the fat dark portion of this circle marked 'escape' have persistently positive normal velocity and will escape. As the figure reveals, such particles have nearly the maximal beam speed as found from adiabatic reflection (Sonnerup, 1969), though the physics is quite different. In reality, a larger portion of the circle will result in escaping particles, and it is uncertain what the centroid of the total population would be, although simulations (Burgess, 1989) provide some indication. Whether it is best to describe this scenario as pitch angle diffusion/scattering is debatable. Scattering due to fluctuations and irregularities in the shock fields (e.g., within the foot, ramp, and/or overshoot regions) almost

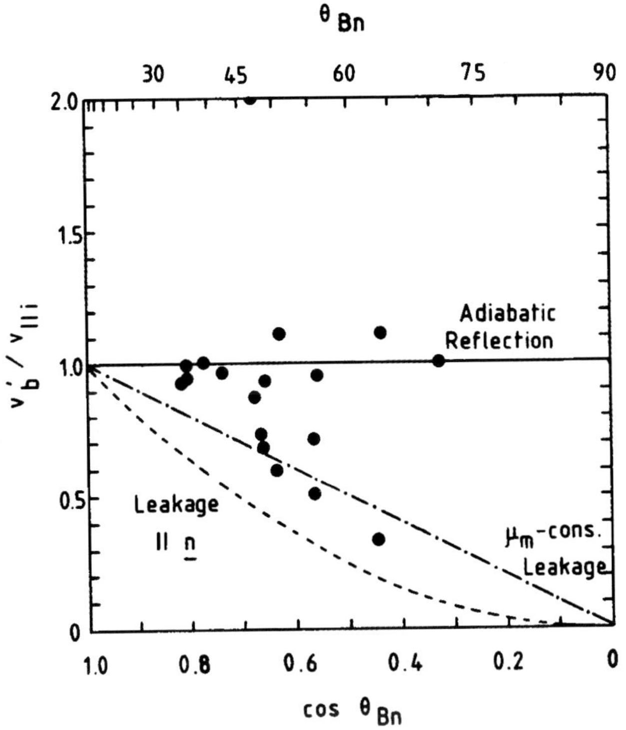

Figure 5.20. Observed velocity of ion beams normalised to the parallel component of the solar wind speed in the dHT frame as a function of the $\cos\theta_{Bn}$. The three curves represent the expected values for the three different ion beam generation models: Adiabatic reflection (Sonnerup, 1969), leakage under conservation of μ (Edmiston et al., 1982), and leakage with acceleration parallel to the shock normal. From Schwartz et al. (1983).

certainly does not preserve kinetic energy in the dHT frame, as such fluctuations propagate at relatively small speeds relative to the bulk plasma flow. Nonetheless, they may play a role, as yet unquantified, in aiding (or hindering) the dc process of multiple shock encounters, or in masking those effects and otherwise diffusing the distribution in velocity space.

A note of caution may be in order. The simple dependence on the local shock normal angle is a direct consequence of the assumption of a planar, featureless, and stationary bow shock. In a way, the predicted and often observed energy dependence on θ_{Bn} may just reflect the necessary escape condition for ion beams. In this picture, it is assumed implicitly that the dHT frame is natural frame of reference, which implies that the reflection and scattering happens in this frame and for a comparison with observations that all parameters for the transformation are known and reasonably constant over the integration period. Any motion of the shock and/or local structures that deviate from a planar shock with the assumed normal may complicate a quantitative comparison with a specific model or even with the simple escape condition.

5.3.2.3 Cluster observations on leakage vs. reflection

The initial results from Cluster have demonstrated that the field-aligned ion beams emerge from the reflected gyrating ion distributions. More recent Cluster studies by Kucharek et al. (2004) suggest a resolution on the source region for the beams. They analysed several quasi-perpendicular shock crossings with CIS and followed the spatial and temporal evolution of the reflected and transmitted ion populations across the shock. Figure 5.21 shows a composite plot during the crossing at 18:48 UT on March 31, 2001, from downstream to upstream, including a snapshot in the shock ramp. The upper panel shows the magnetic field as a function of time, and in the lower panel the ion distributions, parallel and perpendicular to the interplanetary magnetic field (the mean magnetic field orientation is indicated by arrows), are shown for three different locations: downstream, at the ramp, and upstream of the bow shock. The dark blue shaded areas in the magnetic field profile indicate the integration times for the ion distributions. Downstream, the shape of the ion distribution is more elongated perpendicular to the magnetic field. The phase space is filled with ions up to a parallel velocity of $1000\,\mathrm{km\,s^{-1}}$. In the shock ramp, gyrating ions appear whose phase space density extends in parallel velocity, exceeding substantially the limit of $v \approx 1000\,\mathrm{km\,s^{-1}}$. Upstream of the shock (right hand distribution), this part of the distribution decouples from the core and forms a collimated beam along the mean interplanetary magnetic field. It should be noted that the beam occupies a portion of the phase space that is empty downstream.

Such simultaneous observations at different locations at the vicinity of the quasi-perpendicular bow shock indicate that the field-aligned beams most likely result from effective scattering in pitch-angle during reflection in the shock ramp. At least in this low Mach number shock, leakage of thermalized ions from the downstream region does not appear to be the source. While Figure 5.21 presents consecutive observations with Cluster 1, Kucharek et al. (2004) also studied a shock crossing for which simultaneous observations in the shock ramp and upstream are available (their Fig. 6, not shown here). The same pattern as described here is observed in this case.

From their results Kucharek et al. (2004) concluded that processes right in the shock ramp must produce the ion beams. Therefore, scattering in the shock ramp seems to be a major process that is responsible for their generation, which appears to be in accordance with findings from simulations by Scholer et al. (2000).

In their study Kucharek et al. (2004) also find a low intensity field-aligned ion beam upstream of a low Mach number supercritical shock at a rather large shock normal angle ($\theta_{Bn} = 74.5^o$). Interestingly enough it appears as if the basic escape condition might be violated and the conditions are far from reflection under con-

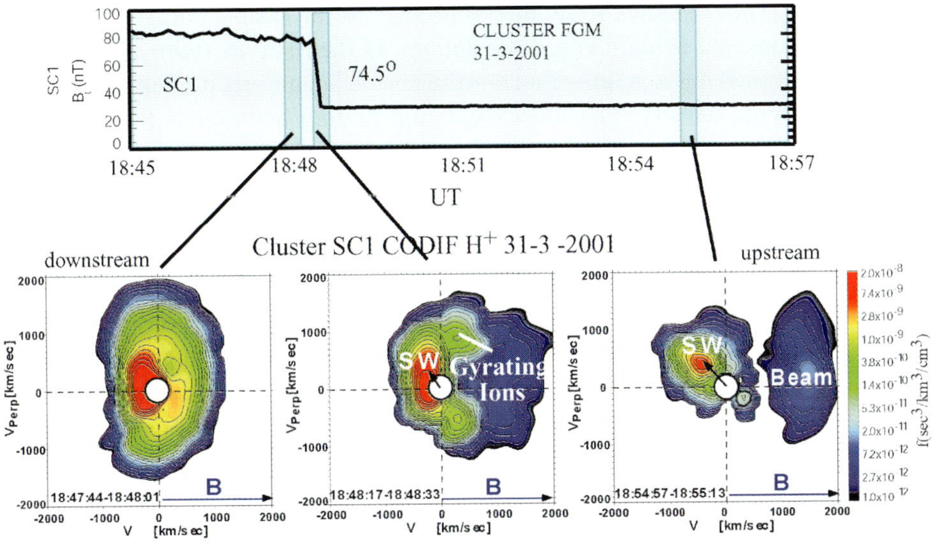

Figure 5.21. Magnetic field (top panel) and ion velocity space distributions (lower panel) downstream, in the ramp, and upstream of a quasi-perpendicular shock. Note that the upstream beam occupies portions of velocity space which are empty downstream, implying that the beams emerge directly from processes within the shock layer. From Kucharek et al. (2004).

servation of the magnetic moment. They point out that the dynamic structure of the bow shock at these large shock normal angles may be important for the visibility of field-aligned beams.

It is commonly known that the Earth's bow shock is a dynamic structure, which responds both locally and globally to changes of the solar wind conditions. During the crossing shown in Figure 5.21 that is associated with the ion beam the bow shock is receding with a velocity of about 30 km s^{-1}toward the Earth. During the preceding crossing, which is not associated with an ion beam (not shown here), the shock is approaching the spacecraft. Both motions result in much larger velocities along the magnetic field, thus necessitating a re-evaluation of the escape condition in a frame that is corrected for the shock motion. With such a correction the escape conditions for the ions are most probably marginally fulfilled for the crossing at 18:47 UT, whereas at 18:30 UT they are not. In none of the models has the dynamic structure of the Earth's bow shock in response to changes of the solar wind been taken into account thus far. The multi-spacecraft capabilities of Cluster provide the tools to include them in the models and to test them with actual observations.

5.3.3 Composition of field-aligned ion beams

Thus far mainly protons in field-aligned beams have been investigated. Compositional studies have been performed by Fuseller and Thomsen (1992) and Ipavich et al. (1988). These studies show that field-aligned ion beams consist almost entirely of protons and contain very little helium. In their survey (with 14 –23 field-aligned ion beams) these authors find that the proton beams reach a density of $< 1\%$ of the solar wind density. The most surprising finding however is the extremely low He^{2+}/H^{+} beam density ratio of $\sim 5 \times 10^{-4}$. This value is about two orders of magnitude lower than the nominal solar wind ratio. This low He^{2+} abundance has implications on the source and generation of field-aligned beams. It appears to be compatible with the finding by Möbius et al. (2001) that beams emerge from the gyrating ion distribution. While solar wind protons are specularly reflected by the shock potential with good efficiency, this seems not to be the case for solar wind He^{2+} (Fuselier and Schmidt, 1994). As a consequence they should be suppressed in their abundance in field-aligned beams as well. For beam events that arise from leakage, higher abundances may be found. It will therefore be an important task in further studies with Cluster to determine He^{2+} abundance values for field-aligned beams and specularly reflected ions.

5.3.4 Discussion on the nature of field-aligned beams

In the past reflected gyrating ions and ion beams at quasi-perpendicular shocks were discussed as two separate and different populations. Recent observations with Cluster have shown that the distribution of the beam ions originates from gyrating ions in the shock ramp. Cluster multi-spacecraft observations support the following, more generalised formation mechanism of field-aligned beams at the quasi-perpendicular Earth's bow shock: Part of the incoming solar wind is reflected and accelerated at the shock to form a gyrating ion distribution. Pitch angle scattering in the shock ramp will produce a small fraction of the gyrating ions which have a high velocity parallel to the magnetic field. Those ions from this scattered distribution that have a velocity component parallel to the shock normal larger than the convection speed of interplanetary magnetic field at the shock ramp will escape upstream and form a field-aligned ion beam. The intensity of the ion beam upstream will then be determined by pitch angle scattering in the shock ramp and not only by the shock geometry. Fluxes of ion beams show significant differences between the spacecraft and vary with time, which seems to reflect spatial and temporal variations.

Surprisingly, Cluster also observed field-aligned ion beams at very high shock normal angles. It appeared as if for this event (see Figure 5.21) the escape condition is violated and that the beam energy is far from that obtained by assuming reflection in the dHT frame under conservation of energy. At this high shock normal angle the beam ions should not be able to escape upstream. However, perpendicular shocks are dynamic and small-scale structures can lead to a deviation of the average θ_{Bn}. Both can modify the critical conditions so that ions can escape upstream. Therefore, it has been pointed out that processes right in the shock ramp and the dynamics of the shock itself seems to be the major production mechanism of these ion beams.

Although Cluster has already provided interesting new insight into processes and sources of ion beams upstream of the Earth's bow shock, many questions are still unanswered. Cluster will contine to contribute to the resolution of these problems with its unique capabilities. For instance: what determines the fraction of reflected ions that forms the gyration ion distribution? What fraction of these ions will escape upstream? What are the critical shock parameters that control the ion beam formation? How important is the internal structure of the shock, and what is the impact of the shock potential on particle reflection and acceleration?

5.4 Summary

The Earth's bow shock represents the most studied example of a collisionless shock. The bow shock under quasi-perpendicular conditions, where the magnetic field is nearly perpendicular to the shock normal, is the most exhaustively scrutinised form due primarily to its relatively clean, characteristic shape and features. Despite a considerable body of pre-existing knowledge, Cluster has contributed several important new results. These results take full advantage of Cluster's unique 4-spacecraft ability to disentangle space and time, and so to infer shape, orientation, static vs. dynamic processes, and simultaneous 3D measurements. These results include:

1. Confirmation that the overall shape of the bow shock, and its motion, are well-represented by previous statistical investigations.

2. Proof that the density and magnetic field ramp widths are governed by a trapped ion gyroscale rather than an ion inertial length. This places severe constraints on the dispersive and dissipative processes which effect the overall shock transition.

3. Identification of small scale electric field spikes within the overall shock transition which appear to be universally present. Their role in the overall shock dynamics remains to be elucidated.

4. Demonstration that much of the variability in the shock profile is confined to the magnetic foot, rather than main ramp, regions. This implies that although ion reflection and dynamics are responsible for the foot appearance, and also for the overall transition scale (as noted above), the main shock ramp and dissipation are not dependent on the fine details of the ion reflection process.

5. Demonstration that energetic field aligned beams of ions observed backstreaming from the bow shock are generated and accelerated at the shock following an initial reflection process.

Other related Cluster results, such as the interaction of field-aligned beams with the upstream wave-field, and the back-reaction of the wave field on the local shock processes, can be found in Eastwood et al. (2005, this issue).

All these results reveal the power of Cluster's multi-spacecraft strategy, and leave open a rich dataset capable of addressing many more outstanding questions on the subject of collisionless shocks under quasi-perpendicular conditions.

Acknowledgements

The contributions of HK and EM were partially supported under NASA Grants NAG5-10131 and NAG5-11804. PPARC (UK) support for this work includes fellowships (TSH and EL) and research grants (DB, SJS, SNW).

References

Asbridge, J. R., S. J. Bame, and I. B. Strong: 1968, 'Outward flow of protons from the Earth's bow shock'. *J. Geophys. Res.* **73**(12), 5777.

Bagenal, F., J. W. Belcher, E. C. Sittler, Jr., and R. P. Lepping: 1987, 'The Uranian Bow Shock: Voyager 2 Inbound Observations of a High Mach Number Shock'. *J. Geophys. Res.* **92**, 8603.

Bale, S. D., F. S. Mozer, and T. S. Horbury: 2003, 'Density-transition scale at quasiperpendicular collisionless shocks'. *Physical Review Letters* **91**(26), 265004.

Balikhin, M., M. Gedalin, and A. Petrukovich: 1995, 'The scales in quasiperpendicular shocks'. *Adv. Space. Res.* **15**, 247.

Balikhin, M., V. Krasnosel'skikh, and M. Gedalin: 1993, 'New mechanism for heating in shocks'. *Phys. Rev. Lett.* **70**, 1259.

Balikhin, M. A., M. Nozdrachev, M. Dunlop, V. Krasnosel'skikh, S. N. Walker, H. S. K. Alleyne, V. Formisano, M. André, A. Balogh, A. Eriksson, and K. Yearby: 2002, 'Observation of the terrestrial bow shock in quasi-electrostatic sub-shock regime'. *J. Geophys. Res.* **107**, 1155, doi: 1029/2001JA000327.

Bame, S. J., J. R. Asbridge, W. C. Feldman, J. T. Gosling, G. Paschmann, and N. Sckopke: 1980, 'Deceleration of the solar wind upstream from the Earth's bow shock and the origin of diffuse upstream ions'. *J. Geophys. Res.* **85**(14), 2981–2990.

Bonifazi, C. and G. Moreno: 1981, 'Reflected and diffuse ions backstreaming from the earth's bow shock. I Basic properties'. *J. Geophys. Res.* **86**(15), 4397–4413.

Burgess, D.: 1987, 'Shock drift acceleration at low energies'. *J. Geophys. Res.* **92**(11), 1119–1130.

Burgess, D.: 1987, 'Simulations of backstreaming ion beams formed at oblique shocks by direct reflection'. *Ann. Geophys.* **5**, 133–145.

Burgess, D.: 1989, 'Alpha particles in field-aligned beams upstream of the bow shock - Simulations'. *Geophys. Res. Lett.* **16**, 163–166.

Burgess, D. and J. G. Luhmann: 1986, 'Scatter-free propagation of low energy protons in the magnetosheath: Implications for the production of field-aligned beams by non-thermal leakage'. *J. Geophys. Res.* **91**, 1439–1449.

Burgess, D. and S. J. Schwartz: 1984, 'The dynamics and upstream distributions of ions reflected at the Earth's bow shock'. *J. Geophys. Res.* **89**(18), 7407–7422.

Dunlop, M. W., A. Balogh, and K.-H. Glassmeier: 2002, 'Four-point Cluster application of magnetic field analysis tools: The discontinuity analyzer'. *J. Geophys. Res.* **107**, 1385, doi:10.1029/2001JA005089.

Dunlop, M. W. and T. I. Woodward: 1998, 'Discontinuity analysis: Orientation and motion'. In: *Analysis Methods for Multispacecraft Data*, Vol. ISSI Sci. Rep. SR-001. Norwell, Mass.: Kluwer Acad., p. 271.

Eastwood, J., E. A. Lucek, C. Mazelle, K. Meziane, Y. Narita, J. Pickett, and R. Treumann: 2005, 'The Foreshock'. *Space Sci. Rev.* **this issue**.

Edmiston, J., C. Kennel, and D. Eichler: 1982, 'Escape of heated ions upstream of quasi-parallel shocks'. *Geophys. Res. Lett.* **9**, 531–534.

Eselevich, V. G.: 1982, 'Shock-wave structure in collisionless plasmas from results of laboratory experiments'. *Space Science Reviews* **32**, 65–81.

Eselevich, V. G., A. G. Eskov, R. C. Kurtmullaev, and A. I. Malyutin: 1971, 'Isomagnetic discontinuity in a collisionless shock wave'. *Soviet Phys. JETP* **33**, 1120.

Farris, M. H. and C. T. Russell: 1994, 'Determining the standoff distance of the bow shock: Mach number dependence and use of models'. *J. Geophys. Res.* **99**(18), 17681.

Formisano, V.: 1979, 'The three-dimensional shape of the bow shock'. *Nuovo Cimento C Geophysics Space Physics C* **2**, 681–692.

Formisano, V.: 1982, 'Measurement of the potential drop across the Earth's collisionless bow shock'. *Geophys. Res. Lett.* **9**, 1033.

Formisano, V.: 1985, 'Collisionless shock waves in space and astrophysical plasmas'. In: *Proc. ESA workshop on future missions in solar, heliospheric and space plasma physics*, Vol. ESA SP-235. p. 83.

Formisano, V. and R. Torbert: 1982, 'Ion acoustic wave forms generated by ion-ion streams at the Earth's bow shock'. *Geophys. Res. Lett.* **9**, 207.

Fuselier, S. and M. F. Thomsen: 1992, 'He^{2+} in field-aligned beams: ISEE results'. *Geophys. Res. Lett.* **19**, 437.

Fuselier, S. A. and W. K. H. Schmidt: 1994, 'H^+ and He^{2+} heating at the Earth's bow shock'. *J. Geophys. Res.* **99**(18), 11539–11546.

Galeev, A. A., C. F. Kennel, V. V. Krasnoselskikh, and V. V. Lobzin: 1988a, 'Quasiperpendicular collisionless high Mach number shocks'. In: *Proc. Joint Varenna-Abastumani Int. School & Workshop on Plasma Astrophysics, Varenna, Italy, 24 Aug - 3 September*. pp. 173–183.

Galeev, A. A., C. F. Kennel, V. V. Krasnoselskikh, and V. V. Lobzin: 1988b, 'The role of whistler oscillations in the formation of the structure of high Mach number collisionless shock'. In: *Proc. Joint Varenna-Abastumani Int. School & Workshop on Plasma Astrophysics, Varenna, Italy, 24 Aug - 3 September*. pp. 165–171.

Galeev, A. A., V. V. Krasnoselskikh, and V. V. Lobzin: 1988c, 'On the fine structure of the quasiperpendicular supercritical collisonless shock,'. *Sov. J. Plasma Phys.* **14**, 697.

Gedalin, M. and E. Griv: 1999, 'Role of overshoots in the formation of the downstream distribution of adiabatic electrons'. *J. Geophys. Res.* **104**(13), 14821–14826.

Giacalone, J., T. P. Armstrong, and R. B. Decker: 1991, 'Effect of magnetic overshoot on shock drift acceleration'. *J. Geophys. Res.* **96**(15), 3621–3626.

Gosling, J. T., J. R. Asbridge, S. J. Bame, G. Paschmann, and N. Sckopke: 1978, 'Observations of two distinct populations of bow shock ions in the upstream solar wind'. *Geophys. Res. Lett.* **5**, 957–960.

Gosling, J. T. and A. E. Robson: 1985, 'Ion reflection, gyration, and dissipation at supercritical shocks'. In: B. Tsurutani and R. Stone (eds.): *Collisionless Shocks in the Heliosphere: Reviews of current research*, Geophys. Monogr. Ser. vol 35. Washington, D.C.: American Geophysical Union, pp. 141–152.

Greenstadt, E. W., C. T. Russell, and M. Hoppe: 1980, 'Magnetic field orientation and suprathermal ion streams in the Earth's foreshock'. *J. Geophys. Res.* **85**(.14), 3473–3479.

Gustafsson, G., R. Bostrøm, B. Holback, G. Holmgren, A. Lundgren, K. Stasiewicz, L. Åéhlen, F. S. Mozer, D. Pankow, P. Harvey, P. Berg, R. Ulrich, A. Pedersen, R. Schmidt, A. Butler, A. W. C. Fransen, D. Klinge, M. Thomsen, C.-G. Falthammar, P.-A. Lindqvist, S. Christenson, J. Holtet, B. Lybekk, T. A. Sten, P. Tanskanen, K. Lappalainen, and J. Wygant: 1997, 'The electric field and wave experiment for the Cluster mission'. *Space Sci. Rev.* **79**, 137–156.

Hada, T., M. Oonishi, B. Lembège, and P. Savoini: 2003, 'Shock front nonstationarity of supercritical perpendicular shocks'. *J. Geophys. Res.* **108**, 1233, doi:10.1029/2002JA009339.

Hellinger, P., P. Trávnícek, and H. Matsumoto: 2002, 'Reformation of perpendicular shocks: Hybrid simulations'. *Geophys. Res. Lett.* **29**, 2234, doi:10.1029/2002GL015915.

Heppner, J. P., N. C. Maynard, and T. L. Aggson: 1978, 'Early results from ISEE-1 electric field measurements'. *Space Sci. Rev.* **22**, 777.

Heppner, J. P., M. Sugiura, T. L. Skillman, B. G. Ledley, and M. Campbell: 1967, 'OGO-A magnetic field observations'. *J. Geophys. Res.* **72**(11), 5417.

Horbury, T. S., P. Cargill, E. A. Lucek, A. Balogh, M. W. Dunlop, T. Oddy, C. Carr, A. Szabo, and K.-H. Fornacon: 2001, 'Cluster magnetic field observations of the bowshock: Orientation, motion and structure'. *Ann. Geophys.* **19**, 1399–1409.

Horbury, T. S., P. J. Cargill, E. A. Lucek, J. Eastwood, A. Balogh, M. W. Dunlop, K.-H. Fornaçon, and E. Georgescu: 2002, 'Four spacecraft measurements of the quasi-perpendicular terrestrial bowshock: Orientation and motion'. *J. Geophys. Res.* **107**(A8), 1208, doi 10.1029/2001JA000273.

Ipavich, F. M., G. Gloeckler, D. Hamliton, and L. Kistler: 1988, 'Protons and alpha particles in field-aligned beams upstream of the bow shock'. *Geophys. Res. Lett.* **15**, 1153.

Karpman, V. I.: 1964, 'Structure of the shock front propagating at an angle of the magnetic field in a low density plasma'. *Sov. Phys. Tech. Phys. Engl. Trans.* **8**, 715.

Kennel, C. F., J. P. Edmiston, and T. Hada: 1985, 'A quarter century of collisionless shock research'. *Washington DC American Geophysical Union Geophysical Monograph Series* **34**, 1–36.

Krasnosel'skikh, V. V.: 1985, 'Nonlinear plasma motions across the magnetic field'. *Sov. Phys. JETP* **62**, 282.

Krasnosel'skikh, V. V., T. Vinogradova, M. A. Balikhin, H. S. C. Alleyne, A. K. Pardaens, L. J. C. Woolliscroft, S. I. Klimov, A. Petrukovich, W. A. C. Mier-Jedrzejowicz, and D. J. Southwood: 1991, 'On the nature of low frequency turbulence in the foot of strong quasi-perpendicular shocks'. *Advances in Space Research* **11**, 15–18.

Krasnosselskikh, V., B. Lemb'ege, P. Savoini, and V. V. Lobzin: 2002, 'Nonstationarity of strong collisionless quasiperpendicular shocks: Theory versus full particle numerical simulations'. *Phys. Plasmas* **9**(4), 1192.

Kucharek, H., E. Möbius, M. Scholer, C. Mouikis, L. Kistler, T. Horbury, A. Balogh, H. Réme, and J. Bosqued: 2004, 'On the origin of field-aligned beams at the quasi-perpendicular bow shock: Multi-spacecraft observations by Cluster'. *Ann. Geophys.* **22**, 2301–2308.

Lembège, B. and J. M. Dawson: 1987a, 'Plasma heating through a supercritical oblique collisionless shock'. *Physics of Fluids* **30**, 1110–1114.

Lembège, B. and J. M. Dawson: 1987b, 'Self-consistent study of a perpendicular collisionless and nonresistive shock'. *Physics of Fluids* **30**, 1767–1788.

Lembège, B. and P. Savoini: 1992, 'Nonstationarity of a two-dimensional quasiperpendicular supercritical collisionless shock by self-reformation'. *Physics of Fluids B* **4**, 3533–3548.

Lembège, B., S. N. Walker, P. Savoini, M. A. Balikhin, and V. Krasnosel'skikh: 1999, 'The spatial sizes of electric and magnetic field gradients in a simulated shock'. *Adv. Space. Res.* **24**, 109–112.

Lepidi, S., U. Villante, and A. J. Lazarus: 1997, 'Single spacecraft identification of the bow shock orientation and speed: A comparison between different methods'. *Nuovo Cimento C Geophysics Space Physics C* **20**, 911.

Leroy, M. M., C. C. Goodrich, D. Winske, C. S. Wu, and K. Papadopoulos: 1981, 'Simulation of a perpendicular bow shock'. *Geophys. Res. Lett.* **8**, 1269–1272.

Leroy, M. M. and D. Winske: 1983, 'Backstreaming ions from oblique Earth bow shocks'. *Annales Geophysicae* **1**, 527–536.

Leroy, M. M., D. Winske, C. C. Goodrich, C. S. Wu, and K. Papadopoulos: 1982, 'The structure of perpendicular bow shocks'. *J. Geophys. Res.* **87**, 5081.

Liewer, P. C., V. K. Decyk, J. M. Dawson, and B. Lembège: 1991, 'Numerical studies of electron dynamics in oblique quasi-perpendicular collisionless shock waves'. *J. Geophys. Res.* **96**, 9455.

Lin, R. P., C. I. Meng, and K. A. Anderson: 1974, '30 -100 keV protons upstream from the Earth bow shock'. *J. Geophys. Res.* **79**, 489.

Livesey, W. A., C. F. Kennel, and C. T. Russell: 1982, 'ISEE-1 and -2 observations of magnetic field strength overshoots in quasi-perpendicular bow shocks'. *Geophys. Res. Lett.* **9**, 1037–1040.

Livesey, W. A., C. T. Russell, and C. F. Kennel: 1984, 'A comparison of specularly reflected gyrating ion orbits with observed shock foot thicknesses'. *J. Geophys. Res.* **89**(18), 6824–6828.

Maksimovic, M., S. D. Bale, T. S. Horbury, and M. André: 2003, 'Bow shock motions observed with CLUSTER'. *Geophys. Res. Lett.* **30**, 1393, doi:10.1029/2002GL016761.

Meziane, K., C. Mazelle, M. Wilber, D. Lequéau, J. Eastwood, H. Rème, I. Dandouras, J. Sauvaud, J. Bosqued, G. Parks, L. Kistler, M. McCarthy, B. Klecker, A. Korth, M. Bavassano-Cattaneo, R. Lundin, and A. Balogh: 2004, 'Bow shock specularly reflected ions in the presence of low-frequency electromagnetic waves: A case study'. *Annales Geophysicae* **22**, 2325–2335.

Möbius, E., H. Kucharek, C. Mouikis, E. Geogescu, L. M. Kistler, M. A. Popecki, M. Scholer, J. M. Bosqued, H. Réme, C. W. Carlson, B. Klecker, A. Korth, G. K. Parks, J. C. Sauvand, H. Balsiger, M.-B. Bavassano-Cattaneu, I. Dandouras, A. M. DiLellis, L. Eliasson, V. Formisano, T. Hobury, W. Lennartson, R. Lundin, M. McCarthy, J. McFadden, and G. Paschmann: 2001, 'Observation of the spatial and temporal structure of field-aligned beam and gyrating ring distributions at the quasi-perpendicular bow shock with Cluster CIS'. *Ann. Geophys.* **19**, 1411.

Morse, D. L.: 1976, 'A model for ion thermalization in the Earth's bow shock'. *J. Geophys. Res.* **81**(10), 6126–6130.

Morse, D. L., W. W. Destler, and P. L. Auer: 1972, 'Nonstationary behavior of collisionless shocks'. *Physical Review Letters* **28**, 13–16.

Newbury, J. A. and C. T. Russell: 1996, 'Observations of a very thin collisionless shock'. *Geophys. Res. Lett.* **23**, 781.

Papadopoulos, K.: 1985, 'Microinstabilities and anomalous transport'. *Washington DC American Geophysical Union Geophysical Monograph Series* **34**, 59–90.

Paschmann, G., N. Sckopke, I. Papamastorakis, J. R. Asbridge, S. J. Bame, and J. T. Gosling: 1981, 'Characteristics of reflected and diffuse ions upstream from the earth's bow shock'. *J. Geophys. Res.* **86**(15), 4355–4364.

Paschmann, G. and P. W. Daly (eds.): 1998, *Analysis methods for multi-spacecraft data*, ISSI Sci. Rep. SR-001. Bern: ISSI.

Paschmann, G., N. Sckopke, S. J. Bame, and J. Gosling: 1982, 'Observations of gyrating ions in the foot of the nearly perpendicular bow shock'. *Geophys. Res. Lett.* **9**, 881.

Paschmann, G., N. Sckopke, I. Papamastorakis, J. Asbridge, S. Bame, and J. Gosling: 1980, 'Energetization of solar wind ions by reflection from the Earth's bow shock'. *J. Geophys. Res.* **85**, 4689.

Peredo, M., J. A. Slavin, E. Mazur, and S. A. Curtis: 1995, 'Three-dimensional position and shape of the bow shock and their variation with Alfvenic, sonic and magnetosonic Mach numbers and interplanetary magnetic field orientation'. *J. Geophys. Res.* **100**(9), 7907–7916.

Quest, K. B.: 1985, 'Simulations of high-Mach-number collisionless perpendicular shocks in astrophysical plasmas'. *Physical Review Letters* **54**, 1872–1874.

Quest, K. B.: 1986, 'Simulations of high Mach number perpendicular shocks with resistive electrons'. *J. Geophys. Res.* **91**(10), 8805–8815.

Russell, C. T. and E. W. Greenstadt: 1979, 'Initial ISEE magnetometer results - Shock observation'. *Space Science Reviews* **23**, 3–37.

Saxena, R., S. D. Bale, and T. S. Horbury: 2004, 'Wavelength and decay length of density overshoot structure at supercritical, collisionless bow shocks'. *Phys. of Plasmas* p. submitted.

Scholer, M., H. Kucharek, and J. Giacalone: 2000, 'Cross-field diffusion of charged particles and the problem of ion injection and acceleration at quasi-perpendicular shocks'. *J. Geophys. Res.* **105**, 18285.

Scholer, M., I. Shinohara, and S. Matsukiyo: 2003, 'Quasi-perpendicular shocks: Length scale of the cross-shock potential, shock reformation, and implication for shock surfing'. *J. Geophys. Res.* **108**, 1014, doi:10.1029/2002JA009515.

Schwartz, S. J.: 1998, 'Shock and discontinuity normals, Mach numbers, and related parameters'. In: G. Paschmann and P. W. Daly (eds.): *Analysis methods for multi-spacecraft data*, ISSI Sci. Rep. SR-001. Bern: ISSI, pp. 249–270.

Schwartz, S. J. and D. Burgess: 1984, 'On the theoretical/observational comparison of field-aligned ion beams in the Earth's foreshock'. *J. Geophys. Res.* **89**, 2381 – 2384.

Schwartz, S. J., M. F. Thomsen, and J. Gosling: 1983, 'Ions upstream of the earth's bow shock: A theoretical comparison of alternative source populations'. *J. Geophys. Res.* **88**, 2039–2047.

Sckopke, N., G. Paschmann, S. J. Bame, J. T. Gosling, and C. T. Russell: 1983, 'Evolution of ion distributions across the nearly perpendicular bow shock - Specularly and non-specularly reflected-gyrating ions'. *J. Geophys. Res.* **88**(17), 6121–6136.

Sckopke, N., G. Paschmann, A. L. Brinca, C. W. Carlson, and H. Lühr: 1990, 'Ion thermalization in quasi-perpendicular shocks involving reflected ions'. *J. Geophys. Res.* **95**, 6337.

Scudder, J. D.: 1995, 'A Review of the Physics of Electron Heating at Collisionless shocks'. *Adv. Space. Res.* **15**, 181.

Scudder, J. D., T. L. Aggson, A. Mangeney, C. Lacombe, and C. C. Harvey: 1986, 'The resolved layer of a collisionless, high beta, supercritical, quasi-perpendicular shock wave. II - Dissipative fluid electrodynamics'. *J. Geophys. Res.* **91**(10), 11053–11073.

Sibeck, D. G., R. E. Lopez, and E. C. Roelof: 1991, 'Solar wind control of the magnetopause shape, location, and motion'. *J. Geophys. Res.* **96**(15), 5489–5495.

Sonnerup, B. U. O.: 1969, 'Acceleration of particles reflected at a shock front'. *J. Geophys. Res.* **74**, 1301.

Spreiter, J. R., A. L. Summers, and A. Y. Alksne: 1966, 'Hydromagnetic flow around the magnetosphere'. *Planet. Space Sci.* **14**, 223–223.

Tanaka, M., C. C. Goodrich, D. Winske, and K. Papadopoulos: 1983, 'A Source of the backstreaming ion beams in the foreshock region'. *J. Geophys. Res.* **88**, 3046.

Terasawa, T.: 1979, 'Energy spectrum and pitch angle distribution of particles reflected by MHD shock waves of the fast mode'. *Planet. Space Sci.* **27**, 193.

Thomsen, M.: 1985, 'Upstream suprathermal ions'. In: B. Tsurutani and R. Stone (eds.): *Collisionless Shocks in the Heliosphere: Reviews of current research*, Geophys. Monogr. Ser. vol 35. Washington, D.C.: American Geophysical Union, pp. 253–270.

Thomsen, M. F., J. T. Gosling, S. J. Bame, W. C. Feldman, G. Paschmann, and N. Sckopke: 1983a, 'Field-aligned beams upstream of the Earth's bow shock: Evidence for a magnetosheath source'. *Geophys. Res. Lett.* **10**, 1207–1210.

Thomsen, M. F., S. J. Schwartz, and J. T. Gosling: 1983b, 'Observational evidence on the origin of ions upstream of the earth's bow shock'. *J. Geophys. Res.* **88**, 7843–7852.

Tsurutani, B. T. and R. G. Stone: 1985, 'Collisionless shocks in the heliosphere: Reviews of current research'. *Washington DC American Geophysical Union Geophysical Monograph Series* **35**.

Vaisberg, O., S. Klimov, G. Zastenker, M. Nozdrachev, A. Sokolov, V. Smirnov, S. Savin, and L. Avanov: 1984, 'Relaxation of plasma at the shock front'. *Adv. Space. Res.* **4**, 265–275.

Vaisberg, O., G. Zastenker, V. Smirnov, Z. Nemecek, and J. Safrankova: 1986a, 'Ion distribution function dynamics near the strong shock front (Project Intershock)'. *Advances in Space Research* **6**, 41–44.

Vaisberg, O. L., G. N. Zastenker, V. N. Smirnov, Z. Nemechek, and I. Shafrankova: 1986b, 'Dynamics of the ion distribution function near the Earth's bow shock (May 11, 1985)'. *Cosmological Research* **24**, 166–176.

Walker, S., H. Alleyne, M. Balikhin, M. Andre, and T. Horbury: 2004, 'Electric field scales at quasi-perpendicular shocks'. *Annales Geophys.* **22**, 2291–2300.

Walker, S. N., M. A. Balikhin, H. S. K. Alleyne, W. Baumjohann, and M. Dunlop: 1999, 'Observations of a very thin shock'. *Adv. Space. Res.* **24**, 47–50.

Wygant, J. R., M. Bensadoun, and F. S. Mozer: 1987, 'Electric field measurements at subcritical, oblique bow shock crossings'. *J. Geophys. Res.* **92**, 11109.

Chapter 6

Quasi-parallel Shock Structure and Processes

D. Burgess[1], E. A. Lucek[2], M. Scholer[3], S. D. Bale[4], M. A. Balikhin[5], A. Balogh[2], T. S. Horbury[2], V. V. Krasnoselskikh[6], H. Kucharek[7], B. Lembège[8], E. Möbius[7 9], S. J. Schwartz[1 10], M. F. Thomsen[11], and S. N. Walker[5]

6.1 Introduction

When the interplanetary magnetic field is oriented such that the angle between the upstream magnetic field and the nominal bow shock normal is small ($\theta_{Bn} <$ 45°), a much more complex shock is observed than in the quasi-perpendicular case. Historically, this has made interpreting single spacecraft data more difficult, so that for a long time the quasi-parallel shock remained relatively poorly understood. The difficulties arise, as we now understand, because the supercritical quasi-parallel shock is a spatially extended and inhomogeneous transition, with smaller length-scale features cyclically reforming within it.

[1]Astronomy Unit, Queen Mary, University of London, London, UK
[2]Space and Atmospheric Physics, The Blackett Laboratory, Imperial College London, London, UK
[3]Max-Planck-Institut für extraterrestrische Physik, Garching, Germany
[4]Department of Physics and Space Sciences Laboratory, University of California, Berkeley, CA, USA
[5]Automatic Control and Systems Engineering, University of Sheffield, Sheffield, UK
[6]LPCE/CNRS, Orléans, France
[7]Space Science Center, Institute for the Study of Earth, Oceans, and Space, University of New Hampshire, Durham, New Hampshire, USA
[8]CETP/IPSL, Velizy, France
[9]Also Department of Physics, University of New Hampshire, Durham, New Hampshire, USA
[10]Now at Space and Atmospheric Physics, The Blackett Laboratory, Imperial College London, London, UK
[11]Los Alamos National Laboratory, Los Alamos, NM, USA

Space Science Reviews 118: 205–222, 2005.
DOI: 10.1007/s11214-005-3832-3

Under the quasi-parallel magnetic geometry ions are able to escape into the region upstream of the shock (e.g., Gosling et al., 1982), where they give rise to and interact with the waves which populate the foreshock (e.g., Le and Russell, 1992), as discussed in Eastwood et al. (2005, this issue). Of particular importance is the association of energetic ions (10–300 keV) with the foreshock and quasi-parallel shock. The role of the quasi-parallel shock as the site of particle acceleration has been fundamental in the development and testing of theories of particle acceleration. The understanding gained has direct implications for other solar system and astrophysical shocks. Recent advances in this area are discussed in Section 6.3.

Before the launch of Cluster a picture of the quasi-parallel shock had been developed from single and dual spacecraft observations, together with results from numerical simulations. Observations of the magnetic field strength changes within the shock showed that they could not all be explained simply by the in-and-out motion of a single shock surface, which would produce nested signatures. A nested signature arises when the boundary crossings observed at one spacecraft are contained, in time, within those observed by a second spacecraft. Instead there were coherent, short scale, magnetic pulsations embedded within the overall transition. These pulsations had sunward directed velocities, but were convected anti-sunwards in the solar wind plasma flow (e.g., Thomsen et al., 1988).

A standard model was developed in which the quasi-parallel shock transition was viewed as being composed of a patchwork of magnetic field enhancements, which grew from the interaction between upstream waves propagating sunward in the plasma frame and a gradient in the supra-thermal particle pressure (e.g., Giacalone et al., 1993; Dubouloz and Scholer, 1995). These field enhancements were characterised by a region where the magnetic field magnitude was a factor of ≥ 4 or so greater than the background field. They were also somewhat separated from surrounding fluctuations, such that they appeared as discrete structures. These magnetic field enhancements were termed SLAMS (short large amplitude magnetic structures) (Schwartz and Burgess, 1991) and were proposed to be the 'building blocks' of the shock. Intrinsic to this picture were the concepts of a spatially extended and patchy shock transition, since the SLAMS collectively caused the thermalisation of the plasma, and temporal or cyclic evolution: new waves growing and steepened as they were convected back toward the shock, replacing those SLAMS which had passed downstream (Burgess, 1989; Schwartz and Burgess, 1991). These pulsations have formed the major focus for Cluster work related to the quasi-parallel shock.

At the time Cluster was launched there were many questions remaining unanswered about the nature of the quasi-parallel shock. Statistical studies of dual spacecraft observations suggested that SLAMS-like pulsations had a shorter correlation length than ULF waves: $\sim$1000 km (Greenstadt et al., 1982) compared with $\sim 0.5 R_E$ (Le and Russell, 1990), but without multi-spacecraft observations it was not possible to determine their overall size or shape perpendicular to the plasma flow direction, over what scale they were coherent, or whether they had internal structure. Their effect on the plasma was not well understood, and although the downstream region showed evidence for variations in the ion reflection properties of the shock (Thomsen et al., 1990), the relative contributions of a spatially extended and temporally varying shock were not well established. Information on SLAMS growth rates was limited by having only two point measurements with an intrinsic temporal/spatial variation ambiguity, and it was not known on what timescales they developed as they approached the shock surface. Simulation results suggested that they had a rapid growth rate, of the order of seconds or less (Giacalone et al., 1994), and that they might be refracted in a direction parallel to the shock as they were convected anti-sunward (Dubouloz and Scholer, 1995). More recently, predictions were also made of the evolution of the shape of the SLAMS structure with time from a ULF wave, to a symmetric magnetic field enhancement and finally to a steepened, asymmetric shape (e.g., Tsubouchi and Lembège, 2004). Simulations also predicted that steepened SLAMS should reflect a portion of the incoming solar wind flow, since their leading edges behave locally as quasi-perpendicular shocks.

6.2 Structure

6.2.1 Overview

With Cluster observations on a range of scales, we can start to address some open questions by exploiting the simultaneous four point measurements of multiple important observables, including magnetic field, electric field and waves, supported by high quality observations of other variables including plasma parameters, ion distributions and energetic particle fluxes. In addition, computer advances have allowed more sophisticated simulations to be made of the parallel shock (e.g., Scholer et al., 2003; Tsubouchi and Lembège, 2004), providing new results for comparison with observations. One difficulty, however, in studying well-developed quasi-parallel shocks using Cluster is the relatively small number of examples which have been observed at each scale. The first season of dayside observations was adversely affected by limited orbital coverage, but even in later years the number of clear quasi-parallel shocks remains small. This might be explained in part by the Cluster orbit, which tends to cross the bow shock at high latitudes near noon and only crosses the bow shock near the equator when apogee is located far on the magnetopause flanks. Cluster consequently samples a much higher proportion of low Mach number shocks than if the spacecraft were in an equatorial orbit and crossed the shock nearer the nose of the magnetopause.

In order to give an overview of a shock crossing, Figure 6.1 shows a quasi-parallel shock observed by Cluster 1 on February 2 (day 33) 2001. Data from the magnetic field instrument (FGM) (Balogh et al., 2001) and the ion instrument (CIS) (Rème et al., 2001) are shown. At the beginning of the interval Cluster was in the

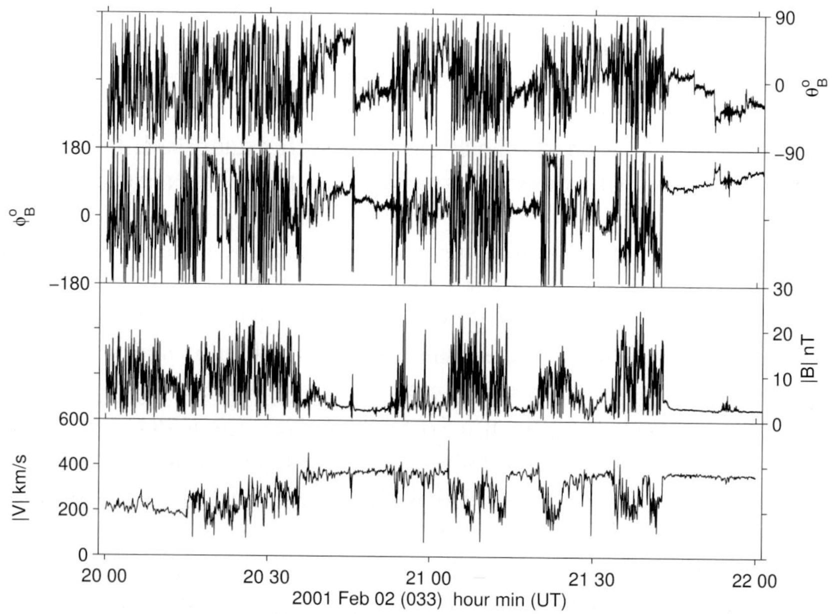

Figure 6.1. Magnetic field and velocity data recorded through a parallel shock crossing on February 2 2001. Data are from Cluster 1. Panels show magnetic field elevation (θ) and azimuthal (ϕ) angles in degrees, plotted in GSE co-ordinates, magnetic field magnitude $|\mathbf{B}|$ (nT), and plasma velocity $|\mathbf{V}|$ (km s^{-1}). Cluster starts the interval in the magnetosheath, observes several shock crossings, and by the end of the interval is in the solar wind. Figure provided by E. A. Lucek.

magnetosheath. The spacecraft then made several crossings of the shock before entering the solar wind. The shock encounters are clearest in the plasma velocity, while the magnetic field data are very 'turbulent' and disturbed over an extended interval. It is worth noting that there are several intervals where the spacecraft returns to undisturbed solar wind, presumably in response to the shock moving due to changes in the upstream conditions.

The association of SLAMS and suprathermal ions is illustrated in Figure 6.2 which shows a close-up of a short interval within the extended transition covered by Figure 6.1. The top set of Figure 6.2 shows data from the CIS HIA sensor. The sub-panels show ion energy flux in counts per second from different directions indicated by the key on the right hand side. In this key, sunwards is toward the top of the page, earthward is toward the bottom of the page, with dusk and dawn toward the left and right respectively. HIA was operating in solar wind mode at this time, so the solar wind beam is measured separately and plotted in the topmost sub-panel. The bottom sub-panel here shows the flux integrated over a 4π solid angle. The bottom set of sub-panels of Figure 6.2 shows magnetic field and plasma velocity data in the same format as Figure 6.1. Just after 2045 UT Cluster encountered an interplanetary discontinuity (which generated a hot flow anomaly, Eastwood et al. 2005, this issue) at which the interplanetary magnetic field turned to an orientation consistent with generating a quasi-parallel shock. The observed magnetic field then remained steady until about 2053 UT when suprathermal ions were seen in the duskward and sunward look directions, associated with enhancements in the magnetic field magnitude and depressions in the plasma flow velocity. Density enhancements can also be seen in the CIS spectra at the times of the magnetic field enhancements. At the very end of the interval Cluster briefly entered a region of shocked plasma.

6.2.2 Pulsations: Structure and scales

The first Cluster magnetic field observations of SLAMS when the tetrahedron scale was 600 km (Lucek et al., 2002) showed that, unexpectedly, significant variations in $|\mathbf{B}|$ occurred on these separations, despite the fact that the overall SLAMS extent exceeded the tetrahedron size as inferred from the overall duration and speed of the signatures.

This is illustrated in Figure 6.3 which shows the overall similarity between the observations at the different spacecraft, but at the same time the significant differences that can occur. In these examples the overall ordering of transitions indicate anti-sunward convection, but sometimes the ordering varies between leading and trailing edges and also between subsequent pulsations. The differences do not appear to be dominated by growth or decay in time: the SLAMS amplitudes are not ordered in time for example, but with only a few samples it is not possible to rule out that time evolution contributes to the observed signatures. Observations of ULF waves during the same pass showed that variations in $|\mathbf{B}|$ typically occur on shorter scales in SLAMS than in ULF waves, as expected from previous observations.

Observations of structures within a quasi-parallel shock when the Cluster spacing was 100 km were reported by Lucek et al. (2004), and shown in the bottom panels of Figure 6.4 which compares SLAMS on both scales. In this case the magnetic field strength enhancements on all spacecraft are better correlated, although with some small differences. Statistical analysis of the differences in $|\mathbf{B}|$ as a function of distance perpendicular to the flow suggests that while the SLAMS extent is much greater than 100 km, the gradient scale appears to be of the order of 100-150 km.

The attempts to use Cluster observations taken at different scales to infer the intrinsic scale of SLAMS has indicated an unexpected paradox. In order use Cluster timings to obtain velocities and orientations (as in the case of the quasi-perpendicular shock), it is necessary that all four spacecraft measure similar profiles. However, determining scale lengths transverse to either flow or motion of structures requires differences between the measured profiles. So on the one hand the requirement is closely positioned observation points, but on the other well separated ones. This suggests that a better separation strategy with unequal separations might have been

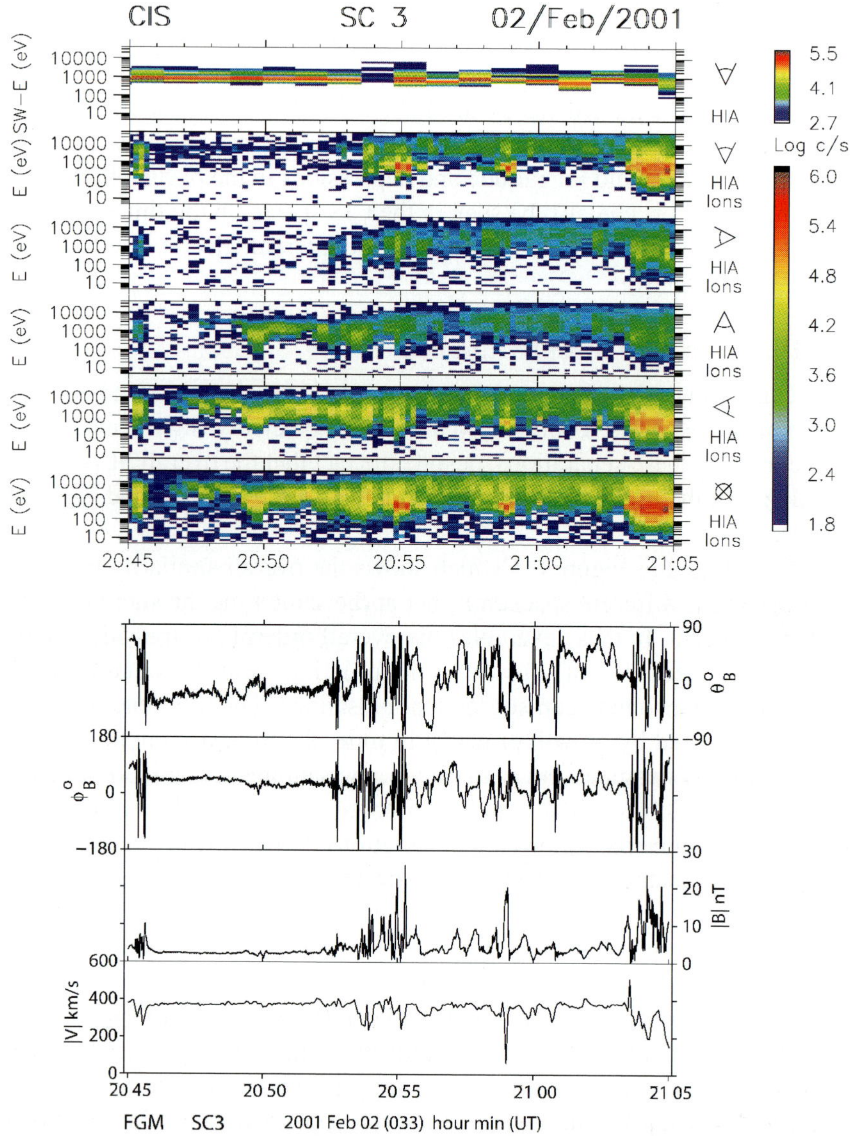

Figure 6.2. Extended transition from clean solar wind (left) to the magnetosheath (far right) showing the appearance and evolution of diffuse ions (top) together with magnetic fluctuations at the bow shock under quasi-parallel conditions. The top panel shows data from the CIS HIA instrument. Sub-panels show ion energy flux in counts per second from different directions indicated by the key on the right hand side: sunwards (up); dusk (left); dawn (right); earthward (down). HIA was operating in solar wind mode at this time, so the solar wind beam is measured separately and plotted in the top sub-panel. The bottom sub-panel shows the flux integrated over a 4π solid angle. The bottom panel shows magnetic field and plasma velocity data in the same format as Figure 6.1. Figure provided by E. A. Lucek, CIS data courtesy of the CIS team.

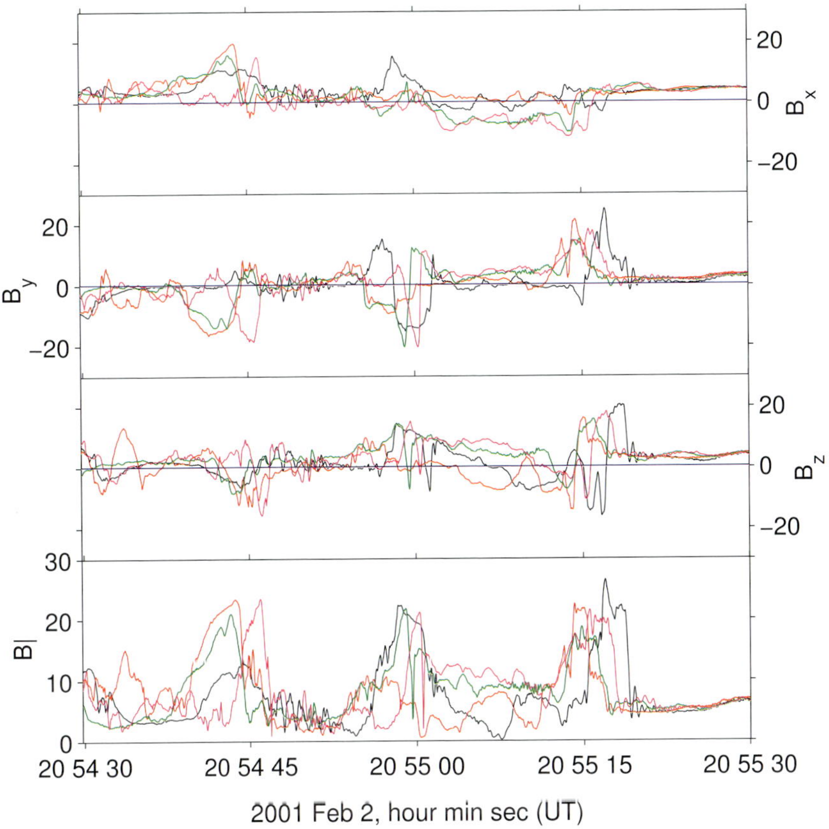

Figure 6.3. Magnetic field data (components in GSE and magnitude) from the four Cluster spacecraft at 22 vectors/s showing magnetic field enhancements within a pulsation region. The spacecraft separations are in the range 400–800 km. From Lucek et al. (2002).

possible for the quasi-parallel shock, although with some loss of information on the orientation of structures.

This paradox is evident in the studies of SLAMS at spacecraft separations of 600 km and 100 km. When the Cluster tetrahedron scale was ∼600 km, there were relatively few SLAMS where the spacecraft profiles were similar enough to calculate the SLAMS orientation; while at a separation of ∼100 km, the method can be used more comprehensively, but there is not any information available over scales larger than the spacecraft separation. A possible future study would be to test statistically whether (single spacecraft) minimum variance analysis is a good measure of the SLAMS orientation locally using timings data for small spacecraft separations. Consequently, SLAMS observations at large tetrahedron scales could be explored for signatures of curvature or acceleration.

6.2.3 Pulsations: Fields and particles

The Cluster EFW experiment measures the spacecraft potential at relatively high cadence (5 samples/second), which serves as a proxy for the electron density. Behlke et al. (2003) presented the first high resolution electric field and spacecraft potential measurements at SLAMS. It was found that the density and magnetic field strength were correlated, confirming the earlier indications that SLAMS were fast mode structures. Timing analysis was used to derive the SLAMS velocity in the spacecraft frame. This differed from the background solar wind velocity and the SLAMS pulsation was found to be propagating sunward in the plasma frame. However, the motional electric field computed using the SLAMS velocity matched the measured electric field within the SLAMS, while a systematic discrepancy was found if the motional electric field was calculated using the background solar wind velocity. Behlke et al. (2003) inferred therefore that, locally, the plasma within the SLAMS moved at the same speed as the SLAMS perturbation, although the SLAMS moved relative to the background solar wind. The authors also noted that in some cases there was a plasma density depletion behind the SLAMS (in the SLAMS frame), indicating some kind of wake structure, and that the presence of differences between the four spacecraft suggested that this wake varied on scales of ~ 600 km.

Examination of high resolution measurements of the electric field within SLAMS showed that on short time scales local discrepancies were observed between the measured and calculated electric fields. Behlke et al. (2003) noted that similar small scale electric field spikes have been observed in simulations of SLAMS (Lembège et al., 2004). These studies of high resolution electric fields at SLAMS were extended by Behlke et al. (2004) who presented evidence of solitary waves within them. The observed waves occurred as bipolar (sometimes tripolar) pulses with a parallel scale of the order of 10 Debye lengths, and peak-to-peak amplitudes of up to 65 $\mathrm{mV\,m^{-1}}$. The velocities of these very small structures were derived using the four probes on a single spacecraft as an interferometer. Solitary waves observed in the auroral acceleration region and at other magnetospheric boundaries have been studied extensively. However the examples observed in the shock pulsations, with a propagation velocity of 400 - 1200 $\mathrm{km\,s^{-1}}$, being higher than the typical ion thermal speed, have properties which make them difficult to explain using any of the current theories of solitary waves. Although they present an interesting problem in terms of plasma physics, these solitary wave structures have no obvious associated net potential drop. Therefore, although they might play a role in electron thermalisation, the authors noted that there was insufficient evidence to conclude that they are important for the structure of SLAMS.

The relationship between magnetic field strength and electron density (as indicated by the spacecraft potential) has been investigated by Stasiewicz et al. (2003). They suggest that some SLAMS have no associated plasma density variations, but that for those with a positive density enhancement, corresponding to a fast mode structure, the density and field changes are not exactly in phase. The maximum in the field magnitude occurs during the density increase. Stasiewicz et al. (2003) present a descriptive model, based on simple momentum conservation and a polytropic equation of state, that is used to explain the profiles in density and field. The explanation is based on assuming compression and then heating in stages through the SLAMS. Although possibly appropriate to some of the observed structures, it is clear from, for example, other work reported here that if there is such variability in pulsation signatures then only a rigorous statistical study will be convincing.

Two examples of SLAMS are shown in Figure 6.4 showing the magnetic field magnitude and the negative spacecraft potential ($-V_{SC}$) which can be used as a proxy for the plasma density (Gustafsson et al., 2001). The top panel shows data for a Cluster separation scale of $\sim$600 km, whilst the bottom panel has data for a separation scale of $\sim$100 km. The correlation between data from different spacecraft as a function of separation scale has already been discussed. Here we note that the correlation between field magnitude and density is not always the same, even within a single event. This is most clearly seen in the top panel where the magnitude of the SLAMS in $|\mathbf{B}|$ is smaller at Cluster 1 than at Cluster 3 or 4, whilst the magnitude of $-V_{SC}$ is approximately the same at these three spacecraft. It can also be noted that the times at which $|\mathbf{B}|$ and $-V_{SC}$ increase differ within a particular SLAMS event. This points to the importance of nonlinear, time non-stationary processes in the SLAMS growth and evolution.

A key issue in understanding the relationship between pulsations and the overall shock transition is their role in the development of ion thermalisation. Before Cluster crossed the SLAMS magnetic field enhancement shown in the lower panel of Figure 6.4, the spacecraft observed a region of enhanced wave activity in the magnetic field, where the correlation between the signals at the different spacecraft was lower; this corresponds to the wake region behind the SLAMS. The implication is that variations occurred on smaller spatial scales in this region: on the scale of the order of the ion inertial length or even less. This might be related to the partial plasma thermalisation, visible in both the velocity moment and CIS ion distributions shown in Figure 6.5. In comparison with the ion distribution in undisturbed solar wind plasma, the plasma on the left-hand side of the SLAMS (downstream in the SLAMS frame) is hotter, slower and deflected slightly. The middle of the three distributions shows a slightly heated solar wind beam, and an additional population of sunward moving ions, consistent with reflection from the SLAMS. Further work is required to relate such observations to the orientation of the pulsations, and also to identify any instability responsible for the thermalization in the wake region.

6.3 Diffuse Ion Acceleration

Since the early observations of ions upstream of Earth's quasi-parallel bow shock by Asbridge et al. (1968) and by Lin et al. (1974) upstream ions with energies just above the solar wind energy up to several hundred keV have been under investiga-

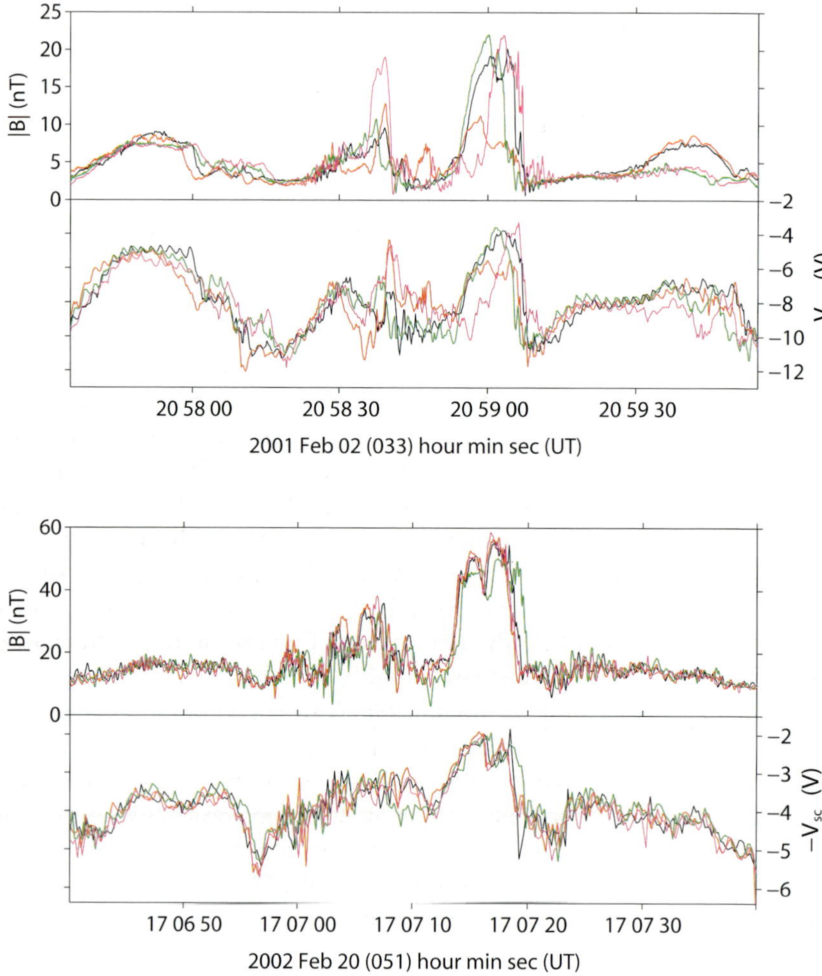

Figure 6.4. Two examples of SLAMS observed in magnetic field magnitude $|\mathbf{B}|$, (top sub-panel) and negative spacecraft potential $-V_{sc}$, (bottom sub-panel). The case in the top panel was observed when the tetrahedron scale was ~600 km, and the lower panel shows an example when the tetrahedron scale was reduced to ~100 km. Figure provided by E. A. Lucek, EFW spacecraft potential data courtesy of R. Behlke.

tion (see also Section 2.3 of Bale et al., 2005). A number of studies have established two different sources for these ions: the so-called magnetospheric bursts at energies above a few hundred keV and the bow-shock associated, so-called diffuse particles. Diffuse ion events last up to several hours and extend in energy up to $\sim$ 150 keV (Ipavich et al., 1981; Scholer et al., 1981). The spectra of protons and alpha particles in these events are generally well described as exponentials in energy per

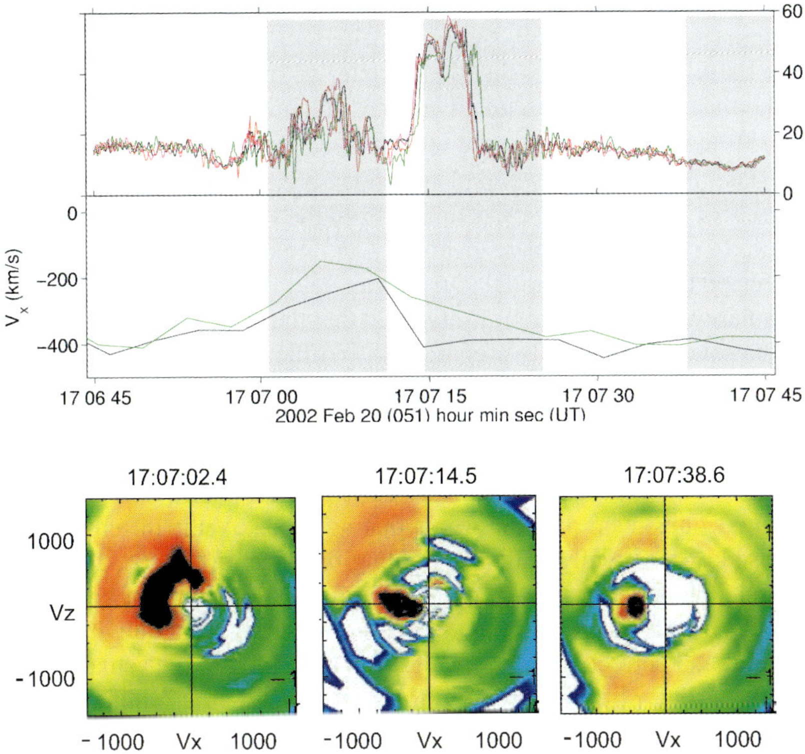

Figure 6.5. Ion reflection associated with SLAMS. The top two panels show magnetic field magnitude and plasma velocity in the X_{GSE} direction. The three shaded areas show the approximate times over which the three ion distributions shown below were accumulated. Each ion distribution shows ion flux plotted on a colour scale where blue and green indicate low fluxes, and black indicates high flux. Each distribution is a cut through $V_Y(GSE) = 0$ and shows V_X on the ordinate and V_Z on the abscissa. Figure provided by E. A. Lucek, CIS data courtesy of I. Dandouras.

charge, and the ratio of the fluxes of the two species is constant as a function of energy per charge (Ipavich et al., 1981). The distribution of diffuse ions is rather isotropic with a shock directed bulk velocity slower than the solar wind speed. Hoppe et al. (1981) demonstrated that there is a one-to-one correlation between the presence of diffuse upstream ions and the occurrence of hydromagnetic waves in the foreshock region. This has led to the widely adopted picture of an intensive interplay between the waves and energetic ions: the waves are thought to constitute scattering centres for the ions, which results in a diffusive transport. If transport is diffusive these particles will experience the shock compression, which leads to first order Fermi acceleration (e.g., Axford et al., 1977, see also reviews by Scholer, 1985 and Forman and Webb, 1985). The diffusive shock acceleration mechanism has often been challenged by the proposition that all upstream ion enhancements at all major magnetospheres, in particular that of Earth's, are exclusively of magnetospheric origin (Sarris et al., 1987). A prerequisite for a first order Fermi mechanism to occur at Earth's bow shock is that the particles do not stream scatter free, as implied by the magnetospheric escape model in which ions from the magnetopause propagate through the magnetosheath upstream against the convecting solar wind.

Steady state diffusive theory at a planar shock predicts that the density of energetic ions (energy E) falls off exponentially from the shock front into the upstream region, with an e-folding distance given by $L(E) = \kappa(E)/v_{sw}$, where $\kappa(E)$ is the spatial diffusion coefficient parallel to the mean field for particles of energy E, and v_{sw} is the solar wind speed (here it is assumed that the shock normal, the magnetic field and the solar wind velocity are aligned). In order to demonstrate that the upstream ions are indeed subject to diffusive transport it is thus of great interest to determine the spatial gradient along the magnetic field. Previous determinations of the spatial variation of upstream particle intensity used single spacecraft data and had therefore to be done on a statistical basis. Ipavich et al. (1981) have shown by analysing 33 upstream particle events that at ~ 33 keV the differential proton flux decreased exponentially with an e-folding distance of $R = 7 \pm 2R_e$ upstream of the shock. Trattner et al. (1994) performed a statistical study of about 300 upstream events. They determined the distance to the shock along the interplanetary magnetic field and corrected the shock position according to the actual solar wind ram pressure. From a linear regression they determined an e-folding distance of $R = 4.8 \pm 0.1R_e$ at 20 keV. The e-folding distance increases with energy and varies from $3.2 \pm 0.2R_e$ at 10 keV to $9.3 \pm 1R_e$ at ~ 67 keV.

Although these statistical results seem to prove the importance of diffusive transport in the upstream region they are hampered by the possibility that the intensity in the upstream region at one location may vary widely from event to event according to interplanetary conditions. The small correlation coefficient of 0.65 of the linear regression analysis of Trattner et al. (1994) may even be taken as evidence that the intensity is more or less constant along the field and that the spread is due to sampling of events with varying intensity. Cluster provides the unique opportunity to investigate the spatial gradient of upstream diffuse ions during times of large spacecraft separation distance. Recent analysis of Cluster data by Kis et al. (2004) provides direct evidence that the upstream particles undergo a diffusive transport.

Kis et al. (2004) have investigated a 10 hour time period on 18 February 2003 when Cluster was upstream of the quasi-parallel bow shock on an inbound orbit. Figure 6.6 shows from top to bottom the solar wind v_x velocity, the three components of the interplanetary magnetic field as measured by FGM on Cluster 1, and the partial density of upstream ions in the energy range 24 - 32 keV as measured by the CIS-HIA instrument at two spacecraft, at Cluster 1 (black line) and at Cluster 3 (green line). Also shown by inserts are projections of the spacecraft orbits and the bow shock onto the GSE $x-y$ and $x-z$ planes. During this time period Cluster 1 is about 1.5 R_e closer to the bow shock than Cluster 3. As can be seen from Figure 6.6, Cluster 1 exhibits a higher partial density during the entire time interval than Cluster 3: the partial density ratio between the two spacecraft varied from a few percent up to 50%. The Cluster spacecraft were magnetically connected to the quasi-parallel bow shock throughout. From the 8 spin (32 sec) averaged magnetic field data, the distance of the spacecraft along the magnetic field to the bow shock intersection point has been determined. For determination of the bow shock position, the Peredo et al. (1995) model has been employed and was expanded to the actual bow shock location by using the spatial position of the bow shock during the first inbound shock crossing of SC1. From this model an average local magnetic field-shock normal angle at the magnetic field-bow shock intersection point of $20° \pm 8°$ has been determined for the whole 10 hour time period. The ratio between the difference of the partial densities at the two spacecraft and the difference between the spacecraft distances along the magnetic field to the bow shock intersection point then gives gradient values at various distances. Figure 6.7 shows these partial density gradients for the 24 - 32 keV energy range in $\mathrm{cm}^{-3}\mathrm{R}_e^{-1}$ versus distance from the bow shock along the magnetic field averaged in seven $1\mathrm{R}_e$ wide distance bins. Since the data points in this logarithmic versus linear representation can be fit by a straight line the gradient is very well represented by an exponential as a function of distance from the shock.

The procedure described above has been performed for the 4 highest energy channels of the CIS-HIA instrument. The resulting e folding distances L for the partial densities in each of the 4 energy channels are shown in Figure 6.8. The error bars indicate the 1σ uncertainties due to the fitting procedure. As can be seen from Figure 6.8 the e-folding distance $L(E)$ of the partial density gradients depends approximately linearly on energy E and increases from $\sim 0.5\,\mathrm{R}_e$ at $\sim$11 keV to $\sim 2.8\,\mathrm{R}_e$ at $\sim$27 keV.

In the simple steady state model of Fermi acceleration at a planar shock, the e-folding distance is given by $L(E) = \kappa(E)/v_{sw}$. Since $\kappa(E) = d(E)v/3$, where $d(E)$ is the diffusion length and v is the particle velocity (in the plasma frame) we can write $d(E) = 3L(E)(E_{sw}/E)^{1/2}$, where E is the particle energy and E_{sw} is the bulk energy of the solar wind. Thus the diffusion length has an energy dependence of $d(E) \propto E^{0.5}$. From the measured e-folding distance and from the solar wind speed Kis et al. (2004) obtained at 30 keV a mean free path of $2.4\,\mathrm{R}_e$. The important result here is that these energetic particles are clearly undergoing a diffuse transport process in the upstream region. It is thus unavoidable that they feel the shock compression and are necessarily accelerated by a first order Fermi acceleration process at the shock. The analysis by Kis et al. (2004) can only be considered as a first step in trying to gain insight into the injection and acceleration process at the quasi-parallel bow shock. For instance it is not clear whether the small e-folding distances found during this event are due to the extremely large solar wind velocity ($\sim 640\,\mathrm{km\,s}^{-1}$). Thus other events with considerably smaller solar wind speed have to be analysed. The strong scattering is due to the waves produced by

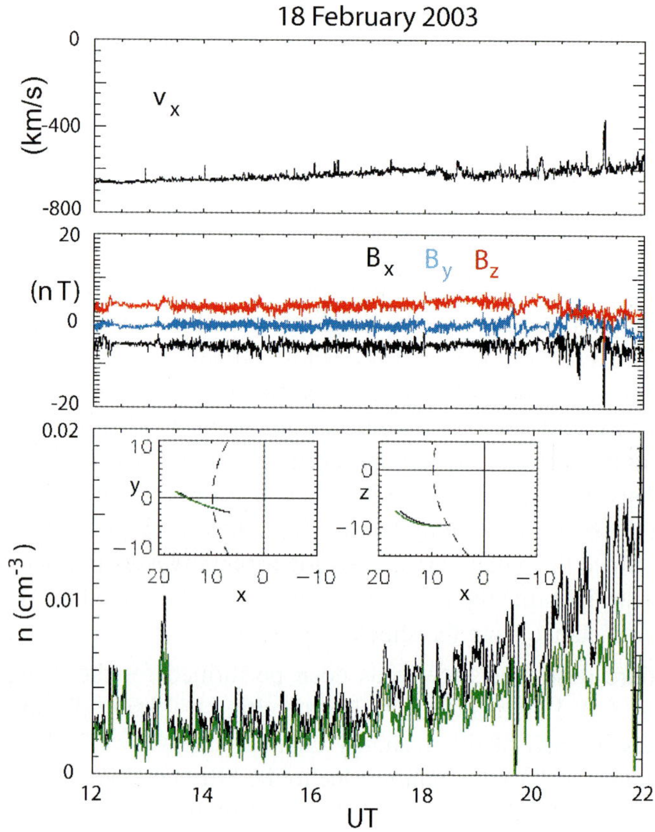

Figure 6.6. Data from which the spatial gradient in diffuse ions, and hence the scattering mean free path, have been determined. From top to bottom: Solar wind velocity component v_x and magnetic field components B_x (black line), B_y (blue line), B_z (red line) as measured on Cluster 1, partial ion density in the 24 - 32 keV energy range as measured at Cluster 1 (black line) and Cluster 3 (green line). Also shown in the lower panel are projections of the spacecraft orbits and bow shock onto the $x-y$ and $x-z$ plane, respectively. From Kis et al. (2004).

the particles themselves. Quasi-linear theory predicts an energy dependence of the diffusion coefficient depending on the slope of the magnetic field power spectrum. Comparison of the spectral exponent of the magnetic field fluctuations with the energy dependence of the diffusion coefficient and of the absolute value of the power in the fluctuations with the magnitude of the diffusion coefficient would provide a test of quasi-linear pitch-angle theory, which is widely applied to other astrophysical settings. The combined data set of magnetic field fluctuations and particle data should provide new insight into injection and acceleration at the quasi-linear bow shock.

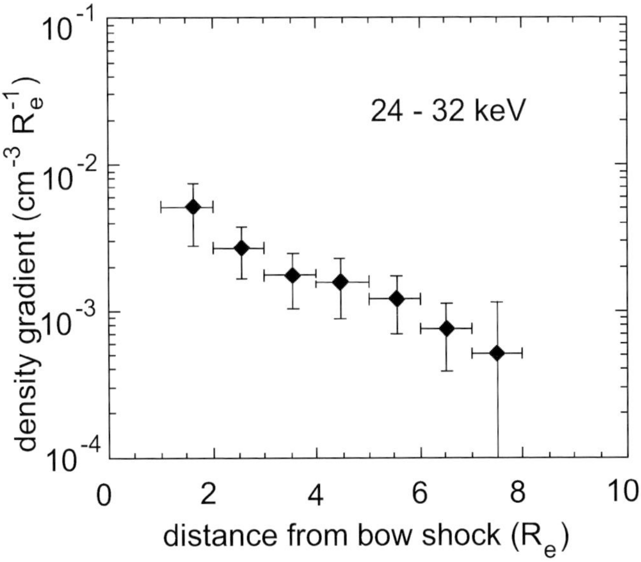

Figure 6.7. Average partial ion density gradient in the 24-32 keV energy range versus distance from the bow shock. From Kis et al. (2004).

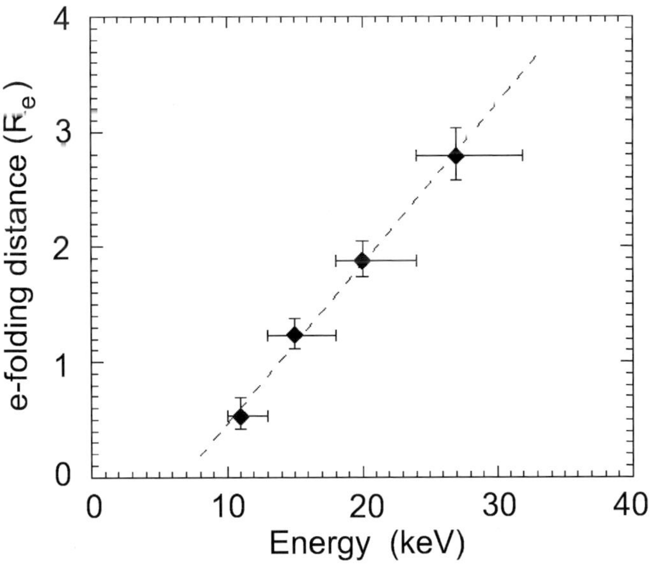

Figure 6.8. The e-folding distance obtained by fitting an exponential to the partial density gradient in 4 energy channels. From Kis et al. (2004).

6.4 Summary

The quasi-parallel regions of the Earth's bow shock, where the interplanetary magnetic field is nearly parallel to the nominal shock normal, provide a complex environment in which multi-streaming collisionless plasma and field components thrive and interact both to effect an overall shock transition, decelerating and heating the incident solar wind, and to accelerate a sub-population of particles to higher energies through the universally invoked Fermi process. The intrinsically unsteady and turbulent nature of this region has made it somewhat less amenable to single spacecraft studies (by comparison to the quasi-perpendicular bow shock reported in Section 2.1 of Bale et al., 2005, this issue), while also shifting the emphasis somewhat from the pure shock transition in terms of dissipation and heating to a more heterogeneous perspective.

Key Cluster results to date as reported in this Chapter include:

1. The revelation that Short Large Amplitude Structures (SLAMS), which prevail under quasi-parallel shock conditions, have considerable sub-structure on scales < 600 km despite their apparent monolithic appearance on scales ~ 6000 km.
2. New results on the electric fields within SLAMS, confirming in some sense their monolithic macroscopic role.
3. Quantitative analysis of the diffusive scattering length in the upstream 'foreshock' region, confirming the operation of a Fermi-like process there.

These results represent only a fraction of the potential of the Cluster dataset, particularly since the later stages of the mission will see the spacecraft separations increase to scales commensurate with the foreshock scales of 10,000's of km.

Acknowledgements

The contributions of HK and EM were partially supported under NASA Grants NAG5-10131 and NAG5-11804. PPARC (UK) support for this work includes fellowships (TSH and EL) and research grants (DB, SJS, SNW).

References

Asbridge, J. R., S. J. Bame, and I. B. Strong: 1968, 'Outward flow of protons from the Earth's bow shock'. *J. Geophys. Res.* **73**(12), 5777.

Axford, W. I., E. Leer, , and G. Skadron: 1977, 'The acceleration of cosmic rays by shock waves'. In: *Proc. Int. Conf. Cosmic Rays 15th*. p. 132.

Bale, S. D., M. A. Balikhin, T. S. Horbury, V. V. Krasnoselskikh, H. Kucharek, E. Möbius, S. N. Walker, A. Balogh, D. Burgess, B. Lembège, E. A. Lucek, M. Scholer, S. J. Schwartz, and M. F. Thomsen: 2005, 'Quasi-perpendicular shock structure and processes'. *Space Sci. Rev.* **this issue**.

Balogh, A., C. M. Carr, M. H. Acuña, M. W. Dunlop, T. J. Beek, P. Brown, K. H. Fornacon, E. Georgescu, K. H. Glassmeier, J. Harris, G. Musmann, T. Oddy, and K. Schwingenschuh: 2001,

'The Cluster magnetic field investigation: Overview of in-flight performance and initial results'. *Ann. Geophys.* **19**, 1207–1217.

Behlke, R., M. André, S. D. Bale, J. S. Pickett, C. A. Cattell, E. A. Lucek, and A. Balogh: 2004, 'Solitary structures associated with short large-amplitude magnetic structures (SLAMS) upstream of the Earth's quasi-parallel bow shock'. *Geophys. Res. Lett.* **31**, L16805, doi:10.1029/2004GL019524.

Behlke, R., M. André, S. C. Buchert, A. Vaivads, A. I. Eriksson, E. A. Lucek, and A. Balogh: 2003, 'Multi-point electric field measurements of Short Large-Amplitude Magnetic Structures (SLAMS) at the Earth's quasi-parallel bow shock'. *Geophys. Res. Lett.* **30**, 1177, doi:10.1029/2002GL015871.

Burgess, D.: 1989, ' Cyclic behaviour at quasi-parallel collisionless shocks'. *Geophys. Res. Lett.* **16**, 345–348.

Dubouloz, N. and M. Scholer: 1995, 'Two-dimensional simulations of magnetic pulsations upstream of the Earth's bow shock'. *J. Geophys. Res.* **100**, 9461–9474.

Eastwood, J., E. A. Lucek, C. Mazelle, K. Meziane, Y. Narita, J. Pickett, and R. Treumann: 2005, 'The Foreshock'. *Space Sci. Rev.* **this issue**.

Forman, M. A. and G. M. Webb: 1985, 'Acceleration of energetic particles'. *Washington DC American Geophysical Union Geophysical Monograph Series* **35**, 91–114.

Giacalone, J., S. J. Schwartz, and D. Burgess: 1993, ' Observations of suprathermal ions in association with SLAMS '. *Geophys. Res. Lett.* **20**, 149–152.

Giacalone, J., S. J. Schwartz, and D. Burgess: 1994, 'Artificial spacecraft in hybrid simulations of the quasi-parallel Earth's bow shock: Analysis of time series versus spatial profiles and a separation strategy for Cluster'. *Ann. Geophys.* **12**, 591–601.

Gosling, J. T., M. F. Thomsen, S. J. Bame, W. C. Feldman, G. Paschmann, and N. Sckopke: 1982, 'Evidence for specularly reflected ion upstream from the quasi-parallel bow shock'. *Geophys. Res. Lett.* **9**, 1333–1336.

Greenstadt, E. W., M. M. Hoppe, and C. T. Russell: 1982, 'Large-amplitude magnetic variations in quasi-parallel shocks: correlation lengths measured by ISEE 1 and 2'. *Geophys. Res. Lett.* **9**, 781–784.

Gustafsson, G., M. André, T. Carozzi, A. Eriksson, C.-G. Fälthammar, R. Grard, G. Homgren, J. Holtet, N. Ivchenko, T. Karlsson, Y. Khotvaintsev, S. Klimov, H. Laakso, G. Marklund, F. Mozer, K. Mursula, A. Pedersen, B. Popielawska, S. Savin, K. Stasiewicz, P. Tanskanen, A. Vaivads, and J.-E. Wahlund: 2001, 'First results of electric field and density observations by Cluster EFW based on initial months of operation'. *Ann. Geophys.* **19**, 1219–1240.

Hoppe, M. M., C. T. Russell, L. A. Frank, T. E. Eastman, and E. W. Greenstadt: 1981, 'Upstream hydromagnetic waves and their association with backstreaming ion populations - ISEE 1 and 2 observations'. *J. Geophys. Res.* **86**, 4471–4492.

Ipavich, F. M., A. Galvin, G. Gloeckler, M. Scholer, and D. Hovestadt: 1981, 'A statistical survey of ions observed upstream of the earth's bow shock - Energy spectra, composition, and spatial variation'. *J. Geophys. Res.* **86**, 4337–4342.

Kis, A., M. Scholer, B. Klecker, E. Möbius, E. Lucek, H. Réme, J. M. Bosqued, L. M. Kistler, and H. Kucharek: 2004, 'Multispacecraft observations of diffuse ions upstream of Earth's bow shock'. *Geophys. Res. Lett.* **31**, L20801, doi:10.1029/2004GL020759.

Le, G. and C. T. Russell: 1990, ' A study of the coherence length of ULF waves in the Earth's foreshock '. *J. Geophys. Res.* **95**, 10703–10706.

Le, G. and C. T. Russell: 1992, 'A study of ULF wave foreshock morphology ? II: Spatial variation of ULF waves'. *Planet. Space Sci.* **40**, 1215–1225.

Lembège, B., J. Giacalone, M. Scholer, T. Hada, M. Hoshino, V. Krasnoselskikh, H. Kucharek, P. Savoini, and T. Terasawa: 2004, 'Selected problems in collisionless-shock physics'. *Space Science Reviews* **110**, 161–226.

Lin, R. P., C. I. Meng, and K. A. Anderson: 1974, '30 -100 keV protons upstream from the Earth bow shock'. *J. Geophys. Res.* **79**, 489.

Lucek, E. A., T. S. Horbury, A. Balogh, I. Dandouras, and H. Rème: 2004, 'Cluster observations of structure at quasi-parallel bow shocks'. *Ann. Geophys.* in press.

Lucek, E. A., T. S. Horbury, M. W. Dunlop, P. J. Cargill, S. J. Schwartz, A. Balogh, P. Brown, C. Carr, K. H. Fornacon, and E. Georgescu: 2002, 'Cluster magnetic field observations at a quasi-parallel bow shock'. *Ann. Geophys.* **20**, 1699–1710.

Peredo, M., J. A. Slavin, E. Mazur, and S. A. Curtis: 1995, 'Three-dimensional position and shape of the bow shock and their variation with Alfvenic, sonic and magnetosonic Mach numbers and interplanetary magnetic field orientation'. *J. Geophys. Res.* **100**(9), 7907–7916.

Rème, H., C. Aoustin, J.-M. Bosqued, and et al.: 2001, 'First multispacecraft ion measurements in and near the Earth's magnetosphere with the identical Cluster ion spectrometry (CIS) experiment'. *Ann. Geophys.* **19**, 1303–1354.

Sarris, E. T., G. C. Anagnostopoulos, and S. M. Krimigis: 1987, 'Simultaneous measurements of energetic ion (50 keV and above) and electron (220 keV and above) activity upstream of Earth's bow shock and inside the plasma sheet - Magnetospheric source for the November 3 and December 3, 1977 upstream events'. *J. Geophys. Res.* **92**(11), 12083–12096.

Scholer, M.: 1985, 'Diffusive acceleration'. *Washington DC American Geophysical Union Geophysical Monograph Series* **35**, 287–301.

Scholer, M., F. M. Ipavich, and G. Gloeckler: 1981, 'Beams of protons and alpha particles greater than approximately 30 keV/charge from the Earth's bow shock'. *J. Geophys. Res.* **86**, 4374–4378.

Scholer, M., H. Kucharek, and I. Shinohara: 2003, 'Short large-amplitude magnetic structures and whistler wave precursors in a full-particle quasi-parallel shock simulation'. *J. Geophys. Res.* **108**, 1273, doi:10.1029/2002JA009820.

Schwartz, S. J. and D. Burgess: 1991, ' Quasi-parallel shocks: A patchwork of three-dimensional structures'. *Geophys. Res. Lett.* **18**, 373–376.

Stasiewicz, K., M. Longmore, S. Buchert, P. K. Shukla, B. Lavraud, and J. Pickett: 2003, 'Properties of fast magnetosonic shocklets at the bow shock'. *Geophys. Res. Lett.* **30**, 2241, doi:10.1029/2003GL017971.

Thomsen, M. F., J. T. Gosling, S. J. Bame, and T. G. Onsager: 1990, 'Two-state heating at quasi-parallel shocks'. *J. Geophys. Res.* **95**, 6363–6374.

Thomsen, M. F., J. T. Gosling, and C. T. Russell: 1988, ' ISEE studies of the quasi-parallel bow shock'. *Adv. Space. Res.* **8**, 9175–9178.

Trattner, K. J., E. Mobius, M. Scholer, B. Klecker, M. Hilchenbach, and H. Luehr: 1994, 'Statistical analysis of diffuse ion events upstream of the Earth's bow shock'. *J. Geophys. Res.* **99**(18), 13389.

Tsubouchi, K. and B. Lembège: 2004, 'Full particle simulations of short large-amplitude magnetic structures (SLAMS) in quasi-parallel shocks'. *J. Geophys. Res.* **109**, A02114, doi: 10.1029/2003JA010014.

Chapter 7

Cluster at the Bow Shock: Status and Outlook

M. Scholer[1], M. F. Thomsen[2], D. Burgess[3], S. D. Bale[4], M. A. Balikhin[5], A. Balogh[6], T. S. Horbury[6], V. V. Krasnoselskikh[7], H. Kucharek[8], E. A. Lucek[6], B. Lembège[9], E. Möbius[8 10], S. J. Schwartz[3 11], and S. N. Walker[5]

When some dynamical energy release occurs in dilute astrophysical plasmas, collisionless shocks arise where the macroscopic flow is regulated by microscopic dissipation and where selective particle acceleration may occur. Collisionless shocks are found in the corona of the Sun, in the solar wind, in front of planetary magnetospheres, and in many other astrophysical settings. The two main questions in collisionless shock physics are: (1) how is dissipation achieved in a plasma where two-body Coulomb collisions are unimportant, and (2) how is part of the thermal plasma accelerated to high energies? The Earth's bow shock is a collisionless shock where these questions have been investigated in great detail for more than three decades by in situ observations. It turned out that physical processes at the bow shock occur on all spatial scales, from the electron inertial scale

[1]Max-Planck-Institut für extraterrestrische Physik, Garching, Germany
[2]Los Alamos National Laboratory, Los Alamos, NM, USA
[3]Astronomy Unit, Queen Mary, University of London, London, UK
[4]Department of Physics and Space Sciences Laboratory, University of California, Berkeley, CA, USA
[5]Automatic Control and Systems Engineering, University of Sheffield, Sheffield, UK
[6]Space and Atmospheric Physics, The Blackett Laboratory, Imperial College London, London, UK
[7]LPCE/CNRS, Orléans, France
[8]Space Science Center, Institute for the Study of Earth, Oceans, and Space, University of New Hampshire, Durham, New Hampshire, USA
[9]CETP/IPSL, Velizy, France
[10]Also Department of Physics, University of New Hampshire, Durham, New Hampshire, USA
[11]Now at Space and Atmospheric Physics, The Blackett Laboratory, Imperial College London, London, UK

Space Science Reviews 118: 223–227, 2005.
DOI: 10.1007/s11214-005-3833-2

in the ramp of the quasi-perpendicular bow shock to many hundreds of ion inertial scales in the foreshock region of the quasi-parallel bow shock. Cluster with its multi-spacecraft capability provides the unique opportunity to unravel many of the outstanding questions in collisionless shock physics. But, as we have seen, when answering one question a plethora of new questions arises. The main achievements of Cluster concerning the Earth's bow shock so far are:

1. Determination of the length scales of the quasi-perpendicular bow shock associated with the shock ramp, overshoot/undershoot, and downstream wavetrain.

2. Detection of small scale electric field structures in the ramp of the quasi-perpendicular bow shock

3. Revealing the origin of field-aligned beams at the quasi-perpendicular bow shock

4. Determination of the internal structure of the short large-amplitude structures in the upstream region of the quasi-parallel bow shock

5. Determination of the mean free path of diffuse energetic ions upstream of the quasi-parallel bow shock

One of the important questions in collisionless shock physics is to which degree the steepening of a magnetosonic wave leading to the shock is balanced by either dissipation or dispersion. The relevant spatial scale associated with dispersion is the electron or ion inertial length, whereas dissipation due to gyroviscosity associated with gradients of the ion stress tensor introduces the gyroscale of the ions in the shock. Bale et al. (2003) have shown that the largest density transition scale of the quasi-perpendicular bow shock is, independent of shock (mgnetosonic) Mach number, the convected ion gyroradius. In contrast, the shock scale in units of the ion inertial scale increases with shock Mach number. The ion inertial scale seems to be appropriate at low Mach numbers only. This demonstrates the importance of gyroviscosity (within the fluid picture) for dissipation in quasi-perpendicular shocks. In astrophysical settings high Mach number, low beta shocks are of particular interest. It remains to be seen whether low beta shocks also scale with the convected ion gyroradius since some particle simulations predict, at least during short times, very small shock transition regions of the order of several electron scales.

The Bale et al. (2003) analysis captures only the largest transition scale at the shock. However, as demonstrated by the electric field measurements by Walker et al. (2004), there are fine-scale features of the electric field in the ramp of the quasi-perpendicular bow shock down to the electron inertial scale. These electric field excursions in the ramp have large amplitudes (up to 70 $\mathrm{mVm^{-1}}$), are layered, with normals in the general shock normal direction, and can contribute up to 50% to the cross shock potential. The origin and nature of these electric field structures is yet unclear and deserves further intense investigation. It will also be interesting to see whether and/or how these electric field structures influence the adiabaticity of the electrons when they pass through the shock ramp.

While it has been known since the ISEE era that the quasi-perpendicular bow shock is a source of field-aligned beams, the Cluster multi-spacecraft capability has now given important clues as to the production mechanism of these beams. During bow shock crossings often one spacecraft has been downstream of the shock while another spacecraft has been upstream and observed a field-aligned beam with speeds of about 1500 km/s. As shown by Kucharek et al. (2004) in the same velocity region the downstream distribution is virtually empty, indicating that these beams do not originate from leakage of downstream heated solar wind ions. In the ramp the gyrating ions are observed which are due to specular reflection. As these ions interact again during their orbital motion with the ramp magnetic field in such a way that their energy is constant in the de Hoffmann-Teller frame they can reach a region in phase space where they have a large velocity parallel to the field and can escape upstream in the form of field-aligned ion beams. However, details of this process have still to be worked out. The picture becomes more complicated due to the three-dimensional structure of the bow shock on smaller scales: for instance, ripples on the shock may change locally the magnetic field-shock normal angle which allows particles to escape which otherwise would have been transmitted downstream. Other open questions are: what determines the intensity of the field-aligned beam and what role does the cross-shock potential play in reflecting the particles? The large data base provided by Cluster constitutes an ideal platform to study these questions in the future in considerable detail.

The upstream region of the quasi-parallel shock is spatially extended and constitutes an inhomogeneous transition region. This region is dominated by diffuse backstreaming energetic ions, supposedly accelerated at the shock out of the solar wind thermal population, ultra-low frequency waves generated by these ions, and large amplitude magnetic pulsations (SLAMS) as the shock is approached. SLAMS are most probably vital for the thermalization process in quasi-parallel collisionless shocks, i.e., understanding their growth, shape and size will contribute to understanding in detail how the inhomogeneous components of the shock cause the thermalization of the plasma. Observations by Cluster upstream of the quasi-parallel bow shock have revealed the internal structure of the SLAMS. It was anticipated that the spatial extent of SLAMS is of the order of one Earth radius. However, it has been found that the SLAMS are highly structured on a much smaller scale: Lucek et al. (2002) reported in SLAMS significant variations of the magnetic field magnitude when the tetrahedron scale was of the order of 600 km. SLAMS orientation has so far been difficult to determine: when the tetrahedron scale was 600 km the structures seen at different spacecraft were rather different, while at a tetrahedron scale of 100 km no information of the orientation over larger scales than the spacecraft separation scale can be made. From timing analysis it was confirmed that SLAMS are fast mode structures propagating sunward in the solar wind frame (Behlke et al., 2003; Stasiewicz et al., 2003). However, the motional electric field within the SLAMS is equal to the measured electric field, indicating that the plasma in the SLAMS moves with the same velocity as the structure itself. The quasi-parallel bow shock clearly contains multiple scale lengths, and further analysis of current Cluster data, together with future Cluster observations at intermediate and large scales can address some of the outstanding questions.

Diffuse energetic ions upstream of the quasi-parallel bow shock are strong evidence for first order Fermi (or diffusive) acceleration at the shock. First order Fermi acceleration requires that the particles cross and recross the shock many times, i.e., they have to be pitch angle scattered in the upstream and downstream medium or, in other words, the mean free path parallel to the magnetic field has to be sufficiently small. When particles diffuse from the shock along the magnetic field upstream against the convection of the solar wind the intensity has to fall off in the steady state exponentially with an e-folding distance given by the diffusion coefficient and the solar wind velocity. Cluster allows for the first time the direct determination of the e-folding distance of the diffuse ion density in the upstream region, and, in turn, a determination of the diffusion coefficient and the mean free path, respectively. During one particular event when the tetrahedron scale was about 5000 km the e-folding distance of the partial diffuse ion densities has been determined in several energy bands in the range between 10 and 32 keV. From the e-folding distances a mean free path from about 0.5 Earth radii at the lowest energy and 3 Earth radii at the highest energy has been determined (Kis et al., 2004). This shows that upstream ions indeed undergo diffusive spatial transport in the region of the quasi-parallel shock and strongly supports the claim that the quasi-parallel bow shock accelerates these ions. An unsolved problem in collisionless shock physics is the so-called injection problem, i.e., how and why a certain part of the thermal solar wind is further accelerated at the shock. It is to be hoped that detailed Cluster studies of the quasi-parallel bow shock also help in unravelling this problem.

The future for Cluster studies of the Earth's bow shock is bright and exciting. The datasets used in the work reported here hold considerably more information waiting to be exploited to the full. Subsequent data will take advantage of novel separation strategies to explore new (mainly larger) scales and to target nearly planar surfaces, such as the bow shock, by correspondingly non-regular tetrahedral spacecraft configurations which take account of the fact that the scales along and perpendicular to the surface are not necessarily the same. The orbit evolution will cause Cluster to encounter the bow shock at lower latitudes and thus closer to the sub-solar regions. Under nominal solar wind conditions, the expected Mach numbers will be higher, as the solar wind velocity and shock normal will be more collinear. Quasi-parallel shock geometries will be more frequently encountered, due to the tendency for 'Parker spiral' interplanetary fields to be predominantly in the plane of the ecliptic. Finally, combined studies with other outer magnetospheric missions will provide new opportunities to apply existing, and new, multi-spacecraft techniques to the many outstanding issues.

Acknowledgements

The contributions of HK and EM were partially supported under NASA Grants NAG5-10131 and NAG5-11804. PPARC (UK) support for this work includes fellowships (TSH and EL) and research grants (DB, SJS, SNW).

References

Bale, S. D., F. S. Mozer, and T. S. Horbury: 2003, 'Density-transition scale at quasiperpendicular collisionless shocks'. *Physical Review Letters* **91**(26), 265004.

Behlke, R., M. André, S. C. Buchert, A. Vaivads, A. I. Eriksson, E. A. Lucek, and A. Balogh: 2003, 'Multi-point electric field measurements of Short Large-Amplitude Magnetic Structures (SLAMS) at the Earth's quasi-parallel bow shock'. *Geophys. Res. Lett.* **30**, 1177, doi:10.1029/2002GL015871.

Kis, A., M. Scholer, B. Klecker, E. Möbius, E. Lucek, H. Réme, J. M. Bosqued, L. M. Kistler, and H. Kucharek: 2004, 'Multispacecraft observations of diffuse ions upstream of Earth's bow shock'. *Geophys. Res. Lett.* **31**, L20801, doi:10.1029/2004GL020759.

Kucharek, H., E. Möbius, M. Scholer, C. Mouikis, L. Kistler, T. Horbury, A. Balogh, H. Réme, and J. Bosqued: 2004, 'On the origin of field-aligned beams at the quasi-perpendicular bow shock: Multi-spacecraft observations by Cluster'. *Ann. Geophys.* **22**, 2301–2308.

Lucek, E. A., T. S. Horbury, M. W. Dunlop, P. J. Cargill, S. J. Schwartz, A. Balogh, P. Brown, C. Carr, K. H. Fornacon, and E. Georgescu: 2002, 'Cluster magnetic field observations at a quasi-parallel bow shock'. *Ann. Geophys.* **20**, 1699–1710.

Stasiewicz, K., M. Longmore, S. Buchert, P. K. Shukla, B. Lavraud, and J. Pickett: 2003, 'Properties of fast magnetosonic shocklets at the bow shock'. *Geophys. Res. Lett.* **30**, 2241, doi:10.1029/2003GL017971.

Walker, S., H. Alleyne, M. Balikhin, M. Andre, and T. Horbury: 2004, 'Electric field scales at quasi-perpendicular shocks'. *Annales Geophys.* **22**, 2291–2300.

III

MAGNETOPAUSE AND CUSP

Chapter 8

Magnetopause and Boundary Layer

J. De Keyser[1], M. W. Dunlop[2], C. J. Owen[3], B. U. Ö. Sonnerup[4], S. E. Haaland[5], A. Vaivads[6], G. Paschmann[7], R. Lundin[8] and L. Rezeau[9]

8.1 Introduction

The Earth's intrinsic dipolar magnetic field is separated from the ambient magnetosheath field by the magnetopause, a thin current-carrying plasma surface layer. The surface current is what is required from Ampère's law to bring about the observed net change in the field across this plasma boundary. A schematic drawing of the magnetopause surface as well as the magnetosphere and its various features is shown in Figure 8.1.

The first unambiguous observations of the magnetopause were made by the Explorer 12 spacecraft, as reported by Cahill and Amazeen (1963). Figure 8.2, taken from their paper, is an excellent illustration of the magnetic field behaviour typically seen in a dayside crossing of the magnetosphere, magnetopause, and magnetosheath. The magnetospheric field magnitude decreases with increasing distance from Earth, but less rapidly than a dipole field in a vacuum. Indeed, the surface current induces a magnetic field that adds to the geomagnetic field inward of the magnetopause, while it prevents the Earth's field lines from penetrating into the

[1]Belgian Institute for Space Aeronomy, Brussels, Belgium
[2]Space Science Division, Rutherford Appleton Laboratory, Chilton, Didcot, Oxfordshire, UK
[3]Mullard Space Science Laboratory, Holmbury St. Mary, Dorking, Surrey, UK
[4]Thayer School of Engineering, Dartmouth College, Hanover, NH, USA
[5]Max-Planck-Institut für extraterrestrische Physik, Garching, Germany
[6]Swedish Institute of Space Physics, Uppsala, Sweden
[7]International Space Science Institute, Bern, Switzerland
[8]Swedish Institute of Space Physics, Kiruna, Sweden
[9]CETP/IPSL/UPMC, Vélizy, France

Space Science Reviews 118: 231–320, 2005.
DOI: 10.1007/s11214-005-3834-1

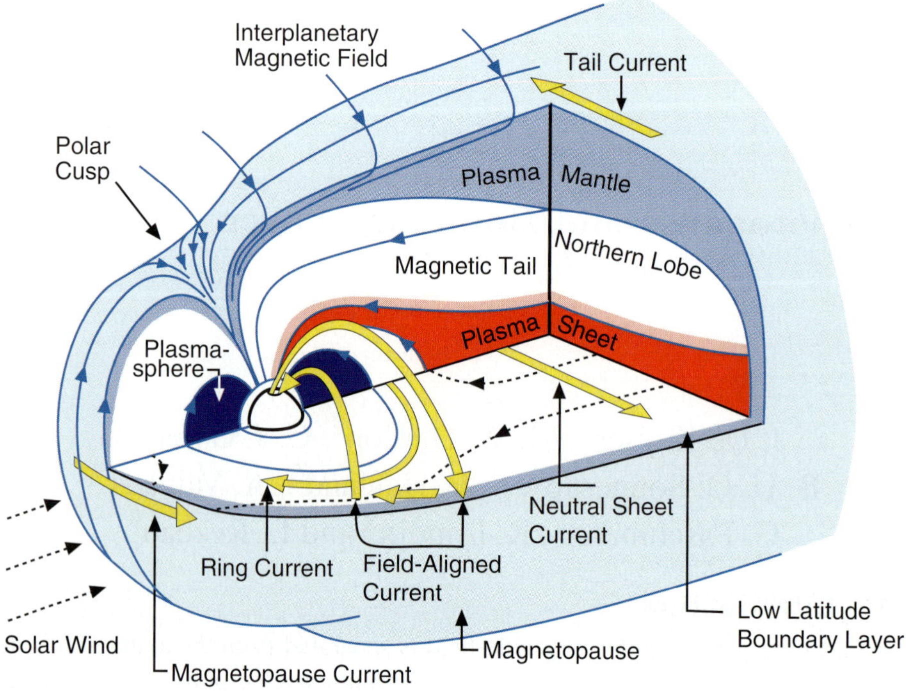

Figure 8.1. Three-dimensional cutaway view of the magnetosphere. The light blue outer surface is the magnetopause, its boundary layers are shown in darker blue. Magnetic field lines are shown in blue, electric currents in yellow. The polar region where the magnetic field lines converge is the polar cusp. The bow shock has been omitted for clarity. (Adapted from Kivelson and Russell, 1995).

magnetosheath, where the field is relatively weak. Just before the magnetopause is encountered, at about $8.2\,R_E$, the measured field is about twice that of the dipole (see e.g., Chapman and Bartels, 1940). In the magnetosheath, the field intensity is usually lower and has larger variability, reflecting the fluctuations in the interplanetary magnetic field. The increase in the field elevation angle, α, just earthward of the magnetopause suggests the presence of a plasma boundary layer, presumably connected to the dayside ionosphere via field-aligned (Region 1) currents. Due to lack of plasma measurements, the actual presence of such a layer in the Explorer 12 pass shown in Figure 8.2 remains uncertain. But measurements by later spacecraft have established the frequent presence of a plasma boundary layer immediately Earthward of the magnetopause. It is now called the 'low-latitude boundary layer' or LLBL (e.g., Eastman et al., 1976; Sckopke et al., 1981).

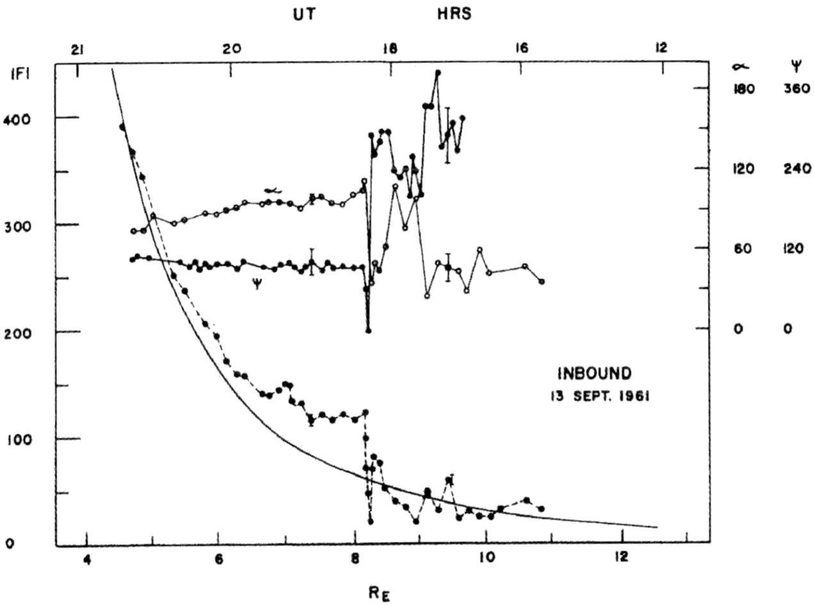

Figure 8.2. Magnetic field measured by Explorer 12 on September 13, 1963. Measured (dashed curve) and dipole (solid curve) field magnitudes are shown, along with measured field elevation angles, α, and longitude angles ψ. (From Cahill and Amazeen, 1963).

8.1.1 Orientation, motion, and thickness

The orientation of the magnetopause surface is of importance, not only for studying its geometry (e.g., surface wave phenomena), but also for the purpose of establishing whether there is a magnetic connection and associated plasma flow across this boundary. Under certain assumptions, the orientation can be deduced from single-spacecraft measurements of the magnetic field alone. If one believes that no normal field and flow components were present, that is, if the magnetopause was a tangential discontinuity (TD), one may obtain the vector normal to the magnetopause simply as the cross product of the field vectors on opposite sides of the layer. If one wants to allow for the possible presence of a normal field component, then one can adopt a one-dimensional model of the local magnetopause structure where the orientation of the current layer is also unchanged during the crossing. Under these assumptions, the condition $\nabla \cdot \mathbf{B} = 0$ requires the normal field component to be strictly constant during the crossing. The orientation can then be found as that direction in space along which the component of the magnetic field has minimum variance (Sonnerup and Cahill, 1967; Sonnerup and Scheible, 1998). Provided the 1D assumption holds and the variance matrix is not near degeneracy, this method, referred to as MVAB (Minimum Variance Analysis of the magnetic field, $\mathbf{B}$), can in principle show whether or not a nonzero normal field component was present during a crossing. With a few exceptions, it was found that this component is so small that it is hidden within its uncertainty estimate; plasma flow across the magnetopause is even more difficult to establish and requires precise knowledge of magnetopause orientation and plasma velocities.

In a time-independent picture, the magnetospheric field magnitude should decrease monotonically with increasing distance from Earth up to the magnetopause. This is not what is seen in Figure 8.2, where magnetic field magnitude fluctuations are found from 6.7 R_E outward, followed by a steep decrease to a more or less constant plateau value that persists out to the magnetopause. These features could be spatial, temporal, or both. In particular, they could be the result of magnetopause motion. From a single-spacecraft magnetic field record alone, one cannot tell for sure, but frequent observations of multiple magnetopause traversals during a single outward or inward pass, by Explorer 12 and later single-spacecraft missions ever since, have provided convincing evidence that the magnetopause is almost always in rapidly changing inward-outward motion with typical speeds of tens of km/s. The magnetopause does not have much mass, so its speed can change abruptly, and by large amounts, in response to even minor changes in the magnetosheath pressure. The amplitude of the radial motion of the magnetopause can be several Earth radii. Surface waves can also lead to multiple crossings of the boundary.

An important parameter is the thickness of the magnetopause, which determines the intensity of the currents in the layer and the steepness of the density gradients and velocities across it. Given the orientation of the magnetopause, its thickness can be inferred from the boundary speed and the crossing duration. The results, however, require that one makes certain assumptions (e.g., planarity of the layer), and they critically depend on a precise determination of the magnetopause speed. Reliable determinations of the magnetopause speed and, from it, the thickness, became possible in the late seventies. The time difference between the passage of the magnetopause over the ISEE-1 and ISEE-2 spacecraft led to speeds ranging from near zero to more than 200 km s^{-1}, with an average around 40 km s^{-1}. The crossing durations then implied thicknesses ranging from 200 km to 1800 km, with an average of about 800 km (Berchem and Russell, 1982). Reliable measurements of the lowest-order plasma moments also became available from ISEE and, later on, from the AMPTE spacecraft. This led to the development of single-spacecraft methods for determination of the magnetopause speed. The earliest of these methods is referred to as MVAB/HT. It is based on the existence of a deHoffmann-Teller (HT) frame, a concept first used by deHoffmann and Teller (1950) in the study of shocks and applied to the magnetopause by Aggson et al. (1983); for a modern version, see (Khrabrov and Sonnerup, 1998a). The HT frame moves with velocity $\mathbf{V}_{HT}$ relative to the spacecraft, in such a manner that, when transformed to this frame, the measured plasma flow becomes as nearly field aligned as the data permit. The component of $\mathbf{V}_{HT}$ along the magnetopause normal then represents the inward/outward motion of the magnetopause, while the tangential components describe motion of field structures along the magnetopause, past the observing spacecraft. Later on, a method referred to as Minimum Faraday Residue (MFR) analysis was developed (Terasawa et al., 1996; Khrabrov and Sonnerup, 1998b). Faraday's law requires the electric-field component tangential to a one-dimensional layer, of fixed structure and moving at constant speed, to be constant. This property, along with $\nabla \cdot \mathbf{B} = 0$, is used in MFR to produce a prediction of both the normal vector and the speed of the magnetopause. Tests of these methods against the timing of magnetopause passages over the two closely spaced spacecraft AMPTE/IRM and AMPTE/UKS revealed substantial uncertainties in the single-spacecraft predictions as well as difficulties in the timing results, the latter caused by difficulties in identifying features in the magnetic field profiles that are needed for an unambiguous determination of the time delays (Bauer et al., 2000).

Data from Cluster taken during the passage of all four spacecraft through the magnetopause, together with certain assumptions, in principle permit an unambiguous determination of magnetopause speed, thickness, and orientation. All methods utilised in this book assume the magnetopause surface to be flat and to maintain a fixed orientation in the entire time interval encompassing the four crossings. In the so-called Discontinuity Analyser, or DA for short (e.g., Dunlop and Woodward, 1998), the magnetopause orientation is taken from a single-spacecraft method such as MVAB. From the time delays between the crossings and the spacecraft separation vectors, the magnetopause speed and its variations can then be determined and, from the duration of each crossing, the corresponding magnetopause thickness can be found. In another method, called the Constant Velocity Approach, or CVA (Russell et al., 1983; Schwartz, 1998), the magnetopause speed is assumed constant during all four crossings. The magnetopause orientation, speed, and a thickness for each crossing, can then be determined from the timing and the crossing durations. In the more recent Constant Thickness Approach (CTA - see Haaland et al., 2004a), the magnetopause thickness is assumed to be the same for all four crossings. This method allows determination of magnetopause orientation and thickness, as well as the speed during each of the four crossings. A combination of CVA and CTA, called Minimum Thickness Variation, or MTV, has also been developed (Paschmann et al., 2004). In this method, the orientation of the magnetopause is a weighted average of the results from CVA and CTA. The magnetopause velocity and thickness are allowed to vary from crossing to crossing but the variation of the latter is minimised. Comparisons of results from various single-spacecraft methods with those from these multi-spacecraft methods are presented in Sections 8.2 and 8.3. Included in these comparisons are two new methods, first tested with Cluster data, that allow determination of the magnetopause normal and speed, either from the conservation of mass, or from the conservation of charge. The former method, referred to as Minimum Massflux Residue (MMR) analysis (Sonnerup et al., 2004a), requires measurements of plasma velocity and density from a single spacecraft. It was made practical, for the first time, by the precision of the Cluster plasma measurements. The latter method, called Minimum Variance of Current (MVAJ) (Haaland et al., 2004b), utilises the curlometer capability of Cluster, i.e., the capability to determine the electric current density from measured magnetic fields on the four spacecraft by use of Ampère's law (also a Cluster 'first', see Section 8.3).

8.1.2 Magnetopause substructure

Over the years, evidence has been mounting that the dominant transport of plasma across the magnetopause is caused by magnetic field reconnection. This process is discussed in detail in Phan et al. (2005). It implies an X-line geometry at the magnetopause. The magnetopause then has a small, non-zero normal magnetic field component with opposite sign on either side of an X-line or reconnection site. Some process must decouple electrons and ions from the magnetic field, but this process needs only be present in the immediate vicinity of the reconnection site. Elsewhere on the magnetopause, it is the normal magnetic field component accompanying reconnection that creates a direct magnetic coupling across the magnetopause and that allows solar-wind plasma to flow into the LLBL. The magnetopause then has a compound structure, with a rotational discontinuity (RD) as its principal ingredient; in a magnetohydrodynamic description, this rotational discontinuity is a large-amplitude standing Alfvén wave. Direct measurements of plasma flow across magnetopause are difficult, as noted earlier. An indirect approach is to perform the so-called Walén test, which is, in effect, a test of the tangential stress balance at the magnetopause (Paschmann et al., 1979). In its modern form, the Walén test consists of plotting the velocity components measured during a magnetopause crossing, and transformed to the HT frame, against the corresponding measured components of the Alfvén velocities. A regression slope of $+1$ indicates Alfvénic flow parallel to the magnetic field while a slope of -1 signifies antiparallel flow at the Alfvén speed. Assuming the flow is directed earthward, the former case indicates an inward, the latter case an outward directed normal magnetic field component. Thus the sign of the slope indicates on which side of the reconnection site the observations were taken. The above results are based on a one-dimensional description of the magnetopause structure during reconnection. Strong 2D effects, such as magnetic islands embedded in the current layer, or, perforce, 3D structures, can lead to Walén slopes that deviate significantly from unity. Here, Cluster can help to identify and assess the existence of additional terms in the tangential stress balance in order to understand such deviations (Section 8.4.1).

It is therefore important to assess the geometry of the current layer. The single-spacecraft methods for the determination of magnetopause orientation and motion are based on the assumption that the current layer is a one-dimensional, time-independent structure with a fixed orientation and moving at constant speed during the crossing. If these assumptions are well satisfied, the one-dimensional field and plasma structure of the layer can be easily reconstructed. If the orientation and 1-D structure are both time-stationary but the velocity is not, then the structure can nevertheless be reconstructed, in the simplest case by allowing for a constant acceleration of the HT frame or, for more complicated motions of the HT frame, by doing the HT analysis with a sliding data window. Local reconstructions of magnetopause structures, based on integration of the Grad-Shafranov equation, have been performed, starting with work by Sonnerup and Guo (1996), and followed by several more recent studies. In this technique, measurements along a single spacecraft trajectory are used as spatial initial values for the integration. Validation of the technique by use of Cluster data is discussed in Section 8.4.2, along with a technique for ingesting data from all four Cluster spacecraft into the reconstruction.

8.1.3 Physical properties

Although much progress has been made since the Explorer 12 epoch, our knowledge of what physical processes operate in the magnetopause current layer and determine its structure remains incomplete. One of the oldest unanswered questions is whether, and in what circumstances, an intrinsic electric field along the magnetopause normal remains after the external electric fields have been removed by, in effect, making a transformation into the HT frame. Ferraro (1952) concluded that, in a TD, i.e., in the absence of a magnetic field component along the magnetopause normal, such an electric field , directed outward from the magnetosphere, is required because of the vastly smaller gyro radii of impinging solar wind electrons compared to those of the impinging ions. This field would bring about an equal penetration depth of the electrons and ions. But such an intrinsic field is not a necessity: particles trapped in the magnetopause layer can in principle nullify the effect. If the magnetopause has a nonzero normal magnetic field component, both solar wind electrons and ions can flow across the magnetopause. An intrinsic electric field may nevertheless be required to assure that the electric current has no component along the magnetopause normal. Prior to Cluster, an intrinsic electric field along the magnetopause normal had not been seen. But in Section 8.5.2, new Cluster observations are presented that provide convincing evidence that such a field is sometimes present.

In an ideal magnetohydrodynamic (MHD) description, which is realistic to lowest order since the plasma is collisionless, the magnetopause is a tangential discontinuity. This means that there should be no particle transfer through the boundary. This is almost true, but not completely: A fraction of the solar wind flux does penetrate the magnetopause. Therefore, some deviation from the tangential discontinuity model should be present: the plasma behaves in a non-ideal way. Diffusive processes based on wave-particle interactions could be responsible for such behaviour. One possibility is that this type of diffusion permits reconnection to occur, so that the boundary becomes a rotational discontinuity with associated direct plasma entry by flow along the normal-field component. Alternatively, if wave-particle interactions are strong enough, they could lead to diffusive plasma entry over large portions of the magnetopause surface. When magnetic field measurements with high time resolution became available in the late sixties and early seventies, it quickly became apparent that the magnetic fluctuation level is high in the magnetopause. Later on, results from ISEE established and quantified the persistent presence of enhanced magnetic and electric field turbulence over a wide range of frequencies in the magnetopause (Gurnett et al., 1979). The nature of the waves and instabilities causing the fluctuations has been studied extensively, but their importance, if any, in producing macro-scale diffusive fluxes of mass, momentum, and energy from the solar wind across the magnetopause into a low-latitude boundary layer remains in doubt. Cluster results concerning these fluctuations, including their origin, relationship to free energy sources such as current density and density gradients, and their effectiveness in producing diffusion in large regions of the magnetopause surface, are presented in Section 8.5.4.

8.1.4 Meso-scale phenomena

On a somewhat larger scale, the magnetopause and the adjacent low-latitude boundary layer are usually not planar. With single-point measurements, one always has to make assumptions about the broader geometrical context. Cluster, by design, produces a picture of the local environment, as it is able to resolve at least part of the spatio-temporal ambiguity. For instance, bipolar signatures in the component of the magnetic field normal to the magnetopause surface were believed by some to be ‘flux transfer events’ (FTEs), caused by brief bursts of magnetic reconnection. Others interpreted these events simply as kinks in the magnetopause surface, travelling past the observing spacecraft. Simultaneous observations by Cluster spacecraft located on the two sides of the magnetopause have unambiguously shown that such FTEs represent ‘bulges’ or local swellings of the magnetopause, rather than magnetopause ripples (Section 8.6). Another example is surface waves (Section 8.7) that travel downtail along the magnetopause surface. Yet another example is the discovery of magnetosheath-like plasma structures inward of the low-latitude boundary layer, but spatially detached from it (Section 8.8). ‘Empirical reconstruction’ techniques, originally developed for single-spacecraft data (Paschmann et al., 1990; De Keyser et al., 2002), can be used in many situations to combine the four-spacecraft data from all instruments into a consistent picture of the topology of the magnetopause and the low-latitude boundary layer (see Sections 8.7 and 8.8).

8.2 Magnetopause Orientation, Motion, and Thickness

As discussed in Section 8.1, knowledge of the magnetopause orientation is important for studying its geometry, but also for the purpose of establishing whether there is a magnetic connection and associated plasma flow across this boundary. Given the orientation of the magnetopause, its thickness can be inferred from the boundary speed and the crossing duration. The thickness in turn determines the intensity of the currents in the layer and the steepness of the density and velocity gradients across it.

The techniques to determine magnetopause orientation, motion and thickness have been described in Section 8.1.1. In this section we will first look at a few magnetopause encounters that illustrate that orientation and speed can change significantly while the magnetopause is moving over the Cluster spacecraft array, which affects the existing analysis methods in different ways. This point will be highlighted by an application of the existing methods, both single- and multi-spacecraft based, to a benchmark case. The message is that there is no single method that is superior to the other under all circumstances, but that the availability of so many different methods greatly increases the confidence in the results. The section will end with the presentation of the statistics of 96 magnetopause crossings.

8.2.1 Case Studies

8.2.1.1 DA and MVAB methods

The basic Discontinuity Analyzer (DA) technique (Dunlop et al., 1997; Dunlop and Woodward, 1998, 1999; Dunlop et al., 2001, 2002a), determines a normal to the boundary for each spacecraft crossing, by using, for example, Minimum Variance Analysis (MVA). The motion and geometry of the boundary can then be calculated by combining crossing times with spacecraft separations and each normal. Figure 8.3 shows the scenario for the case of a planar discontinuity. Parallel normals **n** at each of the 4 spacecraft (1,2,3,4) positions directly imply planar geometry and allow one to characterise boundary motion (v_n^0, a_n) by

$$r_n = v_n^0 t + \frac{1}{2} a_n t^2, \tag{8.1}$$

where $r_n = \Delta\mathbf{r}_{ij} \cdot \mathbf{n}$ and $t = t_{ij}$. As there are three spacecraft pairs but only two unknowns, use of above equation involves fitting a curve to the three pairs of (r_n,t), as indicated in Figure 8.3.

If the speed of the magnetopause is constant over the spacecraft array (a_n is small), Equation 8.1 can be used to compute $\mathbf{n}/v_n^0$ from the timing and position information alone. This is the CVA method introduced earlier. For this method, the crossing times are critical and the result is sensitive to identification errors unless the spacecraft separations are large. If the velocity is actually changing over the spacecraft array, this timing estimate will give a false boundary orientation. In fact, for the DA, if the acceleration of the boundary changes rapidly, a_n may not well represent sudden boundary motion; the timing sequence can easily be used to check the smooth variation of velocity over the spacecraft array.

November 10, 2000 crossing. Figure 8.4 shows data from a high-latitude, duskside crossing on November 10, 2000, early on in the Cluster mission. The lower part shows the magnetic field measurements for the whole pass and confirms that the spacecraft entered the magnetosphere through a series of high shear crossings (dashed vertical lines). Some of the boundary crossings are particularly clear, they

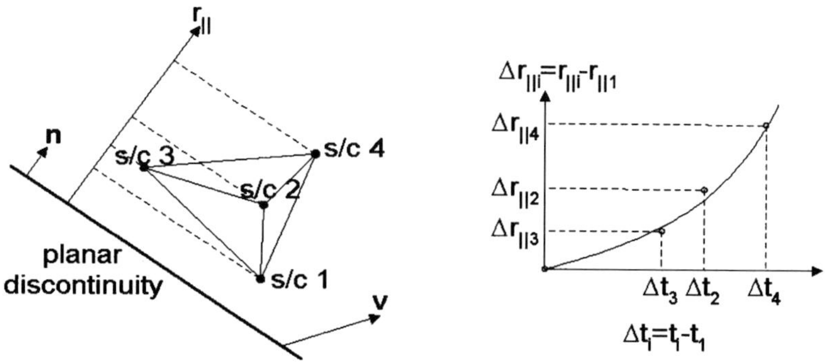

Figure 8.3. Application of the Discontinuity Analyser method to a planar boundary. Timing of boundary crossings together with distance measured along the boundary normal allow determination of boundary speed and acceleration. (After Dunlop and Woodward, 1998).

are all planar and stable and have a variety of both crossing speeds and local boundary orientations, although often the mean normals lie close to the model magnetopause. The orbit track and spacecraft configuration in the top part indicate that the model magnetopause lies close to the C1-C3-C4 plane, while C2 lies someway upstream.

Figure 8.5 shows four selected magnetopause crossings from this pass to illustrate the points made above. In the top crossing, the magnetopause is initially moving outwards so that all spacecraft move from the magnetosheath to the magnetosphere. C2 encounters the magnetopause last while the three others cross almost simultaneously, as expected from the spacecraft configuration. As we move down the panels, the boundary normal **n** changes as indicated by the vector components given in each panel, and the sequence of crossing profiles changes accordingly. Note also that the durations of the traversals through the boundary (extent of each profile from low to high field) at each spacecraft change from crossing to crossing. Until the analysis is performed, it is not known whether changes in traversal times are due to variations in magnetopause thickness or to changes in boundary velocity.

The DA analysis of a few selected crossings is shown in Table 8.1: the time of the crossing, the mean normal, the crossing times Δt with respect to C3, the computed value of $v_n = r_n/\Delta t$ for each pair (from C3 to C1, C2 and C4), and the mean boundary velocity. For comparison, the CVA normals and their deviation from the DA results are given in red. The sign of the mean velocity (positive refers to outward motion) is always consistent with the time sequence, following boundary motion in or out. The change in velocity shows a smooth acceleration over the spacecraft array, matching the sequence of crossing times from spacecraft pair to pair. For example, the in/out crossings at 06:02:00 and 06:03:40 UT show simple reversal of motion at nearly constant outwards acceleration. The range of velocities for each crossing shown, however, implies that 5–10 km s^{-2} acceleration can exist.

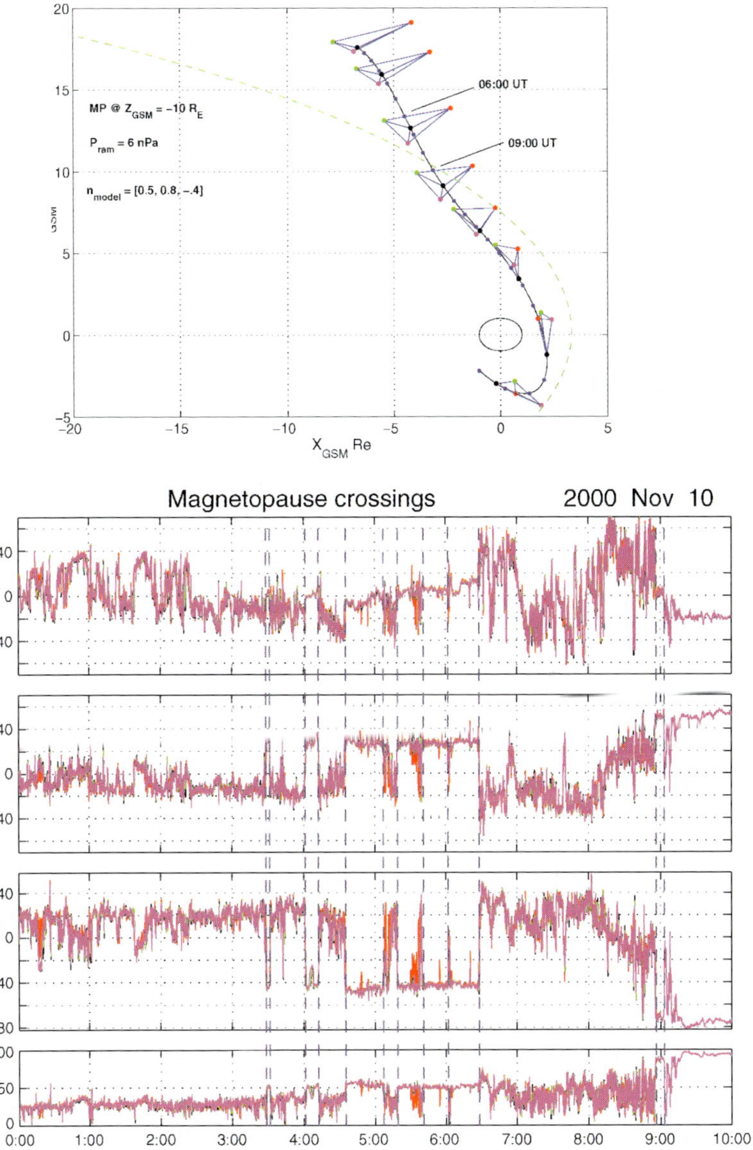

Figure 8.4. A dusk flank pass on November 10, 2000, early in the mission. Top: Cluster orbit and configuration projected on the XY_{GSM} plane (C1 back, C2 red, C3 green and C4 magenta). Bottom: Magnetic field data for the whole pass. (Adapted from Dunlop and Balogh, 2004).

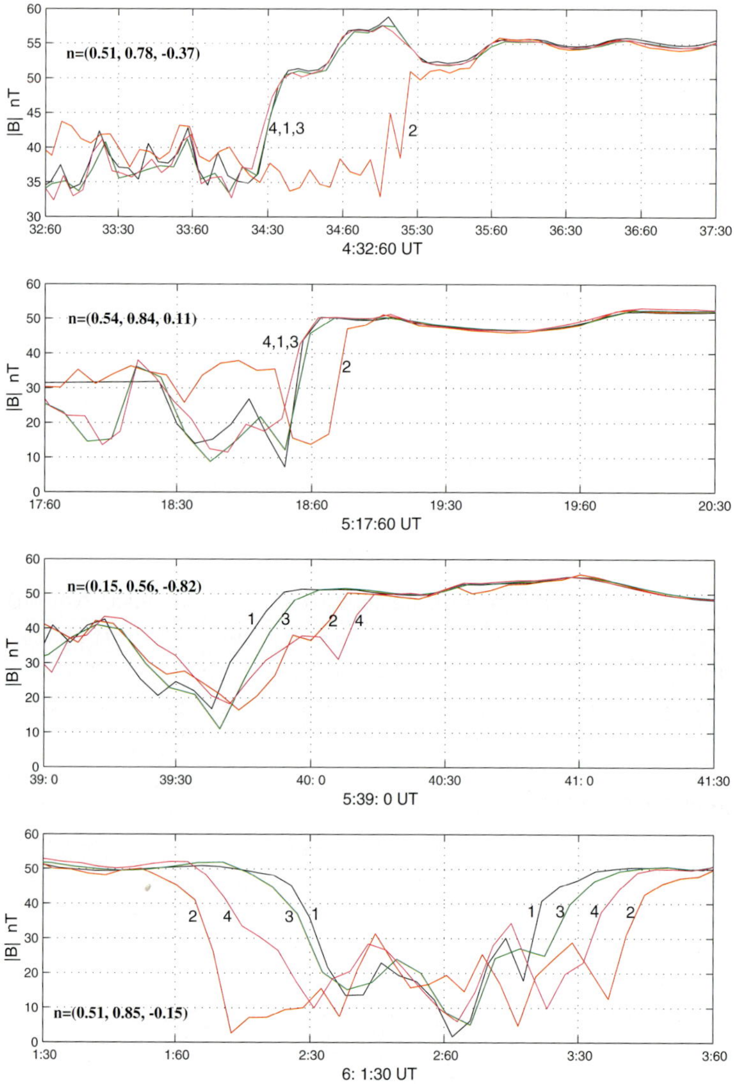

Figure 8.5. Close-up of four crossings from Figure 8.4, with normal directions given in each panel. (Adapted from Dunlop and Balogh, 2004).

The acceleration obtained can be set to zero by changing the common **n**; the result then corresponds to the CVA. The CVA normals are shown in red and can deviate from the mean MVAB solution by up to 30°.

Time UT	$\mathbf{n}_{av}$	$(\Delta t)_{31,32,34}$ s	$(v_n)_{31,32,34}$ km/s	$\langle v_n \rangle$ km/s
08:55:30	0.56 0.77 -0.29	-2 14 5	20 46 38	35
	0.45 0.70 -0.30			29°
06:28:31	0.60 0.74 -0.30	0 -17 -2	-157 -46 -172	-125
	0.50 0.76 -0.10			27°
06:02:00	0.51 0.85 -0.15	3 -20 -11	-6 -36 -10	-17
	0.37 0.83 -0.57			12°
06:03:40	0.51 0.85 -0.15	-4 16 10	5 45 10	20
	0.37 0.83 -0.57			12°
05:40:00	0.14 0.56 -0.82	-2 5 10	139 44 48	77
	0.34 0.29 -0.92			15°

Table 8.1. Planar-DA results for selected magnetopause crossings on November 10, 2000: Time of crossing, average DA normal, time of crossing relative to C3, boundary velocity computed for each pair, and average boundary velocity. In red: CVA normal and its angular deviation from the DA results. (Adapted from Dunlop and Balogh, 2004).

June 11, 2001 crossing. The DA speeds can be combined with the crossing durations to estimate the magnetopause thickness. Figure 8.6 shows a dawn flank event (June 11, 2001) near apogee with a series of crossings that are induced by ripples on the magnetopause moving tailward. The top panel shows the whole interval: The spacecraft are initially in the magnetosheath before repeatedly entering the magnetosphere as indicated. The lower panels zoom in on the individual crossings. The crossing durations Δt (as labelled for two instances) are similar from crossing to crossing, but differ from spacecraft to spacecraft. This suggests the projected motion along each boundary normal is non-constant over the array, but the overall motion of each ripple on the magnetopause is nearly the same. The spacecraft crossing order reveals that the inbound crossings have distinct ordering compared to the outbound crossings, as expected for a rippled boundary (in the manner shown in Figure 8.12, Section 8.3.1.3, where the same event is discussed). The DA results are partially shown in Table 8.6, Section 8.3.3. These results confirm the tilting of the boundary normals and the approximately constant boundary thickness. This constant thickness of ~1200 km is revealed only when the changing velocities over the spacecraft array are used to scale the changing crossing durations. Note that the early dusk-side crossings (e.g., November 10, 2000, Figure 8.4) give a comparable thickness of ~1000 km.

June 30, 2001 crossing. Another dawn flank event, June 30, 2001, provided a number of crossings from high (~10 R_E above the XY_{GSM} plane) to low-latitude regions of the magnetopause, as the orbit skirted the average magnetopause surface for a long period of time. The crossings generally give stable magnetopause orien-

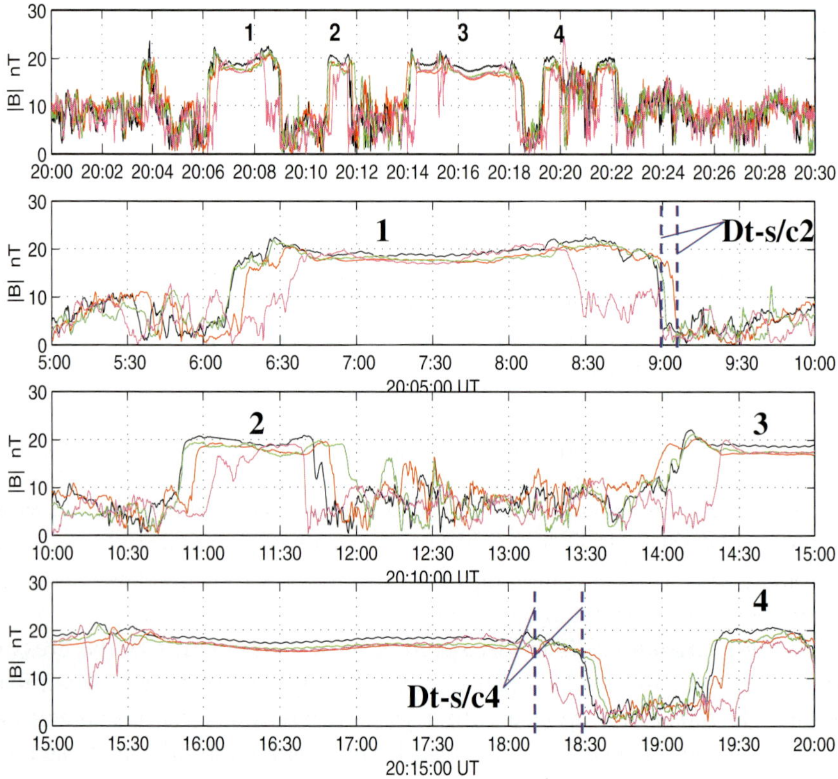

Figure 8.6. Dawn flank pass on June 11, 2001, showing the magnetic field magnitude for a series of magnetopause crossings. (Adapted from Dunlop and Balogh, 2004).

tations, which follow the expected magnetopause shape. A selection of crossings is shown in Table 8.2 and covers 8.5 hours. There is therefore a significant spread in thickness between each encounter. Table 8.2 shows the approximate time of each crossing, the mean MVAB normal, the velocity profile (expressed as before as the mean velocity for each spacecraft pair), the traversal time across the current layer for each of the four spacecraft and the corresponding estimated thickness, D_i, at each spacecraft (according to $D_i = v_i \Delta t_i$, $i = 1, 2, 3, 4$, where the v_i are estimated from $\Delta r_{ij} \cdot \mathbf{n}$). Although there is a significant spread in velocities (ranging from $\sim$1 to 135 km s^{-1}), and significant acceleration (all velocity profiles are smoothly changing over the array), the small differences in D at each spacecraft are all within the uncertainties, for each of the crossings (within $\pm$200 km). If the mean velocity, or the timing result (assuming constant velocity), is used with the Δt_i, the spread of Ds are much larger ($\pm$900 km).

Time UT	$\mathbf{n}_{av}$	v_n km/s	$\Delta t_{1,2,3,4}$ s	$D_{1,2,3,4}$ km
09:15	0.44 -0.87 0.24	72 38 91	8 9 12 10	1314 1128 1308 1282
12:50	0.43 -0.89 0.17	-18 -22 -13	-26 -26 -29 -50	907 1029 1256 1303
13:05	0.46 -0.89 0.05	5 2 7	64 62 57 44	828 811 740 736
17:45	0.53 -0.85 0.01	-113 -46 -135	-20 -33 -35 -20	2484 2620 2778 2732

Table 8.2. DA results for crossing on June 30, 2001: Time, average normal, boundary velocity, crossing times, and corresponding thickness estimates. (Adapted from Dunlop and Balogh, 2004).

8.2.1.2 Benchmark case

Figure 8.7 shows a magnetopause crossing by Cluster on July 5, 2001, in the interval 06:21–06:27 UT, when the spacecraft constellation was located on the dawnside flank of the magnetosphere near $[-6.8,-15.0,6.2]$ R_E GSE. The top three panels show the GSE magnetic field components from FGM at 4 s resolution for the four spacecraft. The following three panels show plasma density, parallel and perpendicular temperatures for C1 and C3, and temperature anisotropy factor, $A_p = (T_{||}/T_{\perp} - 1)$ from CIS/HIA. The bottom three panels show the GSE velocity components at the standard 4 s spin resolution from CIS/HIA for C1 and C3 and from CIS/CODIF for C4. The event displays an unambiguous transition from the hot, tenuous magnetospheric plasma to the cool, dense magnetosheath plasma.

Orientation. The normal vectors, derived from CVA and CTA as well as MTV, are shown in the polar plot in Figure 8.8, together with several single-specraft results. The 'bull's eye' represents the vector $\langle \mathbf{n}_{\text{MVABC}} \rangle = (0.58426, -0.81125, 0.02250)$ (in GSE), which is the average of the four normal vectors obtained by minimum variance analysis (MVAB) of the magnetic field measured in each crossing, using the constraint $\langle \mathbf{B} \rangle \cdot \mathbf{n} = 0$ (MVABC; see Sonnerup and Scheible, 1998). The polar plot represents a projection of the unit hemisphere onto its equatorial plane, i.e., the plane perpendicular to $\langle \mathbf{n}_{\text{MVABC}} \rangle$. The vertical axis in each plot lies in the GSE XZ plane. The horizontal axis points mostly from north to south but with a small dusk-to-dawn component as a consequence of the fact that $\langle \mathbf{n}_{\text{MVABC}} \rangle$ has a small, but positive, GSE Z component. It is seen that the CTA normal falls much closer to the 'bull's eye' normal (MVABC) than the CVA normal. Since one expects the normal magnetic field component to be small, the true normal vector should be close to the MVABC result. Therefore it is reasonable to assume that the CTA normal is closer to the true normal than is the CVA normal. Note that the MVAB normal for C1 falls outside the range of the plots. A closer inspection of the magnetic hodogram indicates a nearly unidirectional current distribution, and hence a breakdown of the conditions for MVAB. It is a situation where some additional constraint on the normal vector is needed: Adding the condition $\langle \mathbf{B} \rangle \cdot \mathbf{n} = 0$ gives a far better result.

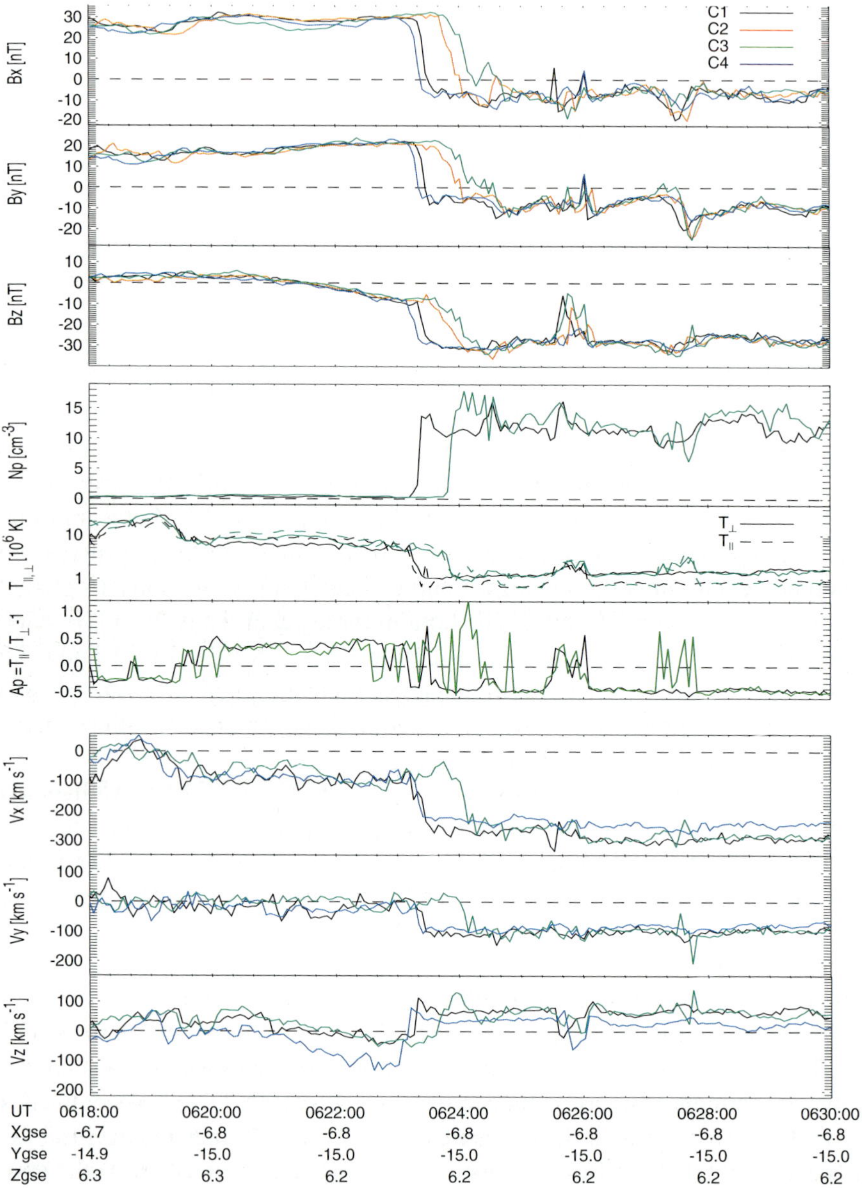

Figure 8.7. Prime parameters measured by Cluster at the magnetopause on July 5, 2001, 06:18–06:30 UT. Top three panels: GSE magnetic field components from FGM. Middle three panels: plasma density N_p, temperatures $T_\parallel$ and $T_\perp$, and anisotropy factor $A_p = T_\parallel / T_\perp - 1$ from CIS/HIA. Bottom three panels: GSE plasma velocity components from CIS/HIA (C1 and C3) and CIS/CODIF (C4). (After Haaland et al., 2004a).

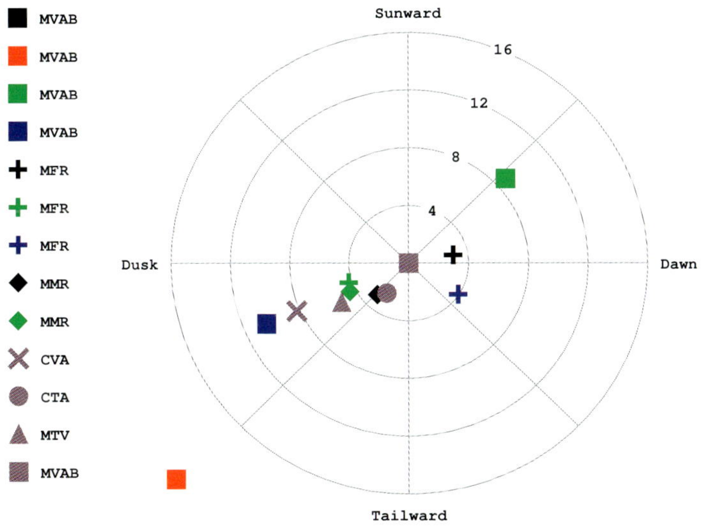

Figure 8.8. Polar plot of magnetopause normals from various single- and multi-spacecraft methods. Each symbol type represents a method (see Section 8.1.1) to calculate a magnetopause normal, and the colours represent the spacecraft; black:C1, red:C2, green:C3, blue:C4. The dark magenta symbols represent normals from the multi-spacecraft methods. Concentric circles indicate the angular deviation from the centre, which corresponds to the normal obtained by averaging the four spacecraft normals derived from MVAB (using 7 nested intervals), with constraint $\langle \mathbf{B} \rangle \cdot \mathbf{n} = 0$. The MVAB normal for C1 falls outside the range of the plot. For C4, MMR (Sonnerup et al., 2004a) is not possible due to the lack of density measurements. (After Haaland et al., 2004a, Figure 5).

Speed. Using the magnetic field profiles for timing, the magnetopause velocity obtained from CVA is $-40\,\mathrm{km\,s^{-1}}$ in this case, the negative sign indicating a transition from the magnetosphere to the magnetosheath, which is consistent with the outbound crossing. The resulting CVA thicknesses are in the range 350–690 km. CTA gives the velocity as a cubic function of time, as shown in Figure 8.9. Also shown in this figure is the velocity curve from the combined (MTV) method.

Benchmarking by Haaland et al. (2004a) and Sonnerup et al. (2004a) showed qualitative agreement between the velocities from CVA, CTA, and the single-spacecraft results from MFR, MMR and HT. Since the latter provides both frame velocity and its acceleration, it can be used to verify acceleration. At C3, the normal HT acceleration for this event is found to be $-0.9\,\mathrm{km\ s^{-2}}$, which, in terms of direction and approximate magnitude, agrees with the CTA result. For C4, the HT acceleration along the normal vector is $-0.6\,\mathrm{km\ s^{-2}}$, corresponding to a small negative slope of the velocity curve at C4. This behaviour is also consistent with the CTA result. At C1, the HT acceleration along the normal is again $-0.6\,\mathrm{km\ s^{-2}}$ whereas the slopes from CTA are seen to be either slightly negative or close to zero.

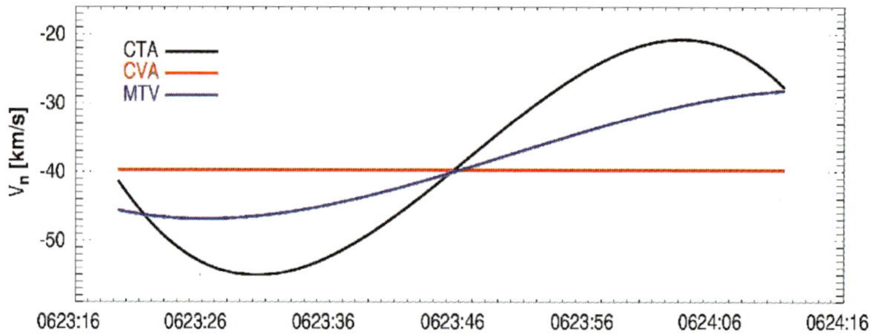

Figure 8.9. Magnetopause velocity as function of time, derived from multi-spacecraft analysis, using the constant velocity approach (CVA), the constant thickness approach (CTA), and the combined method (MTV). CVA gives, as the name suggests, a constant velocity for the entire event. CTA represents the velocity as a quadratic function. The velocity from MTV falls in between the CTA and CVA. Crossing times of the four Cluster spacecraft are indicated along the time axis. (After Haaland et al., 2004a).

It is important to remember that single-spacecraft results, based on either magnetic field or plasma data or a combination of these are calculated from a number of samples; plasma moments from Cluster are available at a time resolution of 4 seconds. Results from the HT, MMR and MFR methods thus represent average velocities over a longer time interval. Since structures in the magnetosphere can evolve rapidly, this must be taken into consideration when comparing multi- and single-spacecraft results.

Thickness. The results from the four-spacecraft thickness determinations are shown in Table 8.3. CVA returns a large range of thicknesses, whereas CTA returns a single thickness. The combined method (MTV), returns thicknesses in between these. A rough idea about the thickness can also be obtained by multiplying the crossing duration with the normal component of the plasma flow velocity. This estimate, denoted CIS in Table 8.3, assumes that there is no plasma flow across the magnetopause.

On the whole, this benchmark event illustrates that great care must be exercised if one wants accurate and consistent answers concerning magnetopause orientation, motion and thickness.

8.2.1.3 More comparisons

In this section, we compare the results from the different methods for a sample of events.

The DA results have indicated that, for separation distances covered by the events here (up to 2000 km), acceleration is common, whereas boundary thickness remains more nearly constant during an encounter. A comparison with CVA and CTA therefore provides a useful crosscheck on the results. To compare CTA

Method	Spacecraft C1	C2	C3	C4
CVA	319	690	667	350
CTA	437	437	437	437
MTV	371	580	490	410
CIS	483	-	918	440

Table 8.3. Magnetopause thickness (in km) for the benchmark event on July 5, 2001, 06:23 UT, obtained with the constant velocity approach (CVA), the constant thickness approach (CTA), a combination of those two (MTV), and the single-spacecraft method consisting of multiplying crossing duration with the normal plasma velocity component from CIS (hence called the CIS method; there are no CIS data for C2). (After Haaland et al., 2004a).

Date	Time UT	$\mathbf{n}_{av}$	$(v_n)_{31,32,34}$ km/s	$\langle v_n \rangle$ km/s	$D_{1,2,3,4}$ km
Nov 10 2000	04:01	0.44 0.89 -0.15	95 70 95	84	2942 3217 2942 3047
	CTA	0.34 0.94 0.05	99 97 97	99	3120 3120 3120 3120
	CVA	0.39 0.92 0.09		99	2710 3295 2709 2709
Jun 11 2001	20:12	0.66 -0.73 0.21	-76 -39 -107	-74	653 619 692 672
	CTA	0.71 -0.56 0.43	-107 -79 -112	-99	661 661 661 661
	CVA	0.68 -0.48 0.55		-106	664 796 1327 464
Jul 05 2001	06:06	0.28 -0.85 0.44	11 10 12	11	206 214 220 210
	CTA	0.32 -0.88 0.34	14 9 13	12	178 178 178 178
	CVA	0.24 -0.82 0.52		11	198 233 208 198
Jul 05 2001	06:23	0.61 -0.78 -0.03	-37 -5 -37	-23	501 457 457 520
	CTA	0.58 -0.81 -0.05	-37 -21 -35	-31	474 474 474 474
	CVA	0.54 -0.83 0.17		-41	335 628 628 337
Mar 02 2002	03:31	0.74 -0.05 0.67	-34 -36 -39	-37	278 269 250 245
	CTA	0.56 0.01 0.83	-46 -46 -47	-46	293 293 293 293
	CVA	0.77 0.25 0.59		-57	289 280 241 236

Table 8.4. Comparison of different methods for determining magnetopause speed and thickness: Planar-DA (black), CTA (green), and CVA (red). Planar-DA and CTA agree reasonably well, CVA only agrees if the velocity indeed turns out to be relatively constant. (Table prepared by M. Dunlop).

to the DA, we compute the mean CTA velocity between each spacecraft pair. In addition, the CTA boundary thicknesses have been scaled to account for different method definitions and fits to the measured durations of the crossing profiles. Table 8.4 summarises the results for planar-DA (black), CTA (green), and CVA (red). The events cover a broad range of spacecraft separations (100–2000 km) as well as cases with both large and small accelerations. The DA and CTA velocities agree to within the analysis sensitivity to the difference in the computed normals. The CVA values deviate significantly more. The CVA thicknesses are computed from

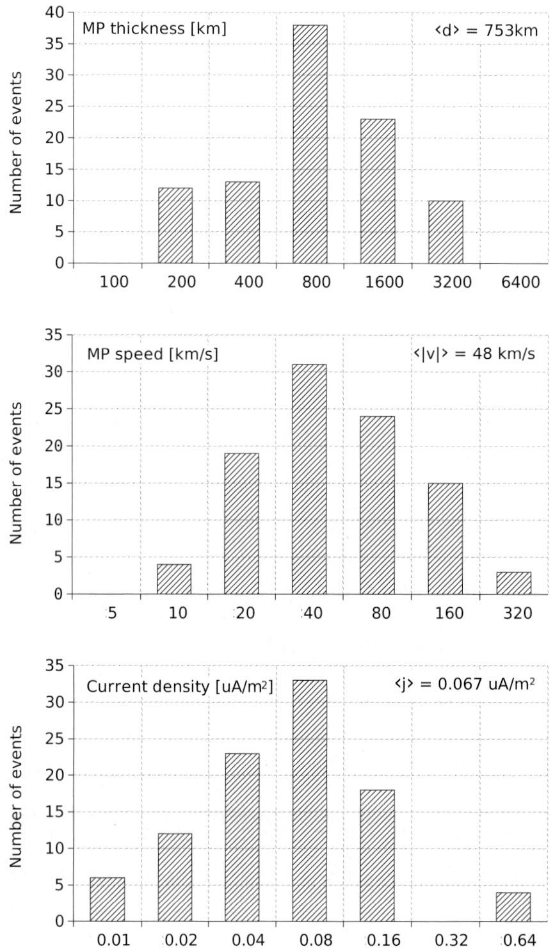

Figure 8.10. Histograms of magnetopause thickness d, speed $|v|$, and average current density j for 24 dawn flank magnetopause crossings by all 4 spacecraft on July 5, 2001, resulting in a total of 96 individual crossings. The average current density in a crossing is obtained from the change in the maximum variance magnetic field component $\Delta B_{\max}$ and the magnetopause thickness by $\mu_0 j = \Delta B_{\max}/d$. (After Paschmann et al., 2004).

the given mean velocities and the crossing durations and result in a much wider range of values from spacecraft to spacecraft (hence the importance of testing for acceleration in order to properly scale planar boundary layer structure). Where the velocity is shown to be constant, all three methods agree.

8.2.2 Statistical study

The July 5, 2001 magnetopause crossing discussed in Section 8.2.1.2 is one of a large number of crossings on that day. Applying the MTV technique to 24 clearly defined crossings by all 4 spacecraft, Paschmann et al. (2004) have obtained a statistical description of magnetopause speeds, thicknesses, and currents for a total of 96 cases.

The top two panels of Figure 8.10 show histograms of speeds and thicknesses respectively. The thickness ranges from 100 to 3000 km, with a peak at 400–800 km. The speeds range from less than $\sim$10 km s^{-1} up to more than 300 km s^{-1}, with a peak around 20–40 km s^{-1}. Both findings are quite consistent with earlier statistics from ISEE (Berchem and Russell, 1982) and from AMPTE-IRM (Phan and Paschmann, 1996). The gyroradii and inertial lengths for these crossings are typically 40–80 km, so scaling the observed thicknesses by those characteristic lengths does not alter the large dynamic range of the magnetopause thicknesses. The current density histogram will be discussed in Section 8.3.3.

8.2.3 Summary and conclusions

Accurate information about the orientation, the motion, and the thickness of the magnetopause has been derived from the Cluster measurements.

- For cases where the magnetopause is planar over a scale size larger than or comparable to the spacecraft separation distance, multi-spacecraft techniques can provide robust estimates of its orientation, motion and thickness, thus augmenting the classical single-spacecraft techniques like MVAB and HT analysis. If there is little or no acceleration, the CVA method will provide reasonable results for both magnetopause orientation and motion. For the more typical cases with non-constant magnetopause motion, the CTA technique is usually better. The Discontinuity Analyser and the MTV methods may also be reasonable choices for many events. Even with multi-spacecraft techniques one should be cautious. For a given event, it is not guaranteed that a particular multi-spacecraft method always produces the more correct answer. For instance, non-time stationary features in the time series are likely to be present at discontinuities, so that correlation techniques must be used with care. In the presence of acceleration, it is vital to determine the boundary normals independently from the timing analysis, since assumptions of constant motion (or constant thickness) can give erroneous results. A consistency check with other multi-spacecraft and/or single-spacecraft methods is therefore always recommended.

- Cluster studies have shown so far that typically the magnetopause boundary is planar on the scale of the spacecraft separation vectors ($\sim$600–2000 km). In contrast, the motion is often not constant while the Cluster array crosses the boundary. For the July 5, 2001 crossings, accelerations of 0.6–2 km s^{-2} were found, but on other occasions accelerations as large as ~5–10 km s^{-2} have been observed.

- From a statistical study the magnetopause thickness is found to vary over a large dynamic range, between 100 and 3000 km, with a peak at 400–800 km. Magnetopause speeds range from less than ~10 km s^{-1} up to more than 300 km s^{-1}, with a peak around 20–40 km s^{-1}. Both findings are quite consistent with earlier statistics from ISEE (Berchem and Russell, 1982) and from AMPTE-IRM (Phan and Paschmann, 1996). The gyroradii and inertial lengths for these crossings are typically 40–80 km, so scaling the observed thicknesses by those characteristic lengths does not significantly alter the large dynamic range of the magnetopause thicknesses.

- While these results agree in a general sense with what was known about the magnetopause from earlier spacecraft, Cluster gives such information with much higher reliability, e.g., by being able to check the planarity hypothesis and/or by including the effect of boundary acceleration. Strong acceleration means that estimates of boundary spatial scales, inferred from dual spacecraft measurements, may in some cases be in error by up to a factor of two or three.

8.3 Magnetopause Currents

Knowledge of the magnetopause currents and their spatial distribution is required for studies of possible relationships between current density and electric/magnetic fluctuation levels within the layer, and for investigating the fine-structure of the currents, both in magnitude and direction.

Individual spacecraft crossing the magnetopause can sample changes in the magnetic field across the current layer, which, in a 1-D geometry, can in principle give an estimate of the current density in the boundary. These estimates, however, depend on accurate knowledge of the orientation and motion of the boundary. With Cluster it has become possible for the first time to determine those currents directly from application of Ampère's law to the magnetometer measurements at the four spacecraft.

We show some key results and illustrate how robust this 'curlometer' technique is, at least in determining the direction of the currents, over a variety of spatial scales and for different locations on the magnetopause (Section 8.3.1.1). Minimum variance analysis of the current density vector obtained in this way provides useful new way of determining boundary orientation and motion, and at the same time can be used to validate the curlometer technique (Section 8.3.2).

8.3.1 Measuring the current with the curlometer

8.3.1.1 Brief method description

The first application of the curlometer technique has been reported by Dunlop et al. (2002b). The method combines simultaneous vector magnetic field data from each spacecraft (spin resolution or high time resolution where appropriate (Balogh et al., 2001)) with the spacecraft positions to calculate the curl of the magnetic field from Ampère's law, thus producing an estimate of the average current density through the spacecraft tetrahedron. The difference approximation

$$\mu_0 \mathbf{j} \cdot (\Delta \mathbf{r}_i \times \Delta \mathbf{r}_j) = \Delta \mathbf{B}_i \cdot \Delta \mathbf{r}_j - \Delta \mathbf{B}_j \cdot \Delta \mathbf{r}_i,$$

with $\Delta \mathbf{r}_i = \mathbf{r}_i - \mathbf{r}_1$ and $\Delta \mathbf{B}_i = \mathbf{B}_i - \mathbf{B}_1$, gives the average current normal to each face $(1, i, j)$ of the spacecraft tetrahedron. Since each face is known by $\Delta \mathbf{r}_i \times \Delta \mathbf{r}_j$, the currents normal to three faces can be re-projected into a Cartesian co-ordinate system. It is also possible to estimate $\nabla \cdot \mathbf{B}$ from

$$\nabla \cdot \mathbf{B} \, |\Delta \mathbf{r}_i \cdot (\Delta \mathbf{r}_j \times \Delta \mathbf{r}_k)| = |\sum_{cyclic} \Delta \mathbf{B}_i \cdot (\Delta \mathbf{r}_j \times \Delta \mathbf{r}_k)|,$$

which usually produces non-zero values, partially as a consequence of non-linear spatial gradients neglected in its estimate. This quantity therefore measures effects of the linear approximation and can be used as a monitor of this error.

8.3.1.2 January 26, 2001 crossing

We first demonstrate that the curlometer technique is able to consistently estimate the magnetopause current direction, using an example taken from January 26, 2001. Figure 8.11(left) shows the spacecraft configurations along the orbit for this outbound pass through the dusk-side magnetopause at high latitude ($Z_{GSM} \sim 9.5\, R_E$). The mean spacecraft separation distance was 600 km (smaller than the boundary thickness). These data, reported by Bosqued et al. (2001), Pu et al. (2003) and Phan et al. (2004), exhibit repeated boundary layer crossings as a result of inward and outward magnetopause motion between 10 and 11 UT. Figure 8.11(right) shows a short segment of the orbit, projected onto the XY_{GSM} plane, along with cuts through a model magnetopause (Sibeck et al., 1991) at 10 and 11 UT, representing the approximate range of sampled positions of the boundary along the orbit. The projections of the estimated current density vectors, plotted along the orbit, show enhanced values within this interval and a clear alignment to the magnetopause orientation. The current bursts between 10 and 11 UT correspond to repeated encounters with the magnetopause and are consistent with a Chapman-Ferraro current. In contrast, enhanced values of current after 11 UT correspond to FTE signatures and exhibit a variety of orientations.

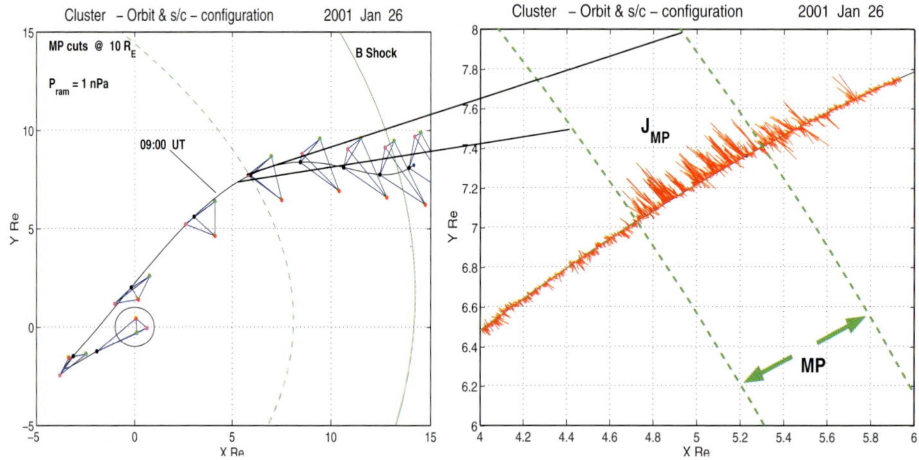

Figure 8.11. Boundary pass on January 26, 2001. Left: The Cluster orbit, projected onto the XY_{GSM} plane, together with the configuration of the four spacecraft, enlarged by a factor of 20, at intervals along the orbit. Magnetopause and bow shock curves for the observed solar wind pressure are shown in green; Right: The computed current estimate from the curlometer, projected into the plane along the orbit track. The magnetopause moves inward and outward several times during the interval between the dashed lines, resulting in the bursts of current which are apparent from the projections. (From Dunlop and Balogh, 2004).

8.3.1.3 June 11, 2001 crossing

Figure 8.12 shows the boundary pass on June 11, 2001 at apogee on the dawn flank, near the equatorial plane. Figure 8.12(bottom) shows the curlometer results. Figure 8.12(top) gives a schematic of the boundary motion, superimposed onto the orbit (projected onto the XY_{GSM} plane, as in Figure 8.11, left). The mean spacecraft separation is $\sim$2000 km. The DA analysis suggests that a series of surface ripples on the magnetopause move tailward past the spacecraft array in the manner indicated, taking them fully into the magnetosphere on each occasion. The lower panel in Figure 8.12 shows the magnetic field magnitude. The transitions from a quiet, 20 nT magnetospheric field to a fluctuating, low magnetosheath field are marked by the dashed lines. Inward and outward crossings, represented by blue and green vertical dashed lines respectively, match two distinct orientations. The boundary normals are depicted schematically by the blue arrows in Figure 8.12(top). The Z_{GSM} components are small. The crossings are irregular and show changing boundary motion at each spacecraft during the crossings. This motion together with transit times for each current layer encounter (measured from the turning in the maximum variance component of the field) can be used to estimate boundary thickness (see below in Table 8.6). It corresponds to a projected tailward motion of ripples along the average magnetopause.

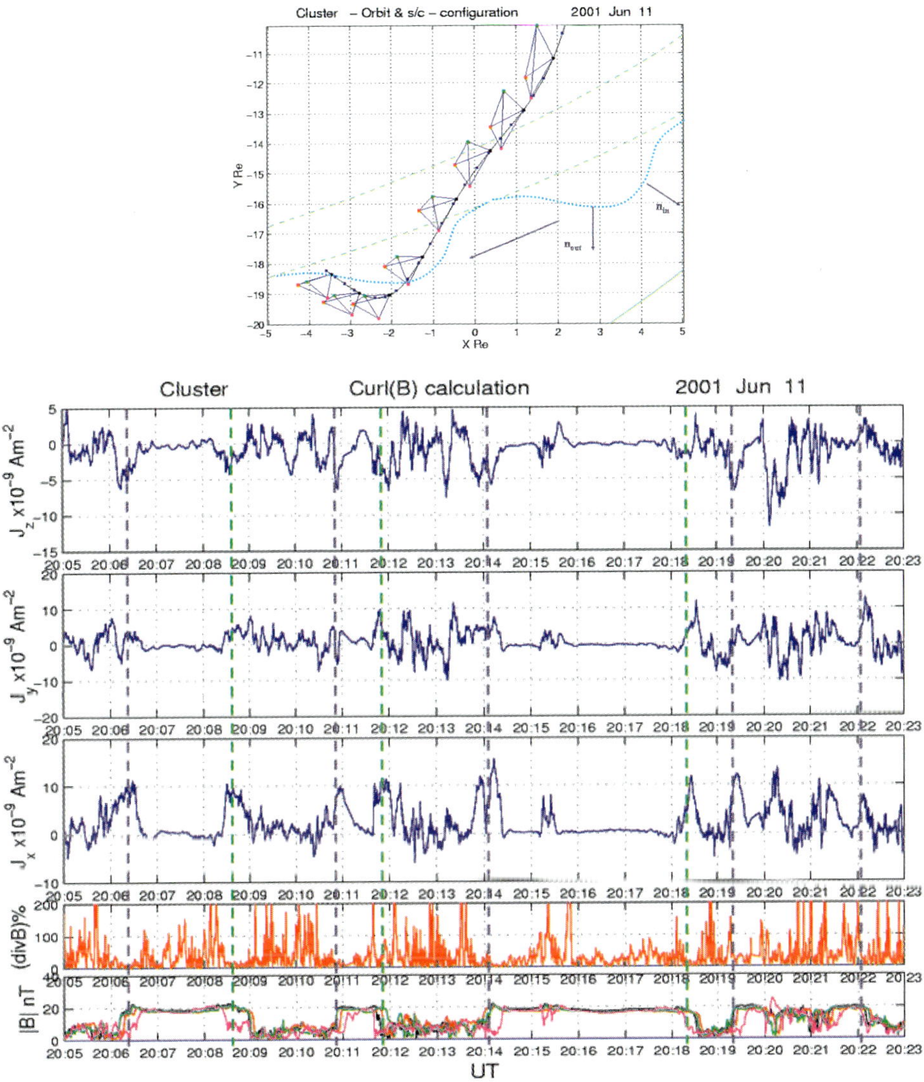

Figure 8.12. Magnetopause crossings on 11 June 2001. Top: Orbit and spacecraft configuration, scaled by a factor 5, projected onto the XY_{GSM} plane, together with a schematic of the boundary ripples implied from the DA analysis. Bottom: Plot of the curlometer estimate of GSM components of the current (top three panels), the estimate of $\nabla \cdot \mathbf{B}$, and the total magnetic field from all four spacecraft, showing the pairs of crossings into (blue) and out of (green) the magnetosphere. (From Dunlop and Balogh, 2004).

The other panels in Figure 8.12 show the estimate of $\nabla \cdot \mathbf{B}$ and the GSM components of $\mathbf{j} = \nabla \times \mathbf{B}/\mu_0$. Although the spacecraft separation distances are slightly larger than the estimated boundary thickness (see Table 8.6), it appears that the current density is still adequately sampled through each crossing to vindicate its direction. In fact, the direction of the current maintains its alignment to the magnetopause orientation as it tilts from crossing to crossing. This can be seen from the components of **j**: The j_x component is maintained for most crossings, whereas the j_y component changes from crossing to crossing so that the majority of the inward crossings (blue dashed lines) have smaller values than the outward crossings (green dashed lines). This follows the orientations expected for the ripple geometry shown by the arrows in Figure 8.12b. The j_z component is less significant for magnetopause alignment since the normals lie nearly in the XY_{GSM} plane. Thus, even for this event, where the magnetopause current layer thickness is larger than the spacecraft separation, the curlometer is giving a good estimate of current direction.

8.3.2 Variance analysis of the current density vector

8.3.2.1 Brief method description

The electric currents obtained from the curlometer can be used as an alternative way to obtain the motion and orientation of the magnetopause (Haaland et al., 2004b). One can estimate the current sheet orientation by performing minimum variance analysis of the current density vector (MVAJ). The underlying physics is that $\nabla \cdot \mathbf{j} = 0$ (just as the basis for MVAB is that $\nabla \cdot \mathbf{B} = 0$). This step provides us with a current sheet aligned coordinate system where the three orthogonal axes are the eigenvectors from the variance analysis. The eigenvector $\hat{\mathbf{x}}_3$ corresponding to the lowest eigenvalue serves as an estimate of the current sheet normal, whereas the two other eigenvectors, $\hat{\mathbf{x}}_2$ and $\hat{\mathbf{x}}_1$, corresponding to the intermediate and maximum eigenvalue, are tangential to the current layer. For a plane one-dimensional layer, MVAJ should give $\langle \mathbf{j} \rangle \cdot \hat{\mathbf{x}}_3 = 0$ (the bracket denotes an average over the data points used). This condition provides a test of the overall consistency of the analysis and can be used as a constraint in MVAJ to improve the prediction of the normal vector. If the discontinuity is a perfect tangential discontinuity (TD), the quantity $\langle \mathbf{B} \rangle \cdot \hat{\mathbf{x}}_3$ should be zero as well. If one has reason to believe that the current sheet is nearly a TD, it is often useful to constrain MVAB and MFR by this condition. This gives a more reliable and stable result for the normal vector.

In the new coordinate system provided by $(\hat{\mathbf{x}}_1, \hat{\mathbf{x}}_2, \hat{\mathbf{x}}_3)$, the two tangential components of Ampère's law can be written as:

$$
\begin{aligned}
\mu_0 j_1 = (\nabla \times \mathbf{B})_1 &= -\frac{\partial B_2}{\partial x_3} = -\frac{dB_2}{dt}\frac{1}{v_3} \\
\mu_0 j_2 = (\nabla \times \mathbf{B})_2 &= \frac{\partial B_1}{\partial x_3} = \frac{dB_1}{dt}\frac{1}{v_3}
\end{aligned}
\tag{8.2}
$$

where v_3 is the velocity, assumed constant, of the magnetopause along the the normal direction $\hat{\mathbf{x}}_3$. The conversion from spatial differentials to time differentials is made by use of v_3. Integrated across the magnetopause, these equations give:

$$\Delta B_2 = -\mu_0 v_3 \int_0^{t_1} j_1 dt$$
$$\Delta B_1 = \mu_0 v_3 \int_0^{t_1} j_2 dt \quad (8.3)$$

where t_1 on the right hand side is the time it takes to cross the current layer. Ideally, the two equations should give the same value for v_3 but, when applied to experimental data, uncertainties and deviations from model assumptions will almost always produce slightly different values.

8.3.2.2 March 2, 2002 crossing

Figure 8.13 shows an example of such an analysis from a magnetopause crossing equatorward of the high-latitude northern cusp on March 2, 2002, around 03:31 UT. The approximate GSE coordinates of Cluster were [7.2, 3.3, 8.4] R_E. The separation distances between the four satellites were only about 100 km on this day. The figure shows hodograms of **j** from MVAJ, the magnetic field in the maximum variance direction, and the current density. From the magnetic field measurements, one sees that there is an interval where all four spacecraft are inside the magnetopause current sheet simultaneously. The conditions for the curlometer are thus satisfied. The current density (lower panel, green curve) reaches a peak value of about 0.16 μA m^{-2} near the outer edge of the current sheet, shortly after 03:31:12 UT, whereas the intermediate variance component (red curve) has an additional peak close to the inner edge of the current sheet around 03:31:05 UT. The minimum variance component of the current (black curve) fluctuates, but is on the whole much smaller than the other components; its average value is 0.0015 ± 0.0029 μA m^{-2}, i.e., it does not differ significantly from zero. This is the expected behaviour.

Applied to the present data set, the above equations give $v_3 = -35.6$ km s^{-1} and -33.4 km s^{-1}, respectively. The average of these numbers, -34.5 km s^{-1}, is the best estimate of the magnetopause velocity from MVAJ for this case. A comparison with magnetopause speeds obtained with other methods is shown in Table 8.5. As one can see, there is very good agreement. This means that the currents used in MVAJ are accurate, which in turn proves that the curlometer technique worked very well. The table also list the normal magnetic field components inferred from the various methods. They are all very small, confirming that the basic assumption of a TD is indeed valid.

Orientation and thickness. Figure 8.14 shows the normal vectors from MVAJ, MVAB, and MFR (constrained by $\langle \mathbf{B} \rangle \cdot \mathbf{n}_{\text{MFR}} = 0$) with their statistical error ellipses, as well as normals from CVA, CTA, MTV. The unconstrained MFR normals and MMR normals fall outside the field of view of the Figure. This plot represents projections of the unit hemisphere onto its equatorial plane, i.e., the plane perpen-

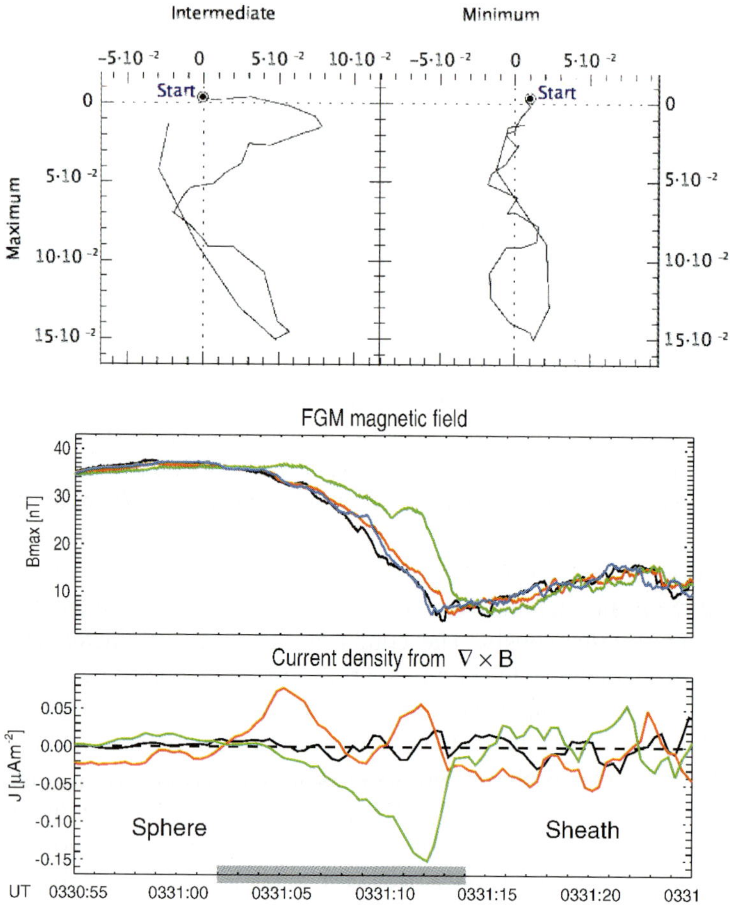

Figure 8.13. Results from MVAJ for the March 2, 2002 crossing. Top: hodogram pair from the variance analysis of the current density [μA m^{-2}] in the interval 03:31:02 - 03:31:14 UT. Middle panel: magnetic field from each spacecraft along the direction of maximum current variation. Lower panel: current density components along $\mathbf{x}_1$ (green, maximum variance direction), $\mathbf{x}_2$ (red) and $\mathbf{x}_3$ (black, minimum variance direction). Shaded bar in lower panel indicates time interval used for the hodogram and MVAJ calculations (After Haaland et al., 2004b).

dicular to the average normal from MVAB ([0.70575, -0.00993, 0.70839] GSE, origin of plot).

As the above example illustrates, minimum variance analysis of the current (MVAJ) provides a novel way to determine the orientation of the current sheet. The condition is that the spacecraft separation is smaller than the magnetopause thickness. The results provide detailed quantitative validation of the curlometer method: Both the velocity (Table 8.5) and orientation (Figure 8.14) are in good agreement

Method	Time interval	No of samples	Speed* [km s^{-1}]	$B_n^\#$ [nT]
MVAJ	03:31:02 - 03:31:14	36	-34.5	-2.94
CVA	-	-	-34.9	-1.04
CTA	-	-	-35.2 to -35.5	-0.51
MTV	-	-	-35.4 to -35.5	-0.93
MVAB/HT	03:30:55 - 03:31:28	16	-33.6 $\pm$ 3.1	-0.58
MFR	03:30:55 - 03:31:28	16	-37.6 $\pm$ 4.5	0.00†
CIS$^+$	03:30:55 - 03:31:28	16	-28.5 $\pm$ 2.1	-0.58

* Statistical uncertainties from Sonnerup and Scheible (1998).
# Average magnetic field along the respective normals.
† Constraint $\langle \mathbf{B} \rangle \cdot \mathbf{n} = 0$ was used.
+CIS/HIA plasma velocity along the normal from MVAB.

Table 8.5. Comparison of velocities from various methods for the March 2, 2002 case. MVAJ, CVA and CTA are based on 36 samples of high resolution data; MVAB/HT, MFR and CIS results are based on 16 samples of 4 s data combined from C1 and C3. (From Haaland et al., 2004b).

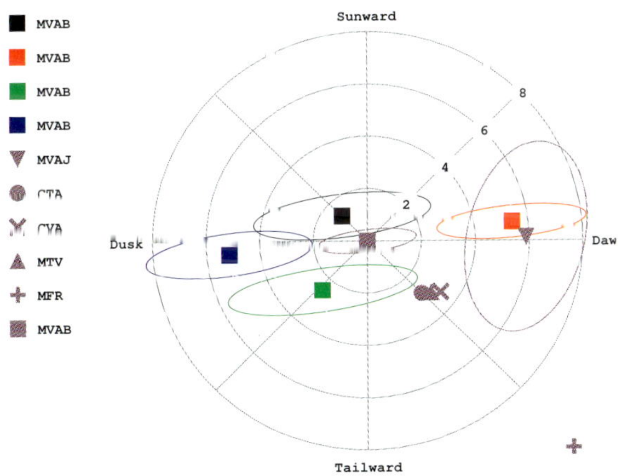

Figure 8.14. Polar plot of the magnetopause normals from various methods; concentric circles indicate the angular deviation from the origin (magenta square), which is the MVAB normal calculated by adding the four individual MVAB covariance matrices, each normalised by their lowest eigenvalue. The four MVAB normals and the multispacecraft normals from CVA and CTA are all within $\simeq 6°$ of the origin. The MFR normal was constrained by $\langle \mathbf{B} \rangle \cdot \mathbf{n}_{MFR} = 0$. Error ellipses (calculated after Sonnerup and Scheible, 1998, Equation 8.23) reflect the statistical uncertainty only. (After Haaland et al., 2004b).

with those from other methods. As is also the case for the other multi-spacecraft methods discussed in this book (CVA, CTA, MTV, DA), MVAJ has an advantage over single-spacecraft methods in cases where the magnetopause duration is short: It does not require plasma data, which usually have a lower time resolution. Note that MVAB by itself does not provide a boundary velocity. Using the speeds determined earlier the thickness has been determined as $\sim$160 km, corresponding to only 2–3 gyro radii of magnetosheath ions for this particular case.

8.3.3 Comparison with average currents

Figure 8.15 shows a series of magnetopause crossings on 5 July, 2001. The Cluster orbit skirted the dawn flank magnetopause while moving from high to low latitudes as shown in Figure 8.15a. This location provided a number of magnetopause crossings for a large fraction of the day; the interval shown in Figure 8.15b represents a few crossings from this set. The Figure shows the magnetic field in the top four panels and the curlometer estimate in the bottom three, all in GSM. These crossings represent small-amplitude ($<1\,R_E$) inward and outward motions of the magnetopause, rather than tailward travelling ripples on the boundary: They produce a sequence of crossings with almost parallel orientations. The bursts of current in the lower three panels reflect this common orientation, aligned to the magnetopause plane. Moreover, the current signature shows both slow crossings, where the spacecraft array remains in the current layer for longer so that the enhanced value of $\nabla \times \mathbf{B}$ remains for minutes, and fast crossings, where the $\mathbf{j}$ profile shows a short spike.

Table 8.6 summarises the results for a set of crossings on June 11, 2001 and on July 5, 2001 that have been analysed in detail. Note that Dunlop and Balogh (2004) used a slightly different method to obtain spacecraft crossing times. The results for the July 5, 2001, 06:23 event in Tables 8.6 and 8.3 are therefore slightly different. The results confirm that the thickness of the current layer is effectively constant while being sampled by the spacecraft array. This thickness can be used to compute an average current density in the magnetopause from $\Delta B_{max}/\mu_0 D$. Comparison of the the two current density columns (j and j_{ave}) in Table 8.6 shows that the absolute values are close for the June 11 crossings, but differ strongly for the July 5 crossings, where the curlometer values are up to a factor 3 smaller. Similarly, computing the angle between $\mathbf{n}_{av}$ and $\mathbf{j}_{GSM}$, shows that the current directions inferred from the curlometer are nearly perpendicular to the normal direction for the June 11 case as required, whereas this is not the case for the July 5 event. This behaviour is expected because the Cluster separation is much larger than the MP thickness in the July 5 cases, and the spatial gradients used by the curlometer are thus underestimated.

For the same 96 crossings already discussed in Section 8.2.2, the distribution of current densities obtained from $\Delta B/D$ has been determined (Paschmann et al., 2004). As shown in the lower panel of Figure 8.10, current densities range from a low value of 0.01 μA m^{-2}up to more than 0.3 μA m^{-2}, with an average value of 0.05 μA m^{-2}. The factor 30 between the highest and lowest value is the same

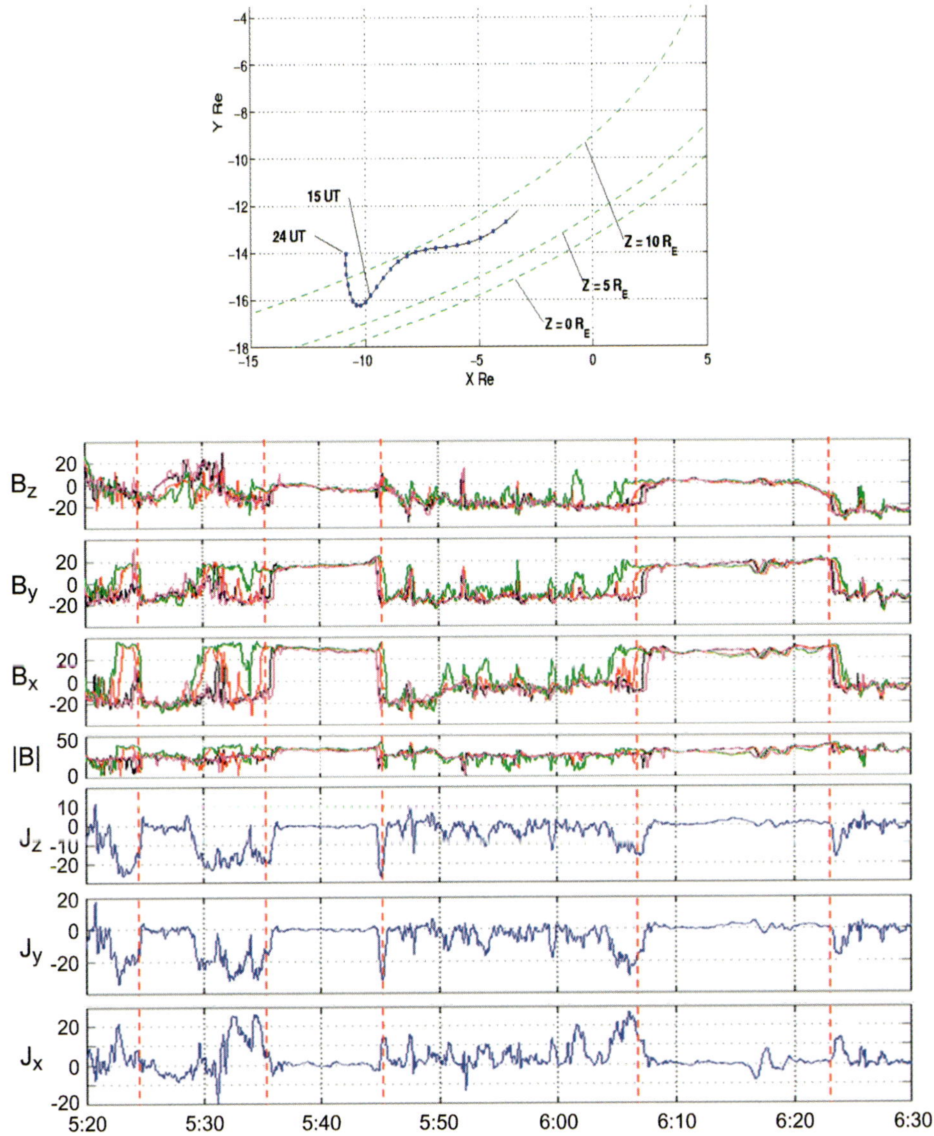

Figure 8.15. Magnetopause crossings on 5 July 2001. Top: The orbit track projected into GSM for the whole day, together with cuts through the magnetopause at the GSE-Z values indicated. Bottom: A plot of the four spacecraft magnetic field data (top four panels) and the GSM components of the current estimate (lower three panels), for a short interval containing a few of the magnetopause crossings observed on this day. (From Dunlop and Balogh, 2004).

Time UT	$\mathbf{n}_{av}$	$\langle \mathbf{V}_n \rangle$ km s^{-1}	$\mathbf{j}_{GSM}$ j nA m^{-2}	j_{ave} nA m^{-2}	$D_{1,2,3,4}$ km
June 11 2001					
20:00:30 out	0.27 -0.94 0.19	120	8 3 -4 10	12	728 746 728 703
in	0.70 -0.62 -0.36	-104	8 6 -2 10	11	778 828 799 863
20:11:00 out	0.46 -0.88 0.13	155	8 3 -4 10	10	908 895 882 869
in	0.66 -0.73 0.21	-74	9 7 -4 11	17	633 619 692 672
20:14:00 out	0.07 -0.99 0.15	32	11 4 -5 13	14	737 927 847 818
in	0.79 -0.61 -0.10	-165	8 7 -2 10	13	601 605 617 618
July 5, 2001					
05:24:00 mp	0.53 -0.84 -0.11	-111	5 -20 -15 27	72	451 357 331 426
05:35:00 mp	0.36 -0.84 0.41	20	24 -26 -18 40	149	165 175 179 191
05:45:00 mp	0.64 -0.76 -0.11	-22	14 -30 -25 40	116	270 235 249 289
06:06:00 mp	0.28 -0.85 0.44	11	22 -22 -15 35	116	206 214 220 210
06:23:00 mp	0.61 -0.78 -0.03	-23	13 -17 -16 25	53	502 457 457 520

Table 8.6. Summary of the boundary analysis for a number of magnetopause crossings. The columns give the the event times, the crossing label (referring to *in* and *out* for the dashed lines in Figure 8.12 and to *mp* for the dashed lines in Figure 8.15), the boundary normal, $\mathbf{n}_{av}$, the average boundary speed, V_n, the current density vector, $\mathbf{j}$ and its magnitude, j, determined with the curlometer, the current density j_{ave}, determined from $\Delta B_{max}/D$, and thickness, D at each spacecraft from the DA technique. (Adapted from Dunlop and Balogh, 2004).

as for the thickness range for this set of crossings, and the current density roughly scales inversely to the thickness, thus preserving a nearly constant net current.

8.3.4 Summary and conclusions

Thanks to Cluster's four-spacecraft configuration, the magnetic field measurements can be combined to determine the electric current density vector in the magnetopause boundary.

- Cluster has enabled, for the first time, a direct measurement, via Ampére's law, of the magnetopause electric current density from in situ measurements. The curlometer estimate is representative even when the spacecraft separation approaches the thickness of the current layer.

- A new method for determining boundary motion and orientation, by performing minimum variance analysis of the current vector (MVAJ), has been constructed. The fact that it works so well is a persuasive argument for the accuracy of the current determinations for small separation distances.

8.4 Structures within the Magnetopause

8.4.1 The magnetopause as an MHD discontinuity

The magnetohydrodynamic (MHD) description of the magnetopause layer should be applicable when its thickness is many ion gyroradii, as is often the case (see the thickness histogram in Figure 8.10). When no reconnection effects are present, the magnetopause can be represented as a tangential discontinuity (TD), in which the normal magnetic field component is identically zero; plasma flow across the layer, from the magnetosheath side to the magnetosphere side, could be present only as a result of diffusion. During reconnection, however, the normal magnetic field component is nonzero, and it is accompanied by plasma flow across the discontinuity from the magnetosheath into the magnetosphere, where it appears as a boundary layer on open field lines adjacent to the magnetopause. The detailed structure of the magnetopause, away from the reconnection site, may involve more than one MHD wave structure: rotational discontinuities (RDs) or intermediate shocks, or slow shocks and expansion fans; the RD is often the dominant feature. For this reason, the magnetopause is commonly designated either as an RD, which indicates the local presence of reconnection signatures, or as a TD, when such signatures are locally absent.

The RD is a large-amplitude Alfvén wave in which the flow, seen in the HT frame, is field aligned and Alfvénic. This feature has led to a simple test of the presence or absence of local signatures of reconnection. This test is referred to as the Walén test. It consists of plotting measured plasma velocity components, transformed into the HT frame, against the corresponding components of the measured Alfvén velocity. It is important that the data be properly chosen and that the entire magnetopause structure be included. A regression line slope of $+1$ or -1, with high correlation coefficient, indicates Alfvénic flow and therefore RD structure. The interpretation is that reconnection is occurring somewhere on the magnetopause, away from the observation point and that, locally, a magnetic field component normal to the magnetopause must be present, even if it is too small to allow direct determination from minimum variance analysis of the magnetic field. This normal component in turn allows local plasma flow across the magnetopause. An example of such a Walén scatter plot, where the slope of the regression line is near $+1$ and the correlation coefficient is high, will be given the in the next Section (Figure 8.21).

The sign of the regression line slope in the Walén scatter plot is significant: A positive slope indicates flow (as seen in the HT frame) that is parallel to the field, whereas a negative slope indicates anti-parallel flow. If one assumes that the flow is always from the magnetosheath into the magnetosphere, then the sign of the slope indicates whether the normal magnetic field points inward (positive slope) or outward (negative slope). This information in turn tells us on what side of the reconnection site (X line) the observations were made. Experience has shown that only

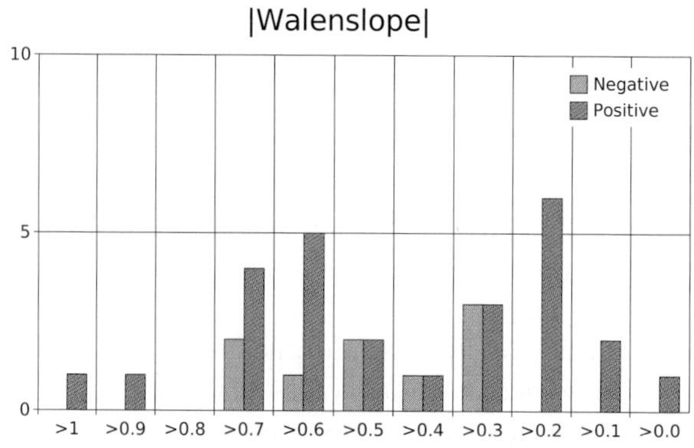

Figure 8.16. Histogram of Walén slopes for 35 Cluster crossings of the magnetopause on July 5, 2001. Only crossings for which the magnitude of the Walén correlation coefficient exceeded 0.85 are included. There were 26 cases with a positive slope, implying reconnection sites tailward of Cluster, and 9 with negative slope, implying reconnection sites sunward of Cluster. (After Paschmann et al., 2004).

fairly rarely is the Walén slope near $+1$ or -1. A far more common situation is that it has smaller magnitude. If the correlation coefficient remains high, such a result indicates that the flow is still field-aligned but is sub-Alfvénic rather than Alfvénic. In the next Section, a number of reasons will be given why the stress balance in directions parallel to the magnetopause can involve, and even be dominated, by terms other than those in a one-dimensional RD, where a simple balance exists between the $\mathbf{j} \times \mathbf{B}$ force and the centrifugal force associated with the field aligned plasma flow. Thus smaller Walén slopes may still be consistent with reconnection, although scenarios are also possible where modest field-aligned flows occur without reconnection. Note also that slow shocks involve exclusively sub-Alfvénic flow, whereas the upstream and downstream states of an intermediate shock are super-Alfvénic and sub-Alfvénic, respectively. Walén slopes significantly larger than unity, with a high correlation coefficient, are essentially never seen.

The histogram in Figure 8.16 shows results of a statistical survey of Walén slopes for all those complete magnetopause crossings by C1 and C3 on July 5, 2001, that yielded a Walén correlation coefficient better than 0.85. A total of 35 crossings meeting this criterion were identified, of which 9 had negative slope and 26 had positive slope. The majority had slope magnitudes in the range 0.2–0.8. To the extent that these various slopes indicate that reconnection-associated plasma flow into the magnetosphere was occurring, the fact that there were many more cases having positive than negative slope would indicate that on this day the reconnection sites were located mostly tailward of Cluster or were originally created sunward, but had moved tailward by the time of the Cluster crossings (the spacecraft location is indicated in Figure 8.15a). In 9 cases the reconnection sites were located sunward of the spacecraft.

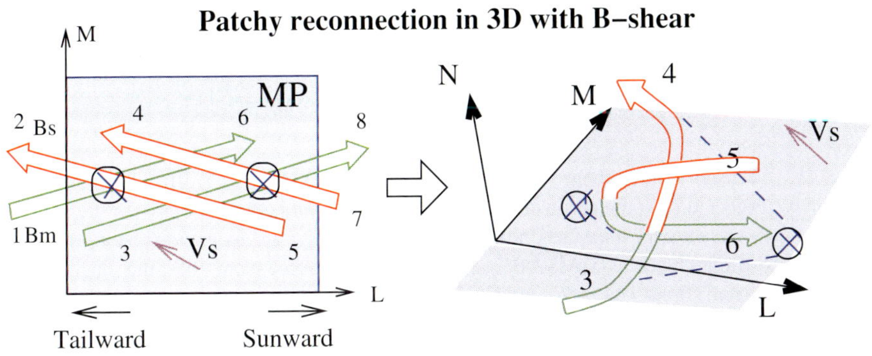

Figure 8.17. Interpretation of Cluster plasma and magnetic field observations for the 05:14 UT magnetopause crossing on June, 30 2001, in terms of a 3-D structure formed by interlinked flux-tubes generated by reconnection at two separate sites, denoted by the X's. (From Louarn et al., 2004).

8.4.2 Examples of two-dimensional structure

The simple one-dimensional view of the magnetopause that underlies most of the results discussed up to this point does not take account of the likely possibility that the layer can have two- and three-dimensional embedded substructures. Three-dimensional features can in principle be detected and, at least qualitatively, described by use of Cluster measurements. Figure 8.17 shows a 3-D structure produced by inter-linked flux tubes, as recently inferred from Cluster plasma and magnetic field observations (Louarn et al., 2004).

For structures that are sufficiently elongated so that they can be considered to be locally two-dimensional, quantitative description of the structures is sometimes possible by use of so-called Grad-Shafranov (GS) reconstruction. This method was initially developed for single-spacecraft data; Cluster has provided an opportunity to validate this version (Hasegawa et al., 2004b). Recently, the method has been adapted to allow for ingestion of multi-spacecraft data (Hasegawa et al., 2004c; Sonnerup et al., 2004a); this version will be used here to discuss two-dimensional structures inside the magnetopause.

GS reconstruction is based on three fundamental assumptions:

- The structure should be approximately two-dimensional, i.e., derivatives along the 'invariant axis' (z) should be much smaller than typical derivatives in the reconstruction plane (xy).
- When viewed in its proper frame, the structure should be approximately time-independent.

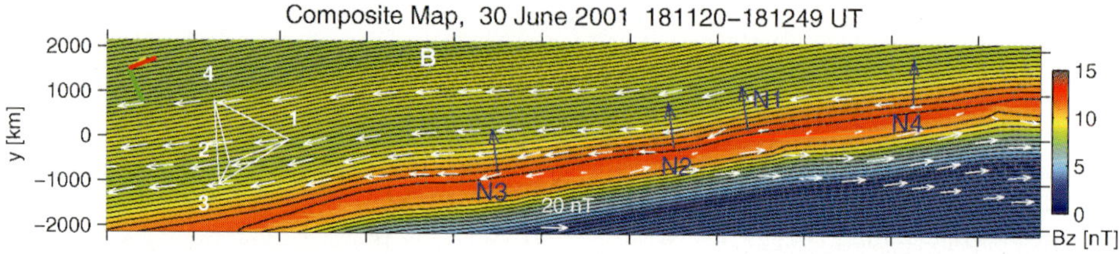

Figure 8.18. Cluster magnetopause crossing on June 30, 2001, 18:12 UT. Black curves represent the field lines, projected onto the reconstruction plane. Background colours represent the axial field B_z. White arrows are actually measured field vectors; blue vectors are magnetopause normals from MVAB with constraint $\langle \mathbf{B} \rangle \cdot \mathbf{n} = 0$ (MVABC). In this map, and subsequent ones, the magnetosphere is towards the bottom. (From Hasegawa et al., 2004c).

- The plasma flow, as seen in the proper frame, should be such that inertia forces can be neglected.

Under these conditions, the one-dimensional vector potential that describes the components of the magnetic field in the reconstruction plane is governed by the plane version of the Grad-Shafranov equation, which describes two-dimensional magnetohydrostatic structures. The reconstruction is obtained by integration of this equation as a spatial initial value problem, using data measured along a spacecraft trajectory as initial values. The details are discussed by Hau and Sonnerup (1999) and Hu and Sonnerup (2003). Cluster data from each spacecraft can be used to produce a map of the magnetic field. The resulting four maps can then be combined into a single map by adding the four maps, each with a properly chosen weight function (Hasegawa et al., 2004c). The composite map, which no longer satisfies the GS equation exactly, is then optimised by choice of invariant axis and weight functions so as to maximise the correlation coefficient between field components predicted from the map and the corresponding measured components. The optimal map may incorporate the effects of inertia forces in a qualitative way.

We now discuss some of the structures that emerge from GS reconstruction of Cluster encounters with the magnetopause.

June 30, 2001, 18:11:20–18:12:49 UT. A composite field map for this event is shown in Figure 8.18. The black curves are the field lines, projected onto the reconstruction plane. The colour represents the strength of the axial field B_z. The red band across the plot is the magnetopause layer, in which the B_z component has a maximum. The magnetosphere is the region below and the magnetosheath the region above the red band. The white arrows anchored at points along the four spacecraft trajectories represent B_x and B_y, the measured field components in the reconstruction plane. The correlation coefficient between the three **B** components predicted by the map at points along the four spacecraft trajectories and the values actually measured at those points is 0.985. The four blue vectors are the magnetopause normals, calculated from MVABC. This figure illustrates that the magne-

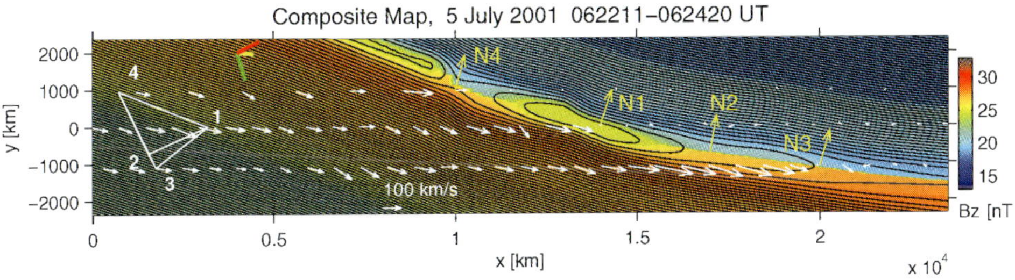

Figure 8.19. Composite field map, based on C4 and C1, for crossing on July 5, 2001, 06:23 UT. White vectors now represent measured plasma velocities (from CIS/HIA for C1 and C3, from CIS/CODIF for C4), transformed to the HT frame; yellow vectors are normals from MVABC. (From Hasegawa et al., 2004c).

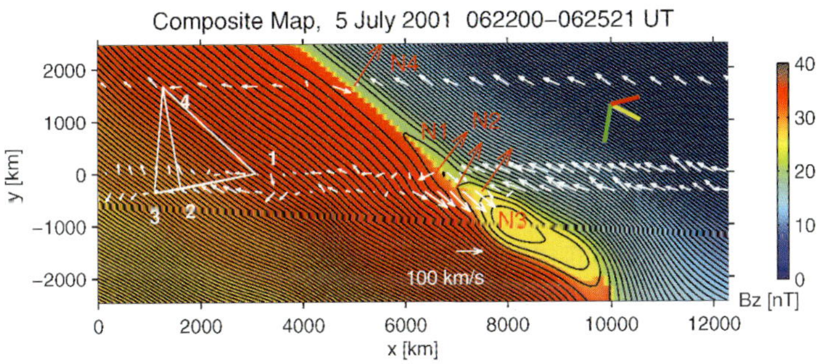

Figure 8.20. Same as in Figure 8.19, except that the map is now based on C2 and C3 instead. Normal vectors from MVABC are red. (From Hasegawa et al., 2004c).

topause sometimes is one-dimensional to a good approximation. Its curvature is rather small on the scale of the Figure, some 20000 km; to model it as a plane layer over the spacecraft separation distance seems justified. The magnetopause thickness is some 500 km, near the middle of the statistical range found in Figure 8.10. Verification of the Walén relation (see also Section 8.4.1) for C1 gives a Walén slope of +0.343 with a relatively poor correlation coefficient $cc = 0.894$, while for C3 the slope is +0.369 with $cc = 0.874$.

July 5, 2001, 06:22:00–06:25:21 UT. This event has been discussed in Section 8.2.1.2 in terms of magnetopause orientation, motion, and speed. An overview of the measurements was shown in Figure 8.7. Haaland et al. (2004a) and Hasegawa et al. (2004b) found evidence that substantial time evolution occurred somewhere in the middle of this magnetopause encounter. For this reason, two composite maps were prepared. The first map, in Figure 8.19, is based on C4 and C1, which were, in the given order, the first two spacecraft to cross the magnetopause. The correlation coefficient for this map is 0.989. The second map, shown in Figure 8.20, is based on the second pair of crossings, by C2 and C3, in that order. It has a correlation coefficient of 0.988. The format in these two Figures is the same as in Figure 8.18, except that the white vectors now represent measured velocities after transformation to the HT frame. The first of the two maps shows two elongated islands embedded in the magnetopause. The fact that the plasma velocities are small in the magnetosheath and large in the magnetosphere indicates that the proper frame (the HT frame from C1 data) of the islands is well anchored in the magnetosheath plasma and that magnetic coupling across the magnetopause is weak or absent. The islands are separated by an X-type null in the transverse field. The Walén test, based on data from C1, gives a slope of $+0.568$ with $cc = 0.976$, suggesting that some reconnection may have been present, or at least in the process of becoming established. The positive sign of the slope indicates that, if plasma crossed the magnetopause from the magnetosheath into the magnetosphere, it would be at a location where the magnetic field component normal to the magnetopause was also earthward. This is the case near, but to the right of, the X point so that the sign of the Walén slope is consistent with what is seen in the map.

In the second map there is evidence that considerable evolution has occurred. The HT frame velocity (now based on C3 data) has decreased significantly and the plasma velocities in the magnetosheath and magnetosphere are now of comparable magnitude, indicating the presence of magnetic coupling across the magnetopause. A magnetic island of substantial width has developed and the Walén slope (from C3 data) has increased to +1.027 with $cc = 0.979$, as shown in Figure 8.21, indicating the presence of full-blown reconnection at the time C3 encounters the magnetopause. It is fully consistent with the reconstructed map in which the reversal of the plasma flow direction at C3, resulting from the $\mathbf{j} \times \mathbf{B}$ force as the plasma crosses the magnetopause, is evident. Again the normal magnetic field component should be negative where the inward plasma flow occurs, as indeed it is in the map. This map also shows a prime reason why the magnetopause thickness is not always constant during an event: Local bursts of reconnection can produce local swelling of the layer. Extreme cases of such swellings are FTEs, which will be discussed in Section 8.6. It must be remembered that, in the event presented here, the model assumption of time stationarity is violated. Nevertheless, the appearance of the large magnetic island in the second half of the event is a qualitatively robust feature, which illustrates that large changes in the magnetopause structure can occur on time scales of 30 s or less.

July 3, 2001, 05:17:04–05:18:17 UT. A final example of magnetopause mesoscale structure encountered by Cluster is shown in Figure 8.22. The GS reconstruction map has a correlation coefficient of 0.994, the highest value found to date (right hand panel). The top panel shows the axial component of the magnetic field in colour and the measured magnetic field as white arrows. In the second and third panels, the colours represent plasma pressure and density, respectively, and the

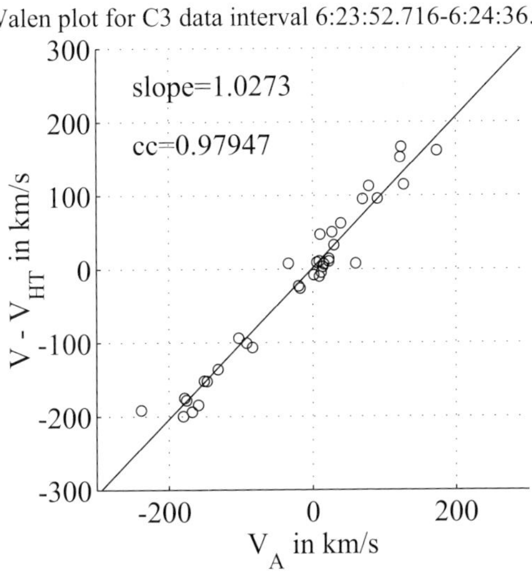

Figure 8.21. Walén scatter plot from C3 for the event in Figure 8.20. (From Hasegawa et al., 2004c).

white arrows are velocities (from CIS/HIA for C1 and C3, and from CIS/CODIF for C4) in the moving frame. The magnetopause, located on the upper right in the three panels, has a large negative B_z component (blue colour). It has very large thickness and a large curvature. It contains two magnetic islands, separated by an X point, but there is no clear indication of ongoing reconnection at that point (the Walén slope from C1 is +0.322 with $cc = 0.891$; for C3 the slope is -0.011 with $cc = 0.051$). This example illustrates an important point. The presence of an X-type null in the transverse field does not necessarily require the presence of reconnection and its associated accelerated plasma flows: In a two-dimensional geometry, tangential stress balance can be maintained by tangential gradients in the transverse pressure $(p + B_z^2/2\mu_0)$. In three-dimensional geometries, additional forces can contribute to the stress balance. For these reasons, it is entirely possible that field-aligned flow across the magnetopause during a reconnection event can be sub-Alfvénic, i.e., have a Walén slope magnitude less than unity. The second panel also shows that the plasma pressure in the magnetopause was lower than on its two sides: As seen in the third panel, there is actually a plasma boundary layer of substantial width on the earthward side of the current layer, the origin of which remains unexplained.

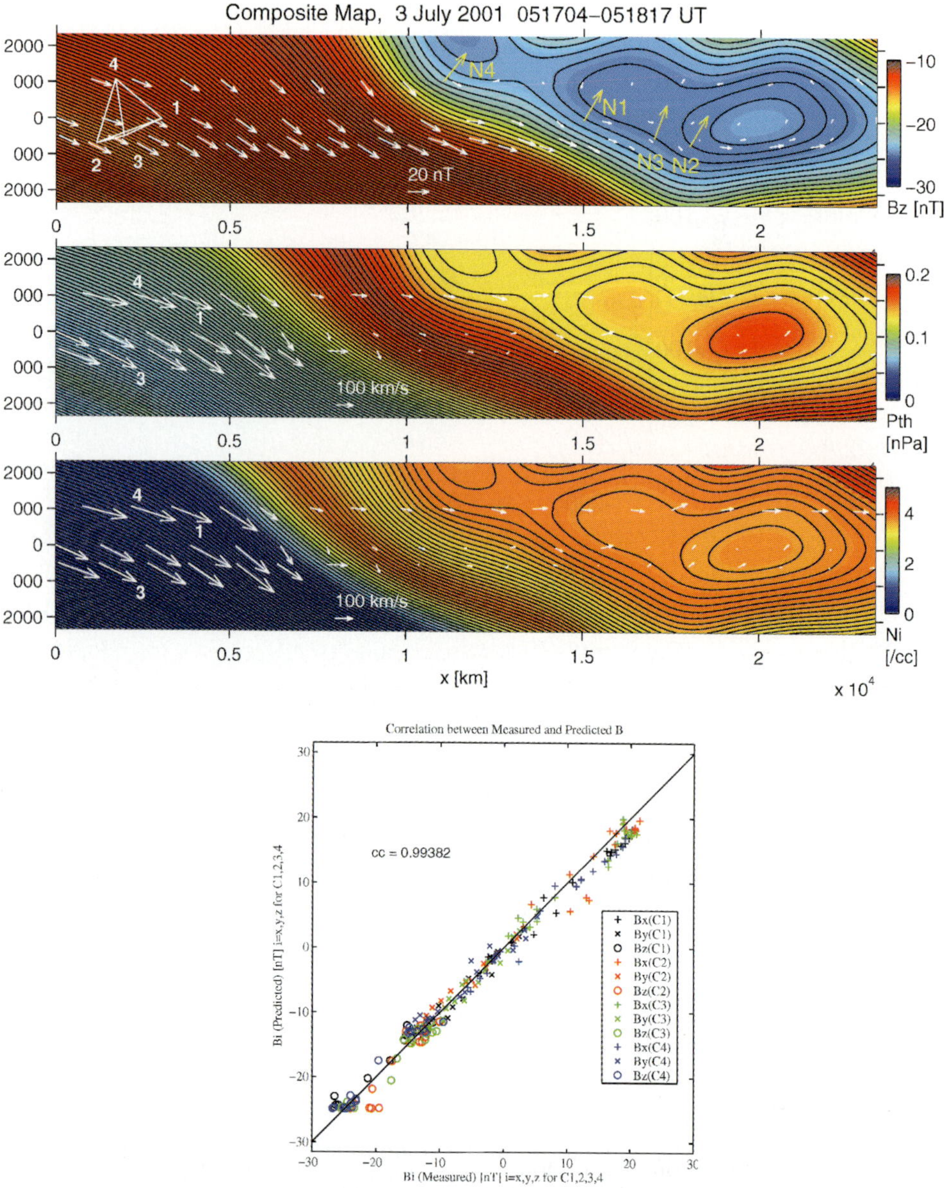

Figure 8.22. Top: Composite maps based on C1, C2, C3, and C4, for the magnetopause crossing on July 3, 2001, 05:17 UT, As before, black lines are the magnetic field lines projected into the reconstruction plane. The top panel shows the axial component of the magnetic field in colour and the measured magnetic field as white arrows. In the second and third panels, the colours represent plasma pressure and density, respectively, and the white arrows are velocities (from CIS/HIA for C1 and C3, and from CIS/CODIF for C4) in the moving frame. Note the presence of an LLBL, earthward of the magnetopause. The scatter plot at the bottom shows the correlation between predicted and measured magnetic field components. (From Hasegawa et al., 2004c).

8.4.3 Summary

The study of Walén slopes is an established way to examine the nature of the magnetopause discontinuity. Even if reconnection is ongoing, the Walén slope magnitude can be less than unity because of spatial structure within the magnetopause. It is therefore important to establish the geometry of the current layer: The Grad-Shafranov reconstruction method, which has been validated and modified to become a true multi-spacecraft technique, can be used for that purpose.

- A statistical study of Walén slopes at the near-tail, dawn magnetopause indicates that reconnection sites are located tailward of the observation site about three times as frequently as sunward of it.

- Using the Cluster observations, Grad-Shafranov reconstruction is capable of producing two-dimensional magnetic field maps that correlate extremely well with the measured field. It accurately determines the invariant direction.

- The magnetopause layer is found to usually contain a wide variety of embedded two-dimensional structures such as magnetic islands, separated by X-type magnetic null points in the transverse field. Since these islands often enclose non-zero axial fields, they are in effect flattened flux ropes.

- The presence of X-type nulls in the magnetopause maps does not necessarily mean that reconnection is occurring. Tangential stress balance can be maintained even in the absence of reconnection-associated jetting of the plasma away from an X line. For the same reason, the Walén slope does not have to be near $+1$ or -1 in order for reconnection to be active.

- The structure of the locally observed magnetopause can change drastically in time intervals between successive spacecraft crossings, i.e., 30 s or less. Specifically, reconnection at a tailward moving X-point can get underway in such a short time, the result being the development of a tailward moving magnetic island that causes substantial localised swelling of the magnetopause layer. This is a possible mechanism for the production of some FTEs.

- Curvature and internal structure of the magnetopause, which appear to be present fairly often, indicate why single- as well as multi-spacecraft methods for determining the magnetopause orientation, most of which assume a planar layer or require identification of common features in successive crossings, can go wrong (Section 8.2). The non-colinear MVABC normal vectors shown in the top panel of Figure 8.22 illustrate this point.

8.5 Electromagnetic Fields in the Magnetopause

8.5.1 Background

Being a plasma discontinuity, the magnetopause has an inner structure that is determined by the distribution of ions and electrons within the boundary and by the corresponding electric and magnetic fields. Many theoretical and numerical studies (see, e.g., Hau and Sonnerup, 1991; Roth et al., 1996, and references therein) have discussed the structure of both the closed and open magnetopause (tangential or rotational discontinuity, respectively). Models of tangential discontinuities show the possibility of strong localised electric fields normal to the discontinuity plane in the regions where two plasmas are in contact (e.g., Sestero, 1966; De Keyser and Roth, 1997). Because of the different magnetospheric and magnetosheath ion and electron gyroradii involved, a small charge imbalance can arise that leads to a polarisation electric field (the Chapman-Ferraro electric field) which is of fundamental importance for understanding the internal structure of the boundary. The open magnetopause studies are usually done in the context of reconnection. Numerical simulations show that also in those cases the dominant electric field component is normal to the discontinuity plane, being strongest along the reconnection separatrices (e.g., Pritchett and Coroniti, 2004). The tangential component is usually much weaker. While there have been numerous observational studies of magnetopause structure based on magnetic field and particle data (e.g., Phan and Paschmann, 1996; Lucek et al., 2001), detailed observations of the electric field have been few (e.g., Aggson et al., 1983; Lindqvist and Mozer, 1990; Mozer et al., 2002). In many cases the observed tangential electric field is close to the instrument precision.

Cluster excels here because of the quality of its instrumentation. As a rule of thumb, Cluster does a very good job at studying the magnetopause at ion scales, spatial scales of a few tens to a few hundreds of kilometers and temporal resolution up to a few Hz. The electric and magnetic field, as well as the plasma density obtained from spacecraft potential measurements, are usually sampled at a sufficiently high rate to resolve the ion scales. These scales correspond to the smallest Cluster separations, so that standard multi-spacecraft analysis techniques such as the curlometer (Section 8.3), measure ion-scale structures. However, the particle instrument time resolution is in most cases insufficient to resolve 3-D distribution functions on ion scales. One can still get particle data at a higher time resolution (particularly when the experiments operate in burst mode), even though these data give only partial information on the distribution functions.

The situation is slightly different for scales between the ion and electron scales or smaller, i.e., scales of only a few km or less and frequencies above a few Hz. From simulations it is known that the electric field tends to structure itself in such narrow layers. There are also strong wave emissions at magnetopause, such as drift lower hybrid and whistler emissions, that fall into this range. To resolve these scales Cluster does a very good job when its instruments are in burst mode, which happens in slightly less than 10 % of all the magnetopause crossings. Sometimes multi-spacecraft analysis techniques can be applied, e.g., for coherent whistler emissions, but in most cases one has to combine multi- and single-spacecraft techniques. Cluster's complement of wave instruments, with their broad frequency range, provide additional information about the highest frequencies and smallest spatial scales.

In this section we present a few examples of Cluster's ability to measure electromagnetic field structure in the magnetopause. Two case studies demonstrate what is possible with burst mode data when the spacecraft are at small separation (100 km): Section 8.5.2 deals with a thin magnetopause without boundary layer, and Section 8.5.3 illustrates the detection of narrow current layers within a thick magnetopause. Section 8.5.4 discusses the properties of electromagnetic wave emissions observed near the magnetopause.

8.5.2 The intrinsic electric field in a thin magnetopause

On 30 March 2002, around 13:11:45 UT, the Cluster spacecraft crossed the magnetopause tailward of the cusp, going from the polar cap into the magnetosheath (Figure 8.23). The magnetopause was moving at $\sim$25 km s^{-1}, and it took only $\sim$2 s to cross it. The magnetopause thickness, $\sim$50 km, is comparable to the gyroradius of a magnetosheath ion. In this case, there was no boundary layer earthward of the magnetopause. The density gradient coincides to a large part with the current sheet.

The magnetic field in Figure 8.23 is plotted in the LMN reference frame, where N is along the magnetopause normal, L is along the maximum variance magnetic field direction, and M completes the right-handed system. The first two panels show the B_L component; panel (a) shows the original times series, while in panel (b) the C2, C3 and C4 observations are shifted so as to align the current sheet crossings. If the magnetopause were moving with a constant velocity, panel (b) would reflect the spatial structure of the magnetopause. The magnetosheath half of the current sheet (13:11:44.5–13:11:45.5) can be nicely aligned, indicating that this part of the current sheet is planar on the spacecraft separation scale and that it moves with approximately constant velocity. The magnetospheric half of the current sheet shows significant variations: B_L variations observed just before the current sheet align in the temporal (panel a) rather than in the spatial domain (panel b), i.e., their phase velocity (roughly estimated at a few hundred km/s) is much higher than the velocity of the magnetopause. In the magnetosphere, the Alfvén velocity is $\sim$1000 km s^{-1}, thus what we observe on the magnetospheric side could be disturbances (e.g., small-scale surface waves) propagating at a fraction of the local Alfvén speed. This demonstrates the importance of multi-spacecraft observations for studying small-scale magnetopause structure. The magnetic shear at the magnetopause is about 110° and thus the B_M component has a constant offset of about -60 nT (panel c). The variations in B_M are indicators of parallel currents within the magnetopause. The constant offset in B_N of about -10 nT indicates that the magnetopause locally is a rotational discontinuity. It has been suggested that thin magnetopause crossings without boundary layer could be encounters not far from the reconnection diffusion region. However, we do not know if this is the case for the event presented here.

Figures 8.23e–f show measurements of the DC and AC electric field normal to the magnetopause. Here DC includes frequencies up to 20 Hz (ten times the pro-

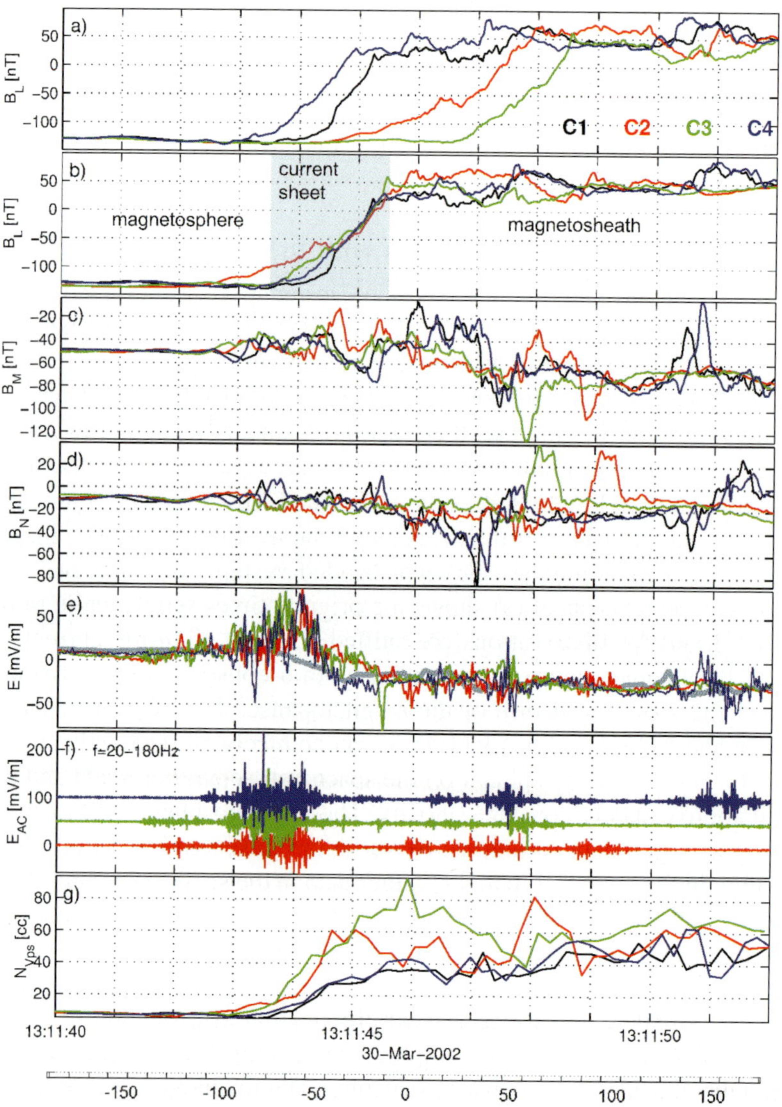

Figure 8.23. Cluster observations of a thin magnetopause. The magnetic field is presented in *LMN* coordinates defined by the local magnetopause orientation. In panels (b–g) the time axis has been shifted for C2, C3, and C4, so as to align the current sheet observations. (a,b) B_L, the maximum variance magnetic field component. (c) B_M. (d) B_N, normal to the magnetopause. (e) The electric field normal to the magnetopause between 0–20 Hz (due to a malfunctioning probe on C1 only data from the other three spacecraft are shown; the grey line is the C4 electric field in the deHoffmann-Teller frame), and (f) in the 20–180 Hz range (the zero levels for Cluster 3 and 4 have been shifted by 50mV m^{-1} and 100mV m^{-1} respectively to improve the visibility). (g) Plasma density derived from the spacecraft potential. (Figure provided by A. Vaivads).

ton gyrofrequency) while AC refers to frequencies of 20 up to 180 Hz (covering the lower hybrid frequency ~70 Hz). The strong DC electric field on the magnetospheric side of the current sheet points from the magnetosphere into the magnetosheath. This is reminiscent of the electric field first discussed for a TD (i.e., $B_n = 0$) by Ferraro (1952, see also Section 8.1.3), although in our case $B_n \neq 0$. One should note that this large DC electric field cannot be transformed away by going to another reference frame. It is the intrinsic field of the magnetopause. To illustrate this, the grey line in Figure 8.23e shows the electric field in the deHoffmann-Teller frame for the time interval 13:11:42–46 UT (C4). The AC component of the electric field (panel f) has amplitudes that are comparable to or larger than the DC field. The strongest AC fields are seen at the density gradients on the magnetospheric side of the current sheet; they are most probably lower hybrid or lower hybrid drift waves. These observations are consistent with Cluster observations at other magnetopause crossings (André et al., 2001). Note that the time series of these waves is available thanks to the experiment being in burst mode; particle data are also available at high time resolution, but we do not discuss them here.

8.5.3 Narrow layers inside a thick magnetopause

In an event on February 6, 2002, around 08:11:58 UT, Vaivads et al. (2004) found at strong localised electric fields within a narrow current layer on the magnetospheric side of a thick magnetopause (see also André et al., 2004). Cluster crossed the magnetopause in the high latitude afternoon sector at ~14 MLT with a spacecraft separation of ~100 km. Figure 8.24 shows C4 data. The top two panels show 70 s of density and magnetic field measurements during the crossing, while the rest of the panels are a 3 s zoom into a narrow layer with the strongest electric fields. The magnetopause/boundary layer crossing lasts ~40 s. The narrow layer that we discuss in more detail is only ~0.2 ion inertial lengths wide and is located at about 08:11:57.5 UT. The electric fields observed within this layer are the strongest of the whole crossing.

The electron sector data in Figure 8.24c and 8.24d correspond to electrons moving perpendicular and parallel to the ambient magnetic field. The high energy electrons of ~1 keV disappear on the magnetosheath side of the layer; there is a pronounced parallel electron beam within the narrow layer (panel d). The electric field normal to the layer is supported by the Hall term in the Generalised Ohm's law, and not by the electron pressure gradient (panel e). This shows that the normal electric field is the Hall electric field. There is a density jump on the same scale as the narrow layer (panel f) accompanied by a corresponding change of the magnetic field strength (panel g). The strong parallel current (panel h) is consistent with the observed electron beam. André et al. (2004) suggest that the narrow layer could be the separatrix of a distant reconnection site. This interpretation is consistent with numerical simulations that show separatrix signatures extending to hundreds of ion inertial lengths from the reconnection site (Shay et al., 2003). Whether there can

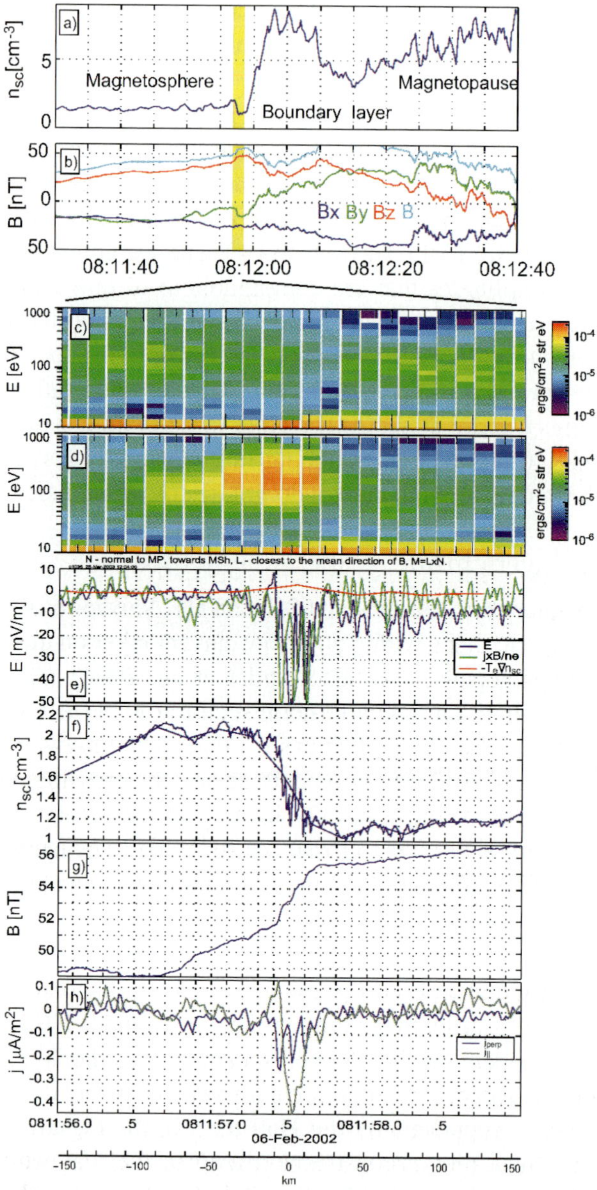

Figure 8.24. Magnetopause crossing on February 6, 2002. (a,b) Overview of density and magnetic field. (c,d) Electron data from sectors 2–3 and 8–9, perpendicular and parallel to the magnetic field. (e) The normal electric field measured by EFW (blue), $\mathbf{j} \times \mathbf{B}/ne$ (green), and $-T_e \nabla n_{sc}$ (red). (f) The plasma density n_{sc} estimated from the spacecraft potential. (g) Magnetic field strength $|\mathbf{B}|$. (h) Current $\mathbf{j}$ calculated assuming that the variation in $\mathbf{B}$ is due to magnetopause motion). Adapted from André et al. (2004) and Vaivads et al. (2004).

be other explanations for the narrow layers, unrelated to reconnection, is not clear. In general, strong and localised electric fields are likely to lead to instabilities and turbulence which would tend to smear out the structures over longer scales. The existence of these small scale features seems to indicate some 'freshness' of the structure.

While Figure 8.24 presents only single-spacecraft data, there are a few important multi-spacecraft aspects to the analysis. First, the estimate of the magnetopause velocity from four spacecraft allows the construction of the spatial scale (at the bottom of the Figure). Secondly, the use of a single-spacecraft technique to estimate currents requires that the phase velocity of magnetic field variations are known. The comparison among all spacecraft is used to confirm the assumption that the narrow current sheets are relatively fixed in the magnetopause reference frame.

8.5.4 Electromagnetic waves

Previous space missions have shown that there is intense wave activity in the low frequency range at the magnetopause. More precisely, the wave activity is low in the magnetosphere, the magnetosheath is strongly turbulent, and inside the magnetopause the power in low frequency waves is on average about 10 times larger than that in the adjacent magnetosheath. Cluster's wave instruments cover a broad frequency range, up to 4 kHz for the magnetic field (STAFF experiment) and up to 80 kHz for the electric field (EFW and WHISPER experiments), and offer an ideal opportunity to study this type of wave activity. Spectrograms for a magnetopause crossing on July 5th, 2001, are given in Figure 8.25. The top panel shows the DC magnetic field to identify the precise magnetopause crossing time. The distinct change in the behaviour of the waves at the magnetopause allows the crossing to be identified from the wave observations directly, without consulting the DC magnetic field. The WHISPER spectrogram shows a cut-off below 10 kHz before the crossing and an emission around 35 kHz after the crossing, indicative of the plasma frequency: The magnetosphere is to the left and the denser magnetosheath to the right. Magnetic wave activity is more intense on the magnetosheath side; electric activity is stronger on the magnetospheric side. In the magnetosphere the wave activity is more electrostatic, while it is electromagnetic in the magnetosheath.

Four-point measurements allow us to spatially localise the wave activity with respect to the magnetopause boundary. The upper panel in Figure 8.26 shows a two-dimensional Grad-Shafranov reconstruction map of the magnetic field, identical to the reconstructions presented in Section 8.4. The lower panel again shows the magnetic field lines in black, with the overlaid colour lines representing the ULF wave power along the trajectories of each spacecraft. Several interesting features can be read off from this plot. First, the magnetic wave power is strong on the magnetosheath side while it ceases upon entering the magnetosphere. Second, the highest fluctuation levels of the order of $\sim$1 nT^2 are found at the magnetopause, in the sheath side magnetic gradient region. Third, when the spacecraft remain close

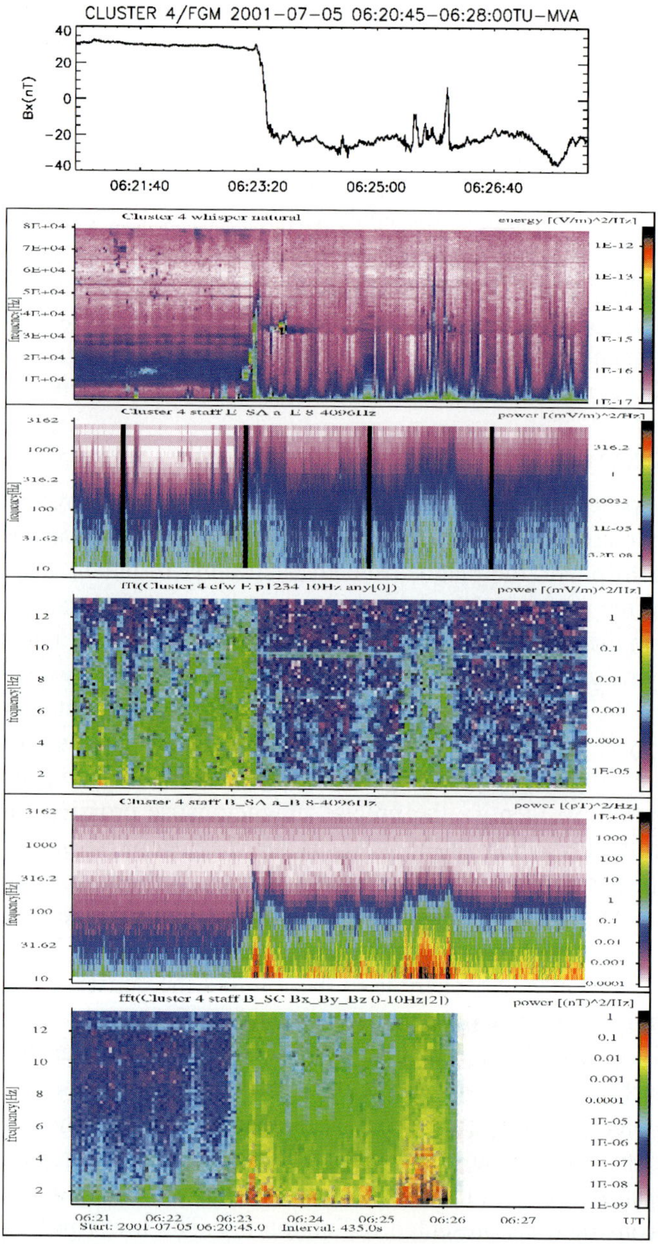

Figure 8.25. Spectrograms from C4 for the magnetopause crossing on July 5th, 2001, around 06:23 UT. The top panel shows the reversal of the maximum variance component of the DC magnetic field. The next 3 panels show the electric field in the frequency ranges 3 kHz–80 kHz (WHISPER), 12 Hz–3 kHz (EFW) and 0–12 Hz (EFW). The bottom two panels show the magnetic field in the frequency bands 0–12 Hz (STAFF-SC) and 12 Hz–3 kHz (STAFF-SA). (Figure provided by D. Attié.)

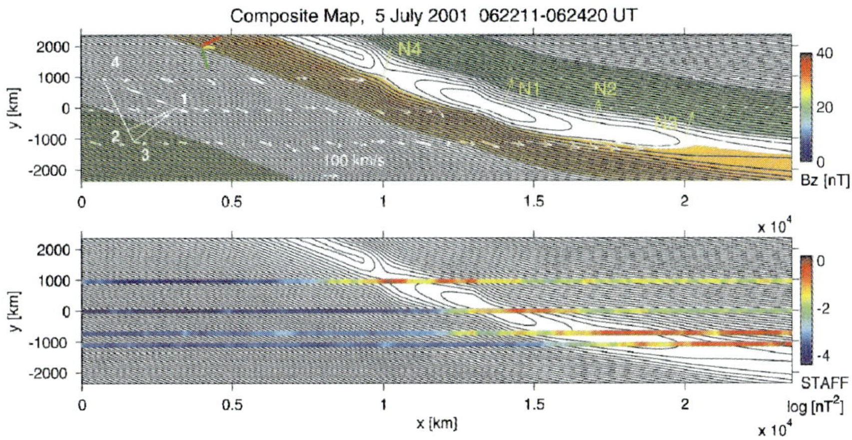

Figure 8.26. The top panel shows magnetic field lines from a Grad-Shafranov reconstruction. The paths of the Cluster tetrahedron are indicated by numbers and the measured flow velocity vectors. The bottom panel shows the reconstructed field contours overlaid by the STAFF low frequency magnetic wave power which is plotted in colour (logarithmic scale on the right) along the 4 spacecraft trajectories. (Data provided by D. Attié and H. Hasegawa.)

to the magnetopause for a long time, as in the right part of the Figure for C2 and C3, high activity levels are observed throughout the entire period of proximity to the magnetopause. Finally, a similarly high activity level is detected close to the presumable X-point detected by C4. From these observations we may conclude that the magnetosheath side of the magnetopause gradient is subject to a high level of magnetic fluctuations at low frequencies. However, a high activity level is also observed near a magnetic X-point.

The nature of these waves could be revealed by looking at the dependence of the integrated wave power measured in different frequency ranges on macroscopic parameters, such as thickness of the current layer, the presence of a boundary layer adjacent to the magnetopause, magnetopause current density, etc. On July 5, 2001, Cluster crossed the flank magnetopause multiple times, allowing for a statistical investigation. There is no apparent correlation of the power of the ULF fluctuations with the thickness of the boundary, neither with the magnetopause current density. A comparison of inbound and outbound crossings (Figure 8.27) shows that the integrated wave power is higher for outbound crossings ($0.18\,\mathrm{nT}^2$ on average) than for inbound ones ($0.08\,\mathrm{nT}^2$). Note that multiple crossings on the flanks of the magnetopause, like on this particular day, have probably been caused by large-scale magnetopause surface waves (see also Section 8.7): Wave activity at smaller scales is related to these macroscopic waves, apparently with an asymmetry between the two edges.

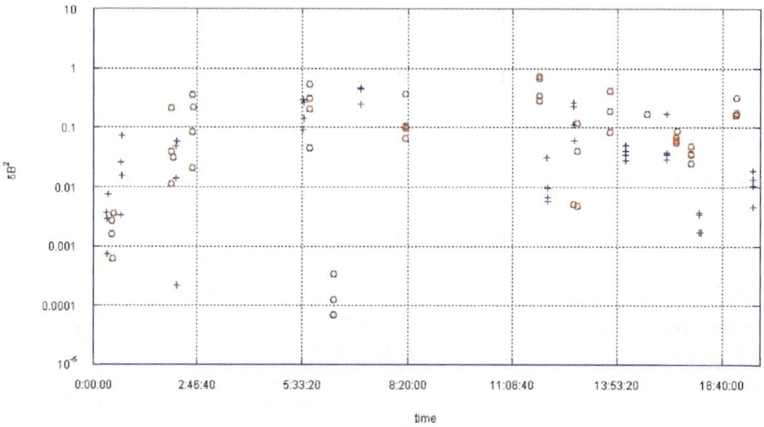

Figure 8.27. Statistics of the power of waves (integrated over frequency, thus in units of nT^2) for the magnetopause crossings observed on July 5, 2001. Outbound crossings (red circles) tend to have higher wave power than inbound crossings (blue pluses). (Figure provided by D. Attié.)

8.5.5 Summary

In its investigation of small-scale structure inside the magnetopause, Cluster draws heavily upon the diversity and the high time resolution of its instruments that measure the electric and magnetic fields. Cluster's major achievements have been

- to directly measure the intrinsic electric field and its spatial distribution inside the boundary, which is fundamental for our understanding of the magnetopause as a plasma discontinuity.
- to locate substructure within the magnetopause/boundary layer, with a tentative identification of the reconnection separatrices.
- to measure the Hall electric field and to demonstrate that it is a key element of the structure of thin current sheets.
- to localise wave emissions relative to the magnetopause.

The investigation of small-scale events involves large amounts of data and has therefore been done only on a case-per-case basis. It would be worthwhile to perform a statistical analysis in the future.

The Cluster data are so detailed that a thorough comparison with models and simulations has become possible. Such a confrontation of theory and observation still has to be done. It would undoubtedly advance our understanding of the physics of plasma boundaries.

Unfortunately, Cluster's capabilities usually do not permit in-depth studies of small-scale structure with the particle instruments, because of the limited time resolution of those instruments (which rely on the spacecraft spin to assemble 3D velocity distributions).

8.6 Flux Transfer Events

8.6.1 Background

Russell and Elphic (1978) reported a regularly occurring characteristic magnetic field signature observed at low latitude by ISEE-1 and -2, which they termed a flux transfer event (FTE). This signature consists of a bipolar fluctuation in the field component normal to the magnetopause over a timescale of a few minutes, corresponding to spatial scales of ~ 1 R_E . Statistical studies of FTEs (e.g., Rijnbeek et al., 1984; Berchem and Russell, 1984; Southwood et al., 1986; Lockwood, 1991; Lockwood and Wild, 1993; Le et al., 1993; Kuo et al., 1995; Russell et al., 1995; Kawano and Russell, 1996, 1997) have shown that the FTE occurrence rate is enhanced for southward IMF, when conditions are optimum for reconnection on the dayside subsolar magnetopause (see Phan et al., 2005), with a mean interval between FTEs of ~ 8 minutes. FTEs are usually associated with a mixed plasma population with components of both magnetospheric and magnetosheath origin (e.g., Paschmann et al., 1982; Farrugia et al., 1988), while the distributions and the composition are modified compared to those found in the bounding regions (Thomsen et al., 1987; Klumpar et al., 1990). Many (but not all) scientists interpret these FTE signatures as the manifestation of transient reconnection.

The original Russell and Elphic (1978) observations led to a picture of a bundle of reconnected flux tubes, formed by a spatially localised patch of reconnection, threading the dayside magnetopause, which are then pulled back over the poles of the Earth under the influence of the $\mathbf{j} \times \mathbf{B}$ force and the external magnetosheath flow. These original observations were of bipolar perturbations directed first in the outward normal and then the inward normal direction. However, Rijnbeek et al. (1982) presented observations of FTEs at equatorial latitudes that were characterised by inward then outward bipolar signatures. These two types of FTE signature, referred to as 'normal' and 'reverse' polarity FTEs, were interpreted as encounters with newly-opened flux tubes connected to the northern and southern hemispheres respectively. This inference was supported by statistical surveys which suggested that normal/reverse polarity FTEs are most commonly observed in the northern/southern hemisphere with a mixture of both near the equator (e.g., Berchem and Russell, 1984; Rijnbeek et al., 1984; Kawano et al., 1992; Russell et al., 1995). Elphic and Southwood (1987) presented simultaneous observations of normal and reverse polarity FTEs by ISEE and AMPTE/UKS, located in opposite hemispheres.

A number of alternate models of FTEs have been proposed. Lee and Fu (1985) suggested that an FTE may be the signature of a magnetic flux rope formed between two or more approximately parallel reconnection lines. Southwood et al. (1986)

and Scholer (1988) suggested that FTEs may be a result of a rapid temporal variation of the reconnection rate at a single reconnection point rather reconnection at multiple sites (see also Scholer, 1995). The transient FTE signature would then be caused by a propagating bulge in a pre-existing reconnection outflow layer moving past the observing spacecraft. These interpretations have been supported by modelling and simulation studies (e.g., Kan, 1988; Hesse et al., 1990; Lee et al., 1993; Ku and Sibeck, 1997; Nakamura and Scholer, 2000). Sibeck (1990, 1992) and Sibeck and Smith (1992) suggested that the magnetospheric FTE signature could also be caused by a transient magnetosheath pressure pulse causing a temporary relocation of the observing spacecraft into the plasma depletion layer and magnetosheath. This would mimic the signature of a 'crater' type FTE (e.g., Paschmann et al., 1982; Farrugia et al., 1988). More detailed studies (e.g., Lockwood, 1991; Smith and Owen, 1992; Elphic et al., 1994; Song et al., 1994) have shown that not all FTE signatures are consistent with the Sibeck (1990, 1992) picture.

FTE signatures are known to map along magnetic field lines into the ionosphere. Ionospheric FTE signatures were first reported by Goertz et al. (1985). Sporadic and spatially limited ionospheric flow bursts moving in the north-west direction were attributed to magnetic merging, and matched a simultaneous FTE observation at the magnetopause. Pinnock et al. (1991, 1993, 1995) presented cases of enhanced convection channels superimposed on the otherwise continuous cusp echoes. Inside these flow channels, the velocity was larger than in the ambient plasma and directed mainly northward, although many also exhibited an azimuthal velocity component, the sign of which depends upon the sign of the IMF B_y component in a manner consistent with the convection of reconnected flux tubes on the dayside magnetopause (e.g., Cowley and Owen, 1989). The longitudinal extent of these flow bursts is generally larger than their latitudinal extent and their average repetition rate is about 7 minutes. Parallel currents flowing on the flanks of the flow burst have been modelled (Southwood, 1987; Lockwood et al., 1990, 2001) and observed (Milan et al., 2000). Results of detailed case studies of corresponding ionospheric and magnetopause features (Elphic et al., 1990; Neudegg et al., 1999, 2000) have been complemented by statistical studies of the location, occurrence, and relation to IMF B_y of ionospheric flow bursts (e.g., Provan and Yeoman, 1999; Provan et al., 1999; McWilliams et al., 2000). Provan et al. (1998), Milan et al. (2000), and Thorolfsson et al. (2000) have used radars and optical observations to determine the size, shape, velocity and recurrence rate of the ionospheric signatures of FTEs. All these observations are consistent with intermittent and/or patchy magnetic reconnection in the form of an FTE driving an ionospheric signature consisting of a localised and sporadic flow burst.

The study of flux transfer events has received renewed interest with the advent of the Cluster mission. The inter-spacecraft separations have generally been appropriate to examine FTEs. Cluster measurements can be used to accurately determine their size and speed of motion across the magnetopause, but for the first time also allow examination of FTE structure, such as the distribution of current flowing within them. Simultaneous observations of FTEs at varying distances from the magnetopause and/or on either side of this boundary are presented below.

8.6.2 Simultaneous multi-point FTE observations

Magnetic field signatures were reported by Dunlop et al. (2001) for a pass where Cluster (spacecraft separation ~600 km) sampled an FTE from both sides of the magnetopause simultaneously. The bipolar normal magnetic field signatures observed on all four spacecraft scaled with distance from the magnetopause. In addition, the traces for $|\mathbf{B}|$ showed a crater-FTE character for the spacecraft nearest the magnetopause. Case studies have also been presented by Bosqued et al. (2001) and Owen et al. (2001). These multi-point analyses directly confirm inferences on the typical size and speed of FTEs from previous missions. For example, the former study presented magnetic field and ion data for an FTE observed by Cluster at the dayside magnetopause under southward IMF conditions on January 26, 2001. Analysis of this FTE indicated that its size along the magnetopause surface was ~2 R_E and that it convected with a velocity of ~200 km s^{-1}. Owen et al. (2001) presented PEACE observations of 10 eV–26 keV electrons just inside the magnetopause on February 2, 2001, together with supporting magnetometer observations. Each of the spacecraft observed a series of transient enhancements of magnetosheath-like electrons within the magnetosphere, together with magnetic field variations typical of magnetospheric FTEs. Timing analysis of one of these FTEs indicated that its size was ~0.8 R_E and that it moved from noon towards dusk with an inferred speed of motion across the magnetopause of ~70 km s^{-1}, again consistent with previous estimates.

These observations from four spatially distinct points also provide new insight into the internal magnetic field and plasma structure of FTEs. Owen et al. (2001) noted that the four observed signatures were not nested, which immediately rules out any explanation for these signatures that involves a simple back and forth motion of the magnetopause. In addition, there were significant differences in the signatures between spacecraft, implying FTE substructure on the scale of the spacecraft separation (~600 km). A delay in and an extended duration of the signature at the spacecraft furthest from the magnetopause led to the conclusion that FTE substructure may include: a compression of the closed flux tubes ahead of the FTE, which cause density and field strength enhancements; a circulation of open flux tubes within the FTE itself, which accounts for the delay in the arrival of the magnetosheath electron populations at locations deepest within the magnetosphere; and a possible trapping of magnetospheric electrons on the most recently opened flux tubes within the FTE.

Robert et al. (2002) have studied an isolated magnetosheath FTE observed on January 26, 2001. These authors use the curlometer to infer the current density within an FTE for the first time. As the spacecraft enter this FTE, data from PEACE

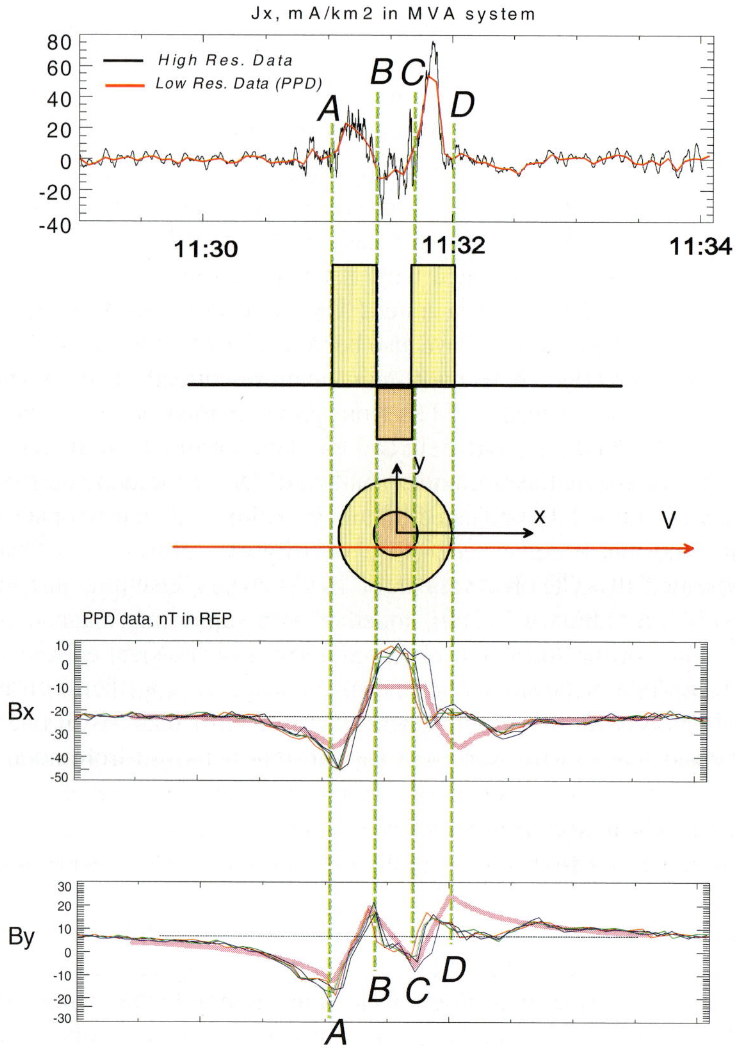

Figure 8.28. FTE observed on January 26, 2001. (a) Field-aligned currents in an FTE computed with the curlometer technique. (b) Schematic of a force-free, double-current, tubular flux rope model. (c) Comparison between the modelled variations and the observations for a sample trajectory through the model. (From Robert et al., 2002).

and CIS provide evidence for enhanced fluxes of energetic electrons and ions, in particular in the directions parallel and anti-parallel to the magnetic field. The ion flow speed was also observed to increase. Curiously, this enhancement in the fluxes of energetic particles lasts much longer in the direction antiparallel to **B** than that of electrons and ions parallel to **B**. Electromagnetic field fluctuations observed by STAFF and EFW are consistent with instabilities driven by field-aligned currents. The analysis reveals a double-current structure, with currents antiparallel to the magnetic field in the outer parts of the FTE, but parallel within the central region (Figure 8.28, top panel). Comparing the observations to a force-free, double-current, tubular flux rope model (central panel), the observed magnetic field profiles are found to be consistent with an outer tube radius of $\sim$5200 km in which the current density is 30 mA km^{-2} (corresponding to a total current of 2300 kA), and an inner tube radius of $\sim$1400 km with a current density of -10 mA km^{-2} (total current 190 kA). The agreement between the observed and modelled magnetic field variations is illustrated in the bottom two panels.

The four-point measurements have also facilitated the simultaneous observation of magnetopause bulges on both sides of the magnetopause, ruling out an explanation for FTEs in terms of a pressure-pulse induced boundary motion. This removes a long-standing ambiguity in the origin of FTEs that could not be resolved with single-spacecraft data. On March 8, 2003, the Cluster spacecraft crossed the high-latitude northern magnetopause in close to an ideal tetrahedron configuration, with separations of $\sim$1 R_E. The plane formed by C1, C2, and C4 lay close to parallel to the magnetopause surface, and these spacecraft led C3 along the orbital track. Data recorded as the spacecraft approached the magnetopause are shown in Figure 8.29. The upper four panels show the fluxes of field-aligned electrons in the energy range 10 eV–26 keV recorded by PEACE on each spacecraft, while the lower four panels show the magnetic field in boundary normal coordinates.

All spacecraft are initially located within the magnetosphere, as indicated by the strong and steady magnetic field and the diffuse, relatively high-energy electron population. At 06:30 and 06:43 UT (marked by the vertical dashed blue lines), each magnetic field dataset shows a positive-negative bipolar signature in the normal component (fifth panel) together with a transient increase in the field strength (bottom panel), consistent with the spacecraft all encountering a magnetospheric FTE. Note that the first of these has no electron signature, while the second shows evidence of the appearance of magnetosheath-like electrons at C1 and C4, indicating that these spacecraft encounter open field lines. At $\sim$06:53 UT (red vertical dashed line) the upstream spacecraft C1, C2, C4 all cross the magnetopause (changes in B_L and B_M, panels 6 and 7); the high energy electrons disappear, and colder, denser electrons start to appear. C3, however, remains within the magnetosphere and observes further magnetospheric FTEs at 07:03, 07:07, 07:11 and 07:16 UT (orange vertical dashed lines).

The three spacecraft in the magnetosheath observe magnetic field and plasma transients at the same times. These transients include an apparent increase in the energies of the magnetosheath electrons and the reappearance of higher energy magnetospheric electrons. However, the magnetic field transients are not simply a return towards the magnetospheric orientation, suggesting that these spacecraft

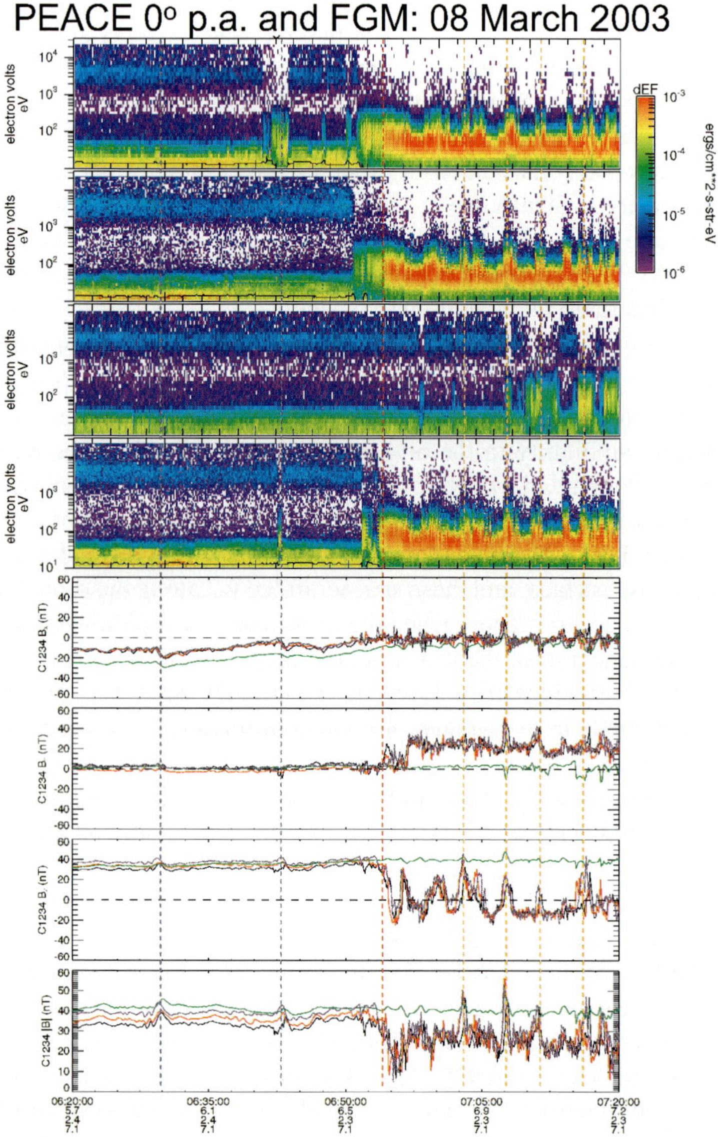

Figure 8.29. Simultaneous observations of magnetospheric and magnetosheath FTEs on March 8, 2003. Panels 1–4: PEACE measurements of the differential energy flux of 10 eV-26 keV electrons from each spacecraft. Panels 5–8: FGM measurements of the magnetic field components and strength in boundary normal coordinates. All spacecraft observe magnetospheric FTE signatures at 06:30 and 06:43 UT (vertical dashed blue lines). C1, C2, C4 cross the magnetopause at 06:55 UT, after which they observe accelerated plasma and magnetic field transients whenever C3 observes magnetospheric FTEs earthward of the magnetopause. (From Owen et al., 2005).

encounter the reconnected field lines outside of the magnetopause. For some of the events, C3 also sees magnetosheath-type electrons. Hence, one can conclude that C3 continues to record the passage of magnetospheric FTEs, while C1, C2, C4 observe their magnetosheath counterpart $\sim$0.7 R_E upstream. There is no evidence for solar wind pressure pulses, and indeed, the simultaneous observations on either side of the magnetopause are not consistent with a simple inward deformation due to a pressure pulse. The observations rather fit a picture of a reconnected flux tube along which magnetosheath and magnetospheric plasmas are mixed, heated and/or accelerated. The orientation and motion of magnetosheath FTEs are consistent with reconnected flux tubes recoiling following component merging from a reconnection site located southward and dawnward of the spacecraft (e.g., Cowley and Owen, 1989). The internal (magnetospheric) FTE signatures thus appear to be caused by the motion of the external portion of a reconnected flux tube over the magnetopause surface, causing at least a partial indentation.

A Grad-Shafranov reconstruction of one of the FTEs within this event supports this interpretation (Sonnerup et al., 2004b, see also Section 8.4.2). The results of this reconstruction in a plane normal to the well-determined axis of this FTE are shown in Figure 8.30. The top left panel indicates the magnetic field within this structure, with contours representing the projection of the field lines onto the plane of the reconstruction and the colours indicating the axial field strength. The white arrows represent measured magnetic field vectors (4 s averages) within the plane for comparison. The lower left panel represents the thermal plasma pressure within the structure (colours and contours) and the plasma flow (white vectors). The magnetosphere is towards the bottom of the Figure and the equator to the right. This result is consistent with a flux rope FTE with a strong core field giving a central region of high magnetic and low plasma pressure, surrounded by a ring of enhanced plasma pressure. This structure represents an outward bulge on both sides of the magnetopause, with a larger extension into the magnetosheath than into the magnetosphere, rather than an indentation of the entire layer. Sonnerup et al. conclude that this is a fossil FTE, created by component merging at an equatorward location, which is relaxing towards a minimum energy state. Time delays between the individual FTEs in this series imply an average reconnection electric field of $\sim$0.18 mV m^{-1}. In addition, the magnetospheric flows observed by C3 in the deHoffmann-Teller frame (evident in the lower half of the second panel of Figure 8.30) are found to be supersonic. Detailed examination of the field and plasma variations seen by C3 are as expected for supersonic field-aligned flow through a converging-diverging nozzle formed by the draping of the magnetospheric field around the bulge in the magnetopause.

The Cluster spacecraft occasionally have been aligned as a 'string of pearls' (during orbital manoeuvres) rather than forming a tetrahedron, which has led to some serendipitous advances. In this case, the spacecraft may pass sequentially through a structure, allowing its temporal development to be assessed, or may sam-

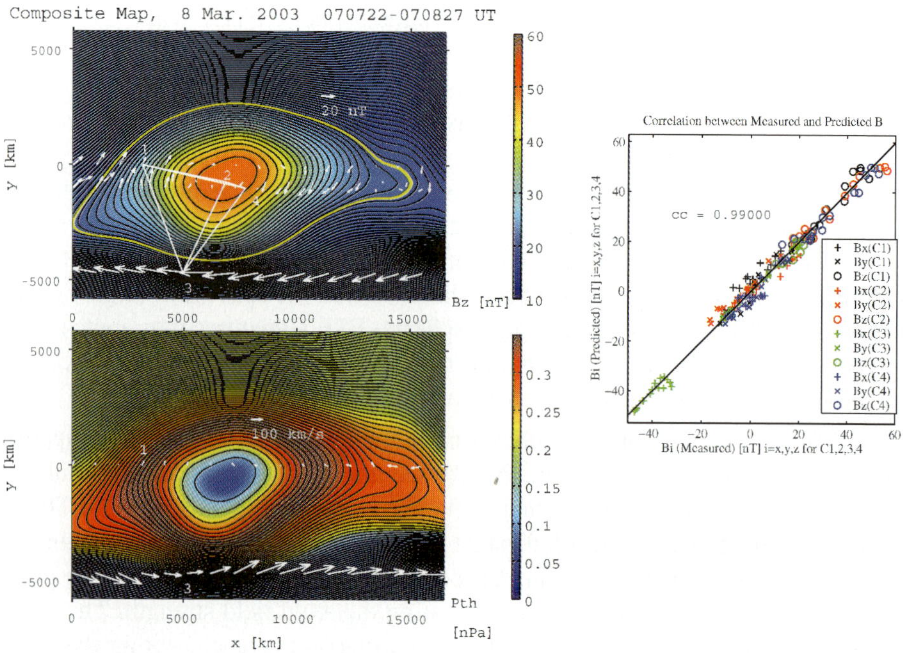

Figure 8.30. A Grad-Shafranov reconstruction in a plane normal to the axis of the FTE observed by Cluster at ~07:07 UT on March 8, 2003 (see Figure 8.29). The upper left panel indicates the axial field strength and the projection of the field lines onto the plane of the Figure (colours and contours respectively). White arrows represent the measured magnetic field vectors within the plane. The lower left panel shows the thermal plasma pressure within the structure (colours and contours) and the plasma flow in the deHoffmann-Teller frame (white vectors). The right-hand panel indicates the high correlation between the predicted and measured magnetic field. The observations are consistent with a bulge in the magnetopause consisting of a region of high magnetic and low plasma pressure, rather than an in-out motion of the entire layer. (From Sonnerup et al., 2004b).

ple a structure passing over the spacecraft at differing depths. An example of the latter is the observation of a crater-type FTE during an outbound pass through the high-latitude post-noon magnetopause at ~07:00 UT on January 11, 2002 (Owen et al., 2004a). Figure 8.31 shows the configuration of the quartet in *LMN* coordinates; *N* is the local normal of a modelled magnetopause boundary , *L* lies parallel to the projection of the terrestrial field on the magnetopause plane, and *M*, pointing dawnwards, makes up the right-handed set. The spacecraft were largely co-linear, with C3 leading the other three by a distance of ~2 R_E along their orbit track. The spacecraft lie along a line of more or less constant *M*, but sample different distances from the magnetopause (which is indicated by the dashed lines in the upper

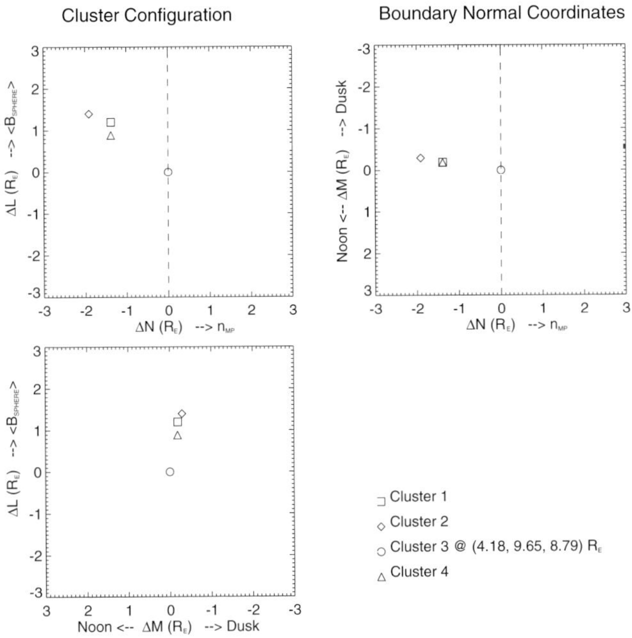

Figure 8.31. Cluster configuration in boundary normal (*LMN*) coordinates at ~06:55 UT on January 11, 2002. The spacecraft are approximately co-linear and sampling different distances from the magnetopause (the projection of which onto the *LN* and *MN* planes is represented by the dashed line in the top 2 panels). (From Owen et al., 2004a).

two panels). During the approach to the magnetopause, C3 was located closest to the boundary, then C1 and C4, and finally C2 at the greatest depth. Prior to crossing into the magnetosheath, the quartet observed a number of FTE-type signatures, an example of which is shown in Figure 8.32 (Owen et al., 2004a). The top four panels show the magnetic field, while the lower four panels contain energy-time spectrograms with the differential energy flux of field-aligned electrons observed at each spacecraft. C1, C2 and C4 see a crater-type FTE. There is a significant $+/-$ bipolar signature in the normal component of the field at these spacecraft, while the predominant B_L component (and thus also the magnitude) shows a double increase in strength on either side of a decrease. These signatures are stronger at C1, closest of the three to the magnetopause, than at C2 and C4, deeper in the magnetosphere. Note, however, that C2, deepest in the magnetosphere, observes a population of accelerated magnetosheath-like electrons (E ~3–400 eV) but little loss of the magnetospheric electrons, which exhibit energies $>$ several keV (6th panel). C4, slightly closer to the magnetopause, observes accelerated magnetosheath electrons at lower fluxes than C2 (bottom panel), and shows evidence of a loss of magnetospheric electrons. C1, again slightly closer to the magnetopause than C4, observes magnetosheath-type electrons (panel 5), but these do not seem

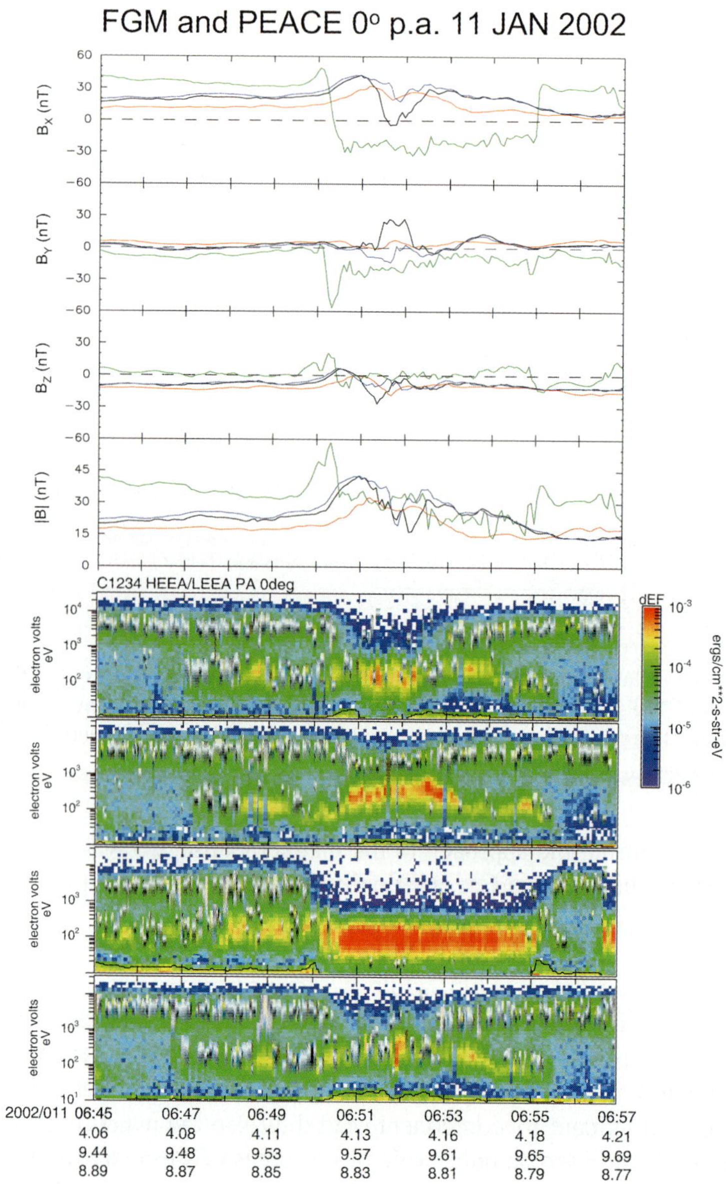

Figure 8.32. A crater-type magnetospheric FTE on January 11, 2002. Panels 1–4: FGM measurements of the magnetic field in boundary normal coordinates. Panels 5–8: PEACE measurements of the differential energy ux of 10 eV–26 keV electrons from each s pacecraft. During this period C1, C2, and C4 sample a reconnection boundary layer inside the magnetopause, as indicated by the appearance of accelerated plasma and magnetic field transients. C3, however, makes a transit into the magnetosheath, indicating the magnetopause moves briefly inward during this period. (Adapted from Owen et al., 2004a).

to have undergone as significant an acceleration as those observed at C2. There is a significant drop-out of the magnetospheric population at C1, which, unlike C2 and C4, also observes a population of accelerated magnetosheath ions (not shown). The final spacecraft, C3, much closer to the magnetopause, makes a very clear and abrupt transition into the magnetosheath proper (indicated by high fluxes of electrons of energy $\sim$100 eV) immediately prior to the event seen at the other spacecraft, before returning to the magnetosphere immediately afterwards. Note that the B_M-component observed by C3 (green trace, second panel) during this period is opposite in sense to the deflection observed by C1 (black trace), confirming that the latter spacecraft does not exit the magnetosphere during this period. Owen et al. (2004a) interpret the C1, C2 and C4 observations inside the magnetosphere as clear evidence of the spacecraft sampling the velocity dispersed reconnection boundary layer. The three spacecraft move onto open field lines (indicated by drop-outs in the $>$ 1 keV electrons) without exiting to the magnetosheath proper. C2 and C4, furthest from the magnetopause, see the electron separatrix boundary layer, while C1, closer to the magnetopause, also samples the ion boundary layer. Acceleration of the observed inflowing magnetosheath populations also suggests a reconnection origin. However, the observations made by C3 indicate that the magnetopause temporarily moved inward during this period. Hence it is not clear whether this FTE signature arose due to, e.g., an unobserved pressure pulse which deflected the pre-existing reconnection layers earthward over the innermost spacecraft, or whether the magnetopause was eroded inwards as a discrete flux tube moved along the boundary.

8.6.3 Coordination with ground-based observations

Lockwood et al. (2001), Moen et al. (2001), Wild et al. (2001, 2003) and Marchaudon et al. (2004) have studied FTE signatures during periods when the Cluster spacecraft were magnetically conjugate with the fields of view of a number of ground-based facilities. Wild et al. (2001, 2003) report observations of four magnetospheric FTEs during an outbound pass through the post-noon high-latitude magnetopause on February 14, 2001. Furthermore, clear FTEs were observed later in the magnetosheath. The magnetic footprint of the Cluster spacecraft during this period mapped centrally within the fields-of-view of the CUTLASS SuperDARN radars, and passed across the field-aligned beam of the EISCAT Svalbard radar (ESR) system. Both the ionospheric flow and the backscatter power observed by CUTLASS at the latitude of the Cluster footprint correlate with the observations of the magnetospheric FTEs. These features are found to propagate poleward, forming classic 'pulsed ionospheric flow' and 'poleward-moving radar auroral form' structures at higher latitudes. Moen et al. (2001) also identified auroral responses corresponding to boundary layer dynamics observed by Cluster on this day, and concluded that an observation of a sequence of three moving auroral forms were likely candidates for flux transfer events. These auroral signatures appear to be examples of accelerated electrons and discrete post-noon aurora on open magnetic field lines.

Lockwood et al. (2001) also report a series of events that have the ion and electron characteristics predicted for a FTE, with allowance for magnetospheric ion reflection at Alfvénic disturbances in the magnetopause reconnection layer. Like FTEs, the events are about 1 R_E in their direction of motion and show a rise in the magnetic field strength, but unlike FTEs, in general, they show no pressure excess in their core and hence no characteristic bipolar signature in the boundary-normal component. However, most of these events were observed when the magnetic field was southward, i.e., on the edge of the interior magnetic cusp, or when the field was parallel to the magnetic equatorial plane. Further evidence that these events correspond to reconnection pulses was obtained from ground-based observations and includes: transient erosion of the noon 630 nm (cusp/cleft) aurora to lower latitudes; transient and travelling enhancements of the flow into the polar cap, imaged by the AMIE technique; and poleward-moving events moving into the polar cap, seen by the EISCAT Svalbard Radar (ESR). The link between the Cluster observations and these ionospheric signatures is strengthened by the observed agreement between the eastward phase speed of the latter with the motion deduced by the cross-correlation of the signatures seen on the four Cluster spacecraft.

Finally, Marchaudon et al. (2004) present observations from March 17, 2001, during a conjunction between Cluster in the high-altitude cusp and the SuperDARN radars. Changes in the convection direction at Cluster are well correlated with IMF B_y variations. The changes in the ionospheric convection near the Cluster footpoint follow those in the magnetosphere with a 2–3 minute delay, although the mapped convection velocity at Cluster is about 1.5 times larger than measured by SuperDARN. In addition, the field and particle observations by Cluster display the characteristic signatures of plasma injections into the magnetosphere suggestive of FTEs. A clear one-to-one correlation is observed between three successive injections at Cluster and impulsive and localised plasma flows in the ionospheric cusp, again with 2–3 minutes delay. For each event, Marchaudon et al. compared the drift velocity of reconnected flux tubes (phase velocity) in the magnetosphere with those in the ionosphere. Figure 8.33 (top panel) shows the projection of the convection velocity observed by C1 onto the GSE XY plane during the period 05:15–06:00 UT. This drift velocity corresponds to flows of the order of 400–600 m s^{-1} when mapped into the ionosphere (Figure 8.33, second panel). The SuperDARN observations (Figure 8.33, panel 3) give the velocity of the ionospheric plasma parallel to the radar look-direction as a function of distance from the radar and time. The Cluster footprint is also indicated (note the 2 minute shift between this panel and the preceding one). The direct comparison between the dawn-dusk convection velocity observed by SuperDARN and inferred from Cluster is shown in the bottom panel: There is a clear qualitative agreement.

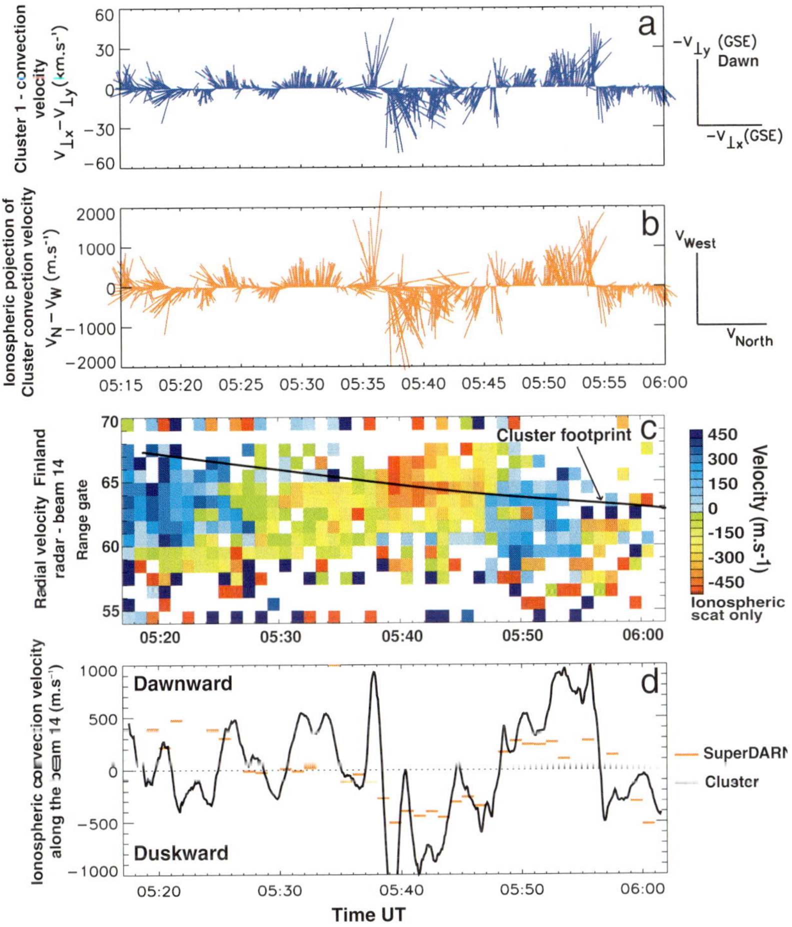

Figure 8.33. Convection velocity measured by Cluster and SuperDARN on March 17, 2001: (a) C1 CIS ion perpendicular velocity vectors. Each vector represents the perpendicular-to-**B** component of the plasma velocity projected onto the XY_{GSE} plane. Because the local magnetic field is close to the Z_{GSE} direction, the $-Y_{GSE}$ component is mainly dawnward. (b) Projection into the ionosphere at 250 km altitude of the CIS velocity vectors, where V_{west} and V_{north} correspond to the velocity components in the magnetic west and north directions respectively in Altitude Adjusted Coorected Geomagnetic Coordinates (AACGM). (c) Range-time plot of line-of-sight velocity for the SuperDARN Hankasalmi beam 14 with the Cluster footprint superimposed (black line). (d) Comparison of velocity measured by Cluster (solid black line) and SuperDARN (red segments). The Cluster CIS velocity is averaged over 1 minute and projected along beam 14 of the Hankasalmi radar. The SuperDARN velocity is the radial velocity measured at the Cluster footprint. (From Marchaudon et al., 2004).

8.6.4 Summary

In summary, it is clear that the advent of the Cluster observations has shed new light on the formation, structure and dynamics of flux transfer events, and will continue to do so.

- Simultaneous observations of FTEs on both sides of the magnetopause clearly demonstrate that FTEs represent a widening of the magnetopause boundary, rather than an indentation, thus discarding the pressure pulse interpretation for these FTE signatures.
- Cluster has provided an unambiguous confirmation of the spatial scales, orientation, and motion of FTE structures. This has been possible because the inter-spacecraft separations were of the order of the FTE scale sizes. None of these parameters can be uniquely determined from single-spacecraft observations.
- The Cluster observations have provided measurements of the internal structure of FTEs, in particular the field-aligned current distribution within an FTE flux tube.
- In conjunction with the unprecedented ground-based coverage of the ionosphere in the vicinity of the Cluster footprints, Cluster has been able to study magnetosphere-scale aspects of FTEs and the associated current system.

The Cluster magnetopause observations provide a rich dataset for continued study of flux transfer events. It should be noted, however, that the Cluster orbits generally intersect the dayside magnetopause at relatively high latitudes, such that the observations of FTEs within the Cluster dataset are limited to a relatively small subsection of the magnetopause surface. Depending on orbit precession and on the mission strategy that will be adopted, the subsolar region, which is of particular importance for studying FTEs since they are believed to originate there for southward IMF, might be sampled in the future as well.

8.7 Surface Waves

8.7.1 Brief history

Variations in the upstream solar wind velocity, in solar wind pressure, and in the interplanetary magnetic field act on the magnetopause boundary so that it is almost continuously moving as the equilibrium between total pressure on either side is dynamically restored. Boundary motion can induce anything from large-amplitude surface waves down to small-scale waves or ripples (Kivelson and Chen, 1995; Seon et al., 1995; Sibeck, 1990; Sibeck et al., 1991). Sckopke et al. (1981) presented in situ observations of boundary waves and discussed different possible topologies depending on the waviness of the magnetopause and of the boundary layer inner edge.

As first suggested by Dungey (1955), the Kelvin-Helmholtz instability (KHI) may also be a mechanism for creating surface waves (Drazin and Reid, 1985). The KHI may occur when two fluids are in motion with respect to one another. The instability of the interface between two magnetised fluids is triggered when the velocity shear exceeds a threshold, while the magnetic field can have a stabilising effect, depending on its orientation (Chandresekhar, 1961; Southwood, 1968). The instability is most likely to occur at the magnetospheric flanks, where the fast-flowing magnetosheath plasma slides past more stagnant magnetospheric or boundary layer populations, thus creating a velocity shear across the boundary. A significant volume of work exists on the role of the KHI at the magnetopause, from a theoretical, an observational, and a simulation point of view (see, for instance, Southwood and Hughes, 1983; Pu and Kivelson, 1983b,a; Fitzenreiter and Ogilvie, 1995; Kivelson and Chen, 1995; Miura, 1995; Fairfield et al., 2000; Otto and Fairfield, 2000; Nykyri and Otto, 2001). Further aspects of the KHI as a physical process will be discussed in Phan et al. (2005), while the present section deals mostly with geometrical aspects of boundary waves.

In a comprehensive case study, Chen et al. (1993) analysed boundary normal orientations from ISEE-1 and -2 data, and concluded that KHI surface waves are non-sinusoidal, with the leading, downtail edges of the waves shallower (i.e. less inclined from a nominal magnetopause orientation) than the trailing, sunward-facing edge. These results are at odds with theory, which predicts steepening at the leading edge (Miura, 1990). More recently, however, Fairfield et al. (2000) compared GEOTAIL observations to MHD simulation results by Otto and Fairfield (2000): Both were consistent with a steepening of the leading edge of the waves. Kivelson and Chen (1995) suggest that the magnetic curvature forces at the low shear boundary could induce steepened trailing edges, while Fairfield et al. (2000) question the methods applied to determine the boundary normal directions in Chen et al. (1993). A number of authors (Sonnerup, 1980; Sckopke et al., 1981; Ogilvie and Fitzenreiter, 1989; Kivelson and Chen, 1995; Farrugia et al., 1998, 2000) have pointed out that the inner edge of the boundary layer might be Kelvin-Helmholtz unstable rather than the magnetopause, as the stabilising effect due to a differently oriented magnetic field on the sunward side of the boundary is absent there.

The predominant way of studying boundary wave structure traditionally consisted of analysing successive boundary crossings to establish boundary orientation and motion; by piecing together this information, a more global sketch of the surface wave arises (e.g., Sckopke et al., 1981). This approach holds promise when applied in a multi-spacecraft context: More crossings are available and/or more precise boundary orientations can be obtained, depending on how the spacecraft separation scale compares to the length scale of the boundary wave, as was already demonstrated in Section 8.3.1.3. Fairly recently, De Keyser et al. (2002) and De Keyser and Roth (2003) have elaborated a new class of analysis techniques, known as empirical reconstruction methods, to study boundary waves.

While originally applied to single-spacecraft observations, these methods are readily extendable to the multi-spacecraft case. If the boundary is simultaneously sampled at several nearby points, there is much more information available than in the single-spacecraft case, which leads to correspondingly more detailed reconstructions. With Cluster, both approaches have been pursued, with success. The ability to deduce a global picture of boundary layer structure opens up new perspectives for studying the physical processes involved.

8.7.2 Tracking the undulating boundary

Empirical reconstruction techniques attempt to interpret observations in terms of stationary structures that are convected across the spacecraft. The observed time-variability then reflects the motion of spatial plasma structure relative to the spacecraft. The spatial information can be extracted from the temporal data only if this motion is known. In a first step, a reference frame is chosen such that the x axis points along the average outward normal direction (see, e.g., De Keyser et al., 2002). The time-varying position $x_{\rm mpbl}(t)$ of the magnetospheric boundary along this normal direction can be obtained from:

$$x_{\rm mpbl}(t) = x(t_0) + \int_{t'=t_0}^{t} v_{\rm mpbl}(t')dt', \tag{8.4}$$

where $x(t_0)$ defines a reference position at an arbitrary time t_0. While the boundary velocity $v_{\rm mpbl}$ is generally not known, Paschmann et al. (1990) proposed to use the locally measured normal plasma velocity v_x as an approximation for it. The one-dimensional spatial profile of any quantity f across the boundary is then obtained from the observed time profile $f(t)$ by plotting the data as $f(x_{\rm sc}(t) - x_{\rm mpbl}(t))$, where $x_{\rm sc}(t)$ is the spacecraft position along its orbit. Plotting the data in this way as a function of the distance of the spacecraft from the boundary, the data points tend to form an essentially single-valued spatial profile. In practice, this technique works only for rather short time intervals, mainly because the integral in 8.4 has an oscillatory integrand $v_{\rm mpbl}$, which leads to increasingly large relative errors of the result as the time interval gets longer, and because of the limited accuracy and time resolution of the plasma data. These difficulties can be overcome by adopting an optimisation approach in which $v_{\rm mpbl}(t)$ is considered a priori as unknown. It is determined such that the spatial profiles of a few relevant physical quantities (so-called guiding variables) are essentially single-valued, by solving an optimisation problem that takes into account the error bars on all observations. The boundary can then be tracked during several hours (De Keyser et al., 2002, 2004b).

As an example, consider the outbound pass on June 6, 2001. The first, the third, and the fifth plot in Figure 8.34 show the electron density obtained by the PEACE instruments on the four spacecraft, the magnetic field strength from the four FGM magnetometers, and the plasma velocity magnitude observed by the CIS/HIA instrument, only operating on C1 and C3, from 01:10 to 03:30 UT. While C1, C2,

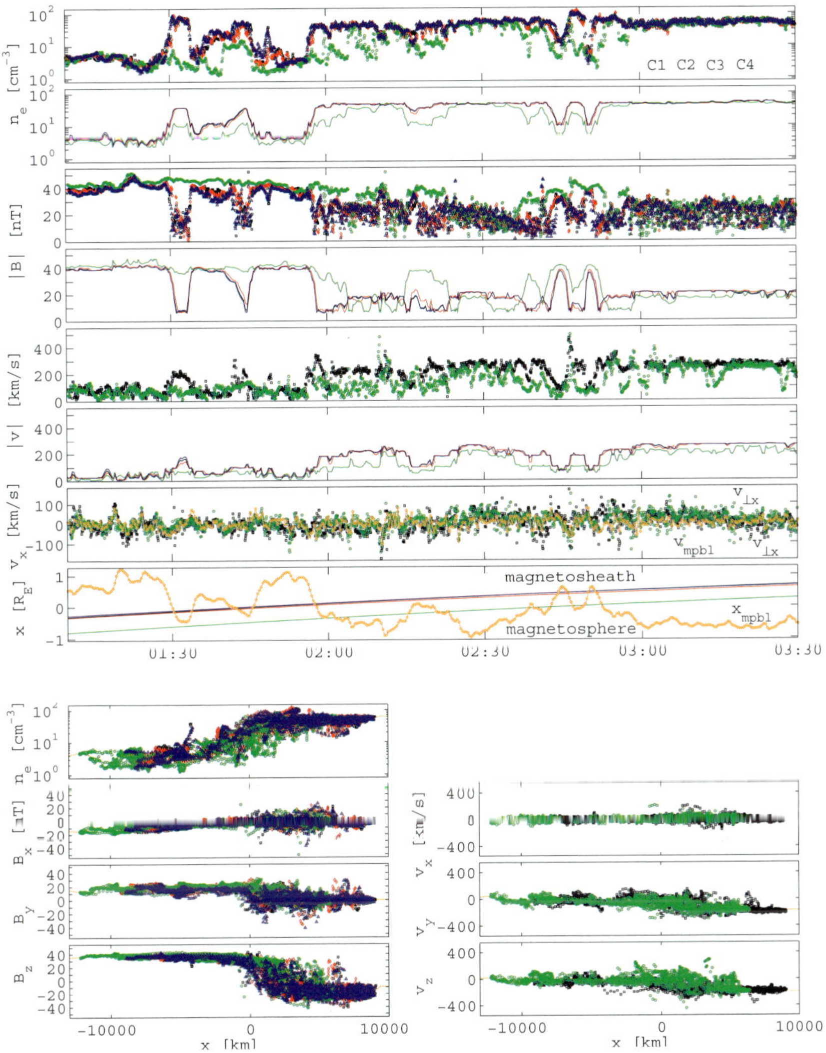

Figure 8.34. Magnetospheric boundary motion during an outbound pass on June 6, 2001, showing the results of an optimisation-based empirical reconstruction with a time resolution of 20 s, using the B_z component and the electron density from the four spacecraft as guiding variables. Top panels: Time profiles of the observed electron density (PEACE, C1–C4), the electron density 'predicted' by the reconstruction (C1–C4), the observed magnetic field strength (FGM, C1–C4), the 'predicted' field strength (C1–C4), the observed plasma velocity magnitude (CIS/HIA, C1 and C3), the 'predicted' plasma velocity magnitude (C1–C4), the boundary velocity proxy $v_{\perp x}$ (from FGM and CIS/HIA on C1 and C3) and the reconstructed boundary velocity v_{mpbl}, and the boundary position x_{mpbl} and spacecraft trajectories. Bottom panels: Spatial plots of electron density, magnetic field, and plasma velocity observations in a co-moving frame. The data points outline single-valued spatial profiles, suggesting the one-dimensional structure of the boundary. (Adapted from De Keyser et al., 2004b).

and C4 are close together and see similar profiles, C3 is about 3000 km earthward, systematically recording lower densities and higher field strength.

De Keyser et al. (2004b) have computed an empirical reconstruction using the optimisation approach. The time resolution with which the motion of the boundary is reconstructed was 20 s. The CIS/HIA perpendicular plasma velocity $v_{\perp x}$ (only on C1 and C3) is used as a proxy for boundary motion. The electron density and the magnetic field z component are used as guiding variables, since density and magnetic field variations are observed by at least one of the spacecraft throughout most of the interval. These guiding variables provide information about boundary position (for instance, the orientation of the magnetic field indicates on which side of the magnetopause a spacecraft is), and this information is exploited by the optimisation technique to determine v_{mpbl} more accurately. Spatial fits can then be constructed for each observed quantity. For instance, plotting the electron density, n_e, as a function of the distance of each spacecraft from the boundary, $x = x_{\mathrm{sc}} - x_{\mathrm{mpbl}}$, produces essentially a single-valued curve. Such curves are shown for electron density, magnetic field, and plasma velocity, in the bottom half of Figure 8.34. There is a certain amount of scatter of the data points around the average spatial profiles, but this scatter is present within each spacecraft data set, while the curves from all the spacecraft are consistent with each other.

Given these spatial profiles and the motion of the boundary, one can 'predict' the time profiles that should be seen along the trajectory of each spacecraft. These 'predictions' are shown in the second, fourth, and sixth panel of the figure. There is a nice overall agreement, but some differences are clearly visible. Interestingly, the method 'predicts' the plasma velocity profiles at C2 and C4, for which no actual observations are available. There are considerable differences between the $v_{\perp x}$ measurements from C1 and C3; v_{mpbl}, being a kind of 20 s average, follows their overall behaviour. The figure also shows the position of the four spacecraft relative to the boundary position x_{mpbl}.

As the back-and-forth boundary motion is combined with the magnetosheath flow, the observed structure is that of a rather gently undulating magnetospheric boundary, with the undulations propagating tailward. To the extent that this propagation speed can be considered constant, the time axis in Figure 8.34, (panel 8 from top), can be rescaled so that the x_{mpbl} profile represents the spatially undulating shape of the boundary. Overall, the amplitude-to-wavelength ratio is relatively small, confirming that the boundary can indeed be regarded as having one-dimensional structure.

In this event, the boundary is continuously moving back-and-forth with an amplitude on the order of an Earth radius. The distribution of v_{mpbl} at 60 s resolution is found to be Gaussian with a standard deviation of 20 km s^{-1}. The average boundary layer thickness is about 5000 km. The structure of the boundary does not seem to be dramatically affected by the motion, although motion by itself implies that the boundary cannot be in equilibrium.

There is a reasonable agreement between the predicted curves and the observations, confirming the validity of the approximations. It can therefore be concluded that the observed time variations and the differences between the time profiles from different spacecraft can be explained to a large extent by the back-and-forth motion of the boundary and by the relative positions of the four spacecraft, but there are differences that may be due to boundary curvature or to time dependent effects (for instance, the higher-than-average density observed at times of rapid boundary motion).

8.7.3 Periodic surface waves

Cluster scientists have also been using the missions novel and unique four-point measuring capabilities to examine the three-dimensional morphology of periodic boundary waves. As it is part of Cluster's mission to resolve spatial-temporal ambiguities, it should shed new light on the question of the steepening of the leading or trailing edges of nonlinear Kelvin-Helmholtz waves (e.g., Chen et al., 1993; Fairfield et al., 2000).

Owen et al. (2004b) presented an example of an extended (>1 hr) interval of Cluster observations in which the quartet made a series of transitions between the magnetosphere and magnetosheath while located on the dawn flank magnetopause at GSE (5.6, 16.3, 4.4) R_E . The PEACE electron spectrometer data for this interval (Figure 8.35, left panel) show a large number of quasi-regular transitions between a cold, dense electron plasma population, typical of the magnetosheath, and a more diffuse population of higher energy, which is typical of the flank magnetosphere/boundary layer. The exact timings of each of these transitions at each of the Cluster spacecraft were determined by identifying the times of maximum gradient in the electron temperature moment. Differences in the transition times at the 4 spacecraft were then used to determine the local orientation and speed of the associated magnetopause boundary as it moves back and forth over the spacecraft locations. The results of this analysis are summarised in the right hand panel of Figure 8.35, which shows the projection of the inferred boundary motion vectors on each of the GSE XY, XZ and YZ planes as a function of position along the spacecraft orbit (black trace). These orientations can be compared with the expectations for the average magnetopause orientation determined from model fits to historical data (Fairfield, 1971), the projection of which is represented by the green trace in each panel. This comparison reveals that both the inbound and outbound crossings of the magnetopause are significantly deflected from the expected orientation, in the senses that are indeed consistent with the passage of a boundary wave past the spacecraft location. As sketched in Figure 8.36, Owen et al. (2004b) also found that the leading edge of the wave was on average significantly steeper in inclination to the model magnetopause than the trailing edge. This observation is consistent with expectations of Kelvin-Helmholtz waves (and possibly waves of other origin), but contradicts the conclusions of the two-spacecraft study by Chen

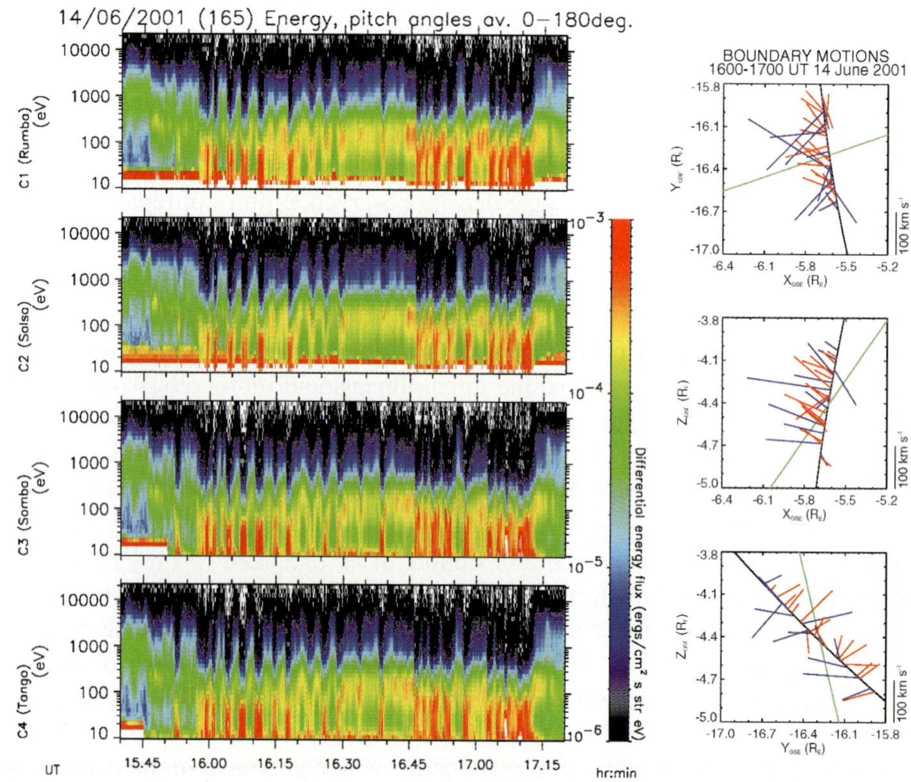

Figure 8.35. Cluster observations of periodic surface waves on June 14, 2001. (left) Energy-time spectrograms for electrons observed by the PEACE instruments in the range 10 eV to 26 keV for the period 15:40–17:20 UT. The four panels show data from C1–C4, respectively. Each spacecraft observes a series of transitions between the dense, low energy electron population, characteristic of the magnetosheath, and the rarer, hotter magnetospheric plasma. (right) Results of a detailed four-spacecraft analysis of the timing of the inbound (red) and outbound (blue) transitions. The projection of the inferred boundary motion vector on each of the GSE XY, XZ and YZ planes is shown as a function of position along the spacecraft orbit (black trace). The projection of an undisturbed model magnetopause surface on each plane is also shown (green line) for comparison. (From Owen et al., 2005).

et al. (1993). The conditions on either side of the magnetopause were used to check a Kelvin-Helmholtz instability criterion based on a simple incompressible plasma model (e.g., Chandresekhar, 1961): The boundary was found to be largely stable during this period. However, there was some suggestion in the results that the sub-intervals which satisfy the instability criterion are closely matched in time with the steepest boundary orientations. The four-spacecraft timing results were also used for the first time to directly determine the wavelength of the surface waves and

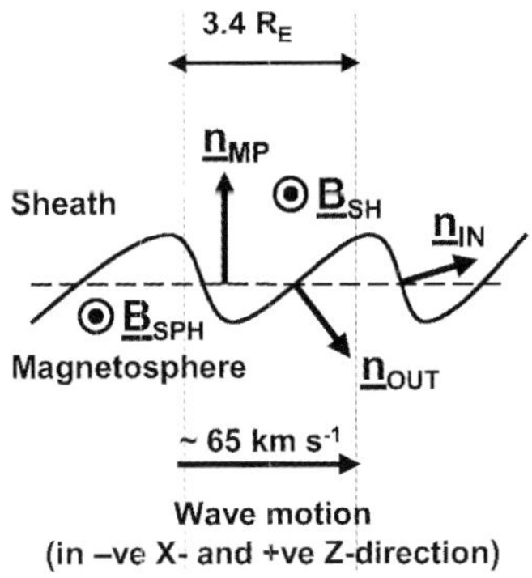

Figure 8.36. Sketch illustrating the average structure of the magnetopause surface wave determined by Owen et al. (2004b). The leading edge of the waves, associated with transitions of the spacecraft from magnetosheath to magnetosphere, were inclined from the expected orientation of the undisturbed magnetopause by a larger angle, on average, than the trailing edge, associated with transitions in the opposite sense. The direction of propagation of the waves, in the negative GSE X- and positive Z-directions, is close to perpendicular to the magnetic fields on either side of the boundary, suggesting that the wave growth and propagation occurs preferentially in the direction in which the wave does not have to bend the magnetic field lines. (From Owen et al., 2004b).

their speed and direction of propagation across the model magnetopause surface. For this event, the average wavelength was $\sim$3.4 R_E , and the waves propagated at an average speed of $\sim$65 km s^{-1} in an anti-sunward and northward direction. The northward component to the motion was unexpected, since the spacecraft were located south of the GSE $Z = 0$ plane at the time of the observations, where the large-scale magnetosheath flow outside the magnetopause is expected to have a southward component at this location. Indeed, the ion flows observed when the spacecraft are in the magnetosheath did not show a significant northward component. However, the propagation direction was close to perpendicular to the average external magnetic field direction, suggesting that these waves may preferentially propagate in the direction which requires no bending of magnetic lines of force. Although this observation is consistent with waves driven by the KHI, which are expected to grow fastest when $\mathbf{k} \cdot \mathbf{B}$ is small, it may also be the energetically most favourable conditions for the growth of other waves. Hence, although the waves show characteristics that might be expected of those driven by the KHI, Owen et al. (2004b) were unable to definitively identify them as such.

Large-scale boundary waves have also been observed by Cluster on 25 November 2001, between 00:00–04:00 UT, as the spacecraft (separation 2500 km) made an exit through the dusk flank magnetopause at GSE location (-4.6, 15.9, 5.8) R_E. Again, an extended series of boundary oscillations could be inferred. EFW observations of the spacecraft potential between 00:40–01:50 UT are shown in Figure 8.37. The left hand panel of this Figure shows variations in the C4 spacecraft

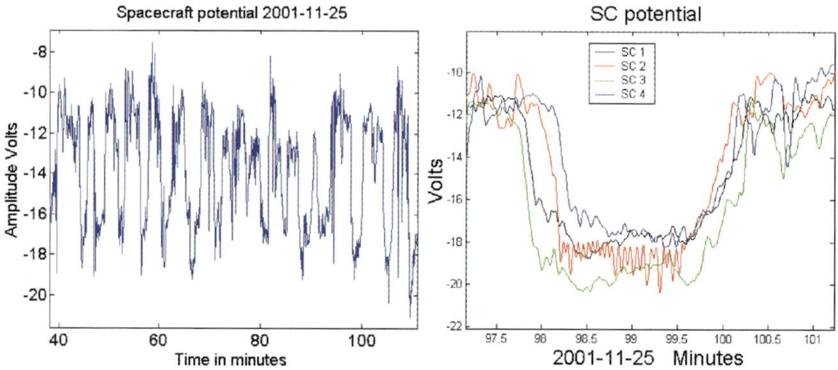

Figure 8.37. Surface wave signatures in the spacecraft potential from EFW, a proxy for electron plasma density, on November 25, 2001. (left) Variations in the C4 spacecraft potential at the dusk flank magnetopause, indicating the presence of a large-scale boundary wave. (right) Four-spacecraft observations during a single cycle of the wave; the nested nature of the signatures indicates the back-and-forth motion of the boundary (Figure provided by G. Gustafsson).

potential, analysis of which shows oscillations with dominant periods of 4.2 and 11.5 minutes. These data correspond to density variations of the ambient electron plasma in the range 0.9–4 cm^{-3}, consistent with the spacecraft making transient movements between the magnetosphere and the magnetosheath. The right hand panel of Figure 8.37 shows four-spacecraft measurements of the spacecraft potential during a single wave period; the nested nature of the four signatures indicate the back-and-forth motion of the magnetopause across the spacecraft tetrahedron. More detailed analysis of the timing of these signatures confirms that a large-scale wave is present that displaces the magnetopause normal back-and-forth from its expected orientation.

De Keyser et al. (2004a) have applied a two-dimensional empirical reconstruction technique to a subset of the data: two wave periods between 01:53 to 02:15 UT (wave period is 11 minutes). The result of this analysis is a two-dimensional picture of the surface waves (the wave structure can be shown to be essentially invariant in one direction, the z direction) in a reference frame that slides along the boundary with the wave, shown in Figure 8.38. The CIS/HIA plasma velocity is directly integrated to obtain x_{mpbl} from Equation 8.4; the long-term overall motion of the boundary is then removed to more clearly show the structure of the periodic wave (De Keyser and Roth, 2003). A constant tangential wave speed of 200 $\mathrm{km\,s}^{-1}$ (about half the magnetosheath flow speed) was assumed. Because of the periodicity, the reconstruction effectively becomes a superposed epoch analysis. The two-dimensional reconstructions in Figure 8.38a–d represent the spatial distribution of a few quantities in the comoving frame (see De Keyser et al., 2004a, for other quantities). Colour-coded reconstructions of ion density (CIS/HIA, C1 and

(a) Ion density (CIS/HIA)

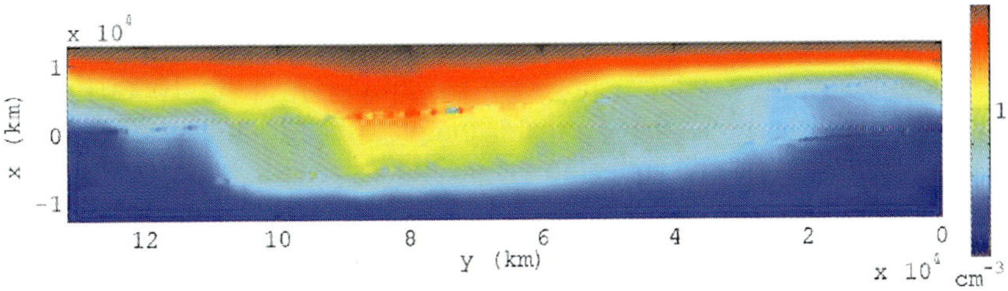

(b) Perpendicular magnetic wave intensity (STAFF, 0.3–10 Hz)

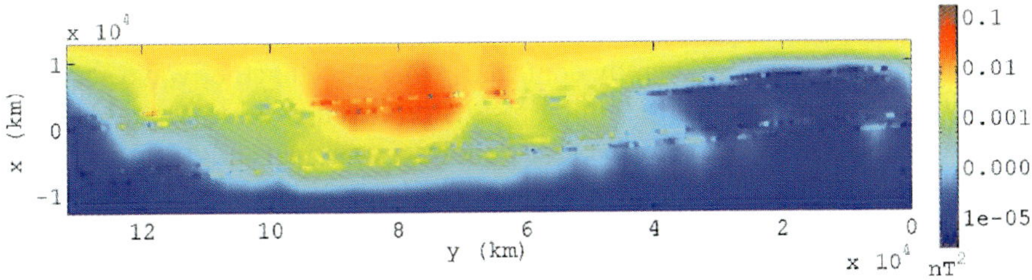

(c) Magnetic field strength and *xy* components (FGM)

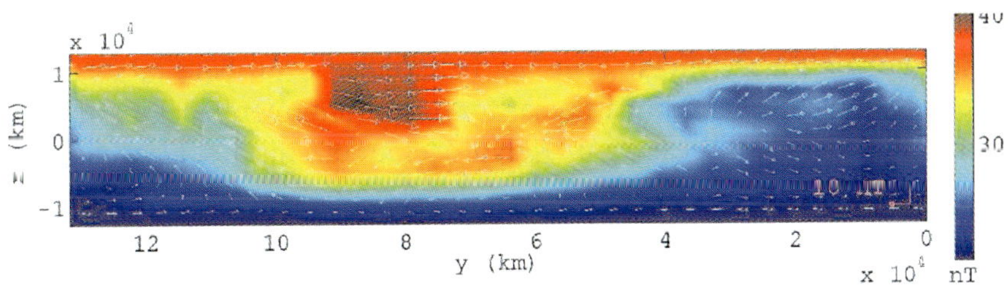

(d) Plasma velocity along *z* and *xy* components (CIS/HIA)

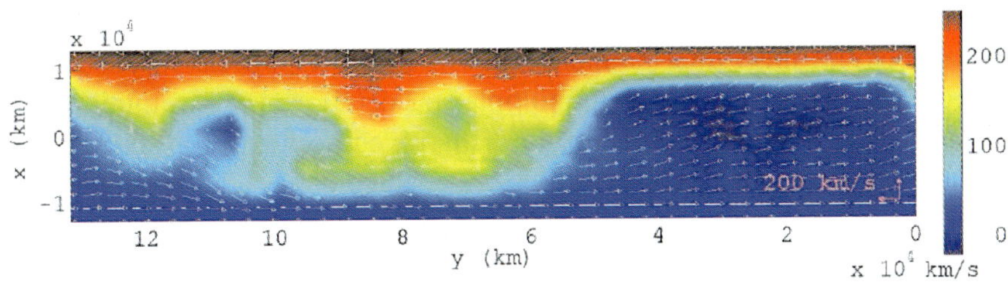

Figure 8.38. Periodic surface waves on November 25, 2001. (a) Reconstruction of ion density in a frame that slides along the boundary with the waves; the magnetosheath is at the top, the magnetosphere at the bottom, the Sun to the right, the tail to the left, with the x scale exaggerated. (b) Perpendicular magnetic wave intensity. (c) Colour-coded magnetic field strength and the magnetic field xy vector field. (d) Colour-coded z velocity and the xy velocity vectors. (From De Keyser et al., 2004a).

C3, Figure 8.38a) and of the perpendicular magnetic ULF wave intensity (STAFF, 0.3-10 Hz, all spacecraft, Figure 8.38b) outline the undulating shape of the wave. The site where the density drops to the magnetospheric level corresponds to the boundary layer inner edge, which shows an amplitude of about 10000 km. The magnetic field strength is enhanced on the magnetosheath side of the indentation (Figure 8.38c). Inside the indentation the field strength is depressed at the site where the magnetic xy vectors reverse direction, thus identifying the magnetopause. The data coverage is not sufficient to trace the magnetopause completely, but the Figure suggests a variable boundary layer width of a few thousand km. Figure 8.38d shows the colour-coded flow in the z direction, as well as the xy velocity vectors in the comoving frame (which depend on the chosen wave speed). The plot demonstrates how the region of maximum flow shear coincides with the reversal of tangential flow direction, thereby confirming that the choice of wave speed was reasonable. In common with the timing results of Owen et al. (2004b), described above, the reconstructed waveform clearly shows a non-sinusoidal waveform, and generally indicates a steepened leading edge. The boundary waves observed in this event are quite complicated because of the simultaneous presence of multiple periodicities.

A recent Cluster result has been the identification of rolled-up plasma vortices at the magnetospheric flanks on November 20, 2001 (Hasegawa et al., 2004a). These vortices are the result of the further nonlinear evolution of large-amplitude Kelvin-Helmholtz surface waves. They are believed to contribute to solar wind entry into the magnetosphere, in particular for northward interplanetary magnetic field. For further details, see Section 'Kelvin-Helmholtz Instability at the Flank Magnetopause' in Phan et al. (2005).

8.7.4 Summary

Cluster's ability to resolve the space-time ambiguity is particularly useful for addressing questions about the structure and dynamics of boundary waves. Both types of surface wave analysis that have been used in single-spacecraft studies, namely, analysis of individual boundary crossings and empirical reconstruction, benefit from the advantage of having multiple spacecraft: fewer assumptions have to be made and the reliability of the results is improved. Generally speaking, multi-spacecraft analysis of Cluster observations has enabled us

- to track the motion of the boundary as a function of time for an extended time interval, and hence to infer information about the spatial shape of the wavy surface; this allows us to put information from the different spacecraft and their various instruments in a common topological context.

- to determine the fundamental parameters that characterise surface waves, such as the wave amplitude, the wavelength, and the wave speed; while not qualitatively different from the values that were inferred from earlier single-spacecraft missions, it has now become possible to significantly reduce the uncertainty margins on these values for any given case, so that a more detailed study of wave morphology becomes possible.

- to confirm, at least from a few events examined this far, that the leading edges of periodic surface waves in their early nonlinear stage of development tend to be steeper than their trailing edges, as predicted by theory.
- to show, again from case studies, that the wave propagation direction lies close to perpendicular to the average magnetic field direction in the external magnetosheath, suggesting that surface waves may preferentially propagate in a direction that requires no bending of these external field lines, thereby confirming the importance of field line curvature pressure for the dynamical development of surface waves.

It can be hoped that, as more Cluster passes through the dusk- and dawn-side flank magnetospheric boundary are examined, we can go beyond individual case studies and build statistics in order to find out whether these findings are typical features of surface waves. Evidently, a better understanding of the morphology of surface waves may lead to a better insight into the physical processes that generate them and that control their time-evolution (see Phan et al., 2005).

8.8 Boundary Layer Structure and Detached Plasma

8.8.1 Background

The existence of a low-latitude boundary layer demonstrates that magnetosheath plasma can penetrate into the magnetosphere. How this actually happens has remained an unresolved question in magnetospheric physics, despite numerous studies based on single-spacecraft observations. Plasma can enter the magnetosphere along open field-lines created by reconnection on the dayside magnetopause. S subsequently convecting to the nightside, and, after reconnecting once more in the magnetotail, entering the dayside magnetosphere on closed field lines again (Dungey, 1961). Alternatively, magnetosheath plasma may perhaps directly enter onto closed magnetic field lines on the dayside by other processes (Haerendel et al., 1978; Eastman, 1979; Olson and Pfitzer, 1985; Lemaire and Roth, 1991, and references therein). Plasma entry may also occur in conjunction with the Kelvin-Helmholtz instability (e.g., Hasegawa et al., 2004a). Each access mechanism would leave its specific imprint on where the penetrating plasma would end up in terms of magnetic topology and in terms of plasma configuration.

The first aspect can be dealt with by checking whether magnetosheath-like plasma found earthward of the magnetopause is situated on open or on closed magnetic field lines, that is, whether it is magnetically connected to the magnetosheath. The reconnection scenario involving open field lines has been observed by Cluster on many occasions (see Phan et al., 2005). However, there are also instances when at least the inner parts of the boundary layer are located on closed field lines. Earlier studies of transient injection of magnetosheath plasma (Carlson and Torbert, 1980; Woch and Lundin, 1992) already established that certain plasma transfer events from open to closed magnetic field lines (PTEs) frequently occur, and Cluster observations confirm this (Lundin et al., 2003).

In order to deal with the issue of plasma configuration, one has to establish whether the penetrating plasma retains a topological connection to the magnetosheath. Experience with single-spacecraft observations has taught us to be very cautious about interpreting a temporally isolated encounter with magnetosheath-like plasma: It could just be a repeated encounter with the moving boundary rather than a spatially isolated feature (see Section 8.7.2). Empirical reconstruction methods that exploit Cluster data help to make that assessment by removing the effect of the boundary motion and reproducing the spatial density profile in the boundary layer. Plasma entering the magnetosphere but still topologically connected to the magnetosheath may show up as substructure inside the boundary layer, so that the transition from magnetosheath to magnetospheric density no longer has to be monotonic, which was believed to be the typical situation (Phan and Paschmann, 1996). Plasma that is able to enter deeper into the magnetosphere should show up as detached plasma entities. In this Section, we present Cluster observations of non-monotonic density structure inside the low-latitude boundary layer (Section 8.8.2). And, more importantly, Cluster has established for the first time that some magnetosheath plasma entities found inside the magnetosphere are indeed topologically disconnected from the boundary layer (Section 8.8.3). Multi-point observations are necessary to make such assessments. In fact, Sibeck et al. (2000) analysed a number of Interball-Tail/MAGION-4 two-spacecraft crossings of the magnetospheric boundary, but they did not find any evidence for a non-monotonic density gradient or for isolated entities. Cluster, being a four-spacecraft mission, has a better spatial coverage and also carries a more complete experiment payload. Additionally, Cluster's variable spacecraft separation strategy covers the length scales relevant for these detached entities, which is crucial to prove or disprove their topological connection to the boundary layer.

8.8.2 Structure inside the boundary layer

Cluster's inbound pass on April 23, 2001, between 14:00 and 17:00 UT (Figure 8.39) and at about 07:30 local time, showed a main magnetopause crossing around 14:26 UT, but transient encounters with magnetosheath-like plasma were observed until more than 2 hours later. The spacecraft separations were of the order of 1000 km; differences between the recorded profiles are seen on a time scale of 1 minute or less. Empirical reconstruction indicates that these transients can be explained in terms of boundary motion, so that Cluster allows us to conclude that these are not spatially isolated structures (as in Section 8.7.2). As an example, consider the largest transient, between 16:15 and 16:40 UT. This time interval is

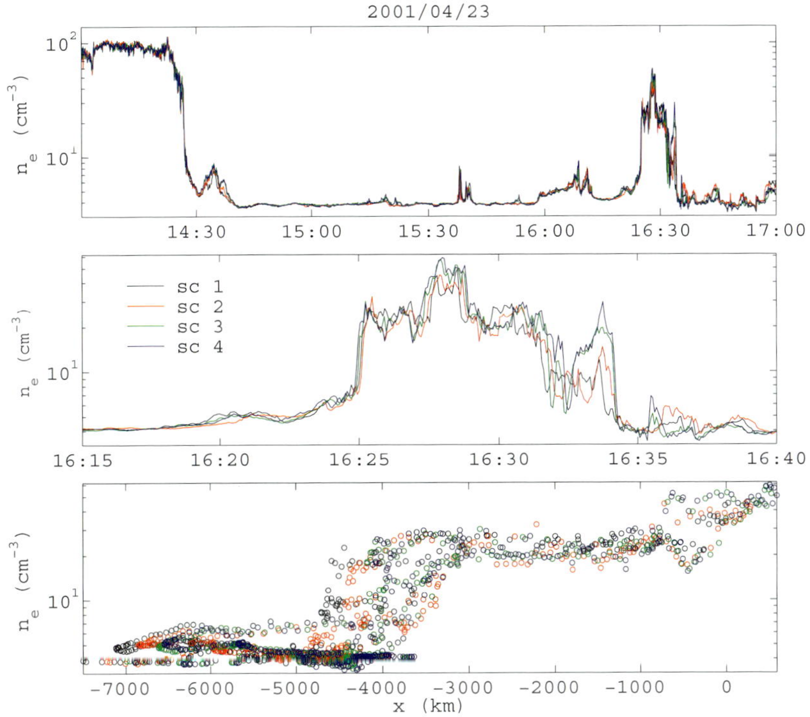

Figure 8.39. Structure of the boundary layer during an inbound pass on April 23, 2001. The electron observations (PEACE) in the top plot show several transient density increases well after the main magnetopause crossing. The last transient, enlarged in the middle plot, is singled out for detailed analysis. The bottom plot shows a one-dimensional empirical reconstruction of the spatial density profile, revealing the existence of a density plateau. (Adapted from De Keyser et al., 2004a).

short enough to reliably compute the boundary position by straightforward integration of the normal component of the CIS/HIA plasma velocity (see Equation 8.4). Plotting all density observations in a reference frame that moves with the boundary, De Keyser et al. (2004a) have shown that a well-defined spatial density profile emerges (Figure 8.39, bottom). The variations from magnetosphere to magnetosheath density and back again, for the four spacecraft, all trace the same profile, adding confidence that this truly reflects the spatial structure of an essentially one-dimensional boundary. Interestingly, the reconstruction reveals a density plateau inside the boundary layer, rather than a strictly monotonically decreasing density. Such substructure places constraints on the mechanisms responsible for the formation of the boundary layer. Not only must the existence of a ~5000 km thick boundary layer be accounted for, but substructure hints at a temporal modulation of the plasma access mechanism. Whether or not most LLBLs have plateau-like density profiles remains to be seen. But for the case discussed here, the plateau profile indicates that the LLBL was not formed by diffusive entry.

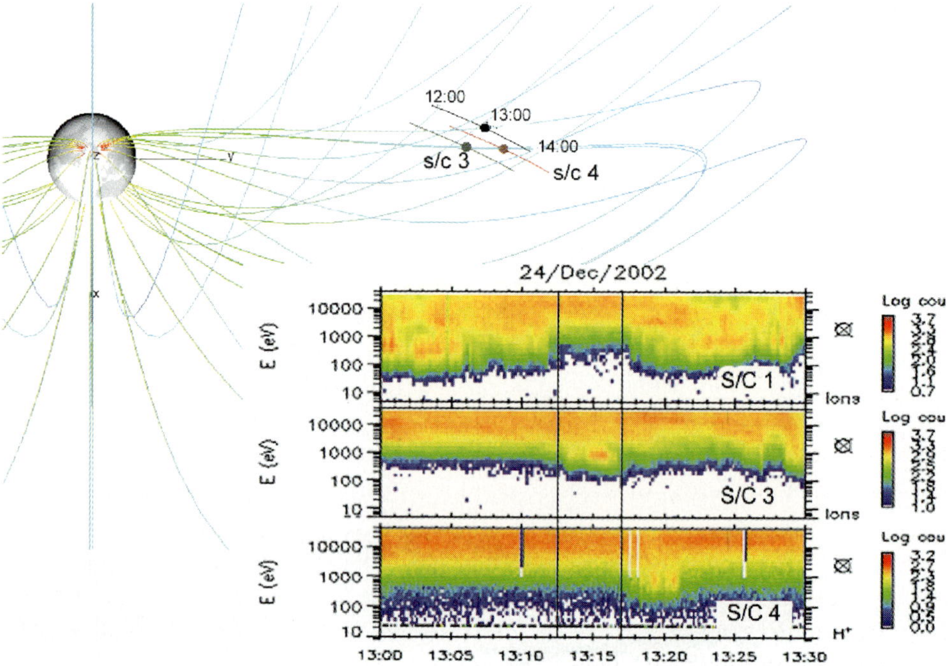

Figure 8.40. Cluster orbital configuration and CIS spectrograms identifying detached magnetosheath plasma near the flank LLBL on December 24, 2002. Red dots mark C1 (upper panel), green dots C3 (middle panel) and black dots C4 (lower panel). (Figure provided by R. Lundin).

8.8.3 Observation of detached plasma

Observational evidence for detached plasma entities was collected between the end of December 2002 and the end of February 2003, when the spacecraft separation distance was of the order of 1 R_E near apogee, mainly by looking at full 3D ion distribution functions and moments measured by CIS with a time resolution up to the spin rate ($\sim$4 s) and by using high time resolution magnetic field measurements from FGM. The FGM instruments on all four spacecraft and the CIS instruments on C1, C3 and C4 were operating during the events. In total we found 8 clear cases of detachment were identified, as well as 6 other cases that may be less clear but that display important aspects of the physical processes involved.

Considering that the width of the plasma structures may be several thousand kilometers, an intermediate to large spacecraft separation is a necessary requirement to determine whether the structure is detached from the boundary layer. Lundin et al. (2003) reported detached plasma observations made at medium spacecraft separation distance ($\sim$600 km). Other detachment events have been identified for large spacecraft separation ($\sim$6000 km).

We present here one large separation example that shows a clear detachment from the boundary layer. On December 24, 2002, Cluster was near the dusk flank low-latitude boundary layer. The Cluster ensemble skims the boundary layer, with C3 located further earthward than the other three spacecraft. In Figure 8.40, the detached plasma blob observed by C3 is marked out. Given the spatial configuration and the fact that C1 and C4 do not observe the blob, we can conclude that there is a topological detachment. The overall signature of the LLBL encounter is that of planar Pc3–5 waves with plasma moving slowly except in the injection structure. The flow ($\sim$50 km/s) is mainly inward, away from the magnetopause. A $-\mathbf{v} \times \mathbf{B}$ analysis shows that the waves are essentially in phase for the three spacecraft and planar, suggesting a ringing motion of the entire boundary layer. The ringing motion seems more pronounced for the innermost spacecraft (C3). This event is also characterized by overlapping energetic ion dispersion signatures.

From the handful of detached plasma entities observed so far, it can be concluded that these injection structures usually have perpendicular dimensions up into the $\sim$1 R_E range at Cluster altitudes. Only relatively small plasma injections are clearly separable by the Cluster spacecraft. We sometimes observe evidence for pressure pulses preceding the injections. From the energetic ion data, we can usually infer that these structures are located on closed field lines. Bipolar perturbations in the magnetic field suggest the presence of downward and upward field-aligned currents, typically in the range 0.01–10 nA m^{-2} (Lundin et al., 2003). Another characteristic feature associated with transient injections is the presence of outflowing energized ions of ionospheric origin (H^+, O^+, but also He^+). All these signatures are consistent with the magnetospheric projection of a dynamo region with a downward current and an electric field structure accelerating ions upward and electrons downward over dayside auroral forms (Lundin et al., 1995).

8.8.4 Summary

Cluster forces us to re-examine our views concerning boundary layer topology, this time without the often questionable assumptions and ambiguities inherent in single-spacecraft studies. We have learned so far

- that the boundary layer may have internal structure, possibly leading to non-monotonic boundary layer density profiles; Cluster allows us to determine those spatial profiles in spite of boundary motion.

- that detached plasma entities inside the magnetosphere do exist.

Both findings affect our ideas about plasma access mechanisms. From the study of a handful of detachment events, we have been able to establish the typical size of detached plasma entities (thousands of kilometers across) and to get a first glimpse of the distribution of field-aligned currents in them. Further examination of these data, and data collected later in the mission, will likely yield clues about

- the source of the electromotive force in the electric current circuit set up by these field-aligned currents;
- how plasma entities evolve in space and time; a first step could be to try to understand the observed energy dispersion structures, which are produced by time-of-arrival and energy diffusion effects;
- how such a plasma entity is coupled to the ionosphere and, in particular, how it deposits energy there;
- how well the observations agree with theoretical models (e.g., Lemaire, 1977).

This is especially interesting from the fundamental physics point of view, since these observations imply a departure from ideal MHD: As all the ions of different origin (cold, magnetosheath, magnetospheric) would experience different drifts, it is not obvious how these plasma entities could retain their identity as single flux tubes; the mechanism would certainly involve local electric fields. Cluster, the only mission able to prove or disprove plasma detachment because of its power to resolve the spatial/temporal ambiguity, is our only hope to make significant progress in this field in the near future.

An obvious question is the following: By which process can magnetosheath plasma manage to penetrate into the magnetosphere that deeply? Although the properties established from the Cluster observations help to constrain the possible answers to that question, Cluster is not likely to catch a plasma entity in the act of penetrating, or if it does, we may not be able to identify it as such. It would take a sequence of global pictures of the topology to unambiguously define the penetration process. Because of the small size of the Cluster configuration relative to the size of the magnetosphere, the observations provide only a local picture of what is going on at a particular place. If we want to draw a global picture of the boundary layer topology or the three-dimensional structure of detached plasma filaments, coordinated observations with other spacecraft and/or with ground-based stations will be necessary. A partial remedy could be to extend the Cluster configuration to even larger spatial separations.

It should not be forgotten that a low-latitude boundary layer is often absent. Cluster observations during 60 crossings by C1 and C3 of the near-tail dawnside magnetopause on 4–5 July 2001, which were analyzed in detail, showed that for 26 of them (i.e., 43 %) there was no evidence of a boundary layer.

8.9 Summary and Conclusions

In this Chapter we have reviewed some of Cluster's main achievements concerning the magnetopause and its major physical characteristics. Cluster has been able to

- determine the orientation, motion, and thickness of the magnetopause at a level of confidence not previously possible;

- establish that the magnetopause usually has a non-zero acceleration during the time it takes a spacecraft to cross the boundary;
- determine the current density distribution inside the magnetopause;
- identify curvature of, and sub-structure in, the magnetopause current layer;
- establish the existence of true flux transfer events that are 'bulges' moving along the magnetopause rather than ripples on the boundary;
- determine the global structure of boundary waves;
- detect plateaus in low-latitude boundary layerdensity profiles which rules out diffusive entry as a formation mechanism in those cases;
- provide evidence for detached magnetosheath-like plasma structures inside the magnetosphere.

These accomplishments result from Cluster's design as a four-spacecraft mission with separation distances varying from small (100 km scale) to moderately large (6000 km scale). These spatial scales are the relevant length scales of the major plasma features near the magnetopause. In addition, Cluster's advanced instrumentation measures plasma response on spin resolution time scales and magnetic and electric fields with a much higher time resolution, thus giving access to scales below the ion inertial length. This has allowed Cluster to

- provide the first direct observation of the Chapman-Ferraro electric field;
- identify the Hall currents.

Some of the key results from Cluster have just confirmed what we already suspected about the magnetopause, by allowing us to verify questionable assumptions that were made in the past, and to quantify our knowledge. But the emerging picture also contains many surprises. On a more macroscopic scale, these are the mapping of substructures and their rapid evolution in the magnetopause and boundary layer; the identification of detached magnetosheath plasma in the magnetosphere. On the microscopic scale, the direct measurements of sub ion-scale features and of the magnetopause electric field present new information that challenges our insight in the fundamental physical properties of boundary layers in collisionless plasmas.

Acknowledgements

JDK acknowledges support by the Belgian Federal Services for Scientific, Technical and Cultural Affairs through ESA/PRODEX Contracts 13127 and 14816. Cluster research in the UK and France is supported by PPARC and CNES, respectively. CJO acknowledges support through a PPARC advanced fellowship. The contributions of TDP and BUÖS were supported by NASA, those of SEH by DLR. Research of AV was supported by the Swedish Research Council.

References

Aggson, T. L., N. C. Maynard, and P. J. Gambardella: 1983, 'Electric field measurements at the magnetopause: Observation of large convective velocities at rotational magnetopause discontinuities'. *jgr* **88**, 10000–10010.

André, M., A. Vaivads, S. C. Buchert, A. N. Fazakerley, and A. Lahiff: 2004, 'Thin electron-scale layers at the magnetopause'. *Geophys. Res. Lett.* **31**, 3803.

André, M., R. Behlke, J.-E. Wahlund, A. Vaivads, A.-I. Eriksson, A. Tjulin, T. D. Carozzi, C. Cully, G. Gustafsson, D. Sundkvist, Y. Khotyaintsev, N. Cornilleau-Wehrlin, L. Rezeau, M. Maksimovic, E. Lucek, A. Balogh, M. W. Dunlop, P.-A. Lindqvist, F. Mozer, A. Pedersen, , and A. Fazakerley: 2001, 'Multi-spacecraft observations of broadband waves near the lower hybrid frequency at the Earthward edge of the magnetopause'. *Ann. Geophys.* **19**, 1471 – 1481.

Balogh, A., C. M. Carr, M. H. Acuna, M. W. Dunlop, T. J. Beek, P. Brown, K.-H. Fornacon, E. Georgescu, K.-H. Glassmeier, J. Harris, G. Musmann, T. Oddy, and K. Schwingenschuh: 2001, 'The Cluster magnetic field investigation: overview of in-flight performance and initial results'. *Ann. Geophys.* **19**, 1207–1217.

Bauer, T. M., M. W. Dunlop, B. U. Sonnerup, N. Sckopke, A. N. Fazakerley, and A. V. Khrabrov: 2000, 'Dual Spacecraft Determinations of Magnetopause Motion'. *Geophys. Res. Lett.* **27**(13), 1835–1838.

Berchem, J. and C. T. Russell: 1982, 'The thickness of the magnetopause current layer - ISEE 1 and 2 observations'. *J. Geophys. Res.* **87**, 2108–2114.

Berchem, J. and C. T. Russell: 1984, 'Flux transfer events on the magnetopause: spatial distribution and controlling factors'. *J. Geophys. Res.* **89**, 6689–6703.

Bosqued, J.-M., T. D. Phan, I. Dandouras, C. P. Escoubet, H. Rème, A. Balogh, M. W. Dunlop, A. D., E. Amata, M.-B. Bavassano-Cattaneo, C. Carlson, and A. M. e. a. DiLellis: 2001, 'Cluster Observations of the High-Latitude Magnetopause and Cusp: First Results from the CIS Ion Instruments'. *Ann. Geophys.* **19**, 1545–1566.

Cahill, L. J. and P. G. Amazeen: 1963, 'The Boundary of the Geomagnetic Field'. *J. Geophys. Res.* **68**, 1835.

Carlson, C. W. and R. B. Torbert: 1980, 'Solar wind ion injections in the morning auroral oval'. *J. Geophys. Res.* **85**, 2903.

Chandrasekhar, S.: 1961, *Hydrodynamic and Hydromagnetic Stability*. Clarendon, Oxford.

Chapman, S. and J. Bartels: 1940, 'Theories of Magnetic Storms and Aurorae'. In: *Geomagnetism, Volume II*. Oxford: Clarendon Press, pp. 850–861.

Chen, S.-H., M. G. Kivelson, J. T. Gosling, R. J. Walker, and A. J. Lazarus: 1993, 'Anomalous Aspects of Magnetosheath Flow and of the Shape and Oscillations of the Magnetopause During an Interval af Strongly Northward Interplanetary Magnetic Field'. *J. Geophys. Res.* **98**, 5727.

Cowley, S. W. H. and C. J. Owen: 1989, 'A Simple Illustrative Model of Open Flux Tube Motion Over the Dayside Magnetopause'. *Planet. Space Sci.* **37**, 1461.

De Keyser, J., F. Darrouzet, and M. Roth: 2002, 'Trying to bring the Magnetopause to a Standstill'. *Geophys. Res. Lett.* **29**, 10.1029/2002GL015001.

De Keyser, J., G. Gustafsson, M. Roth, F. Darrouzet, M. Dunlop, H. Rème, A. Fazakerley, P. Décréau, and N. Cornilleau-Wehrlin: 2004a, 'Reconstruction of the magnetopause and low latitude boundary layer topology using Cluster multi-point measurements'. *Ann. Geophys.* **22**, 2381.

De Keyser, J. and M. Roth: 1997, 'Equilibrium conditions for the tangential discontinuity magnetopause'. *J. Geophys. Res.* **102**, 9513–9530.

De Keyser, J. and M. Roth: 2003, 'Structural analysis of periodic surface waves on the magnetospheric boundary'. *Planet. Space Sci.* **51**, 757–768.

De Keyser, J., M. Roth, M. W. Dunlop, H. Réme, C. J. Owen, and G. Paschmann: 2004b, 'Empirical reconstruction and long-duration tracking of the magnetospheric boundary in single- and multi-spacecraft contexts'. *Ann. Geophys.* p. submitted.

deHoffmann, F. and E. Teller: 1950, 'Magneto-Hydrodynamic Shocks'. *Phys. Rev.* **80**, 692.

Drazin, P. G. and W. H. Reid: 1985, *Hydromagnetic Stability*. Cambridge University Press.

Dungey, J. W.: 1955, 'Electrodynamics of the outer atmosphere'. In: *Proceedings of the ionosphere Conference*. p. p. 225.

Dungey, J. W.: 1961, 'Interplanetary magnetic field and the auroral zones'. *Phys. Rev. Lett.* **6**, 47.

Dunlop, M. W. and A. Balogh: 2004, 'Magnetopause current as seen by Cluster'. *Ann. Geophys.* In press.

Dunlop, M. W., A. Balogh, P. Cargill, R. C. Elphic, K.-H. Fornacon, E. Georgescu, F. Sedgemore-Schultess, and the FGM team: 2001, 'Cluster observes the Earth's magnetopause: Co-ordinated four-point magnetic field measurements'. *Ann. Geophys.* **19**, 1449–1462.

Dunlop, M. W., A. Balogh, and K. H. Glassmeier: 2002a, 'Four-point Cluster application of magnetic field analysis tools: The discontinuity analyzer'. *J. Geophys. Res.* **107**(A11), 1385, doi:10.1029/2001JA0050089.

Dunlop, M. W., A. Balogh, K. H. Glassmeier, and P. Robert: 2002b, 'Four-point Cluster application of magnetic field analysis tools: The Curlometer'. *J. Geophys. Res.* **107**(A11), 1384, doi:10.1029/2001JA0050088.

Dunlop, M. W. and T. I. Woodward: 1998, 'Discontinuity analysis: orientation and motion'. In: G. Paschmann and P. W. Daly (eds.): *Analysis Methods for Multi-Spacecraft Data*, ISSI SR-001. ESA Publications Division, p. 271.

Dunlop, M. W. and T. I. Woodward: 1999, 'Analysis of thick, non-planar boundaries using the discontinuity analyser'. *Ann. Geophys.* **17**(8), 984–995.

Dunlop, M. W., T. I. Woodward, D. J. Southwood, K.-H. Glassmeier, and R. C. Elphic: 1997, 'Merging 4 spacecraft data: concepts used for analysing discontinuities'. *Adv. Space Res.* **20**, 1101.

Eastman, T. E.: 1979, *The plasma boundary layer and magnetopause layer of the Earth's magnetosphere*. Univ. of Alaska, Fairbanks. Ph. D. thesis.

Eastman, T. E., E. W. Hones, Jr., S. J. Bame, and J. R. Asbridge: 1976, 'The Magnetospheric Boundary Layer: Site of Plasma, Momentum and Energy Transfer from the Magnetosheath into the Magnetosphere'. *Geophys. Res. Lett.* **3**, 685.

Elphic, R. C., W. Baumjohann, C. A. Cattell, H. Lühr, and M. F. Smith: 1994, 'A search for upstream pressure pulses associated with flux-transfer events - an AMPTE/ISEE case-study'. *J. Geophys. Res.* **99**, 13521.

Elphic, R. C., M. Lockwood, S. W. H. Cowley, and P. E. Sandholt: 1990, 'Flux-transfer events at the magnetopause and in the ionosphere'. *Geophys. Res. Lett.* **17**, 2241.

Elphic, R. C. and D. J. Southwood: 1987, 'Simultaneous measurements of the magnetopause and flux-transfer events at widely separated sites by AMPTE UKS and ISEE-1 and ISEE-2'. *J. Geophys. Res.* **92**, 13666.

Fairfield, D. H.: 1971, 'Average and unusual locations of the Earths magnetopause and bow shock'. *J. Geophys. Res.* **76**, 6700.

Fairfield, D. H., A. Otto, T. Mukai, S. Kokubun, R. P. Lepping, J. T. Steinberg, A. J. Lazarus, and T. Yamamoto: 2000, 'GEOTAIL observations of the Kelvin-Helmholtz instability at the equatorial magnetotail boundary for parallel northward fields'. *J. Geophys. Res.* **105**, 21159.

Farrugia, C. J., F. T. Gratton, L. Bender, H. K. Biernat, N. V. Erkaev, J. M. Quinn, R. B. Torbert, and V. Dennisenko: 1998, 'Charts of joint Kelvin-Helmholtz and Rayleigh-Taylor instabilities at the dayside magnetopause for strongly northward interplanetary magnetic field'. *J. Geophys. Res.* **103**, 6703–6727.

Farrugia, C. J., R. P. Rijnbeek, M. A. Saunders, D. J. Southwood, D. J. Rodgers, M. F. Smith, C. P. Chaloner, D. S. Hall, P. J. Christiansen, and L. J. C. Woolliscroft: 1988, 'A multi-instrument study of flux transfer event structure'. *J. Geophys. Res.* **93**, 14,465–14,477.

Farrugia, C. T., F. T. Gratton, J. Contin, C. C. Cocheci, R. L. Arnoldy, K. W. Ogilvie, R. P. Lepping, G. N. Zastenker, M. N. Nozdrachev, A. Federov, J.-A. Sauvaud, J. T. Steinberg, and G. Rostoker: 2000, 'Coordinated Wind, Interball/tail, and ground observations of Kelvin-Helmholtz waves at the near-tail, equatorial magnetopause at dusk: January 11, 1997'. *J. Geophys. Res.* **105**, 7639–7667.

Ferraro, V. C.: 1952, 'On the Theory of the First Phase of a Geomagnetic Storm in a New Illustrative Calculation Based on an Idealized (Plane not Cylindrical) Model Field Distribution'. *J. Geophys. Res.* **57**, 15.

Fitzenreiter, R. J. and K. W. Ogilvie: 1995, 'Kelvin-Helmholtz Instability at the Magnetopause: Observations'. In: P. Song, B. U. Ö. Sonnerup and M. F. Thomsen (eds.): *Physics of the Magnetopause*, Geophys. Monograph 90, Washington, DC, USA: American Geophysical Union, pp. 277–284.

Goertz, C. K., E. Nielsen, A. Korth, K.-H. Glassmeier, C. Haldoupis, P. Hoeg, and D. Hayward: 1985, 'Observations of a possible ground signature of flux transfer events'. *J. Geophys. Res.* **90**, 4069–4078.

Gurnett, D. A., R. R. Anderson, B. T. Tsurutani, E. J. Smith, G. Paschmann, G. Haerendel, S. J. Bame, and C. T. Russell: 1979, 'Plasma Wave Turbulence at the Magnetopause: Observations from ISEE'. *J. Geophys. Res.* **84**, 7034.

Haaland, S., B. U. Ö. Sonnerup, M. W. Dunlop, A. Balogh, H. Hasegawa, B. Klecker, G. Paschmann, B. Lavraud, and H. Rème: 2004a, 'Four-Spacecraft Determination of Magnetopause Orientation, Motion and Thickness: Comparison with Results from Single-Spacecraft Methods'. *Ann. Geophys.* **22**, 1347.

Haaland, S., B. U. Ö. Sonnerup, E. Georgescu, B. Klecker, G. Paschmann, and A. Vaivads: 2004b, 'Orientation and Motion of a Discontinuity from Cluster Curlometer Capability: Minimum Variance of Current Density'. *Geophys. Res. Lett.* **31**(10).

Haerendel, G., G. Paschmann, N. Sckopke, H. Rosenbauer, and P. C. Hedgecock: 1978, 'The frontside boundary layer of the magnetosphere and the problem of reconnection'. *J. Geophys. Res.* **83**, 3195.

Hasegawa, H., M. Fujimoto, T.-D. Phan, H. Rème, A. Balogh, M. W. Dunlop, C. Hashimoto, and R. TanDokoro: 2004a, 'Transport of solar wind into Earth's magnetosphere through rolled-up Kelvin-Helmholtz vortices'. *Nature* **430**, 755–758.

Hasegawa, H., B. U. Sonnerup, M. W. Dunlop, A. Balogh, S. E. Haaland, B. Klecker, G. Paschmann, B. Lavraud, I. Dandouras, and H. Rème: 2004b, 'Reconstruction of Two-Dimensional Magnetopause Structures from Cluster Observations: Verification of Method'. *Ann. Geophys.* **22**, 1251–1266.

Hasegawa, H., B. U. Sonnerup, B. Klecker, G. Paschmann, M. W. Dunlop, and H. Rème: 2004c, 'Optimal Reconstruction of Magnetopause Structures from Multi-Spacecraft Data'. Submitted to Annales Geophysicae.

Hau, L. and B. U. Sonnerup: 1999, 'Two-Dimensional Coherent Structures in the Magnetopause: Recovery of Static Equilibria from Single-Spacecraft Data'. *J. Geophys. Res.* **104**, 6899–6917.

Hau, L.-N. and B. U. O. Sonnerup: 1991, 'Self-consistent gyroviscous fluid model of rotational discontinuities'. *J. Geophys. Res.* **96**, 15767.

Hesse, M., J. Birn, and K. Schindler: 1990, 'On the topology of flux-transfer events'. *J. Geophys. Res.* **95**, 6549.

Hu, Q. and B. U. Sonnerup: 2003, 'Reconstruction of Two-Dimensional Structures in the Magnetopause: Method Improvements'. *J. Geophys. Res.* **108**(A1) 1011, doi: 10.1029/2002JA009323.

Kan, J.: 1988, 'A theory of patchy and intermittent reconnections for magnetospheric flux-transfer events'. *J. Geophys. Res.* **93**, 5613.

Kawano, H., S. Kokubun, and K. Takahashi: 1992, 'Survey of transient magnetic field events in the dayside magnetosphere'. *J. Geophys. Res.* **97**, 10 677–10 692.

Kawano, H. and C. T. Russell: 1996, 'Survey of flux transfer events observed with the ISEE 1 spacecraft: Rotational polarity and the source region'. *J. Geophys. Res.* **101**, 27299.

Kawano, H. and C. T. Russell: 1997, 'Survey of flux transfer events observed with the ISEE 1 spacecraft: Dependence on the interplanetary magnetic field'. *J. Geophys. Res.* **102**, 11307.

Khrabrov, A. V. and B. U. Ö. Sonnerup: 1998a, 'DeHoffmann-Teller Analysis'. In: G. Paschmann and P. W. Daly (eds.): *Analysis Methods for Multi-Spacecraft Data*, ISSI SR-001. ESA Publications Division, pp. 221–248.

Khrabrov, A. V. and B. U. Ö. Sonnerup: 1998b, 'Orientation and motion of current layers: Minimization of the Faraday residue'. *Geophys. Res. Lett.* **25**, 2373–2376.

Kivelson, M. G. and S.-H. Chen: 1995, 'The Magnetopause: Surface waves and Instabilities and their possible Dynamical Consequences'. In: P. Song, B. U. Ö. Sonnerup and M. F. Thomsen (eds.): *Physics of the Magnetopause*, Geophys. Monograph 90, Washington, DC, USA: American Geophysical Union, pp. 257–268.

Kivelson, M. G. and C. T. Russell: 1995, *Introduction to space physics*. Cambridge ; New York : Cambridge University Press, 1995.

Klumpar, D. M., S. A. Fuselier, and E. G. Shelley: 1990, 'Ion composition measurements within magnetospheric flux-transfer events'. *Geophys. Res. Lett.* **17**, 2305.

Ku, H. C. and D. G. Sibeck: 1997, 'Internal structure of flux transfer events produced by the onset of merging at a single X line'. *J. Geophys. Res.* **102**, 2243.

Kuo, H., C. Russell, and G. Le: 1995, 'Statistical studies of flux-transfer events'. *J. Geophys. Res.* **100**, 3513.

Le, G., C. T. Russell, and H. Kuo: 1993, 'Flux transfer events - spontaneous or driven?'. *Geophys. Res. Lett.* **20**, 791.

Lee, L. C. and Z. F. Fu: 1985, 'A Theory of Magnetic Flux Transfer at the Earth's Magnetopause'. *Geophys. Res. Lett.* **12**, 105–108.

Lee, L. C., Z. W. Ma, Z. F. Fu, and A. Otto: 1993, 'Topology of magnetic-flux ropes and formation of fossil flux-transfer events and boundary-layer plasmas'. *J. Geophys. Res.* **98**, 3943.

Lemaire, J.: 1977, 'Impulsive penetration of filamentary plasma elements into the magnetospheres of the Earth and Jupiter'. *Planet. Space Sci.* **25**, 887.

Lemaire, J. and M. Roth: 1991, 'Non-steady-state solar wind–magnetosphere interaction'. *Space Sci. Rev.* **57**, 59–108.

Lindqvist, P.-A. and F. S. Mozer: 1990, 'The average tangential electric field at the noon magnetopause'. *J. Geophys. Res.* **95**, 17137–17144.

Lockwood, M.: 1991, 'Flux-transfer events at the dayside magnetopause - transient reconnection or magnetosheath dynamic pressure pulses?'. *J. Geophys. Res.* **96**, 5497.

Lockwood, M., S. W. H. Cowley, P. E. Sandholt, and R. P. Lepping: 1990, 'The ionospheric signatures of flux-transfer events and solar-wind dynamic pressure changes'. *J. Geophys. Res.* **95**, 17113.

Lockwood, M., A. Fazakerley, H. J. Opgenoorth, J. Moen, A. P. van Eyken, B. Dunlop, M., J.-M., G. Lu, C. Cully, P. Eglitis, I. W. McCrea, M. A. Hapgood, M. N. Wild, R. Stamper, W. Denig, M. Taylor, J. A. Wild, G. Provan, O. Amm, K. Kauristie, T. Pulkkinen, A. Stromme, P. Prikryl, F. Pitout, A. Balogh, H. Réme, R. Behlke, T. Hansen, R. Greenwald, H. Frey, S. K. Morley, D. Alcayde, P.-L. Blelly, E. Donovan, M. Engebretson, M. Lester, J. Waterman, and M. F. Marcucci: 2001, 'Coordinated Cluster and ground-based instrument observation of transient changes in the magnetopause boundary layer during an interval of predominantly northward IMF: relation to reconnection pulses and FTE signatures'. *Ann. Geophys.* **19**, 1613–1640.

Lockwood, M. and M. N. Wild: 1993, 'On the quasi-periodic nature of magnetopause flux-transfer events'. *J. Geophys. Res.* **98**, 5935.

Louarn, P., A. Fedorov, E. Budnik, G. Fruit, J. A. Sauvaud, C. C. Harvey, I. Dandouras, H. Rème, M. C. Dunlop, and A. Balogh: 2004, 'Cluster observations of complex 3D magnetic structures at the magnetopause'. *Geophys. Res. Lett.* **31**, 19805.

Lucek, E. A., P. Cargill, M. W. Dunlop, L. M. Kistler, A. Balogh, W. Baumjohann, K.-H. Fornacon, E. Georgescu, and G. Haerendel: 2001, 'Equator-S magnetopause crossings at high time resolution'. *J. Geophys. Res.* **106**, 25409–25418.

Lundin, R., J.-A. Sauvaud, H. Rème, and A. B. et al.: 2003, 'Evidence for impulsive solar wind plasma penetration through the dayside magnetopause'. *Ann. Geophys.* **21**, 457–472.

Lundin, R., J. Woch, M. Yamauchi, and G. Marklund: 1995, 'Boundary layer polarization and voltage in the 14 MLT region'. *J. Geophys. Res.* **100**, 7587.

Marchaudon, A., J.-C. Cerisier, J.-M. Bosqued, M. W. Dunlop, J. A. Wild, P. M. E. Décreau, M. Foerster, D. Fontaine, and H. Laakso: 2004, 'Transient plasma injections in the dayside magnetosphere: one-to-one correlated observations by Cluster and by SuperDARN'. *Ann. Geophys.* **22**, 141–158.

McWilliams, K. A., T. K. Yeoman, and G. Provan: 2000, 'A statistical survey of dayside pulsed ionospheric flows as seen by the CUTLASS Finland HF radar'. *Ann. Geophys.* **18**, 445–453.

Milan, S. E., M. Lester, S. W. H. Cowley, and M. Brittnacher: 2000, 'Convection and auroral response to a southward turning of the IMF: Polar UVI, CUTLASS, and IMAGE signatures of transient magnetic flux transfer at the magnetopause'. *J. Geophys. Res.* **105**, 15,741–15,755.

Miura, A.: 1990, 'Kelvin Helmholtz instability for supersonic shear flow at the magnetopause boundary'. *Geophys. Res. Lett.* **17**, 749.

Miura, A.: 1995, 'Kelvin-Helmholtz Instability at the Magnetopause: Computer Simulations'. In: P. Song, B. U. Ö. Sonnerup and M. F. Thomsen (eds.): *Physics of the Magnetopause*, Geophys. Monograph 90, Washington, DC, USA: American Geophysical Union, pp. 285–291.

Moen, J., J. Holtet, A. Pedersen, B. Lybekk, K. Svenes, K. Oksavik, W. Denig, E. Lucek, F. Søraas, and M. André: 2001, 'Cluster boundary layer measurements and optical observations at magnetically conjugate sites'. *Ann. Geophys.* **19**, 1655–1668.

Mozer, F. S., S. D. Bale, and T. D. Phan: 2002, 'Evidence of Diffusion Regions at a Subsolar Magnetopause Crossing'. *Phys. Rev. Lett.* 89, doi: 10.1103/Phys.Rev.Lett.89.015002.

Nakamura, M. and M. Scholer: 2000, 'Structure of the magnetopause reconnection layer and of flux transfer events: Ion kinetic effects'. *J. Geophys. Res.* **105**, 23179.

Neudegg, D. A., S. W. H. Cowley, S. E. Milan, T. K. Yeoman, M. Lester, G. Provan, G. Haerendel, W. Baumjohann, B. Nikutowski, J. Büchner, U. Auster, K.-H. Fornacon, and E. Georgescu: 2000, 'A survey of magnetopause FTEs and associated flow bursts in the polar ionosphere'. *Ann. Geophys.* **18**, 416–435.

Neudegg, D. A., T. K. Yeoman, S. W. H. Cowley, G. Provan, G. Haerendel, W. Baumjohann, U. Auster, K.-H. Fornacon, E. Georgescu, and C. J. Owen: 1999, 'A flux transfer event observed at the magnetopause by the Equator-S spacecraft and in the ionosphere by the CUTLASS HF radar'. *Ann. Geophys.* **17**, 707–711.

Nykyri, K. and A. Otto: 2001, 'Plasma transport at the magnetospheric boundary due to reconnection in Kelvin-Helmholtz vortices'. *Geophys. Res. Lett.* **28**, 3565.

Ogilvie, K. W. and R. J. Fitzenreiter: 1989, 'The Kelvin-Helmholtz instability at the magnetopause and the inner boundary layer surface'. *J. Geophys. Res.* **94**, 15113–15123.

Olson, W. P. and K. A. Pfitzer: 1985, 'Magnetospheric responses to the gradient drift entry of solar wind plasma'. *J. Geophys. Res.* **90**, 10823.

Otto, A. and D. H. Fairfield: 2000, 'Kelvin-Helmholtz instability at the magnetotail boundary: MHD simulation and comparison with GEOTAIL observations'. *J. Geophys. Res.* **105**, 21175.

Owen, C. J., M. W. Dunlop, A. N. Fazakerley, J.-M. Bosqued, J. P. Dewhurst, R. C. Fear, S. A. Fuselier, A. Balogh, and H. Rème: 2004a, 'Cluster observations of 'crater' flux transfer events at the dayside high-latitude magnetopause'. *Ann. Geophys.* Submitted.

Owen, C. J., A. N. Fazakerley, P. J. Carter, A. J. Coates, I. C. Krauklis, S. Szita, M. G. G. T. Taylor, P. Travnicek, G. Watson, R. J. Wilson, A. Balogh, and M. W. Dunlop: 2001, 'CLUSTER PEACE Observations of Electrons During Magnetospheric Flux Transfer Events'. *Ann. Geophys.* **19**, 1509.

Owen, C. J., R. C. Fear, A. N. Fazakerley, M. W. Dunlop, and A. Balogh: 2005, 'Simultaneous Cluster observations of flux transfer events both up- and downstream of the magnetopause'. *Ann. Geophys.* Submitted.

Owen, C. J., M. G. G. T. Taylor, I. C. Krauklis, A. N. Fazakerley, M. W. Dunlop, and J.-M. Bosqued: 2004b, 'Cluster observations of surface waves on the dawn flank magnetopause'. *Ann. Geophys.* **22**, 971.

Paschmann, G., S. Haaland, B. U. Ö. Sonnerup, E. Georgescu, B. Klecker, A. Vaivads, H. Réme, T. D. Phan, and H. Hasegawa: 2004, 'Characteristics of the near-tail dawn magnetopause and boundary layer'. *Ann. Geophys.* p. submitted.

Paschmann, G., G. Haerendel, I. Papamastorakis, N. Sckopke, S. J. Bame, J. T. Gosling, and C. T. Russell: 1982, 'Plasma and magnetic-field characteristics of magnetic-flux transfer events'. *J. Geophys. Res.* **87**, 2159.

Paschmann, G., B. U. Sonnerup, I. Papamastorakis, N. Sckopke, G. Haerendel, S. J. Bame, J. R. Asbridge, J. T. Gosling, C. T. Russell, and R. C. Elphic: 1979, 'Plasma Acceleration at the Earth's Magnetopause: Evidence for Reconnection'. *Nature* **282**, 243.

Paschmann, G., B. U. Ö. Sonnerup, I. Papamastorakis, W. Baumjohann, N. Sckopke, and H. Lühr: 1990, 'The magnetopause and boundary layer for small magnetic shear: Convection electric fields and reconnection'. *Geophys. Res. Lett.* **17**, 1829–1832.

Phan, T., M. W. Dunlop, G. Paschmann, B. Klecker, J.-M. Bosqued, H. Rème, A. Balogh, C. Twitty, F. S. Mozer, C. W. Carlson, C. Mouikis, and L. M. Kistler: 2004, 'Cluster Observations of Continuous Reconnection at the Magnetopause under Steady Interplanetary Magnetic Field Conditions'. *Ann. Geophys.* in press.

Phan, T. D., C. P. Escoubet, L. Rezeau, R. Treumann, A. Vaivads, G. Paschmann, S. A. Fuselier, D. Attié, B. Rogers, and B. U. Ö. Sonnerup: 2005, 'Magnetopause Processes'. *Space Sci. Rev.* this issue.

Phan, T. D. and G. Paschmann: 1996, 'Low-latitude dayside magnetopause and boundary layer for high magnetic shear: 1. Structure and motion'. *J. Geophys. Res.* **101**, 7801–7815.

Pinnock, M., A. S. Rodger, J. R. Dudeney, K. B. Baker, P. T. Newell, R. A. Greenwald, and M. E. Greenspan: 1993, 'Observations of an enhanced convection channel in the cusp ionosphere'. *J. Geophys. Res.* **98**, 3767–3776.

Pinnock, M., A. S. Rodger, J. R. Dudeney, R. A. Greenwald, K. B. Baker, and J. M. Ruohoniemi: 1991, 'An ionospheric signature of possible enhanced magnetic field merging on the dayside magnetopause'. *J. Atmos. Terr. Phys.* **53**, 201–212.

Pinnock, M., A. S. Rodger, J. R. Dudeney, F. Rich, and K. B. Baker: 1995, 'High spatial and temporal resolution observations of the ionospheric cusp'. *Ann. Geophys.* **13**, 919–925.

Pritchett, P. L. and F. V. Coroniti: 2004, 'Three-dimensional collisionless magnetic reconnection in the presence of a guide field'. *J. Geophys. Res.* **109**, 1220.

Provan, G. and T. K. Yeoman: 1999, 'Statistical observations of the MLT, latitude and size of pulsed ionospheric flows with the CUTLASS Finland radar'. *Ann. Geophys.* **17**, 855–867.

Provan, G., T. K. Yeoman, and S. W. H. Cowley: 1999, 'The influence of the IMF By component on the location of pulsed flows in the dayside ionosphere observed by an HF radar'. *Geophys. Res. Lett.* **26**, 521–524.

Provan, G., T. K. Yeoman, and S. E. Milan: 1998, 'CUTLASS Finland radar observations of the ionospheric signatures of flux transfer events and the resulting plasma flows'. *Ann. Geophys.* **16**, 1411.

Pu, Z.-P. and M. G. Kivelson: 1983a, 'Kelvin-Helmholtz instability at the magnetopause: Energy flux into the magnetosphere'. *J. Geophys. Res.* **88**, 853–861.

Pu, Z.-P. and M. G. Kivelson: 1983b, 'Kelvin-Helmholtz instability at the magnetopause: Solution for compressible plasma'. *J. Geophys. Res.* **88**, 841–852.

Pu, Z. Y., Q. G. Zong, T. Fritz, Z. Xiao, Y. Huang, S. Y. Fu, Q. Q. Shi, M. W. Dunlop, K.-H. Glassmeier, A. Balogh, P. Daly, J. B. Cao, Z. X. Liu, H. Rème, and I. Dandouras: 2003, 'Multiple Flux Rope Events at the High-latitude Magnetopause: Cluster/Rapid Observation on January 26, 2001'. *Ann. Geophys.* in press.

Rijnbeek, R. P., S. W. H. Cowley, D. J. Southwood, and C. T. Russell: 1982, 'Observations of reverse polarity flux-transfer events at the earths dayside magnetopause'. *Nature* **300**, 23.

Rijnbeek, R. P., S. W. H. Cowley, D. J. Southwood, and C. T. Russell: 1984, 'A survey of dayside flux-transfer events observed by ISEE-1 and ISEE-2 magnetometers'. *J. Geophys. Res.* **89**, 786.

Robert, P., A. Roux, D. Fontaine, P. Canu, . Le Contel, S. Perraut, N. Cornilleau-Wehrlin, M. W. Dunlop, A. Balogh, K.-H. Glassmeier, J.-M. Bosqued, A. Fazakerley, R. Behlke, and P. Décréau: 2002, 'Cluster observations of a flux transfer event (FTE)'. In: *COSPAR, Plenary Meeting*.

Roth, M., J. De Keyser, and M. M. Kuznetsova: 1996, 'Vlasov Theory of the Equilibrium Structure of Tangential Discontinuities in Space Plasmas'. *Space Sci. Rev.* **76**, 251–317.

Russell, C. T. and R. C. Elphic: 1978, 'Initial ISEE magnetometer results: Magnetopause observations'. *Space Sci. Rev.* **22**, 681–715.

Russell, C. T., G. Le, and H. Kuo: 1995, 'The occurrence rate of flux transfer events'. *Adv. Space Res.* **18**, 197–205.

Russell, C. T., M. M. Mellott, E. J. Smith, and J. H. King: 1983, 'Multiple observations of interplanetary shocks: four spacecraft determination of shock normals'. *J. Geophys. Res.* **88**, 4739–4748.

Scholer, M.: 1988, 'Magnetic flux transfer at the magnetopause based on single X-line bursty reconnection'. *Geophys. Res. Lett.* **15**, 291–294.

Scholer, M.: 1995, 'Models of Flux Transfer Events'. In: P. Song, B. U. Ö. Sonnerup, and M. F. Thomsen (eds.): *Physics of the Magnetopause*, Geophys. Monograph 90, Washington, DC, USA: American Geophysical Union, pp. 235–245.

Schwartz, S. J.: 1998, 'Shock and Discontinuity Normals, Mach Numbers, and Related Parameters'. In: G. Paschmann and P. W. Daly (eds.): *Analysis Methods for Multi-Spacecraft Data*, ISSI SR-001. ESA Publications Division, pp. 249–270.

Sckopke, N., G. Paschmann, G. Haerendel, B. U. Ö. Sonnerup, S. J. Bame, T. G. Forbes, E. W. Hones, and C. T. Russell: 1981, 'Structure of the low-latitude boundary layer'. *J. Geophys. Res.* **86**(15), 2099.

Seon, J., L. A. Frank, A. J. Lazarus, and R. P. Lepping: 1995, 'Surface waves on the tailward flanks of the Earth's magnetopause'. *J. Geophys. Res.* **100**, 11,907–11,922.

Sestero, A.: 1966, 'Vlasov equation study of plasma motion across magnetic fields'. *Phys. Fluids* **9**, 2006.

Shay, M. A., J. F. Drake, M. Swisdak, W. Dorland, and B. N. Rogers: 2003, 'Inherently three dimensional magnetic reconnection: A mechanism for bursty bulk flows?'. *Geophys. Res. Lett.* **30**, 1345, doi:10.1029/2002GL016267.

Sibeck, D. G.: 1990, 'A Model for the Transient Magnetospheric Response to Sudden Solar Wind Dynamic Pressure Variations'. *J. Geophys. Res.* **95**, 3755–3771.

Sibeck, D. G.: 1992, 'Transient events in the outer magnetosphere - boundary waves or flux-transfer events?'. *J. Geophys. Res.* **97**, 4009.

Sibeck, D. G., R. E. Lopez, and E. C. Roelof: 1991, 'Solar wind control of the magnetopause shape, location and motion'. *J. Geophys. Res.* **96**, 5489.

Sibeck, D. G., L. Prech, J. Safrankova, and Z. Nemecek: 2000, 'Two-point measurements of the magnetopause: Interball observations'. *J. Geophys. Res.* **105**, 237–244.

Sibeck, D. G. and M. F. Smith: 1992, 'Magnetospheric plasma flows associated with boundary waves and flux transfer events'. *Geophys. Res. Lett.* **19**, 1903–1906.

Smith, M. F. and C. J. Owen: 1992, 'Temperature anisotropies in a magnetospheric FTE'. *Geophys. Res. Lett.* **19**, 1907.

Song, P., G. Le, and C. T. Russell: 1994, 'Observational differences between flux-transfer events and surface-waves at the magnetopause'. *J. Geophys. Res.* **99**, 2309.

Sonnerup, B. U. and L. J. Cahill, Jr.: 1967, 'Magnetopause Structure and Attitude from Explorer 12 Observations'. *J. Geophys. Res.* **72**, 171.

Sonnerup, B. U. O.: 1980, 'Theory of the low-latitude boundary layer'. *J. Geophys. Res.* **85**(14), 2017.

Sonnerup, B. U. Ö. and M. Guo: 1996, 'Magnetopause Transects'. *Geophys. Res. Lett.* **23**, 3679–3682.

Sonnerup, B. U. Ö., S. Haaland, G. Paschmann, B. Lavraud, and M. W. Dunlop: 2004a, 'Orientation and Motion of a Discontinuity from Single-spacecraft Measurements of Plasma Velocity and Density: Minimum Massflux Residue'. *J. Geophys. Res.* **109**, A03221, doi:10.1029/2003JA010230.

Sonnerup, B. U. Ö., H. Hasegawa, and G. Paschmann: 2004b, 'Anatomy of a flux transfer event seen by Cluster'. *Geophys. Res. Lett.* **31**, L11803. doi:10.1029/2004GL020134.

Sonnerup, B. U. Ö. and M. Scheible: 1998, 'Minimum and Maximum Variance Analysis'. In: G. Paschmann and P. W. Daly (eds.): *Analysis Methods for Multi-Spacecraft Data*, ISSI SR-001. ESA Publications Division, pp. 185–220.

Southwood, D. J.: 1968, 'The hydromagnetic stability of the magnetospheric boundary'. *Planet. Space Sci.* **16**, 587.

Southwood, D. J.: 1987, 'The ionospheric signature of flux-transfer events'. *J. Geophys. Res.* **92**, 3207.

Southwood, D. J. and W. J. Hughes: 1983, 'Theory of hydromagnetic waves in the magnetosphere'. *Space Sci. Rev.* **35**, 301–366.

Southwood, D. J., M. A. Saunders, M. W. Dunlop, W. A. C. Mierjedrzejowicz, and R. P. Rijnbeek: 1986, 'A survey of flux-transfer events recorded by the UKS spacecraft magnetometer'. *Planet. Space Sci.* **34**, 1349.

Terasawa, T., H. Kawano, I. Shinohara, T. Mukai, Y. Saito, M. Hoshino, A. Nishida, S. Machida, T. Nagai, T. Yamamota, and S. Kokubun: 1996, 'On the Determination of a Moving MHD Structure: Minimization of the Residue of Integrated Faraday's Equation'. *J. Geomagn. Geoelectr.* **48**, 603–614.

Thomsen, M. F., J. A. Stansberry, S. J. Bame, S. A. Fuselier, and J. T. Gosling: 1987, 'Ion and electron velocity distributions within flux-transfer events'. *J. Geophys. Res.* **92**, 12127.

Thorolfsson, A., J.-C. Cerisier, M. Lockwood, P. E. Sandholt, C. Senior, and M. Lester: 2000, 'Simultaneous optical and radar signatures of poleward-moving auroral forms'. *Ann. Geophys.* **18**, 1054–1066.

Vaivads, A., M. André, S. Buchert, J.-E. Wahlund, A. Fazakerley, and N. Cornilleau-Wehrlin: 2004, 'Cluster observations of lower hybrid turbulence within thin layers at the magnetopause'. *Geophys. Res. Lett.* **31**, L0380H. doi:10.1029/2003GL018142.

Wild, J. A., S. W. H. Cowley, J. A. Davies, H. Khan, S. E. Milan, G. Provan, T. K. Yeoman, A. Balogh, M. W. Dunlop, K.-H. Fornacon, and E. Georgescu: 2001, 'First simultaneous observations of flux transfer events at the high-latitude magnetopause by the Cluster spacecraft and pulsed radar signatures in the conjugate ionosphere by the CUTLASS and EISCAT radars'. *Ann. Geophys.* **19**, 1491–1508.

Wild, J. A., S. E. Milan, S. W. H. Cowley, M. W. Dunlop, C. J. Owen, J. M. Bosqued, M. G. G. T. Taylor, J. A. Davies, M. Lester, N. Sato, A. S. Yukimatu, A. N. Fazakerley, A. Balogh, and H. Rème: 2003, 'Coordinated interhemispheric SuperDARN radar observations of the ionospheric response to flux transfer events observed by the Cluster spacecraft at the high-latitude magnetopause'. *Ann. Geophys.* **21**, 1807–1826.

Woch, J. and R. Lundin: 1992, 'Signatures of transient boundary layer processes observed with Viking'. *J. Geophys. Res.* **97**, 1431.

Chapter 9

Cluster at the Magnetospheric Cusps

P. J. Cargill[1], B. Lavraud[2], C. J. Owen[3], B. Grison[4], M. W. Dunlop[5], N. Cornilleau-Wehrlin[4], C. P. Escoubet[6], G. Paschmann[7], T. D. Phan[8], L. Rezeau[4], Y. Bogdanova[3], and K. Nykyri[1]

9.1 Introduction

The existence of the Earth's magnetospheric cusp was discussed originally by Chapman and Ferraro (1930). Although their work pre-dates the concept of a magnetosphere confined by the solar wind, they argued that there would be a singular magnetic field line extending from the Earth's surface to a boundary, now known as the magnetopause. Considering for example the noon-midnight plane, field lines slightly displaced from this line separate at the magnetopause, and close either on the dayside or nightside. In this picture, the cusp can be defined as the singular field line, a separatrix, which spreads over the whole magnetopause. This definition carries over to a modern magnetosphere where the magnetopause connects to the cusp, and can be seen in models of a closed magnetosphere (Tsyganenko, 1989).

Of course the real cusp is much more complex, and its properties are determined by the nature of the interaction of the solar wind, and especially the interplanetary

[1]Space and Atmospheric Physics, The Blackett Laboratory, Imperial College London, London, UK
[2]Los Alamos National Laboratory, Los Alamos, NM, USA
[3]Mullard Space Science Laboratory, Holmbury St. Mary, Dorking, Surrey, UK
[4]CETP/IPSL, Vélizy, France
[5]Space Science Division, Rutherford Appleton Laboratory, Chilton, Didcot, Oxfordshire, UK
[6]ESTEC, Noordwijk, The Netherlands
[7]International Space Science Institute, Bern, Switzerland
[8]Space Sciences Laboratory, University of California, Berkeley, CA, USA

Space Science Reviews 118: 321–366, 2005.
DOI: 10.1007/s11214-005-3835-0

magnetic field (IMF), with the magnetosphere. An all-inclusive definition of the cusp is difficult, and indeed may not be entirely desirable, but a working definition is that it is (a) part of the magnetosphere in the vicinity of the polar region at high magnetic latitudes, where a significant quantity of magnetosheath plasma is detected inside the nominal magnetopause position and (b) a region that encompasses the demarcation between dayside and nightside field lines. It should be noted that when examining in-situ particle and field measurements, the cusp tends to identify itself readily by its location and by the presence of magnetosheath plasma, and electromagnetic turbulence.

At low altitudes, where it was first discovered (Heikkila and Winningham, 1971; Frank, 1971), the cusp is located near magnetic noon and extends 1-2° in latitude and 1-2 hours in magnetic local time, MLT (Newell and Meng, 1988). Its position responds to both external changes like the solar wind ram pressure and IMF direction, and to internal changes like the dipole tilt angle and the geomagnetic activity, and can be found between 73° and 80° magnetic latitude and between 10:30 and 13:30 MLT (Newell and Meng, 1994; Yamauchi et al., 1996; Newell et al., 1989, and references therein). Since Cluster does not sample the low-altitude cusp, it is not discussed further in this chapter. At mid-altitudes, Cluster samples a very complex region of precipitating, mirrored and upwelling ions and electrons, with associated magnetic and electric field perturbations and turbulence. This is a reflection of the key role that the cusp plays in the transport of energy from the magnetopause to the ionosphere, as well as reflecting the ionospheric response to this energy input.

However, understanding what is seen in the cusp at low and middle altitudes requires knowledge of what is happening at the high-altitude cusp and associated magnetopause. This is a complex region of the magnetosphere, with the complexity arising from both the global interaction of the IMF with the magnetosphere, and the special local nature of the terrestrial magnetic field at the high-altitude cusp. It is now accepted that magnetic reconnection has a major influence on magnetospheric dynamics. In simple terms, for southward IMF, reconnection may occur in the vicinity of the sub-solar region, and for northward IMF, it can occur at the lobes. The cusp plays a special role since it is the reversal of the terrestrial magnetic field with respect to the IMF around the cusp that permits both southward and northward IMFs to reconnect.

For southward IMF, the high-altitude cusp will be influenced by the reconnection process as field lines reconnected at low latitudes are swept tailwards with the magnetosheath flow (Shelley et al., 1976; Reiff et al., 1977, 1980). How reconnected pulses and field lines evolve as they move along the magnetopause is unknown (see Smith and Lockwood, 1990, for suggestions). However, major effects in the cusp should be the presence of tailward convection and associated plasma mantle (Rosenbauer et al., 1975) and the necessity that the magnetic boundary with the magnetosheath be a rotational discontinuity (RD) (e.g., Vasyliunas, 1995). For a northward IMF, lobe reconnection will influence the cusp from the tailward side (Gosling et al., 1991; Kessel et al., 1996). The effect here is less clear, since for large enough magnetosheath plasma flows (i.e., super-Alfvénic), reconnected field lines should be swept tailward.

In the above scenarios, the cusp simply is a region that responds to what is happening elsewhere on the magnetopause: in other words a transit point in the flow of magnetic flux and particles in the magnetosphere. However, the cusp may also have its own intrinsic properties. For example, at the cusp, the Earth's magnetic field will have a region of relative weakness, leading to the suggestion that the magnetopause can sag inwards there, and that the magnetosheath flow needs to be deflected accordingly (Haerendel et al., 1978; Haerendel, 1978). A complex system of shock and rarefaction waves would be needed to accomplish this, as has been discussed by, for example, Walters (1966); Yamauchi and Lundin (2001); Taylor and Cargill (2002), and could lead to the injection of solar wind plasma independent of reconnection elsewhere on the magnetopause.

A related issue is that if one considers the cusp to be a funnel, with magnetic field lines converging from 360°, then local magnetic reconnection of anti-parallel fields would appear to be inevitable in the cusp vicinity for any IMF orientation, not just at the low-latitude magnetopause or at the lobes (see also Chapter 8). Also, a number of studies have shown the probable occurrence of multiple reconnections in the cusp region, eventually leading to turbulent mixing and entry penetration (e.g., Savin et al., 2004).

9.1.1 Previous work

9.1.1.1 *Mid-altitude*

The mid-altitude cusp region has been the subject of previous studies based on missions such as ISEE, Dynamics Explorer, Viking and Polar. This region is located near 12 MLT, and is characterised by direct penetration of magnetosheath plasma into the Earth's magnetosphere/ionosphere. Under southward IMF conditions, energy-latitude dispersions in ion data are the most pronounced cusp signature, and are usually interpreted as evidence of magnetosheath plasma entry following reconnection. The anti-sunward convection of open field lines through the polar ionosphere, and finite and different velocities of the magnetosheath ions entering from a sub-solar reconnection site give rise to the velocity filter effect.

Under steady northward IMF it seems likely that reconnection can occur at the high-latitude magnetopause and the observed 'reversed' ion dispersion signatures (Woch and Lundin, 1992b,a) have been interpreted as a consequence of this 'lobe reconnection' (Bosqued et al., 1985; Fuselier et al., 2000a). There is also evidence that component merging at the low-latitude magnetopause leads to cusp injections (Chandler et al., 1999). Finally, the IMF B_y component influences plasma convection in the polar region (e.g., Gosling et al., 1990; Cowley et al., 1991). For example, for IMF $B_y > (<)0$, flows in the cusp show a strong dawnward (duskward) bias.

More complex ion behaviour is often seen, such as steps in the cusp ion dispersion signatures (e.g., Escoubet et al., 1992; Lockwood and Smith, 1992). Models in which particle motions are traced from a magnetopause reconnection site down cusp field lines and into the ionosphere have been developed (e.g., Onsager et al., 1993; Smith and Lockwood, 1996) and such models reproduce the observed ion signatures including the cusp ion energy steps (e.g., Lockwood et al., 1995). This was interpreted as a signature of variations in the reconnection rate. However, Trattner et al. (1999, 2002) have suggested that some of these features can also be explained by sampling of spatially separated structures in some cases.

Another interesting phenomenon is the possibility of 'double reconnection', which is believed to produce two auroral forms that can be observed simultaneously on the equatorward and poleward side of the cusp (Øieroset et al., 1997; Sandholt et al., 1998, 2001) as well as the so-called 'double cusp' (Wing et al., 2001; Pitout et al., 2002). In support of this conjecture, modelling carried out by Wing et al. (2001) showed that when there was a strong IMF B_y with either a small negative or positive B_z, reconnection can take place both on the dayside and near the lobes, and so produce simultaneously two cusp injections at different locations. Although quite rare, such observations do exist in the DMSP database.

Lockwood et al. (1985) noted persistent outflows of low-energy O^+ ions at altitudes of 2000 - 5000 km. It was suggested that they had their origin in the dayside cleft/cusp region and manifested themselves as spatially-dispersed field-aligned flows at higher altitude (e.g., Moore et al., 1986). Such outflows have been suggested as a major source of magnetospheric heavy ion populations (André and Yau, 1997). Observations have shown strong localised perpendicular energisation of these ions (Moore et al., 1999), which can be attributed to wave-particle interactions. For example, the energisation has been associated with enhanced broad band extra low frequency (BBELF) wave power in the frequency range between 1 Hz and 1 kHz (André et al., 1990, 1998; Norqvist et al., 1998), as well as with electromagnetic ion cyclotron (EMIC) waves and/or lower hybrid waves (Moore et al., 1999). However, the free energy source for the wave growth and the identification of the wave modes responsible for ion heating are still open issues.

9.1.1.2 High-altitude

There have been five previous major missions that have led to progress in understanding the high-altitude cusp: HEOS-1 and -2 (1969 - 1974), Hawkeye (1973 - 1975), Polar (1996 - present) and Interball (1997 - 2000). All had different orbits from Cluster, and the precise orbit plays a critical role in determining which parts of the high-altitude cusp and its surroundings are sampled. The HEOS and Hawkeye spacecraft had apogees many R_E outside the bow shock near to the positive GSE-Z axis, and so cut through the high-altitude cusp. Polar has an apogee of 8.9 R_E , and so skims through the lower part of this region. Interball had a highly eccentric orbit with a 63° inclination, leading to the crossing of the cusp and plasma mantle regions mostly at very high magnetic latitudes.

There was recognition from the time of the HEOS mission that the cusp represented a region where magnetosheath plasma was located inside a 'nominal' magnetopause, and that the plasma flow in this region was weaker and more disordered than in the neighbouring magnetosheath (Rosenbauer et al., 1975; Paschmann et al., 1976; Haerendel, 1978; Haerendel et al., 1978). However, the lack of a solar wind monitor upstream of the Earth at these times rendered the identification of the regions and boundaries complicated. Further evidence of extensive solar wind plasma penetration through the magnetopause and into the cusp came from analysis of data from the Hawkeye (Eastman et al., 2000) and Polar (Zhou et al., 1999) missions.

The location of the magnetopause was defined by Haerendel et al. (1978) as the innermost sharp change in the magnetic field vector. Despite extensive analysis of data from the HEOS and Hawkeye spacecraft (e.g., Dunlop et al., 2000; Zhou and Russell, 1997; Eastman et al., 2000), the question of whether such a magnetopause was 'indented' has been of enduring interest. Moreover the definition of the magnetopause at the cusp is ambiguous because the region is likely to have boundaries with both the magnetosheath and magnetosphere. In this chapter a working definition of the magnetopause is the outermost magnetic boundary, usually identified as an abrupt deviation of the field direction from a magnetospheric orientation. An additional complication is that the cusp boundaries and properties are likely to be influenced by the occurrence of magnetic reconnection at the magnetopause either on the dayside or at the lobes (e.g., Lockwood and Smith, 1992; Gosling et al., 1991; Vasyliunas, 1995). In particular, the nature of the boundaries and associated plasma flows play a key role in establishing the penetration mechanisms of solar wind plasma into the cusp.

9.1.2 How Cluster investigates the cusp

Cluster is very well suited to investigate the cusp because of its comprehensive suite of instruments, and the ability of multiple spacecraft to obtain accurate measurements of the motion and structure of boundaries and waves, as well as of small-scale plasma micro-processes. The orbit is such that Cluster encounters the high-altitude cusp in the late winter and spring, and the mid-altitude cusp in the late summer and fall. Not only does Cluster sample the cusp over a large range of latitudes within a single orbit, but the inertial orbit implies that the cusp is also sampled widely in local time, hence building up a very extensive data base of cusp measurements.

The high-altitude cusp was defined as a major target of the Cluster mission. With solar wind monitors being continually available, the orbit of the spacecraft allows an investigation of the detailed structure and dynamics of this region as a function of the solar wind conditions, whose exact role is still to be established, in particular for northward IMF.

When analysing data in the different cusp regions, an important parameter is the speed of the spacecraft with respect to the speed of the cusp boundaries (Lockwood and Smith, 1994). In the high-altitude cusp the spacecraft speed is roughly 2-3 $\mathrm{km\,s^{-1}}$, while the speed of the boundaries can be more than 10 times larger. The spacecraft can then be considered as almost stationary with respect to motion of the cusp. At very low altitude ($< 1\,R_E$), the opposite situation arises since the spacecraft speed (a few $\mathrm{km\,s^{-1}}$) is much larger than the convection speed. Here the spacecraft give a snapshot of the cusp.

At mid-altitude, the spacecraft have a speed comparable to the convection and boundary speeds and therefore the observations are a mixture of spatial and temporal effects. These can be disentangled since, when the four Cluster spacecraft cross the mid-altitude cusp in their 'string-of-pearls' configuration, three spacecraft follow each other within a few minutes while the fourth arrives later (depending on the spacecraft separation). This configuration permits the investigation of spatial and temporal variations there.

9.2 The High-Altitude Cusp

Cluster encounters the high-altitude cusp in late winter and spring. In the years 2001 and 2002, data was taken for most outbound cusp crossings whereas over half the inbound crossings had no data due to telemetry restrictions on the spacecraft. By 2003 Cluster was able to return data from its entire orbit, so that inbound coverage is much enhanced.

The nature of the Cluster orbit is such that different parts of the cusp are sampled by the inbound and outbound crossings. Figure 9.1 shows the spacecraft configuration and location of the centroid with respect to the Tsyganenko (1989) magnetospheric magnetic field model for quiet solar wind conditions on March 17, 2001. For outbound crossings Cluster cuts through the cusp initially well inside the nominal magnetopause, exiting into the magnetosheath after the cusp traversal and often numerous or lengthy dayside magnetosphere encounters. For the inbound crossings in the Southern hemisphere, Cluster tends to cut through the tailward edge of the high-altitude cusp, giving quite short encounters (as for the February 13, 2001 event of Section 9.2.3). Depending on the solar wind conditions and dipole tilt angle, the spacecraft may remain in the high-altitude cusp for a long time before a direct exit into the magnetosheath, such as in the event of February 4, 2001 (Section 9.2.2). These conjectures are verified by examination of many individual crossings.

The primary focus of this section is thus on a series of case studies. The events analysed were chosen because they occurred under different (and well-defined) solar wind conditions and have been well-studied (Sections 9.2.1 - 9.2.3). Ultimately a survey of many more crossings will be required to assess how the cusp proper-

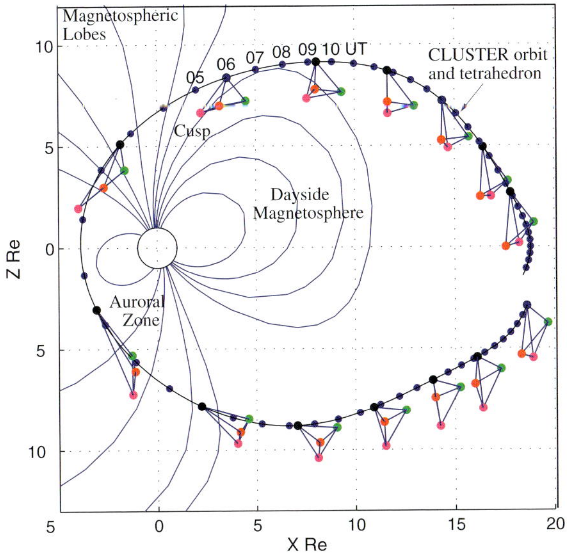

Figure 9.1. The orbit and tetrahedron configuration of the Cluster fleet during the cusp traversals in the two hemispheres on March 16 - 17, 2001. The interval 05 to 10 UT on March 17, 2001, to be discussed in Section 9.2.1, is marked along the orbit. The background field lines have been computed by use of the Tsyganenko (1989) magnetic field model. The tetrahedron configuration is magnified by a factor of 20 for clarity. (Figure provided by M. W. Dunlop).

ties depend on solar wind conditions and seasonal effects. This is some way from completion, although a statistical survey has been carried out (Section 9.2.4). Another important advance with Cluster is the ability to study small-scale cusp plasma processes, especially electromagnetic waves, and this is discussed in Section 9.2.5.

The results of the cusp geometry can be summarised as shown in the two panels of Figure 9.2. For southward IMF (left panel), magnetic reconnection occurs at the subsolar magnetopause, which then leads to a large scale convection electric field in the cusp region, producing a velocity filter effect on the entering plasma. The plasma mantle, at the poleward edge of the cusp, is formed by the tail of the dispersed cusp plasma and so contains mainly tailward and up-flowing plasma. For northward IMF (right panel), the convection in the cusp reverses and becomes sunward, due to magnetic reconnection occurring at the high-latitude magnetopause tailward of the cusp, as discussed in Phan et al. (2005). Under such conditions, the magnetosheath and magnetospheric fields have anti-parallel components in the lobe region. This northward IMF scenario should lead to the absence of a plasma mantle and a different large-scale plasma flow behaviour in the whole region.

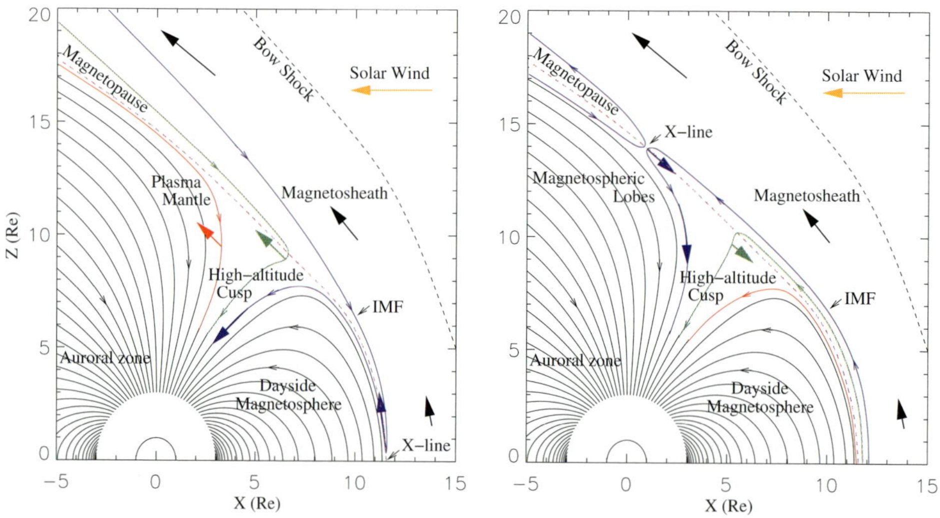

Figure 9.2. A schematic overview of the large-scale structure of the magnetic field topology and of the plasma flows in the high-altitude cusp and surrounding regions for southward (left) and northward (right) IMF directions. The location of the X-line is indicative of an approximate location of the reconnection process for both IMF geometries. The dashed purple (black) line is the approximate location of the magnetopause (bow shock) and the coloured arrows are indicative of the plasma flow directions expected for each case (orange: solar wind; black: magnetosheath; blue: reconnection-associated; green and red: large-scale convection). The blue, green, and red field lines are meant to show the time evolution of reconnected field lines. (Adapted from Lavraud and Cargill, 2005).

9.2.1 March 17, 2001

The first example is the outbound cusp crossing on March 17, 2001, 05:00 - 10:00 UT. The Cluster position and configuration during this event is marked in Figure 9.1. This is a case with relatively stable, weak, predominately northward IMF, and quiet solar wind conditions (density of $5\,\mathrm{cm}^{-3}$, velocity under $300\,\mathrm{km\,s}^{-1}$and dynamic pressure of 0.65 nPa). It is selected because of the extensive presence of solar wind plasma well inside the nominal magnetopause, the clean plasma boundaries on both entry and exit, the evidence for plasma flows associated with lobe reconnection, and the copious magnetic turbulence around the ion cyclotron frequency discussed further in Section 9.2.5. Figure 9.3 shows a summary plot of this event.

The cusp structure is seen most clearly in the ion data. Cluster leaves the inner magnetospheric lobe region at ~05:08 UT and enters the cusp, as characterised by a broadband signature of ions at magnetosheath energies. Cusp entry is abrupt, indicating the absence of a plasma mantle, as would be expected for a crossing with northward IMF and no tailward convection. These ions persist until approximately 06:25 UT, when Cluster enters the dayside magnetosphere, characterised by low-

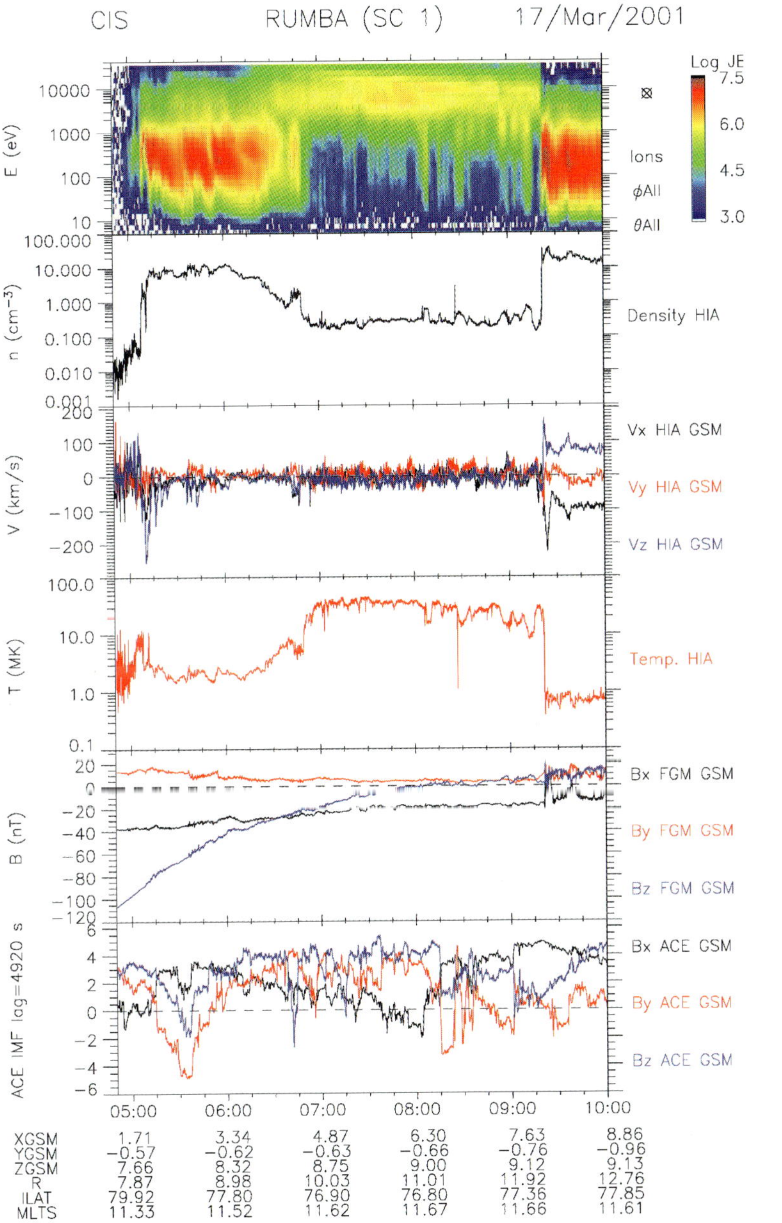

Figure 9.3. A summary plot of the Cluster data during the cusp interval 05:00 - 10:00 UT on March 17, 2001. The panels show, from top to bottom, the HIA ion omni-directional energy fluxes, the HIA ion density, velocity and temperature, the FGM magnetic field, all from spacecraft 1, and the lagged IMF from the ACE spacecraft. (Figure provided by B. Lavraud).

density, high energy ion populations of plasma sheet origin. Cluster remains in this region until it crosses the dayside magnetopause at 09:20 UT, equatorward of the cusp.

The cusp is less recognisable in the magnetic field, with B_z and B_y following the prediction of the Tsyganenko T89 model closely until the spacecraft nears the dayside magnetopause. The magnetic field is initially directed southward and slightly duskward, but after about 08:00 UT, the field turns from southward to northward, corresponding to passage through the dayside magnetosphere. The cusp is however recognisable by the presence of magnetic field fluctuations which commence immediately on entry at 05:08 UT (see Section 9.2.5).

The ion density and temperature in the central cusp remain relatively constant at $\sim 10\,\text{cm}^{-3}$and 2-3 MK respectively. However, the strength of the magnetic field is such that the plasma-β is of order 0.03, so that the plasma structure must be determined by either local field dynamics or flows parallel to the magnetic field. A magnetically-dominated state persists throughout this encounter indicating that Cluster is well inside the nominal magnetopause.

There are two clear boundaries: a sharp one on cusp entry at 05:08 UT and a more extended one on exit at 06:25 UT. As would be expected from a low-beta plasma, there is almost no change in the magnetic field strength across either, nor is there any identifiable change in the field orientation. Figure 9.4 shows the PEACE energy spectra between 05:00 and 05:20 UT. The boundary on cusp entry is sharp, with no evidence of cusp electrons leaking into the lobes, although of course the four spacecraft each see the boundary at different times. This boundary is likely to reflect the presence of newly reconnected lobe field lines (thus containing magnetosheath plasma) directly adjacent to empty non-reconnected lobe field lines, and may be interpreted as a tangential discontinuity (TD).

However, analysis of multi-spacecraft data (Vontrat-Reberac et al., 2003; Holland et al., 2004) revealed that this boundary is corrugated on the scale of the spacecraft separation, and is also moving, presumably in response to the IMF and solar wind dynamic pressure. Hence, the standard assumption that boundaries between plasma regimes are infinite, planar structures is unlikely to be valid in the present case.

Cusp entry is also characterised by the immediate onset (at 05:08 UT) of downward field-aligned plasma flows with magnitudes between $\sim$20 and $\sim 200\,\text{km}\,\text{s}^{-1}$. These begin at different times at each of the spacecraft, and are associated with the immediate onset of magnetic field turbulence. The flows weaken the further into the cusp Cluster progresses, giving way to occasional stagnant regions, and are transient, appearing and vanishing on timescales of a few minutes. It is believed that these are the remains of plasma jets generated by quasi-continuous lobe reconnection, evidence for which has also recently been given by Frey et al. (2003) on the basis of proton auroral precipitation, and in conjunction with Cluster data

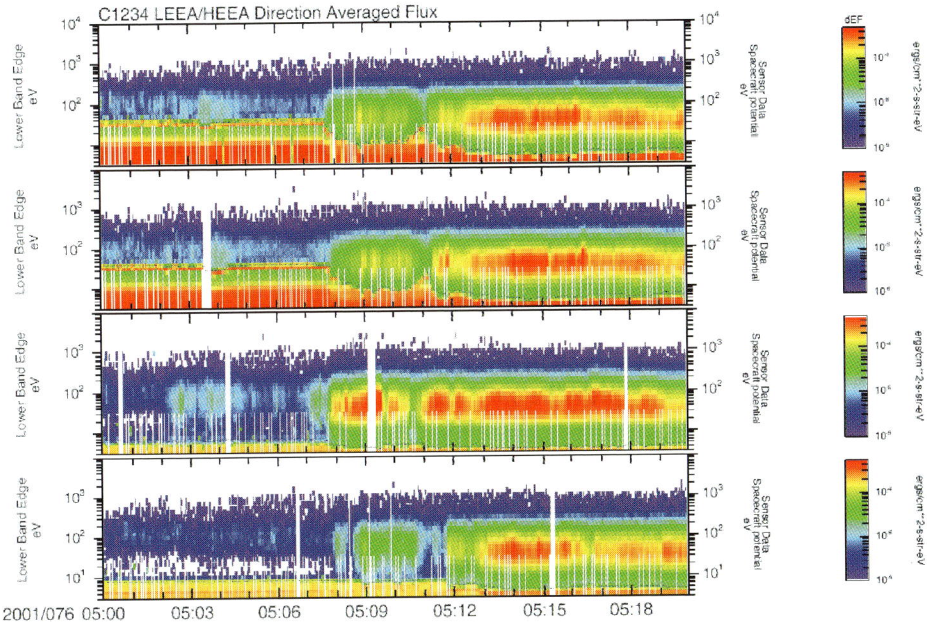

Figure 9.4. The direction-averaged electron energy flux from the four PEACE instruments with spacecraft 1 - 4 shown from top to bottom. The interval shown, between 05:00 and 05:20 UT, corresponds to the entry of Cluster into the cusp from the lobes. (Figure provided by C. J. Owen).

(Phan et al., 2003). The flows are sometimes sheared on the scale of the spacecraft separation (about 600 km for this event).

Further evidence for dynamic lobe reconnection comes from transient signatures in the ion distribution functions. Some relate to ion dispersion (Vontrat-Reberac et al., 2003) and others to transient, impulsive bursts characterised by short, purely temporal, pitch angle- and energy-time dispersions, which can be correlated with plasma flows in the ionosphere (Marchaudon et al., 2004). The global picture is consistent with ion dispersion for Sunward convection and subsequent mirroring at low altitudes.

The cusp exit is more extended. The density falls slowly, starting at around 06:00 UT, to the low densities expected for the dayside magnetosphere (0.2 - 0.3 cm^{-3}) at 06:25 UT, which is observed as an onset in the flux of the high-energy band of ions (between $\sim$5 - 10 keV). Between 06:42 and 06:48 UT there is a burst of magnetosheath plasma that can be associated with a weak transient southward turning of the IMF. This is interpreted as a transient motion (back and forth) of the boundary between the dayside magnetosphere and the cusp. This boundary is often extended and characterised by a mixture of solar wind and magnetospheric plasma (the low-latitude boundary layer: LLBL).

The orbit of Cluster allows one to successively sample the poleward boundary of the cusp, its equatorward boundary with the dayside magnetosphere and finally the dayside magnetopause at low latitudes (equatorward of the cusp). This type of encounter is observed quite frequently in Cluster data. While the crossing of March 17, 2001 described here is by far the most studied, other good examples can be found on March 2 and March 9, 2002.

9.2.2 February 4, 2001

The second example is an outbound crossing on February 4, 2001, 19:00 - 23:00 UT. The IMF was again predominantly directed northward during the whole interval. The solar wind was relatively stable with a velocity of $\sim$310 km s^{-1}, a density of $\sim$3 cm^{-3}and a dynamic pressure of $\sim$0.5 nPa. This event is shown because it demonstrates the presence of a large volume of stagnant plasma inside the magnetopause, the dynamic nature of the boundary between the cusp and the magnetosphere, as well as a magnetopause crossing that shows evidence for plasma inflow, and highlights the role of the plasma depletion layer (PDL) under northward IMF. Figure 9.5 shows a summary plot of Cluster data.

Cluster entered the high-altitude cusp at approximately 19:55 UT. As in the case of March 17, 2001, entry was characterised by a sudden onset of magnetosheath-like plasma, moving downwards with velocities occasionally greater than 200 km s^{-1}, and an onset of magnetic field turbulence, but no major change in field intensity or direction. Cluster remained in this region of magnetosheath-like plasma until its exit into the magnetosheath proper at 22:02 UT. Thus, unlike the encounter on March 17, there was no complete entry into the dayside magnetosphere, although there was a partial one at 21:41 UT. So, while the magnetopause on March 17 separates the dayside magnetosphere from the magnetosheath, the boundary observed at 22:02 UT on February 04 directly separates cusp plasma of magnetosheath origin from the magnetosheath itself. An extended PDL is observed outside the magnetopause between 22:02 and $\sim$22:30 UT and there is possibly a partial magnetopause encounter at 22:15 - 22:30 UT.

Considering the ion properties first, after cusp entry the mean energy of the ions gradually decreases. The observed behaviour is characteristic of reversed energy-time dispersion that is also consistent with the measured Sunward plasma convection. Also, on small-scales within the cusp structure (e.g., at 20:00 and 20:10 UT), short plasma injections are observed, similar to those on March 17. The particles are initially mainly field aligned, but with increasing time (and thus decreasing latitude), more up-flowing ions are detected and the ion distribution functions become more isotropic (Lavraud et al., 2002).

From around 21:20 UT until the exit into the magnetosheath, the distributions are very isotropic, the plasma flows are small, and the magnetic field strength is very low ($<$20 nT). This region was descriptively called the stagnant exterior cusp by Lavraud et al. (2002). The weak magnetic field in this region leads to a plasma

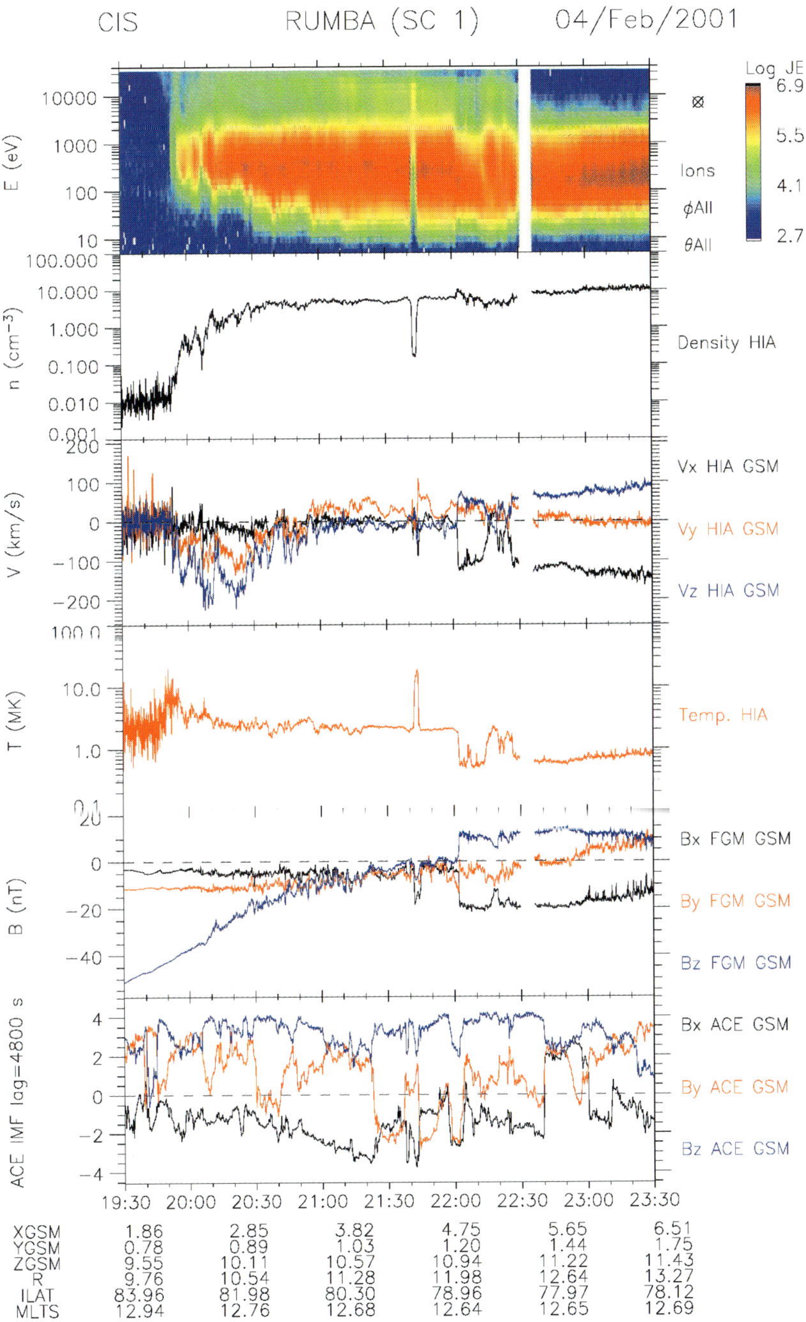

Figure 9.5. Summary plot of Cluster data during the high-altitude cusp interval 19:30 - 23:30 UT on February 4, 2001. See the caption of Figure 9.3 for details. (Figure provided by B. Lavraud).

beta of order 10, indicating that the field properties are determined by the plasma behaviour. The isotropic distributions observed there may not be the simple result of the mixing of the two populations, up and down-going (Fuselier et al., 2000a), together with the low convection (perpendicular velocity). It seems possible that wave-particle interaction would be needed to provide substantial ion scattering and isotropisation (Nykyri et al., 2003, 2004; Grison et al., 2004, see next sections), and such fluctuations are present there. Throughout the whole interval, from cusp entry up to the start of the low magnetic field exterior cusp, all variations look gradual without showing the clear boundaries that might be expected if a detached, stagnant region existed. This suggests that the stagnant exterior cusp is probably non-detached, rather being the extension of the cusp at high altitudes under northward IMF conditions (Russell et al., 2000; Lavraud et al., 2004b).

There are three encounters with distinct cusp boundaries. Cusp entry is similar to that on March 17, so is not discussed further. The crossing to and from the dayside magnetosphere at 21:41 UT can be identified as a sudden drop in the ion fluxes at magnetosheath energies, but as an increase in the fluxes at plasma sheet energies at all spacecraft, as well as a change in the magnetic field (Figure 9.6). Analysis of the magnetic field shows that this is approximately a'nested' structure, which is to be contrasted with a'convected' structure. The latter is characteristic of a transient feature passing through the spacecraft array while the former indicates that the boundary is moving back and forth across the spacecraft. Minimum variance analysis (MVA) of the magentic field data shows that the boundary has a negligible normal field component on both entry and exit (Lavraud et al., 2002, 2004c). The upper panel of Figure 9.6 further shows that the total pressure is conserved across the boundary. Together with the failure of the Walén test (Lavraud B., PhD Thesis, University of Toulouse, France), these facts suggest that the boundary is a Tangential Discontinuity (TD). Application of the discontinuity analyser (Dunlop et al., 2002, 2004) reveals a small motion of about 10 km s^{-1}back and forth, which may be associated with a rotation of the IMF seen in Figure 9.5.

The magnetosheath and the exterior cusp are separated by a sharp boundary across which the plasma velocity, as well as the magnetic field strength and direction, change abruptly. Also, both the ion and electron temperatures increase drastically (factor 3 for the ions) upon entry into the exterior cusp, revealing a possible heating mechanism at the boundary. Figure 9.7 shows the details of the boundary with magnetic field data (8 Hz) from all 4 spacecraft, and the normal ion velocity from spacecraft 1, 3 and 4 (in the MVA coordinates of spacecraft 1). In GSM coordinates the normal is (0.61, 0.33, 0.72), indicating a significant deviation from the nominal value, and the normal speed of the boundary is 8.8 km s^{-1}inward (indicated in Figure 9.7) and with a crossing duration of 25 s, one can obtain a boundary thickness of 220 km, which corresponds to 5 gyroradii of the thermal magnetosheath ions.

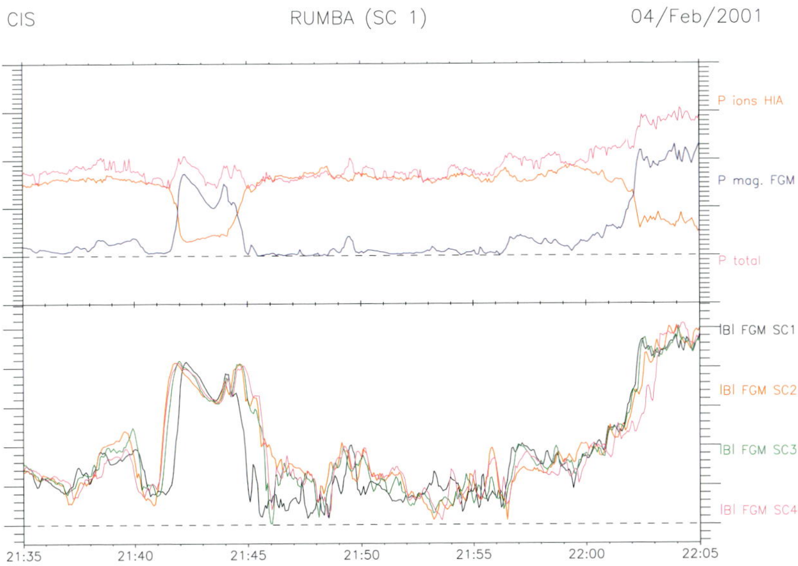

Figure 9.6. Ion plasma and magnetic pressure (upper panel) from spacecraft 1 and four-spacecraft magnetic field data (lower panel) for the interval 21:35 - 22:05 on February 4, 2001. The nested feature observed at 21:41-21:45 UT is an excursion into the dayside magnetosphere. The structure at 22:02 UT is the boundary between the exterior cusp and magnetosheath. (From Lavraud et al., 2004c).

The normal components of the plasma velocity, of order -20 to -30 km s^{-1}at the boundary, is to be compared with the normal boundary speed of -8.8 km $s^{-1}$and allows one to infer a net plasma inflow of about 10 to 20 km s^{-1}(Figure 9.7). A determination of the deHoffmann-Teller (HT) frame gave a small Sunward and downward transformation velocity (Lavraud et al., 2002). In addition, the magnetic field shows an inward normal component of $\sim$3 nT. The two latter arguments are consistent with a reconnection site poleward of the spacecraft while the former suggests plasma transport across the boundary. An application of the Walén test revealed that the ion flow speed in the HT frame was 91% of the Alfvén speed, in good agreement with expectations from a rotational discontinuity (RD).

The PDL plays an important role in sustaining this configuration. Figure 9.8 shows CIS and FGM data at and outside the magnetopause. The magnetopause is indicated by the first vertical black line and the extent of the PDL is shown by a green arrow. Without considering the complex structure observed between 22:15 and 22:30 UT (not analysed here), the PDL is characterised by a decrease in density and increase in magnetic field strength on approach to the magnetopause (as indicated by the magnitude of the Alfvén speed in Figure 9.8). Additionally, there are smaller flows than in the outer parts of the magnetosheath, in contrast to the sub-

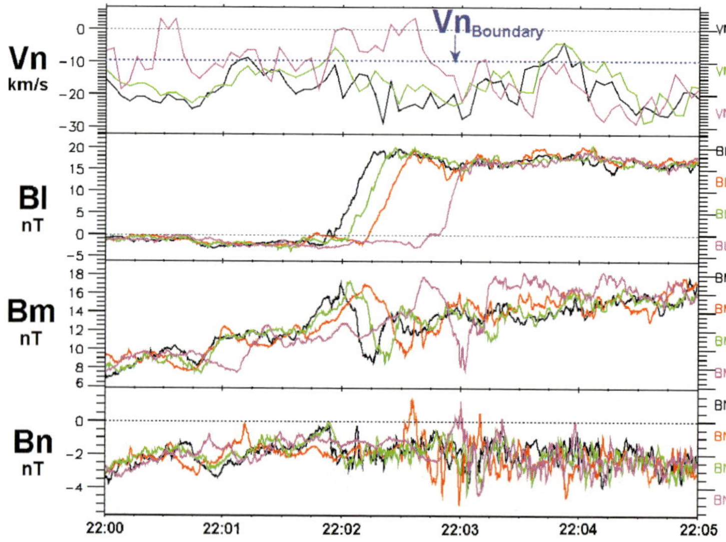

Figure 9.7. Blow-up of the boundary between exterior cusp and magnetosheath on February 4, 2001. The panels show, from top to bottom, the normal ion velocity from spacecraft 1, 3 and 4, and four spacecraft FGM magnetic field data components in the MVA coordinates (maximum, intermediate, and minimum variance, respectively). (Adapted from Lavraud et al., 2002).

solar PDL where the tangential flow speeds up on approach to the magnetopause (Phan et al., 1994). The combination of the decrease of flow speed and density, and the increase of magnetic field strength renders the flow sub-Alfvénic in this layer (see also Petrinec et al., 2003).

This analysis therefore suggests that the boundary may be a RD propagating in a direction opposite to the external magnetosheath flow in the Sunward and downward directions at a speed close to the local Alfvén speed. However, the RD can propagate Sunward only because of the presence of the sub-Alfvénic PDL. In addition, it may be that the field lines forming the discontinuity have been generated at the high-latitude magnetopause at a lobe reconnection site also located in sub-Alfvénic flows, in the PDL. The overall stability of the high-altitude cusp and of the reconnection site may therefore be highly dependent on the presence, and effects, of the PDL (Fuselier et al., 2000a; Avanov et al., 2001; Lavraud et al., 2004b).

The observation of stagnant plasma in the exterior cusp appears to be quite common under northward IMF, as well as the presence of a PDL on the outside of the magnetopause. More events are discussed by Lavraud et al. (2004b), see also Phan et al. (2003).

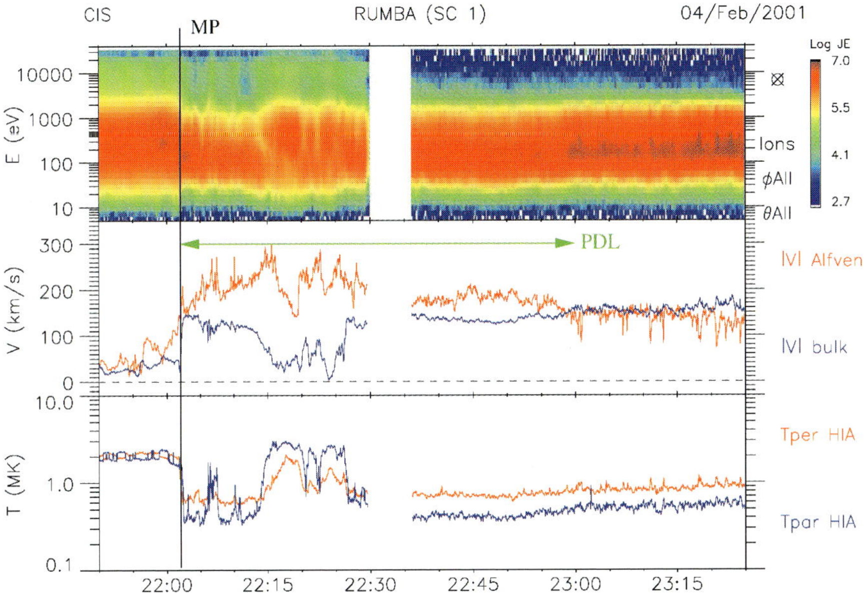

Figure 9.8. Data for an interval focused on the magnetopause boundary between 21:50 and 23:25 UT on February 4, 2001. The three panels show the omni-directional ion energy fluxes, the magnitude of the plasma and Alfvén velocities, and the parallel and perpendicular temperatures obtained from HIA. (Figure provided by B. Lavraud).

9.2.3 February 13, 2001

An inbound crossing occurred on February 13, 2001, 19:45 - 20:30 UT. The IMF was strongly southward, with a strong solar wind (velocity $530\,\mathrm{km\,s^{-1}}$, density $8\,\mathrm{cm^{-3}}$ and a dynamic pressure of 3.8 nPa). This event is shown to contrast with the two previous northward IMF cases. It is very dynamic with no evidence for stagnant plasma and has distinct magnetic field and plasma boundaries. Also, it demonstrates that, as we noted above, the inbound and outbound crossings sample the cusp at different altitude and latitude, thus perhaps sampling different regions. Figure 9.9 summarises the Cluster data for this encounter (Cargill et al., 2004).

The cusp is entered abruptly at 20:00 UT with a large decrease in the magnetic field strength, and rotation of the field. The cusp exit occurs at 20:08 UT with an enhancement in the magnetic field: there is a cusp re-entry seen at 20:12 UT, after which the spacecraft are in the plasma mantle region. The plasma parameters show no such abrupt change, and decay slowly to and after 20:30 UT. This encounter differs considerably from the two discussed above. It is clearly dominated by the southward IMF and high solar wind speed which are likely to be driving strong low-latitude magnetopause reconnection. This in turn leads to convection of magnetic field lines through the cusp so that there is no evidence for stagnant plasma as

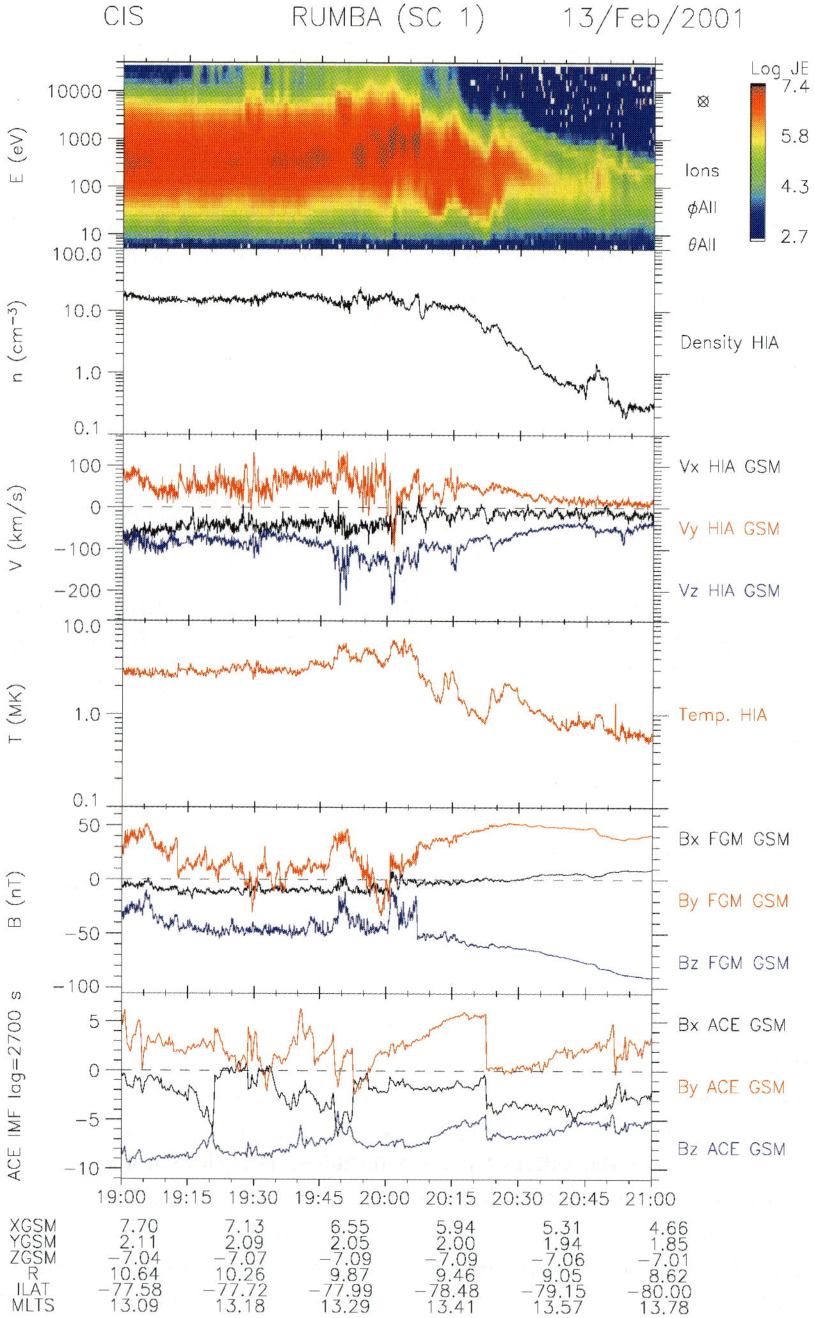

Figure 9.9. Summary plot of the Cluster data during the high-altitude cusp interval 19:00 - 21:00 UT on February 13, 2001. See the caption of Figure 9.3 for details. (Figure provided by B. Lavraud).

discussed above for the cases of northward IMF. The plasma flows seen are indeed large (>100 km s^{-1}) and in overall accordance with those observed in the magnetosheath, although there is an acceleration and deflection in the y-direction on entry (at 20:00 UT) and exit (at 20:07 UT).

The extent of the global anti-Sunward convection is demonstrated by the consistently negative values of $V_{\perp xz}$ in Figure 9.10. This leads to the presence of an extended plasma mantle, appearing as the low energy tail of the ion dispersion structure after 20:30 UT in Figure 9.9 and as the extended density profile in 9.10. An important feature to note is the large-amplitude fluctuations superimposed on the average convection velocities. These fluctuations die out with elapsed time (or distance) from the magnetopause crossing. Towards the end of the crossing the convection velocities become very small, only 2 km s^{-1}.

High-resolution (22.4 vectors per second) FGM data from the interval 19:59 - 20:10 UT are shown in Figure 9.11. Consider first the field magnitude. We show the start and finish of cusp entry (as defined by the decrease in field magnitude) by the first two solid lines. Cusp entry is through a boundary of duration just over a minute, occurs to within a few seconds at all spacecraft, and the y- and x-components of the magnetosheath field are greatly diminished. A minimum variance analysis was performed over the interval 19:59:30 - 20:01:30 UT (Cargill et al., 2004). Good normals (minimum to intermediate eigenvalue ratios of order 10 or greater) were found for spacecraft 1, 3 and 4, but not for spacecraft 2. This appears to be due to large fluctuations in the shape of the boundary. The average normal from spacecraft 1, 3 and 4 was **n** = (0.96, -0.21, -0.16), with an average normal field of +3.6 nT, consistent with a field directed outward through the boundary (note this case is in the southern hemisphere). The boundary is oriented both Sunward and dawnward, which should be contrasted with the normal of a model magnetopause, **n** = (0.6, 0.15, -0.75). Thus the magnetopause has undergone considerable distortion. A good deHoffman-Teller frame was found for spacecraft 1 and 3, but tests of the Walén relation did not show an exact linear relation. Any identification as a RD is suggestive rather than definitive.

Examination of the timing of the spacecraft passage through the magnetopause reveals a complex geometry. For entry through the centre of a stationary, planar cusp-magnetosheath boundary, spacecraft 1 would be expected to encounter the boundary ahead of the others by 2 - 3 minutes. This does not happen. From the viewpoint of the field magnitude, all spacecraft see the cusp entry at approximately the same time, and spacecraft 1 only occasionally is the first to see the transition (Figure 9.11). As we noted above, spacecraft 2 - 4 see the rotation of B_y before spacecraft 1. This is likely to be due to a strong motion of the cusp boundary in the y-direction.

This type of crossing, short and traversing the very high altitude and latitude parts of the cusp, appears to be quite common on the inbound orbits. The absence of stagnation, but rather the occurrence of large tailward convection, is also a common

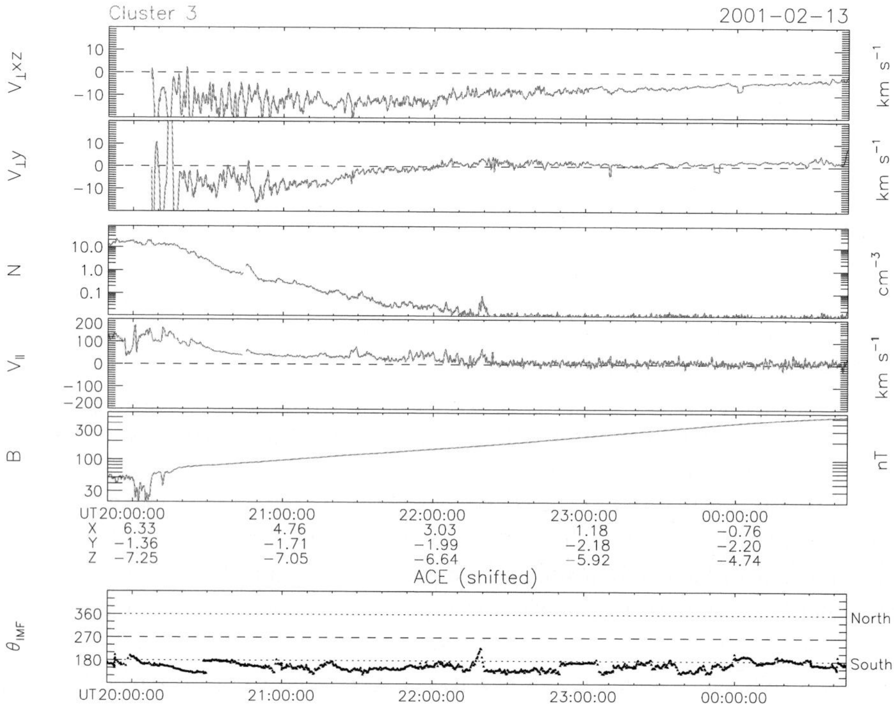

Figure 9.10. Cluster 3 measurements during the inbound pass over the southern polar cap on February 13, 2001. The top two panels show the convection velocities measured by the electron drift instrument (EDI), after applying a 30 s-smoothing to the full-resolution data; $V_{\perp xz}$ is the component in the GSE (X,Z)-plane (with negative values implying anti-sunward convection), $V_{\perp y}$ the component along GSE-Y. The third and fourth panels show the ion density, N, and the parallel component of the bulk velocity, $V_{\|}$, measured by CIS/HIA, the fifth panel the magnetic field strength measured by FGM. The bottom panel shows the lagged IMF clock-angle, confirming the consistently southward direction of the IMF. (From Vaith et al., 2004).

feature under southward IMF. There is an almost identical case on February 20, 2001, and others are also evident in the first two years of the mission.

9.2.4 Statistical survey

While the case studies presented in the previous three sections can highlight important physics issues, a comprehensive analysis of a large number of crossings is required to really understand the cusp. An alternative approach is to carry out statistical surveys that discuss the cusp in terms of a number of key parameters to highlight general trends.

Lavraud et al. (2004a) carried out such a statistical survey of three years of high-altitude cusp encounters (2001 - 2003 inclusive). They discussed (1) the global

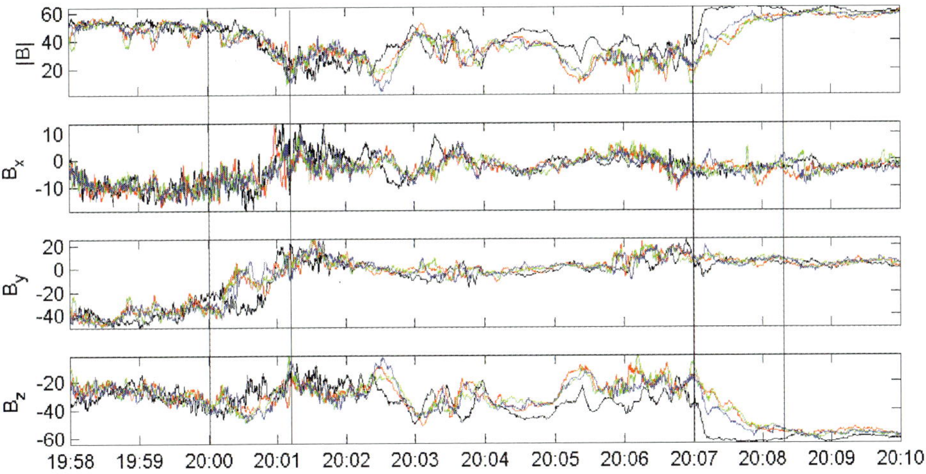

Figure 9.11. The magnetic field with a resolution of 22.4 vectors per second from 19:58 - 20:10 UT on February 13, 2001. The six panels show the magnetic field magnitude and components in GSE coordinates. (Figure provided by P. J. Cargill).

magnetic field and plasma properties of the high-altitude cusp diamagnetic cavity (the 'exterior cusp') and its surrounding regions, (2) the identification of the various boundaries surrounding the exterior cusp and (3) the dependence of cusp plasma flows on the IMF orientation. Motion of the cusp and magnetopause positions in response to solar wind conditions requires the use of appropriate transformations in order to ensure that averages are calculated from data that is spatially concordant. This can be achieved by the use of the Tsyganenko and Stern (1996): T96 magnetic field model to account for cusp angle deviations, and the magnetopause model of Shue et al. (1997) to include radial magnetopause variations. The technique is presented fully in Lavraud et al. (2004a).

Figures 9.12a and 9.12b show distributions of the average magnetic field vector and the ratio of measured magnetic pressure to the magnetic pressure estimated from the T96 model. Figures 9.12c and 9.12d show, respectively, the distribution of the average ion density normalised to that in the solar wind, and the ion temperature normalised to that predicted near the magnetopause at high latitudes, using the solar wind parameters as input in the Spreiter et al. (1966) magnetosheath model. Figures 9.13a and 9.13b show the distributions of the parallel plasma flows for southward and northward IMF orientations respectively, and Figures 9.13c and 9.13d the distributions of the perpendicular velocity components in the X direction of the normalised frame respectively for southward and northward IMF (Lavraud et al., 2005). In all plots, three distinct boundaries are shown as guides, and are discussed below.

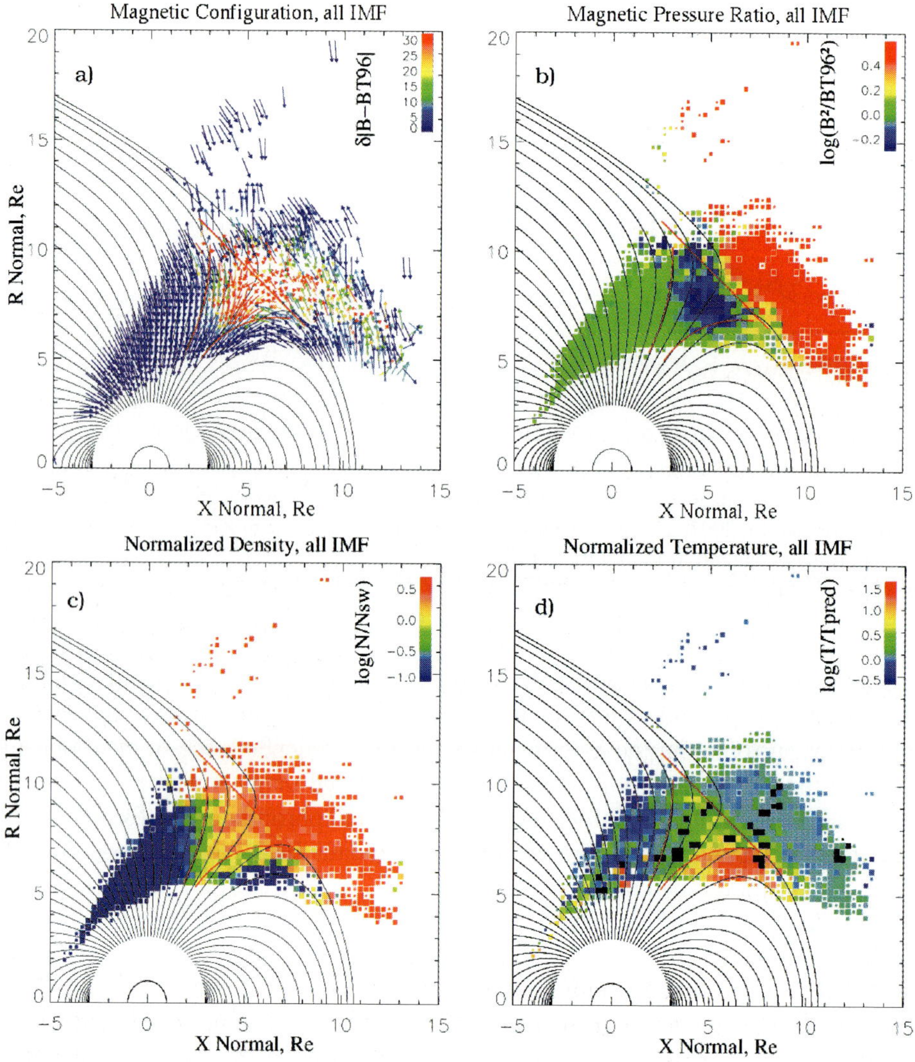

Figure 9.12. The spatial distribution of (a) the magnetic field vectors, where the size of each vector is the field magnitude in logarithmic scale. The colour of the vectors corresponds to the deviation of the measured magnetic field to the T96 model field ($B_{meas} - B_{T96}$). (b) The ratio of the measured magnetic pressure to that calculated from the T96 model. (c) The ratio of the measured density to that monitored in the solar wind and (d) the ratio of the measured temperature to the temperature predicted in the magnetosheath. In all panels, data are averaged over bins of 0.3 R_E. In panels (b) - (d) the sizes of the squares are proportional to the number of samples but are saturated at the maximum (0.3 R_E) for more than 20 samples. The background field lines are calculated from the T96 model, and colour palettes are used to illustrate the amplitude of the parameters studied. In this figure all IMF conditions are taken into account. (Adapted from Lavraud et al., 2004a).

9.2.4.1 Global properties for all IMF conditions

The distribution of the magnetic field vector highlights the presence of an intermediate region between the magnetosheath and the magnetosphere which is the exterior cusp. In this region, both magnetic field direction and strength are variable (Figure 9.12a). The exterior cusp extends a few R_E inside the nominal magnetopause location and is characterised by the presence of cold (Figure 9.12d) and dense (Figure 9.12c) plasma of solar wind origin. Comparison with the measured magnetic pressure (which shows a deficit) demonstrates that this region is diamagnetic.

The magnetic field and plasma distributions allow one to establish the unambiguous presence of three distinct boundaries surrounding the exterior cusp region: inner boundaries with the lobes at the poleward edge and the dayside magnetosphere at the equatorward edge, and an external boundary with the magnetosheath. These results also show that as one travels from the magnetosheath to the exterior cusp, the external boundary is characterised by a density decrease and temperature increase.

9.2.4.2 Dependence of large-scale plasma flows on IMF orientation

For southward IMF, solar wind plasma is statistically observed to be flowing Earthward (field-aligned) primarily at the equatorward side of the cusp (Figure 9.13a), and Figure 9.13c shows that these precipitating ions are characterised by large tailward convection. Overall, the convection in the cusp and plasma mantle is consistently directed tailward.

By contrast, under northward IMF conditions, no downward flows are seen at all at the equatorward edge of the cusp (Figure 9.13b). However, there is evidence for field-aligned downward flows at higher latitudes, near the boundary with the lobes. Consistent with this, no large-scale convection is observed in the cusp but a slight Sunward convection is present near the poleward boundary.

While convection is large and tailward in the exterior cusp for southward IMF, it is very small (and perhaps Sunward) for northward IMF conditions. This is compatible with the idea that convection is a consequence of reconnection and more specifically that the reconnected field lines need to travel opposite to the magnetosheath flow in the case of high-latitude reconnection. All of the above statistical findings suggest that the whole high-altitude cusp region is structured, at large scales, by magnetic reconnection occurring near the lobes under northward IMF, but at the lower latitude magnetopause for southward IMF.

9.2.5 Waves and turbulence

The cusp is also now recognised as being the site of small-scale plasma processes, manifested in particular by electric and magnetic field fluctuations (e.g., Bahnsen et al., 1975; Gurnett and Frank, 1978; Pottelette et al., 1990; Pickett et al., 1999,

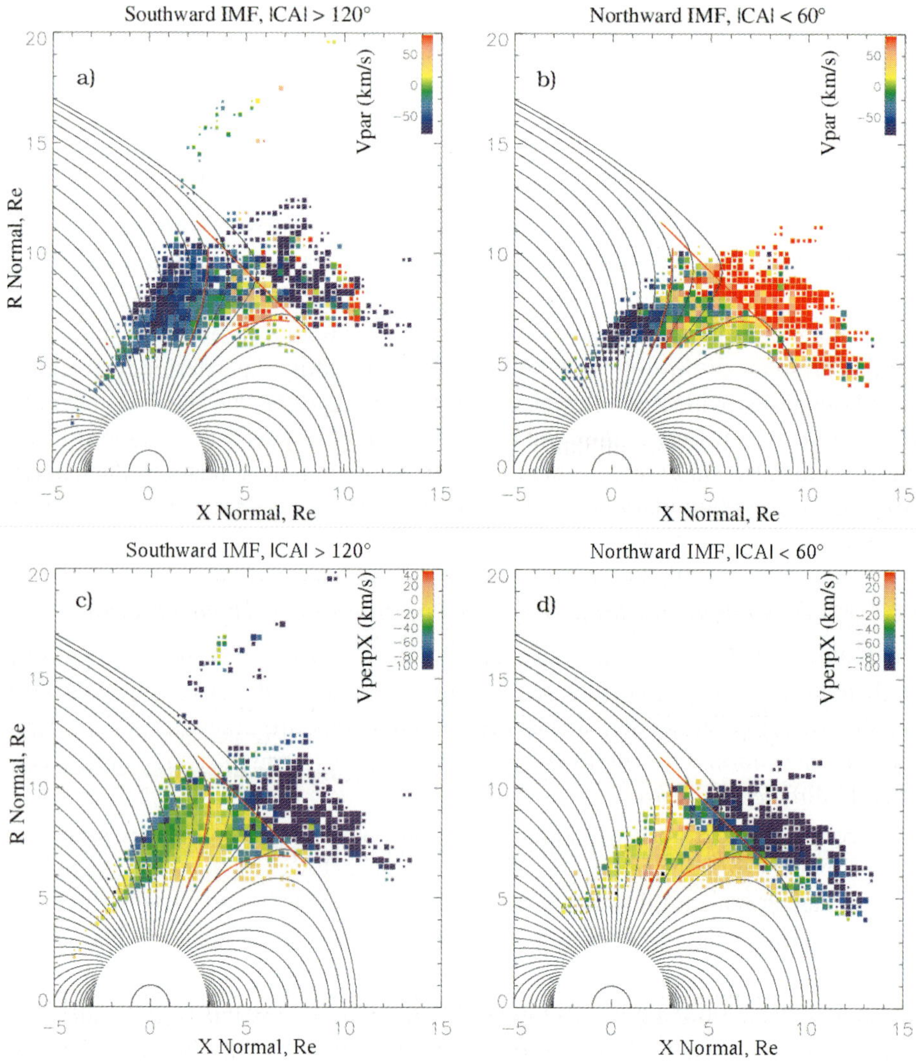

Figure 9.13. Spatial distributions of plasma flows in the high-altitude cusp region when IMF conditions are restricted to southward (IMF clock angle CA >120°) and northward (CA < 60°) orientations. (a) Field-aligned components of the plasma velocity for southward IMF. Its magnitude (in $\mathrm{km\,s^{-1}}$) is shown by the colour palette. (b) The parallel flow in the case of northward IMF. The lower panels show the spatial distribution of the X component of the perpendicular flow velocity for (c) southward IMF and (d) northward IMF. Other details are as in Figure 9.12. (Adapted from Lavraud et al., 2005).

2001; Le et al., 2001; Savin et al., 2004). Such fluctuations are important because they can lead to particle energisation and/or scattering. Electric field fluctuations

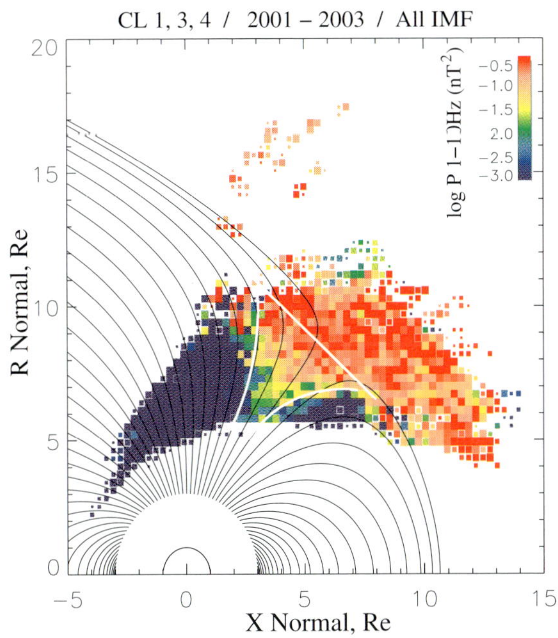

Figure 9.14. The spatial distribution of the magnetic wave power between 1 and 10 Hz measured by the STAFF experiment between 2001 and 2003. All IMF orientations are included. Other details may be found in Figure 9.12 caption. (Figure provided by N. Cornilleau-Wehrlin).

can also provide the localised diffusion needed to, for example, initiate and sustain the magnetic reconnection process.

A good summary of the extent of cusp magnetic field turbulence can be obtained from a statistical survey using the method discussed in Section 9.2.4. Figure 9.14 shows the magnetic wave power in the high-altitude cusp measured in the frequency range between 1 and 10 Hz by the Cluster STAFF experiment. High intensities are seen in both the magnetosheath and in the outer regions of the exterior cusp. A decreasing level of wave power is seen as one moves further into the cusp. The funnel-shape of wave power above $10^{-2.5}$ nT^2 resembles the distribution of both the density and magnetic field cavity shown in Figure 9.12. These distributions could either reflect the transport of waves into the cusp, or their in-situ generation, but to resolve this issue requires detailed case-study analysis of the waves.

A good example of the nature of cusp magnetic field turbulence can be found during the March 17, 2001 encounter discussed above. This shows bursty turbulence, an association of wave power with sheared plasma flows (see Figure 9.3), and the evidence of major power enhancement at the ion cyclotron frequency and its harmonics. Figure 9.15 shows the power in the magnetic field fluctuations in the

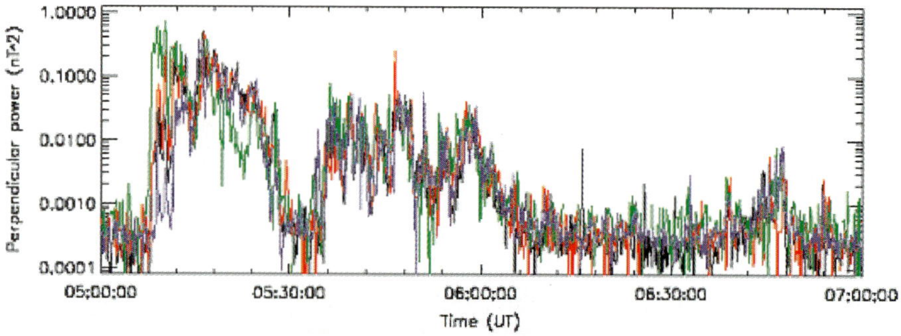

Figure 9.15. The power in perpendicular magnetic field fluctuations between 05:00 and 07:00 UT on March 17, 2001. The power is integrated over the frequency range 0.5 - 7 Hz. (Figure provided by K. Nykyri).

interval 05:00 - 07:00 UT as determined from the four FGM instruments (Nykyri et al., 2003, 2004). The level of wave power changes considerably as Cluster moves through the cusp, and varies by up to an order of magnitude between the spacecraft at some locations. At cusp entry the onset of the waves is extremely rapid, and can be directly linked to the sudden appearance of reconnection flows (Holland et al., 2004): Section 9.2.1). Indeed CIS data indicates that the power level is in general closely associated with the presence of field-aligned plasma flows. This is especially striking at 05:30 UT where the rapid drop-out and subsequent enhancement in wave power corresponds to a vanishing and subsequent enhancement of the bulk flows, despite the continued presence of magnetosheath plasma.

The detailed wave properties such as polarisation, propagation direction and correlation between spacecraft have also been analysed (Nykyri et al., 2004). An example of the power spectrum at four different times is shown in Figure 9.16. Particularly important points to note are the large differences in power between the spacecraft, the differences in the peak power at different times and the peaks in the spectrum at the first and second harmonics of the ion cyclotron frequency. Indeed, examination of STAFF data in the interval around 05:27 UT reveals up to five cyclotron harmonics between 1 and 10 Hz (Figure 9.17). A correlation analysis between the four spacecraft (which are separated by approximately 600 km) reveals no correlation between the various signals. A similar analysis has been carried out for two cusp crossings in 2002 (March 2 and March 9) and show broadly similar results, including sometimes a lack of correlation when the separation is only 100 km.

Whether these waves are generated remotely and convected to the spacecraft, or are generated in situ is unclear at present. Clearly the sheared bulk flows are a possible energy source for the waves, and there exists an extensive literature on the generation of ion cyclotron waves and harmonics in such a situation. However, the

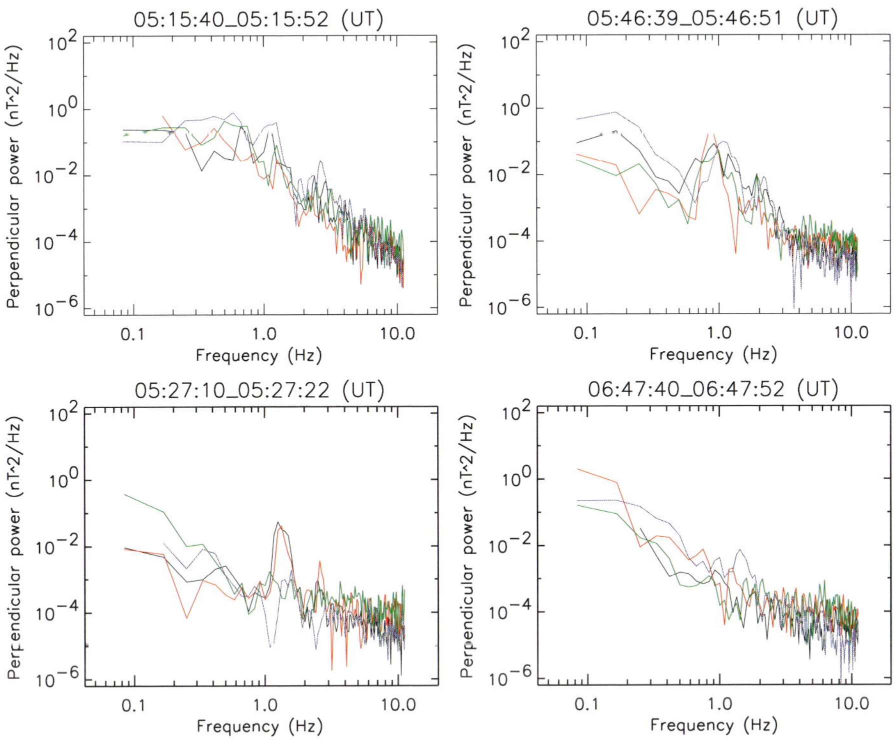

Figure 9.16. The power spectrum of perpendicular fluctuations in the frequency range 0.5 - 7 Hz from the four FGM instruments during four different intervals between 05:00 and 07:00 UT on March 17, 2001. (Figure provided by K. Nykyri).

low plasma-β would appear to favour the generation of electrostatic waves, so their electromagnetic nature needs to be explored further.

It is highly desirable to understand how the waves interact with the particles in the cusp, and the outbound crossing of March 23, 2002 provides an excellent laboratory for this. In early 2002 the spacecraft configuration had a 100 km separation which is better suited for application of the k-filtering method (see below). Initially there was a northward IMF and quiet solar wind conditions (a density of 3 cm^{-3} and velocity of 400 $km\,s^{-1}$). However, a strong interplanetary shock arrives at the approximate lagged time of 11:52 UT, once Cluster is already in the magnetosheath.

Figure 9.18 shows an overview of the ion properties observed by CIS during the crossing. The top panel shows the bulk velocity and the bottom panel the ion pitch angle spectra. Three different plasma regimes can be identified. First (until 10:55 UT), the plasma moves Earthward approximately parallel to the magnetic field. Later (until 11:20 UT), there is a disturbed plasma with the perpendicular

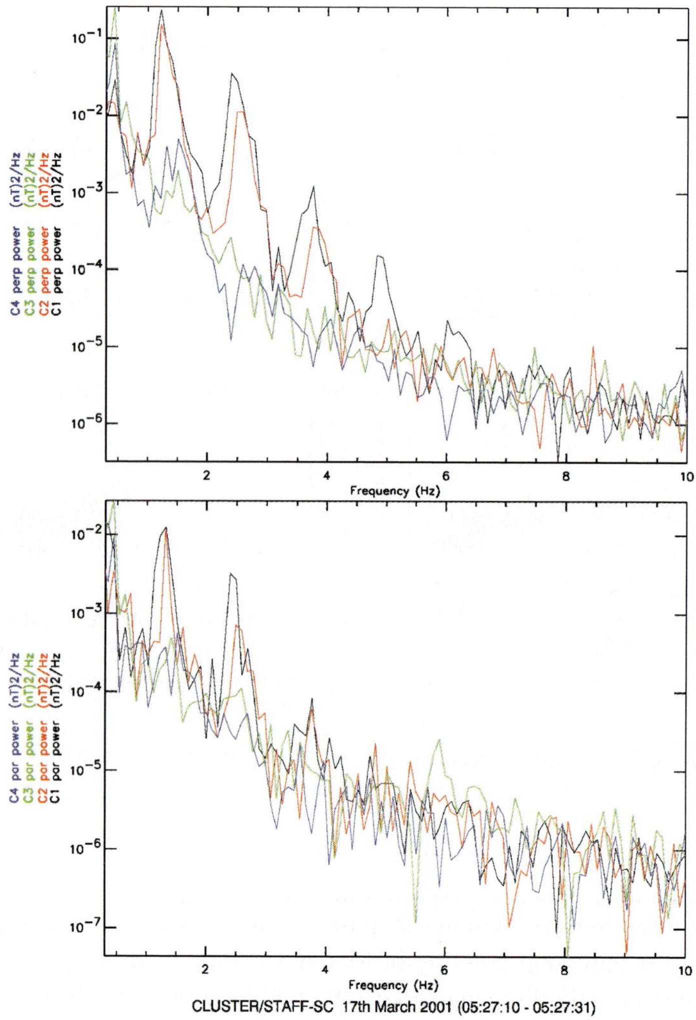

Figure 9.17. The power spectrum for magnetic field fluctuations obtained by the STAFF instrument in the frequency range 0.3 - 10 Hz between 05:27:10 and 05:27:31 UT on March 17, 2001. (Figure provided by N. Cornilleau-Wehrlin).

and parallel components of the flow fluctuating with nearly the same magnitude. Finally, between 11:20 and 11:40 UT the plasma is almost stagnant after which the spacecraft enter the magnetosheath.

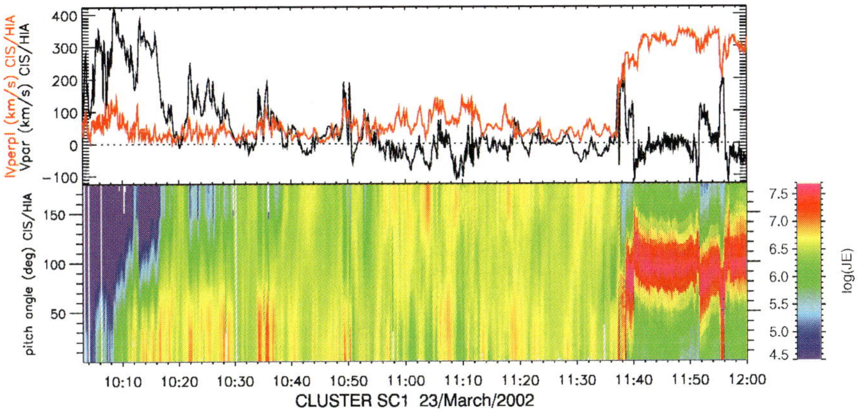

Figure 9.18. The bulk flow velocity and the ion pitch angle spectra (calculated in the spacecraft frame) measured by CIS/HIA instrument onboard spacecraft 1 between 10:03 and 12:00 on March 23, 2002. (Figure provided by B. Grison).

Figure 9.19 presents a more comprehensive picture of the first part of this crossing. The two top panels show the energy and pitch angle spectra of the ions from CIS and the bottom panel shows the power density spectra of the ULF magnetic waves measured by STAFF. Plasma injections are clearly observed, and are characterised by time-of-flight energy dispersions (top panel), pitch angle dispersions and a lack of upward particles (centre panel). The injections appear as purely temporal phenomena and the ULF electromagnetic wave bursts are observed to be associated with them. In distinction from March 17, 2001, the waveforms show a generally good correlation between the spacecraft, although there is occasional evidence for sub-100 km scales. There is also no clear wave polarisation, nor evidence of any monochromatic wave, as the frequency spectra have a power law behaviour.

Figure 9.20 focuses on the multiple injection events around 10:50 UT. The bulk flow velocity and the magnetic wave power between 1 and 10 Hz are shown in the top left and right corners of the figure, respectively. The bottom part of the figure shows cuts of the ion distribution functions at three selected times. This (multiple) injection event is sub-divided into 4 periods, as shown in the figure. During the first interval (A) the waves start to develop and the ion distribution function is mainly isotropic (not shown). An initial injection is then seen at about 10:49:15 UT (interval B and distribution (a)) as the wave activity reaches a maximum. The ions are scattered in velocity space as the wave activity starts to decrease (see the distribution function (b) at 10:50:03 UT). During the next interval (C), a second (less intense) injection is rapidly recorded, and the decrease of the wave activity slows down (see interval C in the top right-hand plot). In this interval, as the distribution function becomes more isotropic (distribution (c)), the wave activity

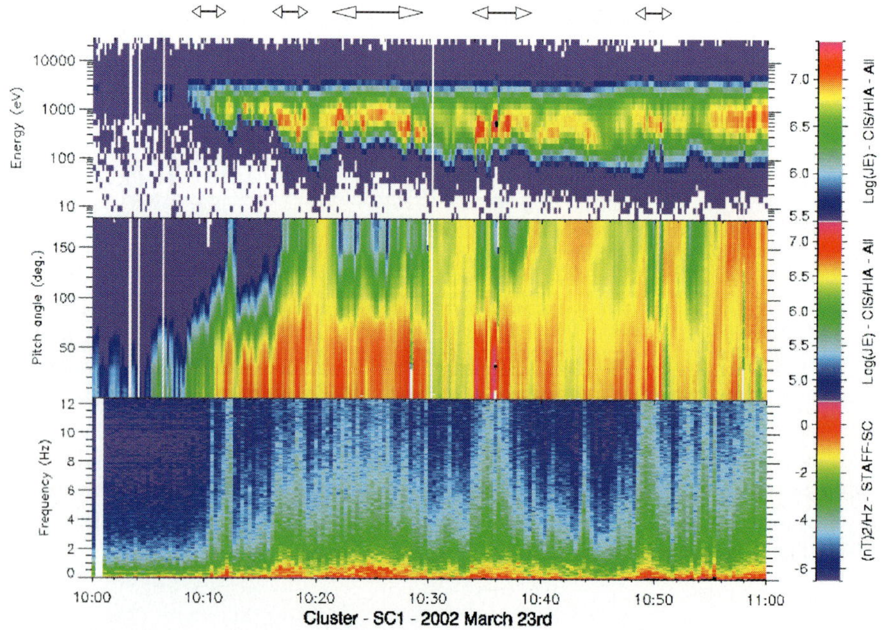

Figure 9.19. The ion energy spectra, the ion pitch angle spectra (calculated in the spacecraft frame) both measured by CIS/HIA, and the ULF magnetic wave spectra measured by STAFF-SC between 10:00 and 11:00 UT on March 23, 2002. The horizontal arrows at the top correspond to the location of injection events. (Figure provided by B. Grison).

decreases to a minimum. During the last interval (D), there is a slow scattering of downward-moving particles . The time of flight and pitch angle dispersion effects allow us to describe the evolution of the distribution function as being purely temporal. These observations thus show that wave activity increases simultaneously with the intensity of the injections and that the ion distribution functionis isotropised very quickly as the wave activity decreases. This isotropisation can be interpreted either by particle ballistic effects or by pitch angle diffusion by waves, or even by a combination of both. Further studies are needed to disentangle between these two effects.

Further information on these waves can be obtained by using the k-filtering technique (Pinçon and Lefeuvre, 1991; Pinçon and Motschmann, 1998), as adapted by Sahraoui et al. (2003) for an investigation of magnetosheath turbulence. The method requires an optimal tetrahedron geometry of the spacecraft as occurs here. Figure 9.21 shows the energy density for f_{sc} = 0.46 Hz ($\sim f_{ci}$) using an isocontour representation. The most intense peak is identified at k_z = -0.0011 rad km^{-1} and its wave vector makes an angle of 88° with respect to the ambient magnetic field. The corresponding frequency in the plasma frame ($f_{plas} = f_{sc} - k \cdot v$) is about 0.21 Hz.

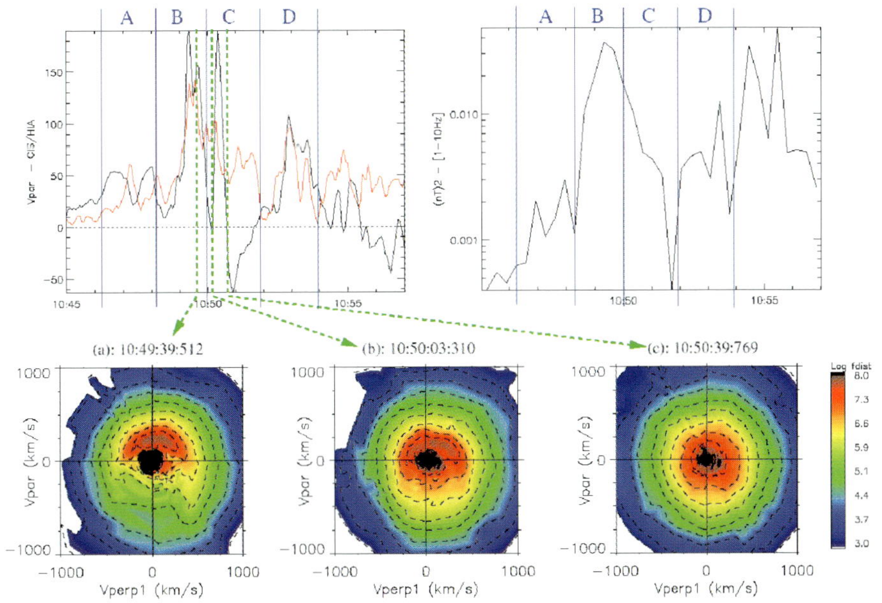

Figure 9.20. An example of a multiple ion injection event. The top two panels the bulk flow velocity (CIS/HIA) on the left, and the ULF magnetic wave power (STAFF-SC) on the right. Four intervals, A, B, C and D, are separated by vertical lines. Ion distribution functions are shown at the bottom of the figure for three selected times in the interval. The times are centered in the 12 s interval during which the distribution functions are measured. The distribution functions show cuts in the parallel-perpendicular plane. (Adopted from Grison et al., 2004).

Using plasma parameters measured by CIS, the WHAMP program (Rönmark, 1982) can be used to calculate the theoretical dispersion relations and it can be seen that the general form of the energy density spectrum of the peak is consistent with the dispersion curve of the Alfvén mode (over-plotted red lines). The identification of an Alfvén wave during this injection event is consistent with the observation of the strong perpendicular magnetic fluctuations mentioned above. Other results (not shown here) show also the presence of other weaker peaks, which seem to be linked to the Bernstein mode (Grison et al., 2004). This case study of a typical turbulent wave event in the cusp has allowed us to show the close link between wave amplitude and particles injections. Thanks to the small separation of the Cluster tetrahedron and the use of the *k* filtering method, a first identification of the wave modes has been possible. It is shown that due to Doppler effect, at a given frequency in the spacecraft frame, there is sometimes a superimposition of the dominant Alfvén mode together with Bernstein modes, that are at different frequencies in the plasma frame. Clearly, more events have to be studied to complement those first new results.

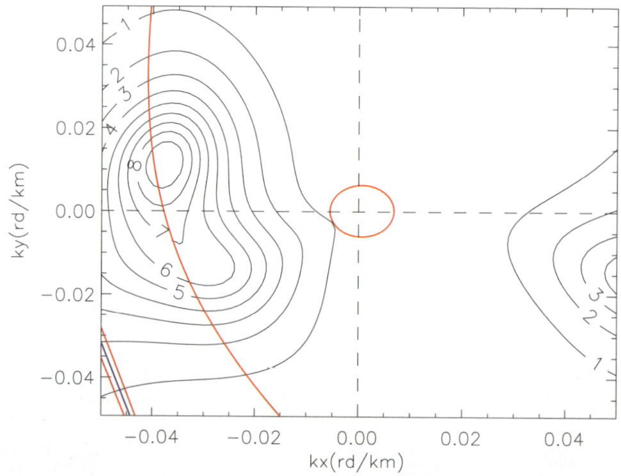

Figure 9.21. Superposition of the experimental magnetic energy density isocontour (thin black lines) and the theoretical dispersion relations of the low frequency modes (red thick lines) in the spacecraft frame for a frequency of 0.46 Hz. The blue line represents the Doppler shift $\omega = \mathbf{k} \cdot \mathbf{v}$ (or mirror mode). The main peak is consistent with an Alfvén wave of frequency 0.21 Hz in the plasma frame. (From Grison et al., 2004).

9.3 The Mid-Altitude Cusp

9.3.1 Structure: Case study

Cluster has had numerous encounters with the mid-altitude cusp during the Northern hemisphere summer months. Figure 9.22 shows the mid-altitude cusp encounter of August 30, 2001 (Escoubet et al., 2004). The upper two panels show the ion energy spectra observed by spacecraft 4 and 1 respectively, and the lowest panel shows the lagged IMF. The IMF is strongly southward until 15:40 UT, and then fluctuates around zero, while B_y is persistently negative. Between 15:00 UT and 15:30 UT spacecraft 4 crosses the dayside magnetosphere, characterised by ions above 5 keV located on closed field lines equatorward of the polar cusp. Spacecraft 4 then entered the polar cusp at 15:31:30 UT and detected the typical poleward ion dispersion (energy decrease as the spacecraft move poleward) observed when the IMF is southward. The dispersion lasts up to 16:00 UT. After that time a low flux of ions gradually decreases down to background levels in the plasma mantle. Starting at about 15:33:05 UT, spacecraft 1 observed a similar cusp structure.

This encounter is particularly interesting because of what is seen by spacecraft 3 (third panel on Figure 9.22) which entered the cusp at 16:18 UT and also observed an initial poleward dispersion that persisted until about 16:48 UT. A second poleward ion dispersion was then observed as the energy of the ions increased abruptly to a maximum of a few keV at 16:48 UT, subsequently decreasing slowly until 17:25 UT. The first dispersion observed by spacecraft 3 is very similar to those ob-

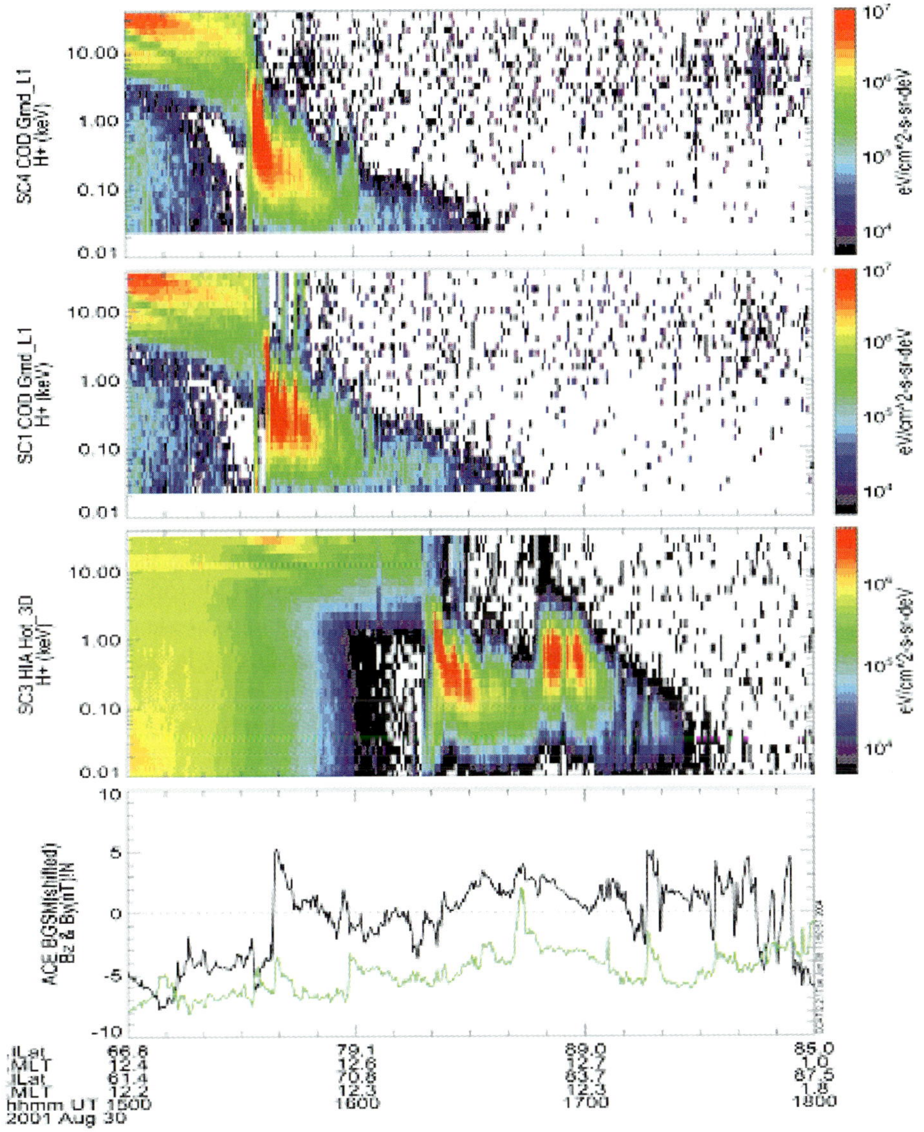

Figure 9.22. A single cusp observed by spacecraft 4 and 1, and a double cusp observed by spacecraft 3 on 30 Aug. 2001. The top three panels show the CIS ion energy spectra from spacecraft 4, 1 and 3 respectively. The last panel shows the IMF B_z (black) and B_y (green) components from ACE lagged by 63 min. The invariant latitude and magnetic local time of spacecraft 1 and 3 are given at the bottom. (From Escoubet et al., 2004).

served by Cluster-4 and Cluster-1, suggesting that the cusp did not change during the intervening 45 minutes. The fact that these dispersions are located at approximately the same invariant latitude (bottom of Figure 9.22) also suggests that the cusp was not moving significantly before 16:48 UT. On the other hand, the second dispersion observed by spacecraft 3 was not observed by the earlier crossing of Cluster-4 and Cluster-1.

The north/south IMF component was small during the first dispersion seen by spacecraft 3 but was northward during the second dispersion. DMSP satellite data, SuperDarn radar observations and a global MHD simulation model (none shown here) all support the idea that the cusp moved poleward due to the reconnection point moving from the dayside magnetosphere (when B_z was negative) to the dawn flanks (when B_z was slightly positive and B_y dominant). This would explain why the cusp was encountered twice by spacecraft 3 as it moves slowly poleward, as well as demonstrating the extreme sensitivity of the cusp properties to changes in the solar wind (Escoubet et al., 2004).

9.3.2 Structure: Statistical survey

A preliminary statistical study of the Cluster cusp crossings during the summer of 2002 has been conducted by Pitout et al. (2004). The cusp has been classified into five categories according to the shape of the precipitating ion distributions: (1) normal dispersion with energy decreasing with invariant latitude (Figure 9.22, upper panels), (2) reverse dispersion with energy increasing as latitude increases, (3) no dispersion, (4) irregular when the precipitation is highly variable with no clear dispersion and (5) discontinuous if made up of several dispersions (Figure 9.22, third panel). The distribution of cusp encounters as a function of IMF property is shown in Figure 9.23 and confirms that normal dispersions are occurring when the IMF is South, and reversed ones when the IMF is North. However these are not the majority of the crossings as irregular cusps occur about 50% of the time and discontinuous ones about 25%. Most of these crossings occurred for variable IMF direction.

9.3.3 Ionospheric ions

Cluster has been able to examine the nature and spatial extent of the heating region driving ionospheric ion outflow from the mid- and high-altitude cusps. Bogdanova et al. (2004a) showed that local transverse heating of O^+ ions may occur at 3.5-5 R_E in the 'electron-only cleft' region located equatorward of the main cusp. Here, electron injections were observed without any corresponding magnetosheath ions and there was also localised low-frequency magnetic field wave power with a broadband spectrum.

The four Cluster spacecraft also give multiple samples of the same section of the cusp with a few minute time delay between spacecraft. Bogdanova et al. (2004a) used this property to demonstrate that the location of the ion heating region in

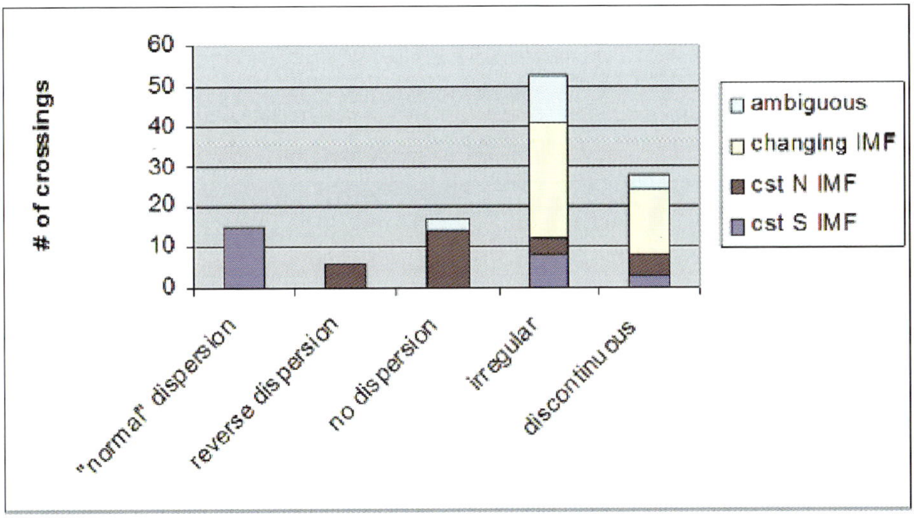

Figure 9.23. Distribution of ion signatures in the cusp observed by the Cluster spacecraft during summer 2002. (From Pitout et al., 2004).

the mid-altitude cusp coincides with anisotropic electron injections into the cleft. Figure 9.24 shows a typical example of oxygen outflows observed by spacecraft 4 as it crossed the mid-altitude cusp on August 23,2001. The open-closed field line boundary (OCB), defined by the earliest injection of magnetosheath-like electrons in the top panel, is marked by the first dashed vertical line. The oxygen outflow (panels 3 and 4) begins at this boundary and also correlates with a sharp enhancement in the electrostatic and electromagnetic wave power (panel 6).

The O^+ ions have a conic velocity distribution, with a significant perpendicular component, indicating the spacecraft has entered an ion heating region. Inside this region, the O^+ field-aligned outflow velocity (panel 5) increases from 0 km/s near the equatorward boundary of the heating region to 100-200 km s^{-1}(1 - 2 keV) at its most poleward extent (marked by the second dashed line). The extremely low frequency (ELF) electromagnetic waves are localised inside the ion heating region, while the ELF electrostatic waves are detected throughout the cleft/cusp/mantle regions, suggesting that the former are more likely responsible for the ion heating. As localised ion heating and electromagnetic waves were observed in electron-only cleft, the electrons must play an important role in wave generation.

Comparison of the data shown in Figure 9.24 with that from Cluster 1 (not shown), which crossed the same heating region 4 minutes later, shows that there is a correlation between ion outflow and the injected electron fluxes. Based on these observations, Bogdanova et al. (2004a) further suggested that many properties of the ion outflow, such as relative fluxes, location, longitudinal and latitudinal extent,

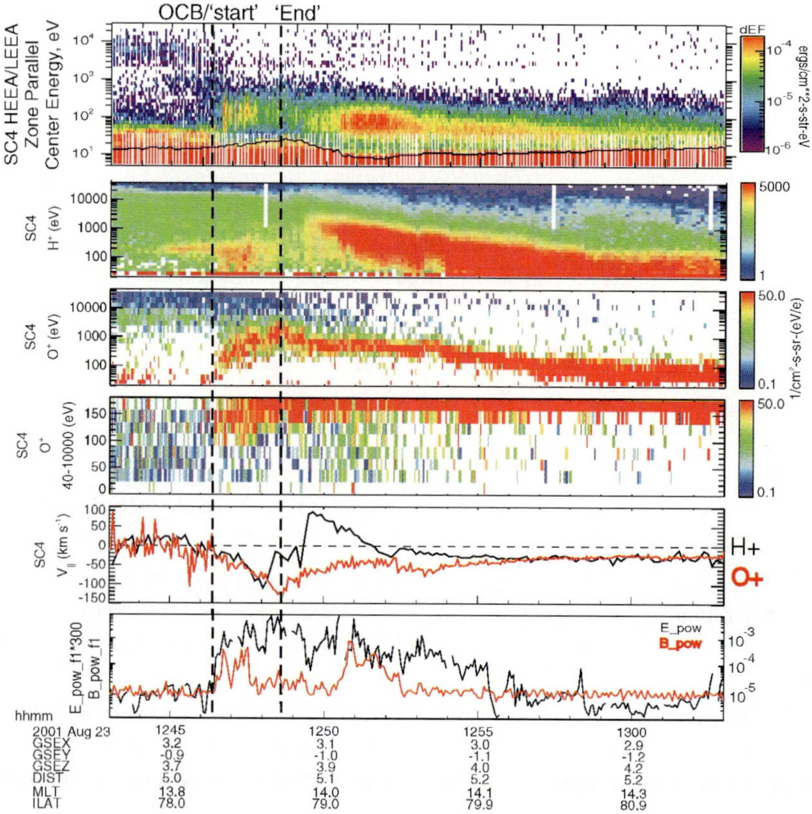

Figure 9.24. An example of oxygen outflow near the equatorward boundary of the mid-altitude cusp observed by Cluster 4 on 23 August 2001. From top to bottom the figure shows the energy-time spectrogram of the field-aligned (injected) electrons. [The black trace at ~10 eV marks the spacecraft potential.] The four middle panels show energy-time spectrograms of H^+ and O^+, the pitch-angle spectrogram of O^+ and the parallel velocity component of these ions, (H^+ in black and O^+ in red). The bottom panel shows the spectral power density of electric field fluctuations (increased by a factor 300 for comparison, black line) and magnetic field fluctuations (red line) in the frequency band 1-10 Hz. The first dashed line marks the Open-Closed Boundary which coincides with the equatorward boundary of the ion local perpendicular heating region. The second dashed line indicates the poleward boundary of this region. Between ~12:46 UT and ~12:49 UT the spacecraft observes localised O^+ heating associated with an electron-only injection into the cleft region and an increase in the broadband magnetic field wave power. (From Bogdanova et al., 2004a).

may be related to the properties of the electron injections from the reconnection site.

Cluster data has been used to put limits on the spatial extents of the cleft/cusp ion heating regions. Bogdanova et al. (2004c) presented a statistical study of 10 events with strong perpendicular energisation of O^+ ions within the localised heating region at altitudes of 4.5-6 R_E . For each event, the heating region is marked by a sharp enhancement of energy and parallel ion outflow velocity. This is followed by a long-lived oxygen beam showing gradual decline of energy and velocity in the cusp and mantle. The size and location of the ionospheric heating region was estimated by using the T89 and T96 Tsyganenko models to trace the Cluster locations corresponding to the equatorward and polarward edges of the heating region down to the level of the ionosphere. This demonstrates that the ion heating region is typically located near the equatorward boundary of the cleft/cusp region and that ion outflow occasionally begins equatorward of the observed magnetosheath proton injections. The region with strong perpendicular heating is very localised in latitude, with average width of 1.5°, but can extend to >27° in longitude (the greatest Cluster separation).

Cluster data can also be used to examine the small scale structure of the ion outflows. Bouhram et al. (2004) presented a statistical study of the coherence of observed ion outflows between different Cluster spacecraft for 18 events in the mid-altitude cusp between July and November 2001. A cross correlation analysis revealed significant variability in the transport patterns of the ion outflow. This lack of coherence is consistent with changes in the convection patterns, which in turn depend on the IMF magnitude and orientation. However, the total oxygen outflow rate observed at each spacecraft appears to be related to smaller scale processes. In addition, Nilsson et al. (2004) presented a case study of the oxygen outflow observed by two of the Cluster spacecraft in the high-altitude cusp at 5-9 R_E . At high altitude, oxygen outflow is seen as a beam with narrow energy spread, showing increasing field-aligned velocity and temperature. Cross-correlation analysis of the oxygen moments from the two spacecraft show that structures in ion density, temperature and field-aligned flow appear to drift with the observed field-perpendicular drift. Much of this structure can be also explained by transverse heating below the spacecraft.

9.3.4 Mid-altitude signatures of the LLBL

Cluster passes through the mid-altitude cusp have also enabled insight to be obtained into the properties of the low latitude boundary layer (LLBL), as observed at mid-altitudes. One study has involved estimating the temporal variation of the thickness of the LLBL as a function of IMF condition. Data was taken from the northern hemisphere passes and the study originally included 12 events (Bogdanova et al., 2004b), but only one event on August 5, 2002 is shown here.

The upper panels of Figure 9.25 show the energy-time spectrograms of the differential energy flux of 1 eV to 30 keV electrons moving parallel to magnetic field observed by the PEACE instruments on each spacecraft. The lower panels on the left show the three components of the lagged IMF from the ACE spacecraft in the GSM coordinate system while the lower right panel shows the position and motion of the spacecraft in the ILAT-MLT coordinate system for the 30 minutes after

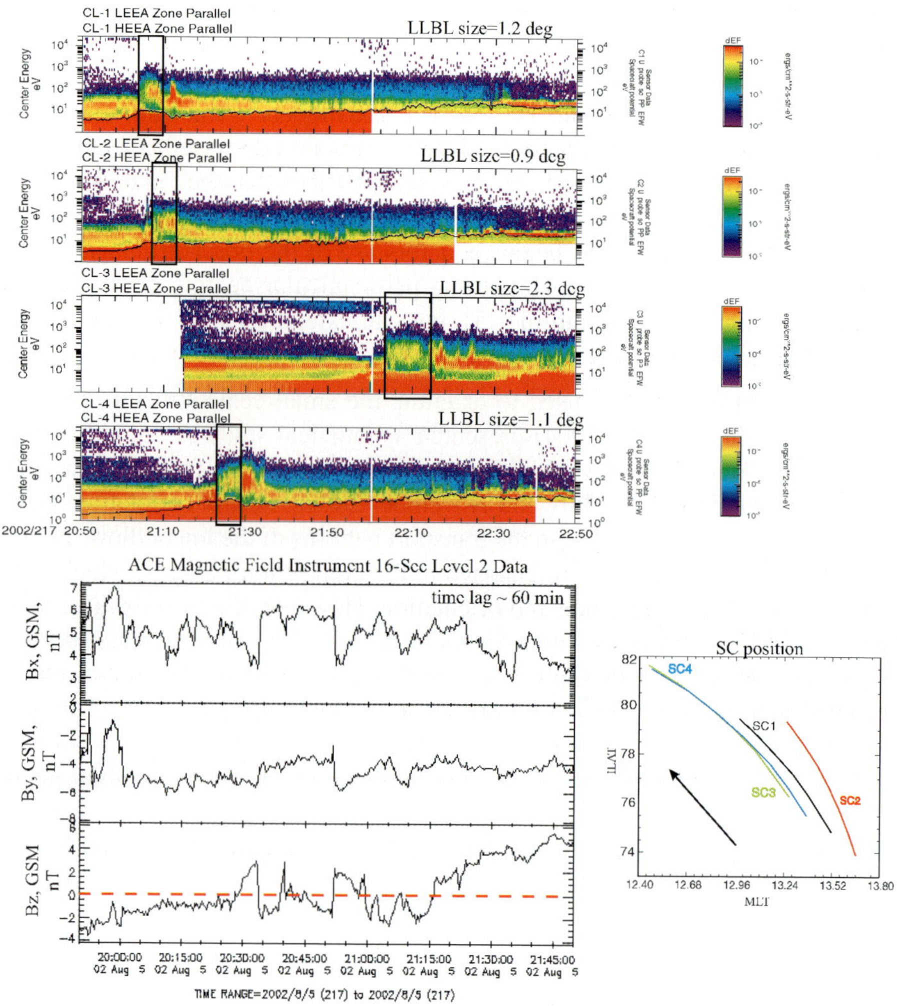

Figure 9.25. Four-spacecraft PEACE electron spectra for the mid-altitude cusp pass on August 5, 2002. The black trace in each of the upper panels indicates the spacecraft potential. The extent of the LLBL is marked by the black boxes in these panels. The three components of the lagged ACE IMF (lower left), and spacecraft position and motion (lower right) for the 30 minutes after the OCB crossing are also shown. (From Bogdanova et al., 2004c).

the crossing of the OCB. Note that Cluster-3 has almost the same trajectory as Cluster-4 but follows with a 40 minutes delay.

As seen by PEACE, we define the cleft/LLBL as having a spectral flux peak about 10 times lower than the cusp, with lower ion bulk flow velocities and higher electron temperatures (the cusp has T_e = 30 - 100 eV while the cleft/LLBL has T_e = 70 - 200 eV (Newell and Meng, 1988; Onsager and Lockwood, 1997)). In addition, the average LLBL density is approximately 20% that in the cusp. During this pass, spacecraft 1, 2 and 4 pass through a cleft region of latitudinal size in the range 0.9-1.2° (as measured from the open-closed boundary crossing to the cusp proper injections). However, Cluster 3, following some 40 minutes later, observes a much broader cleft/LLBL region which we estimate has a latitudinal extent of 2.3°. The IMF was rather unstable during the time between these observations and exhibits 3 transient northward turnings. It thus appears that the cleft/LLBL region expands under these conditions.

The statistical study of Bogdanova et al. (2004c) shows that thickness of the cleft/LLBL on open field lines remains approximately constant under relatively stable, but southward IMF. It generally increases under more variable conditions for which multiple reconnection events may be expected to occur. This study illustrates the potential of using Cluster multi-spacecraft mid-altitude measurements to provide new information on the temporal variations and spatial structure of the LLBL as a function of the IMF.

9.4 Discussion

In the early part of the mission, Cluster has made major advances in understanding the physics of the mid- and high-altitude cusp. Cluster has:

- demonstrated the fundamental role played by the IMF in determining cusp properties;
- characterised the global, large-scale plasma and field properties of the cusp;
- determined the location and dynamics of the various cusp boundaries;
- characterised the small-scale plasma and field microprocesses occurring in the cusp.

We now address these four results in turn.

The Role of the IMF: Cluster has demonstrated beyond any reasonable doubt that the strength and especially the orientation of the IMF determine what is seen in the cusp. Cluster has confirmed the expectation that subsolar and high-latitude (lobe) reconnection would be important for southward and northward IMF respectively. Cluster has also demonstrated the strong sensitivity of the cusp position to the y-component of the IMF, with very small changes leading to almost immediate motion of the cusp boundaries at mid- and high latitudes. This implies that the interaction of the solar wind and terrestrial field is a very 'sticky' process for all IMF directions, i.e., the coupling is effective.

Global Properties: Cluster has demonstrated the presence of solar wind plasma well inside the nominal magnetopause position for a wide range of IMF conditions.

In both cases, the cusp is populated by plasma entry via magnetic reconnection. As a result of that presence, the region is a diamagnetic cavity, most clearly observable in the exterior part of the region at high altitudes. However, for northward (southward) IMF, the cusp is often stagnant (convecting).

Cusp boundaries: The cusp has two distinct boundaries, one curved (or indented) boundary between it and the inner magnetospheric region, separating the region from the dayside magnetosphere at its equatorward edge and from the lobes at its poleward edge, and another between the cusp and the magnetosheath. It appears as if the cusp-magnetosphere boundary may best be described as a tangential discontinuity, and that between cusp and magnetosheath has some form of rotational discontinuity for both northward and southward IMF directions. For northward IMF it appears as if a plasma depletion layer exists outside the magnetopause, and may play an important role in sustaining quasi-steady lobe reconnection. The boundaries all show considerable dynamic behaviour. The question of indentation of the outer boundary is unclear. Some events show a tilted magnetopause, whereas the wider statistical survey is inconclusive.

Small-scale processes: The cusp has proved to be an excellent laboratory for studying plasma turbulence and wave-particle interactions. It has been demonstrated that magnetic turbulence in the vicinity of the ion cyclotron frequency is linked to plasma jets associated with lobe reconnection under northward IMF. The interaction of these waves with particles has also been strongly suggested. There is also evidence of electric and magnetic field turbulence at mid-altitudes.

This chapter has been limited to results from the first two years of the Cluster mission, implying spacecraft separations of approximately 600 and 100 km in 2001 and 2002 respectively. In the subsequent years of the mission, larger separations have been available, and one has been able to make estimates of the global extent of a single cusp encounter. However, the principal future work in the cusp must focus on the analysis of many crossings, as opposed to case studies as have been predominately discussed here.

Acknowledgements

Cluster research in the UK and France is supported by PPARC and CNES, respectively. PJC acknowledges support through a PPARC senior research fellowship and CJO through a PPARC advanced fellowship. Work at Los Alamos was conducted under the auspices of the US Department of Energy, with support from the NASA Cluster program. The contributions of CPE and TDP were supported by ESA and NASA, respectively.

References

André, M., G. B. Crew, W. K. Peterson, A. M. Persoon, and C. J. Pollock: 1990, 'Ion heating by broadband low-frequency waves in the cusp/cleft'. *J. Geophys. Res.* **95**(14), 20809–20823.

André, M., P. Norqvist, L. Andersson, L. Eliasson, A. I. Eriksson, L. Blomberg, R. E. Erlandson, and J. Waldemark: 1998, 'Ion energization mechanisms at 1700 km in the auroral region'. *J. Geophys. Res.* **103**(12), 4199–4222.

André, M. and A. Yau: 1997, 'Theories and Observations of Ion Energization and Outflow in the High Latitude Magnetosphere'. *Space Sci. Rev.* **80**, 27–48.

Avanov, L. A., S. A. Fuselier, and O. L. Vaisberg: 2001, 'High-latitude magnetic reconnection in sub-Alfvénic flow: Interball tail observations on May 29, 1996'. *J. Geophys. Res.* (15), 29491–29502.

Bahnsen, A., N. Dangelo, and A. M. Hansen: 1975, 'On double current layers in the polar cusp'. *J. Geophys. Res.* **80**(9), 201.

Bogdanova, Y. V., A. N. Fazakerley, C. J. Owen, B. Klecker, N. Cornilleau-Wehrlin, B. Grison, M. André, P. Cargill, H. Réme, J. M. Bosqued, L. M. Kistler, and A. Balogh: 2004a, 'Correlation between suprathermal electron bursts, BBELF waves and ion heating in the mid-altitude cleft/LLBL observed by Cluster'. *J. Geophys. Res.* **109**, accepted.

Bogdanova, Y. V., A. N. Fazakerley, C. J. Owen, B. Klecker, H. Réme, M. Dunlop, and A. Balogh: 2004b, 'Cleft/LLBL topology derived from Cluster multi-point measurements'. In: *IUGG 2003 Proceedings*, in preparation.

Bogdanova, Y. V., B. Klecker, G. Paschmann, L. M. Kistler, C. Mouikis, E. Moebius, H. Réme, J. M. Bosqued, I. Dandouras, J. A. Sauvaud, N. Cornilleau-Wehrlin, H. Laakso, A. Korth, M. B. Bavassano.Cattaneo, T. Phan, C. Carlson, G. Parks, J. P. McFadden, M. McCarthy, and R. Lundin: 2004c, 'Investigation of the source region of ionospheric oxygen outflow in the cleft/cusp using multi-spacecraft observations by CIS onboard Cluster'. *Adv. Space. Res.* **34**(11), 2459 2646.

Bosqued, J. M., J. A. Sauvaud, H. Reme, J. Crasnier, and I. I. Galperin: 1985, 'Evidence for ion energy dispersion in the polar cusp related to a northward-directed IMF'. *Adv. Space. Res.* **5**, 149–153.

Bouhram, M., B. Klecker, G. Paschmann, H. Réme, A. Blagau, L. Kistler, P. Puhl-Quinn, and J.-A. Sauvaud: 2004, 'Multipoint analysis of the spatio-temporal coherence of dayside O^+ outflows with Cluster'. *Ann. Geophys.* **22**, 2507–2514.

Cargill, P. J., M. W. Dunlop, B. Lavraud, R. C. Elphic, D. L. Holland, K. Nykyri, A. Balogh, I. M. Dandouras, and H. Réme: 2004, 'CLUSTER encounters with the high altitude cusp: boundary structure and magnetic field depletions'. *Ann. Geophys.* **22**, 1739–1754.

Chandler, M. O., S. A. Fuselier, M. Lockwood, and T. E. Moore: 1999, 'Evidence of component merging equatorward of the cusp'. *J. Geophys. Res.* **104**(13), 22623–22634.

Chapman, S. and V. Ferraro: 1930, 'A new theory of magnetic storms'. *Nature* **126**, 129–130.

Cowley, S. W. H., J. P. Morelli, and M. Lockwood: 1991, 'Dependence of convective flows and particle precipitation in the high-latitude dayside ionosphere on the X and Y components of the interplanetary magnetic field'. *J. Geophys. Res.* **96**(15), 5557–5564.

Dunlop, M. W., A. Balogh, and K.-H. Glassmeier: 2002, 'Four-point Cluster application of magnetic field analysis tools: The discontinuity analyzer'. *J. Geophys. Res.* **107**, 1385, doi:10.1029/2001JA005089.

Dunlop, M. W., P. J. Cargill, T. J. Stubbs, and P. Woolliams: 2000, 'The high-altitude cusps: HEOS 2'. *J. Geophys. Res.* **105**, 27509–27518.

Dunlop, M. W., B. Lavraud, P. J. Cargill, M. G. G. T. Taylor, A. Balogh, H. Réme, P. Décreau, K.-H. Glassmeier, R. C. Elphic, J.-M. Bosqued, A. Fazakerley, I. Dandouras, and A. Marchaudon: 2004, 'Cluster observations of the cusp: magnetic structure and dynamics'. *Surv. Geophys.* in press.

Eastman, T. E., S. A. Boardsen, S.-H. Chen, S. F. Fung, and R. L. Kessel: 2000, 'Configuration of high-latitude and high-altitude boundary layers'. *J. Geophys. Res.* (14), 23221–23238.

Escoubet, C. P., J. M. Bosqued, J. Berchem, J. Baker, P. C. Anderson, M. Fehringer, H. Laakso, M. Taylor, M. Dunlop, H. Réme, Y. Bogdanova, B. Klecker, P. Puhl-Quinn, A. Fazakerley, J. C. Cerisier, R. Greenwald, and K. J. Trattner: 2004, 'Doule cusp observed by Cluster on 30 August 2001'. *Ann. Geophys.* submitted.

Escoubet, C. P., M. F. Smith, S. F. Fung, P. C. Anderson, R. A. Hoffman, E. M. Basinska, and J. M. Bosqued: 1992, 'Staircase ion signature in the polar cusp - A case study'. *Geophys. Res. Lett.* **19**, 1735–1738.

Frank, L. A.: 1971, 'Plasma in the Earth's polar magnetosphere'. *J. Geophys. Res.* **76**, 5202.

Frey, H. U., T. D. Phan, S. A. Fuselier, and S. B. Mende: 2003, 'Continuous magnetic reconnection at Earth's magnetopause'. *Nature* **426**, 533–537.

Fuselier, S. A., K. J. Trattner, and S. M. Petrinec: 2000, 'Cusp observations of high- and low-latitude reconnection for northward interplanetary magnetic field'. *J. Geophys. Res.* (14), 253–266.

Gosling, J. T., M. F. Thomsen, S. J. Bame, R. C. Elphic, and C. T. Russell: 1990, 'Plasma flow reversals at the dayside magnetopause and the origin of asymmetric polar cap convection'. *J. Geophys. Res.* **95**(14), 8073–8084.

Gosling, J. T., M. F. Thomsen, S. J. Bame, R. C. Elphic, and C. T. Russell: 1991, 'Observations of reconnection of interplanetary and lobe magnetic field lines at the high-latitude magnetopause'. *J. Geophys. Res.* **96**(15), 14097.

Grison, B., F. Sahraoui, B. Lavraud, T. Chust, N. Cornilleau-Wehrlin, H. Réme, and A. Balogh: 2004, 'Wave particle interactions in the distant polar cusp: A Cluster case study'. *Ann. Geophys.* submitted.

Gurnett, D. A. and L. A. Frank: 1978, 'Plasma waves in the polar cusp - Observations from Hawkeye 1'. *J. Geophys. Res.* **83**(12), 1447–1462.

Haerendel, G.: 1978, 'Microscopic plasma processes related to reconnection'. *J. Atmos. Terr. Phys.* **40**, 343–353.

Haerendel, G., G. Paschmann, N. Sckopke, and H. Rosenbauer: 1978, 'The frontside boundary layer of the magnetosphere and the problem of reconnection'. *J. Geophys. Res.* **83**(12), 3195–3216.

Heikkila, W. J. and J. D. Winningham: 1971, 'Penetration of magnetosheath plasma to low altitudes through the dayside magnetospheric cusps'. *J. Geophys. Res.* **76**, 883.

Holland, D. L., K. Nykyri, P. J. Cargill, M. W. Dunlop, B. Lavraud, E. A. Lucek, T. S. Horbury, A. Balogh, I. Dandouras, H. Réme, Y. Bogdanova, M. G. G. T. Taylor, and A. Fazakerley: 2004, 'The cusp during a period of weak northward IMF: The Cluster encounter on March 17 2001'. *Ann. Geophys.* submitted.

Kessel, R. L., S.-H. Chen, J. L. Green, S. F. Fung, S. A. Boardsen, L. C. Tan, T. E. Eastman, J. D. Craven, and L. A. Frank: 1996, 'Evidence of high-latitude reconnecting during northward IMF: Hawkeye observations'. *Geophys. Res. Lett.* **23**, 583–586.

Lavraud, B. and P. J. Cargill: 2005, 'Cluster reveals magnetospheric cusps'. *Astronomy and Geophysics* **46**, in press.

Lavraud, B., M. W. Dunlop, T. D. Phan, H. Rème, J.-M. Bosqued, I. Dandouras, J.-A. Sauvaud, R. Lundin, M. G. G. T. Taylor, P. J. Cargill, C. Mazelle, C. P. Escoubet, C. W. Carlson, J. P. McFadden, G. K. Parks, E. Moebius, L. M. Kistler, M.-B. Bavassano-Cattaneo, A. Korth, B. Klecker, and A. Balogh: 2002, 'Cluster observations of the exterior cusp and its surrounding boundaries under northward IMF'. *Geophys. Res. Lett.* **29**, 56–1.

Lavraud, B., A. Fedorov, E. Budnik, A. Grigoriev, P. J. Cargill, M. W. Dunlop, H. Rème, I. Dandouras, and A. Balogh: 2004a, 'Cluster survey of the high-altitude cusp properties: A three-year statistical study'. *Ann. Geophys.* **22**(8), 3009–3019.

Lavraud, B., A. Fedorov, E. Budnik, M. F. Thomsen, A. Grigoriev, P. J. Cargill, M. W. Dunlop, H. Rème, I. Dandouras, and A. Balogh: 2005, 'High-altitude cusp flow dependence on IMF orientation: A three-year Cluster statistical study'. *J. Geophys. Res.* **110**, in press.

Lavraud, B., T. D. Phan, M. W. Dunlop, M. G. G. T. Taylor, P. J. Cargill, J.-M. Bosqued, I. Dandouras, H. Rème, J.-A. Sauvaud, C. P. Escoubet, A. Balogh, and A. Fazakerley: 2004b, 'The exterior cusp and its boundary with the magnetosheath under northward IMF: Cluster multi-event analysis'. *Ann. Geophys.* **22**(8), 3039–3054.

Lavraud, B., H. Rème, M. W. Dunlop, J. M. Bosqued, I. Dandouras, J.-A. Sauvaud, A. Keiling, T. D. Phan, R. Lundin, P. J. Cargill, C. P. Escoubet, C. W. Carlson, J. P. McFadden, G. K. Parks, E. Moebius, L. M. Kistler, E. Amata, M.-B. Bavassano-Cattaneo, A. Korth, B. Klecker, and A. Balogh: 2004c, 'Cluster observes the high-altitude cusp region'. *Surv. Geophys.* in press.

Le, G., X. Blanco-Cano, C. T. Russell, X.-W. Zhou, F. Mozer, K. J. Trattner, S. A. Fuselier, and B. J. Anderson: 2001, 'Electromagnetic ion cyclotron waves in the high altitude cusp: Polar observations'. *J. Geophys. Res.* (15), 19067–19080.

Lockwood, M., S. W. H. Cowley, P. E. Sandholt, and U. P. Løvhaug: 1995, 'Causes of plasma flow bursts and dayside auroral transients: An evaluation of two models invoking reconnection pulses and changes in the Y component of the magnetosheath field'. *J. Geophys. Res.* **100**(9), 7613–7626.

Lockwood, M. and M. F. Smith: 1992, 'The variation of reconnection rate at the dayside magnetopause and cusp ion precipitation'. *J. Geophys. Res.* **97**(16), 14841.

Lockwood, M. and M. F. Smith: 1994, 'Low and middle altitude cusp particle signatures for general magnetopause reconnection rate variations. 1: Theory'. *J. Geophys. Res.* **99**, 8531–8553.

Lockwood, M., J. H. Waite, T. E. Moore, C. R. Chappell, and J. F. E. Johnson: 1985, 'A new source of suprathermal O+ ions near the dayside polar cap boundary'. *J. Geophys. Res.* **90**(9), 4099–4116.

Marchaudon, A., J. Cerisier, J. Bosqued, M. Dunlop, J. Wild, P. Décréau, M. Förster, D. Fontaine, and H. Laakso: 2004, 'Transient plasma injections in the dayside magnetosphere: one-to-one correlated observations by Cluster and SuperDARN'. *Ann. Geophys.* **22**, 141–158.

Moore, T. E., R. Lundin, D. Alcayde, M. André, S. B. Ganguli, M. Temerin, and A. Yau: 1999, 'Chapter 2-Source Processes in the High-Latitude Ionosphere'. *Space Sci. Rev.* **88**, 7–84.

Moore, T. E., C. J. Pollock, R. L. Arnoldy, and P. M. Kintner: 1986, 'Preferential O(+) heating in the topside ionosphere'. *Geophys. Res. Lett.* **13**, 901–904.

Newell, P. T. and C. Meng: 1988, 'The cusp and the cleft/boundary layer - Low-altitude identification and statistical local time variation'. *J. Geophys. Res.* **93**(12), 14549–14556.

Newell, P. T. and C. Meng: 1994, 'Comment on "Unexpected features of the ion precipitation in the so-called cleft/low-latitude boundary layer region: Association with sunward convection and occurrence on open field lines" by A. Nishida, T. Mukai, H. Hayakawa, A. Matsuoka, K. Tsuruda, N. Kaya, and H. Fukunishi'. *J. Geophys. Res.* **99**(18), 19609–19614.

Newell, P. T., C. Meng, D. G. Sibeck, and R. Lepping: 1989, 'Some low-altitude cusp dependencies on the interplanetary magnetic field'. *J. Geophys. Res.* **94**(13), 8921–8927.

Nilsson, H., S. Joko, R. Lundin, H. Réme, J.-A. Sauvaud, I. Dandouras, A. Balogh, L. M. Carr, L. M. Kistler, B. Klecker, C. W. Carlson, M. B. Bavassano-Cattaneo, and A. Korth: 2004, 'The structure of high altitude O^+ energization and outflow: a case study'. *Ann. Geophys.* **22**, 2497–2506.

Norqvist, P., T. Oscarsson, M. André, and L. Blomberg: 1998, 'Isotropic and perpendicular energization of oxygen ions at energies below 1 eV'. *J. Geophys. Res.* **103**(12), 4223–4240.

Nykyri, K., P. J. Cargill, E. Lucek, T. Horbury, B. Lavraud, A. Balogh, M. W. Dunlop, Y. Bogdanova, A. Fazakerley, I. M. Dandouras, and H. Réme: 2004, 'Cluster observations of magnetic field fluctuations in the high-altitude cusp'. *Ann. Geophys.* **22**(7), 2413–2429.

Nykyri, K., P. J. Cargill, E. A. Lucek, T. S. Horbury, A. Balogh, B. Lavraud, I. Dandouras, and H. Réme: 2003, 'Ion cyclotron waves in the high altitude cusp: CLUSTER observations at varying spacecraft separations'. *Geophys. Res. Lett.* **30**, 2263, doi:10.1029/2003GL018594.

Øieroset, M., P. E. Sandholt, W. F. Denig, and S. W. H. Cowley: 1997, 'Northward interplanetary magnetic field cusp aurora and high-latitude magnetopause reconnection'. *J. Geophys. Res.* **102**(11), 11349–11362.

Onsager, T. G., C. A. Kletzing, J. B. Austin, and H. Mackiernan: 1993, 'Model of magnetosheath plasma in the magnetosphere - Cusp and mantle particles at low-altitudes'. *Geophys. Res. Lett.* **20**, 479–482.

Onsager, T. G. and M. Lockwood: 1997, 'High-Latitude Particle Precipitation and its Relationship to Magnetospheric Source Regions'. *Space Sci. Rev.* **80**, 77–107.

Paschmann, G., G. Haerendel, N. Sckopke, H. Rosenbauer, and P. C. Hedgecock: 1976, 'Plasma and magnetic field characteristics of the distant polar cusp near local noon - The entry layer'. *J. Geophys. Res.* **81**(10), 2883–2899.

Petrinec, S. M., K. J. Trattner, and S. A. Fuselier: 2003, 'Steady reconnection during intervals of northward IMF: Implications for magnetosheath properties'. *J. Geophys. Res.* **108**, 1458, doi:10.1029/2003JA009979.

Phan, T., H. U. Frey, S. Frey, L. Peticolas, S. Fuselier, C. Carlson, H. Rème, J.-M. Bosqued, A. Balogh, M. Dunlop, L. Kistler, C. Mouikis, I. Dandouras, J.-A. Sauvaud, S. Mende, J. McFadden, G. Parks, E. Moebius, B. Klecker, G. Paschmann, M. Fujimoto, S. Petrinec, M. F. Marcucci, A. Korth, and R. Lundin: 2003, 'Simultaneous Cluster and IMAGE observations of cusp reconnection and auroral proton spot for northward IMF'. *Geophys. Res. Lett.* **30**, 1509, doi:10.1029/2003GL016885.

Phan, T. D., C. P. Escoubet, L. Rezeau, R. Treumann, A. Vaivads, G. Paschmann, S. A. Fuselier, D. Attié, B. Rogers, and B. U. Ö. Sonnerup: 2005, 'Magnetopause Processes'. *Space Sci. Rev.* **this issue**.

Phan, T.-D., G. Paschmann, W. Baumjohann, N. Sckopke, and H. Luehr: 1994, 'The magnetosheath region adjacent to the dayside magnetopause: AMPTE/IRM observations'. *J. Geophys. Res.* **99**(18), 121–141.

Pickett, J. S., J. R. Franz, J. D. Scudder, J. D. Menietti, D. A. Gurnett, G. B. Hospodarsky, R. M. Braunger, P. M. Kintner, and W. S. Kurth: 2001, 'Plasma waves observed in the cusp turbulent boundary layer: An analysis of high time resolution wave and particle measurements from the Polar spacecraft'. *J. Geophys. Res.* **106**, 19081–19100.

Pickett, J. S., J. D. Menietti, J. H. Dowell, D. A. Gurnett, and J. D. Scudder: 1999, 'Polar spacecraft observations of the turbulent outer cusp/magnetopause boundary layer of Earth'. *Nonlinear Processes in Geophysics* **6**, 195–204.

Pinçon, J. L. and F. Lefeuvre: 1991, 'Local characterization of homogeneous turbulence in a space plasma from simultaneous measurements of field components at several points in space'. *J. Geophys. Res.* **96**(15), 1789–1802.

Pinçon, J. L. and U. Motschmann: 1998, 'Multi-Spacecraft Filtering: General Framework'. In: G. Paschmann and P. W. Daly (eds.): *Analysis Methods for Multi-Spacecraft Data*, ISSI SR-001. ESA Publications Division, pp. 65–78.

Pitout, F., C. P. Escoubet, H. Rème, and A. Balogh: 2004, 'Cluster survey of the mid-altitude cusp: morphology and plasma properties'. *Ann. Geophys.* submitted.

Pitout, F., P. T. Newell, and S. C. Buchert: 2002, 'Simultaneous high- and low-latitude reconnection: ESR and DMSP observations'. *Ann. Geophys.* **20**, 1311–1320.

Pottelette, R., M. Malingre, N. Dubouloz, B. Aparicio, and R. Lundin: 1990, 'High-frequency waves in the cusp/cleft regions'. *J. Geophys. Res.* **95**(14), 5957–5971.

Reiff, P. H., J. L. Burch, and T. W. Hill: 1977, 'Solar wind plasma injection at the dayside magnetospheric cusp'. *J. Geophys. Res.* **82**, 479–491.

Reiff, P. H., R. W. Spiro, and J. L. Burch: 1980, 'Cusp proton signatures and the interplanetary magnetic field'. *J. Geophys. Res.* **85**, 5997–6005.

Rönmark, K.: 1982, 'WHAMP: Waves in homogeneous anisotropic multicomponent plasmas'. Report 179, Kiruna Geophys. Inst., Kiruna, Sweden.

Rosenbauer, H., H. Gruenwaldt, M. D. Montgomery, G. Paschmann, and N. Sckopke: 1975, 'Heos 2 plasma observations in the distant polar magnetosphere - The plasma mantle'. *J. Geophys. Res.* **80**(9), 2723–2737.

Russell, C. T., G. Le, and S. M. Petrinec: 2000, 'Cusp observations of high-and low-latitude reconnection for northward IMF: An alternative view'. *J. Geophys. Res.* (14), 5489–5496.

Sahraoui, F., J. L. Pinçon, G. Belmont, L. Rezeau, N. Cornilleau-Wehrlin, P. Robert, L. Mellul, J. M. Bosqued, A. Balogh, P. Canu, and G. Chanteur: 2003, 'ULF wave identification in the magnetosheath: The k-filtering technique applied to Cluster II data'. *J. Geophys. Res.* **108**, 1335, doi:10.1029/2002JA009587.

Sandholt, P. E., C. J. Farrugia, S. W. H. Cowley, and M. Lester: 2001, 'Dayside auroral bifurcation sequence during B_y-dominated interplanetary magnetic field: Relationship with merging and lobe convection cells'. *J. Geophys. Res.* (15), 15429–15444.

Sandholt, P. E., C. J. Farrugia, J. Moen, S. W. H. Cowley, and B. Lybekk: 1998, 'Dynamics of the aurora and associated convection currents during a cusp bifurcation event'. *Geophys. Res. Lett.* **25**, 4313–4316.

Savin, S., L. Zelenyi, S. Romanov, I. Sandahl, J. Pickett, E. Amata, L. Avanov, J. Blecki, E. Budnik, J. Büchner, C. Cattell, G. Consolini, J. Fedder, S. Fuselier, H. Kawano, S. Klimov, V. Korepanov, D. Lagoutte, F. Marcucci, M. Mogilevsky, Z. Nemecek, B. Nikutowski, M. Nozdrachev, M. Parrot, J. Rauch, V. Romanov, T. Romantsova, C. Russell, J. Safrankova, J. Sauvaud, A. Skalsky, V. Smirnov, K. Stasiewicz, J. Trotignon, and Y. Yermolaev: 2004, 'Magnetosheath-cusp interface'. *Ann. Geophys.* **22**, 183–212.

Shelley, E. G., R. D. Sharp, and R. G. Johnson: 1976, 'He^{++} and H^{+} flux measurements in the day side cusp - Estimates of convection electric field'. *J. Geophys. Res.* **81**, 2363–2370.

Shue, J.-H., J. K. Chao, H. C. Fu, C. T. Russell, P. Song, K. K. Khurana, and H. J. Singer: 1997, 'A new functional form to study the solar wind control of the magnetopause size and shape'. *J. Geophys. Res.* **102**(11), 9497–9512.

Smith, M. F. and M. Lockwood: 1990, 'The pulsating cusp'. *Geophys. Res. Lett.* **17**, 1069–1072.

Smith, M. F. and M. Lockwood: 1996, 'Earth's magnetospheric cusps'. *Reviews of Geophysics* **34**, 233–260.

Spreiter, J. R., A. L. Summers, and A. Y. Alksne: 1966, 'Hydromagnetic flow around the magnetosphere'. *Planet. Space Sci.* **14**, 223.

Taylor, M. G. G. T. and P. J. Cargill: 2002, 'A magnetohydrodynamic model of plasma flow in the high-altitude cusp'. *J. Geophys. Res.* **107**, 1084, doi:10.1029/2001JA900159.

Trattner, K. J., S. A. Fuselier, W. K. Peterson, and C. W. Carlson: 2002, 'Spatial features observed in the cusp under steady solar wind conditions'. *J. Geophys. Res.* **107**, 1288, doi:10.1029/2001JA000262.

Trattner, K. J., S. A. Fuselier, W. K. Peterson, J.-A. Sauvaud, H. Stenuit, N. Dubouloz, and R. A. Kovrazhkin: 1999, 'On spatial and temporal structures in the cusp'. *J. Geophys. Res.* **104**(13), 28411.

Tsyganenko, N. A.: 1989, 'A magnetospheric magnetic field model with a warped tail current sheet'. *Planet. Space Sci.* **37**, 5–20.

Tsyganenko, N. A. and D. P. Stern: 1996, 'Modeling the global magnetic field of the large-scale Birkeland current systems'. *J. Geophys. Res.* **101**(10), 27187–27198.

Vaith, H., G. Paschmann, J. M. Quinn, M. Förster, E. Georgescu, S. E. Haaland, B. Klecker, C. A. Kletzing, P. A. Puhl-Quinn, H. Réme, and R. B. Torbert: 2004, 'Plasma convection across the polar cap, plasma mantle and cusp: Cluster EDI observations'. *Ann. Geophys.* **22**, 2451–2461.

Vasyliunas, V. M.: 1995, 'Multiple-branch model of the open magnetopause'. *Geophys. Res. Lett.* **22**, 1145–1147.

Vontrat-Reberac, A., J. M. Bosqued, M. G. G. T. Taylor, B. Lavraud, D. Fontaine, M. W. Dunlop, H. Laakso, N. Cornilleau-Werhlin, P. Canu, and A. Fazakerley: 2003, 'Cluster observations of the high-altitude cusp for northward interplanetary magnetic field: A case study'. *J. Geophys. Res.* **108**, 1346, doi:10.1029/2002JA009717.

Walters, G. K.: 1966, 'On the Existence of a Second Standing Shock Wave Attached to the Magnetosphere'. *J. Geophys. Res.* **71**(10), 1341.

Wing, S., P. T. Newell, and J. M. Ruohoniemi: 2001, 'Double cusp: Model prediction and observational verification'. *J. Geophys. Res.* **106**(15), 25571–25594.

Woch, J. and R. Lundin: 1992a, 'Erratum: "Magnetosheath plasma precipitation in the polar cusp and its control by the interplanetary magnetic field"'. *J. Geophys. Res.* **97**(16), 3225–3225.

Woch, J. and R. Lundin: 1992b, 'Magnetosheath plasma precipitation in the polar cusp and its control by the interplanetary magnetic field'. *J. Geophys. Res.* **97**(16), 1421–1430.

Yamauchi, M. and R. Lundin: 2001, 'Comparison of various cusp models with high- and low-resolution observations'. *Space Sci. Rev.* **95**, 457–468.

Yamauchi, M., H. Nilsson, L. Eliasson, O. Norberg, M. Boehm, J. H. Clemmons, R. P. Lepping, L. Blomberg, S.-I. Ohtani, T. Yamamoto, T. Mukai, T. Terasawa, and S. Kokubun: 1996, 'Dynamic response of the cusp morphology to the solar wind: A case study during passage of the solar wind plasma cloud on February 21, 1994'. *J. Geophys. Res.* **101**(10), 24675–24688.

Zhou, X. W. and C. T. Russell: 1997, 'The location of the high-latitude polar cusp and the shape of the surrounding magnetopause'. *J. Geophys. Res.* **102**(11), 105–110.

Zhou, X.-W., C. T. Russell, G. Le, S. A. Fuselier, and J. D. Scudder: 1999, 'The polar cusp location and its dependence on dipole tilt'. *Geophys. Res. Lett.* **26**, 429–432.

Chapter 10

Magnetopause Processes

T. D. Phan[1], C. P. Escoubet[2], L. Rezeau[3], R. A. Treumann[4], A. Vaivads[5], G. Paschmann[6], S. A. Fuselier[7], D. Attié[8], B. Rogers[9], and B. U. Ö. Sonnerup[10]

In its simplest description, the magnetopause is an impenetrable boundary that separates the solar wind from Earth's magnetosphere. The well-established existence of plasma of dominantly solar wind origin inside the magnetosphere does, however, imply that solar wind plasma is able to penetrate the magnetopause. Establishing the mechanisms by which this occurs is one of the greatest challenges of magnetospheric physics. Candidate processes include magnetic reconnection, the Kelvin-Helmholtz instability, diffusion, and impulsive penetration. In this chapter we provide highlights of Cluster's achievements in understanding some of these processes during the first three years of its mission. In Sections 10.1 and 10.2 we describe large-scale characteristics of magnetic reconnection and the Kelvin-Helmholtz instability at the magnetopause. In Section 10.3 we illustrate findings concerning microphysical properties of the magnetopause processes. It will be clear, from the examples discussed in this chapter, that the Cluster mission has made important contributions to the understanding of solar wind entry processes at the magnetopause. Some of the findings were possible only because of the high-

[1]Space Sciences Laboratory, University of California, Berkeley, CA, USA
[2]ESTEC, Noordwijk, The Netherlands
[3]CETP/IPSL/UPMC, Vélizy, France
[4]Max-Planck-Institut für extraterrestrische Physik, Garching, Germany
[5]Swedish Institute of Space Physics, Uppsala, Sweden
[6]International Space Science Institute, Bern, Switzerland
[7]Lockheed-Martin Advanced Techn. Center, Palo Alto, CA, USA
[8]CETP/IPSL, Vélizy, France
[9]Physics Department, Dartmouth College, Hanover, NH, USA
[10]Thayer School of Engineering, Dartmouth College, Hanover, NH, USA

Space Science Reviews 118: 367–424, 2005.
DOI: 10.1007/s11214-005-3836-z

resolution multi-point measurements, while others are due to the Cluster orbits that provide ideal coverage of the high-latitude and flank magnetopause and the magnetospheric cusps.

10.1 Magnetopause Reconnection

In this section, we present highlights of Cluster studies regarding large-scale magnetopause properties of magnetic reconnection (or merging). The next four sections discuss (1) the temporal and spatial nature of reconnection, (2) the evidence for component merging, (3) the occurrence and properties of tailward-of-the-cusp reconnection, and (4) the quantitative fluid and kinetic tests of reconnection.

10.1.1 Intermittent vs. quasi-steady reconnection

While single-spacecraft measurements have provided ample in-situ evidence for the occurrence of reconnection at the magnetopause (e.g., Paschmann et al., 1979, 1986; Sonnerup et al., 1981; Gosling et al., 1991; Smith and Rodgers, 1991; Fuselier et al., 1991), the large-scale spatial and temporal nature of reconnection is still not well understood. One of the unanswered questions is whether reconnection at the magnetopause is necessarily intermittent, or can be continuous, or even quasi-steady. Reconnection is termed intermittent if it turns on and off. Continuous reconnection may operate at a variable rate but never ceases; if the fluctuation is a small fraction of the average rate then the reconnection is termed 'quasi-steady'.

The distinction between intermittent and quasi-steady reconnection may depend on the variations of the solar wind and magnetic field. If there are frequent changes in the solar wind, one would not expect reconnection to be steady. An interesting and fundamental question is whether reconnection is intermittent even when the solar wind conditions are stable.

Before the launch of Cluster, this question had been investigated either by in-situ observations at the magnetopause by single spacecraft or deduced remotely by evaluating the ion dispersion signatures in the mid-altitude cusp by single spacecraft or by two spacecraft separated by large distances. Radar observations from the ground have also been used to infer the large-scale temporal and spatial nature of reconnection.

Repeated encounters of reconnection jets at multiple magnetopause crossings have been interpreted as evidence for continuous reconnection (e.g., Sonnerup et al., 1981; Gosling et al., 1982). On the other hand, frequent observations at the magnetopause of flux transfer events (FTEs) (Russell and Elphic, 1978), poleward moving auroral forms (e.g., Farrugia et al., 1998) or steps in the ion dispersion signatures seen in the magnetospheric cusps at low-altitudes (e.g., Newell and Meng, 1991; Escoubet et al., 1992; Lockwood and Smith, 1992; Lockwood et al., 2003) have been interpreted as signatures of intermittent reconnection.

The advantage of in-situ magnetopause observations is that one can quantitatively deduce the occurrence of reconnection by examining its fluid and kinetic signatures. However, because signatures of reconnection are localised in a thin layer along the magnetopause, single-spacecraft can observe reconnection signatures only during the short time the spacecraft crosses that layer. Typical crossing times are only a few minutes and the crossings are often far apart in time. Therefore one could never be sure whether reconnection had actually stopped in between successive magnetopause crossings. Furthermore, the time-variability of the reconnection rate is not known from in-situ observations because of the difficulties in determining the reconnection rate accurately.

On the other hand, observing ion dispersion signatures in the mid-altitude or low-altitude cusp has the advantage that the spacecraft samples in a short time many flux tubes which map to large areas of the magnetopause. With modelling, one can deduce in principle the variation of the reconnection rate (e.g., Lockwood and Smith, 1992; Onsager et al., 1993). The disadvantage is that variations in the ion dispersion may not be due entirely to temporal variations of reconnection. A spacecraft path which crosses different parts of the convection cells could observe irregular ion dispersion signatures as well (e.g., Trattner et al., 2002).

With Cluster multi-point measurements, some of the ambiguities associated with in-situ observations at the magnetopause as well as remote-sensing observations within the cusp can be removed. For example, if all spacecraft crossing the magnetopause consecutively detect reconnection signatures, it would indicate that reconnection is continuously active during those times. In the cusp, the timing of ion dispersion signatures observed by multi-spacecraft will distinguish temporal from spatial variations. Both magnetopause and cusp observations are complemented in an important way by the availability of simultaneous observations of proton auroras by the IMAGE spacecraft and by radar measurements from the ground that provide global views of reconnection (but at a lower time resolution).

In this Section, we will first address the cusp ion dispersion observations by Cluster to demonstrate that some cusp dispersion features are indeed temporal and consistent with intermittent reconnection while others are spatial variations of the reconnection pattern. We then show examples of continuous reconnection at the local magnetopause for both southward and northward IMF when the solar wind conditions are stable. When the IMF changes, the reconnection site moves and FTEs (presumably due to intermittent reconnection) can be formed. The accumulated evidence suggests that the key controlling factor of the steadiness of the reconnection process is the steadiness of the solar wind and IMF. Thus intermittency does not appear to be a fundamental property of the reconnection process at the magnetopause.

10.1.1.1 Intermittent reconnection inferred from mid-altitude cusp and radar observations

Cluster has made a number of observations in the polar cusp region indicating intermittent reconnection at the magnetopause (Lockwood et al., 2001b,a; Wild

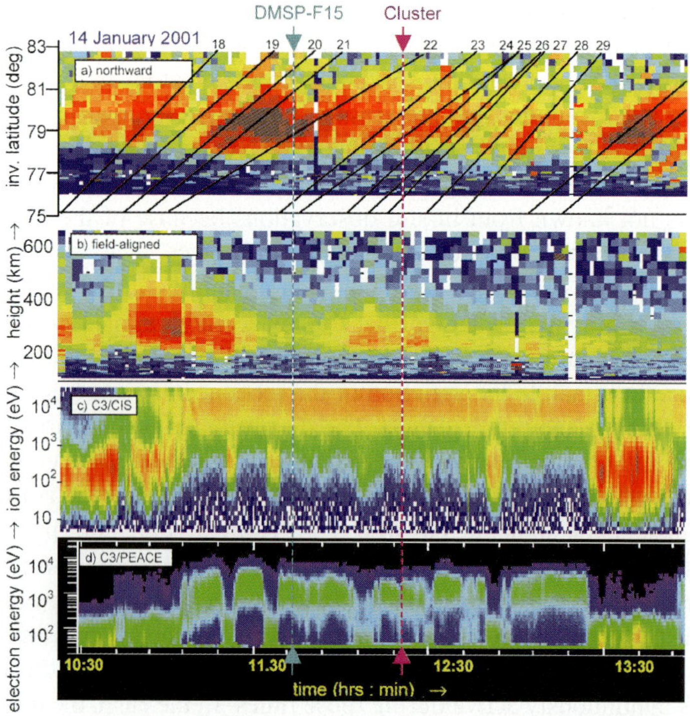

Figure 10.1. Observation of transient events seen by Cluster and the EISCAT Svalbard Radar(ESR) on January 14, 2001. (a) The plasma concentrations seen in the ionosphere along the low elevation (30 °) poleward beam of the ESR are colour-coded as a function of time and latitude. The centre of poleward-moving events are marked with a black line which is numbered in continuation of the events earlier on the same day. (b) The observations along the field-aligned ESR beam are shown as a function of time and altitude. (c) An energy-time spectrogram of differential energy flux of ions, integrated over all pitch angles, as observed by the CIS instrument on Cluster C3. (d) An energy-time spectrogram of the count rate of electrons observed by the HEEA detector of the PEACE instrument on Cluster C3 in zone 11 (electrons moving in the +Z GSE direction) (From Lockwood et al., 2001a).

et al., 2001; Owen et al., 2001; Bosqued et al., 2001; Moen et al., 2001; Vontrat-Reberac et al., 2003; Marchaudon et al., 2004). In this section we will show three examples of such transient reconnection events.

Lockwood et al. (2001a) studied a series of entries into the Low Latitude Boundary Layer (LLBL) observed on all four spacecraft during an outbound pass through the mid-afternoon magnetopause. At the same time the EISCAT Svalbard Radar (ESR) detected poleward moving events in the ionosphere. Figure 10.1 shows the radar data and Cluster 3 data. Between 11 and 13 UT, Cluster was in the dayside magnetosphere on closed field lines, characterised by high fluxes of high energy (>3 keV) ions (third panel) and intermediate energy (>500 eV) electrons (fourth panel). In this region, many transient occurrences of magnetosheath plasma are observed, characterised by ions below 3 keV and electrons below 300 eV. During the same time interval, the ESR radar observed poleward moving events (panel 1, marked with solid lines). Using the four spacecraft, Lockwood et al. (2001a) could determine that these transients were moving poleward at a speed around 26 $\mathrm{km\,s^{-1}}$, consistent with the ground-based observations. From these combined observations, Lockwood et al. (2001a) proposed that these events are the magnetospheric signatures of FTEs produced by intermittent reconnection at the magnetopause.

Trattner et al. (2005) reported Cluster observations of a series of temporal structures observed in the cusp (see Figure 10.2). On September 23, 2001, the spacecraft crossed the mid-altitude polar cusp (at about 5 R_E altitude). C4 was leading, followed by C2, 1 minute later, and C1, 2 minutes later. C3 crossed the cusp 40 minutes later. The IMF was dominated by a strong but variable negative B_y component ranging from -3 nT to -15 nT. The B_x component was positive and centered around 5 nT while the B_z component was less then 5 nT and switched several times between northward and southward. The cusp observations were made in a Magnetic Local Time (MLT) range from 11:30 to 13:00, an Invariant Latitude (ILAT) range from 75° to 83° and a geocentric distance from 4.5 R_E to 5.4 R_E. C1 entered the cusp at about 11:08 UT and subsequently observed two typical step-up cusp structures in the ion energy dispersion at about 11:18 UT and 11:25 UT. Cluster spacecraft C4 entered the cusp about 1 minute before SC1 at 11:07 UT. Like C1, C4 encountered two step-up cusp structures at about 11:19 and 11:27 UT. The entry in the polar cusp is clearly observed at 11:07-11:08 UT, where C4 first entered the cusp, then C2 and C1. On the other hand the two structures around 11:20 and 11:28 show the reversed order, first C1 detects the structure, then C2 and finally C4. This suggests that these structures are temporal, moving poleward through the spacecraft. Both structures seem to have convected 1° poleward in the direction of the convection path within about 1 minute, which represents a convection velocity of about 1.5 $\mathrm{km\,s^{-1}}$, in agreement with observed convection speed in the ionosphere of about 1.2 $\mathrm{km\,s^{-1}}$, as measured by the SuperDARN radars. This Cluster cusp event was characterised by strong variations in solar wind density and IMF direction that are likely to introduce temporal changes in the reconnection location, and probably, temporal changes in the reconnection rate.

Another example of signatures of intermittent reconnection has been studied by Bosqued et al. (2004). Using Cluster multi-point data at $\sim$5 R_E altitude and global dayside imaging data provided by the IMAGE-FUV-SI12 proton aurora imager, they showed that large variations in solar wind dynamic pressure were producing intermittent reconnection. The IMF was pointing southward/strongly duskward and, for more than 30 minutes, the solar wind pressure P_{sw} was highly variable and reached 13 nPa. Simultaneous observations in the northern 1300-1450 MLT sector, at 71-75° latitude, reveal intense cusp activity, characterised by multiple, impulsive energy-dispersed ion injections, with a recurrence time of $\sim$8-10 minutes, and

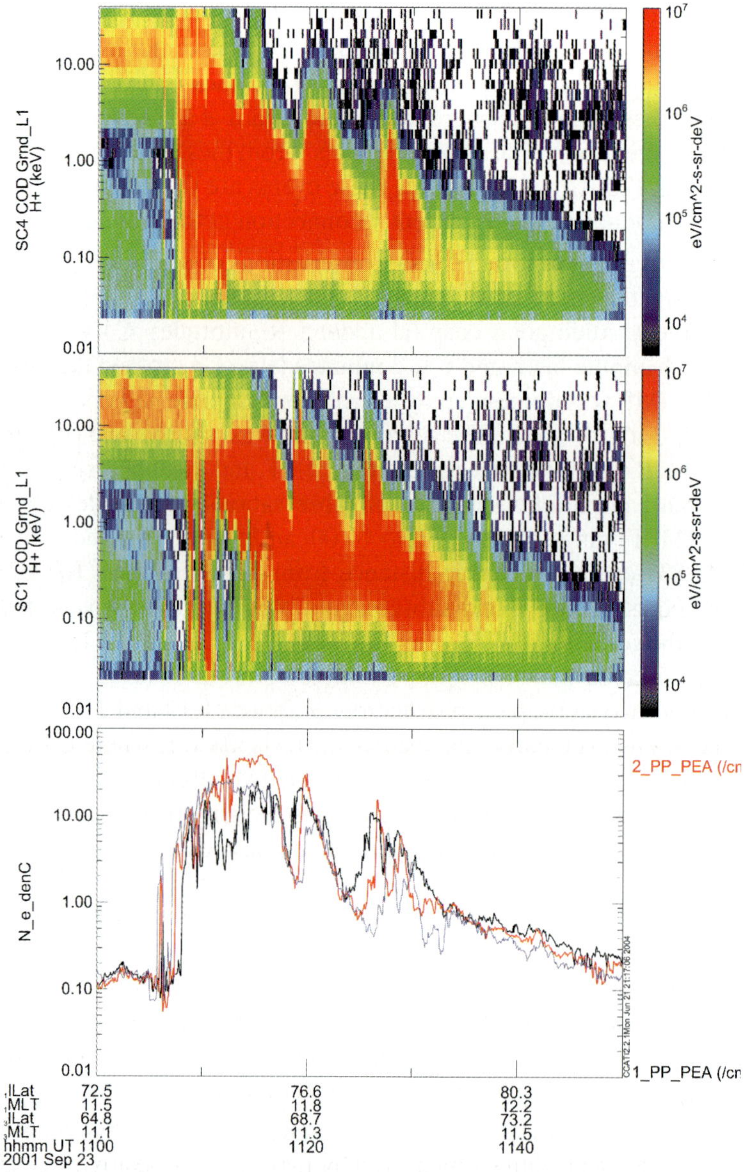

Figure 10.2. Observations of temporal ion structures in a Cluster cusp crossing on September 23, 2001. The top and middle panels show energy-time spectrograms of differential energy flux of ions, integrated over all pitch angles, as observed by the CIS instrument on Cluster C4 and C1 respectively. The bottom panel shows the electron density (in cm^{-3}) measured by PEACE on C1 (black), C2 (red) and C4 (blue). At the bottom are given the Invariant Latitude and Magnetic Local Time of C1 and C3. (Adapted from Trattner et al., 2005).

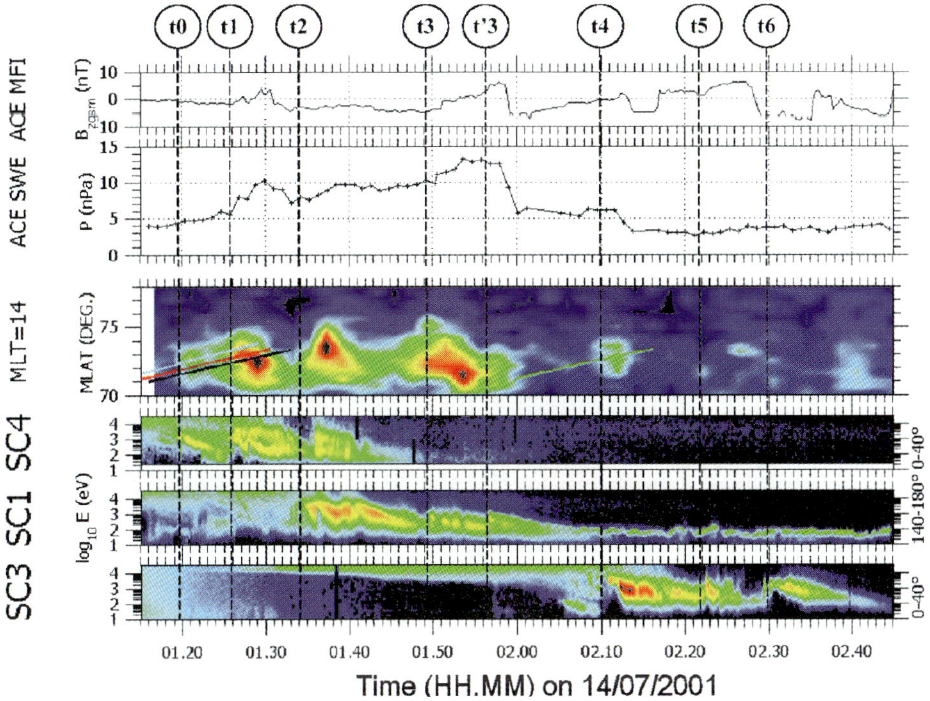

Figure 10.3. Simultaneous interplanetary data, IMAGE-FUV-SI12 keogram and Cluster CIS data for the period 01:15- 02:45 UT on July 14, 2001. From top to bottom: IMF B_z (nT), P_{sw} (nPa), MLAT-time-intensity keogram for 14.00 MLT (integration from 14.00 to 14.50) and, in the three bottom panels, Cluster C4, C1, and C3 CIS ion spectrograms (downgoing ions for C4 and C3, upgoing ions for C1). The tracks of each spacecraft are given by coloured segment lines in the appropriate keograms (C1: black, C2: red, C3: green, C4: magenta). Vertical dashed lines mark the main injection times. IMAGE-SI12 data are not shifted by the ion travel time, 60-100 s, from the Cluster altitude to the ionosphere (Adapted from Bosqued et al., 2004).

perfectly correlated in space and time with an intense doppler-shifted Ly-α brightening, i.e., proton aurora produced by precipitating solar wind protons in the cusp (Figure 10.3). These two coupled transient signatures were one-to-one related to repeated P_{sw} enhancements.

The proton aurora brightness measured by IMAGE was positively related to P_{sw} but not to the IMF clock angle, and was remarkably similar (within a factor of 2) to the expected count rate deduced from the total precipitating ion energy flux measured by Cluster on the same field line. A multi-point analysis revealed that a newly reconnected flux tube was moving westward with a very high velocity, up to $\sim$50 km s^{-1}, controlled by magnetic tension forces when IMF $B_y \gg 0$. Upward field-aligned currents, carried by 10-200 eV electrons, are identified on the poleward edge of the tube, in agreement with FTE models. The authors inferred a

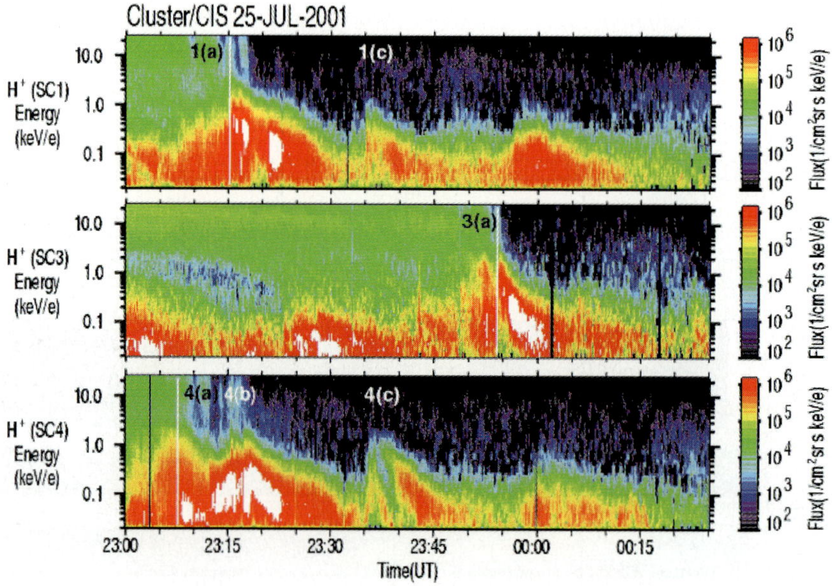

Figure 10.4. Cluster-CIS observation for a cusp crossings on July 25, 2001. Plotted are H^+ omnidirectional flux measurements (1 cm^{-2} s^{-1} sr^{-1} keV^{-1} e) for satellites C1, C3 and C4. All satellites encounter distinctive structures, sudden jumps in the ion energy dispersion that are similar on C1 and C4, but different on the later arriving C3 satellite. (From Trattner et al., 2003).

rather stable reconnection site located at ~10-12 R_E from Cluster, i.e., on the dusk flank of the compressed magnetosphere, around 1700-1800 MLT. They interpreted these very dynamic and transient features as clear signatures of pulsed magnetic reconnection operating in a localised region of the magnetopause centered on the preferential anti-parallel merging site. Bosqued et al. (2004) suggested that the reconnection rate at the reconnection site was not spontaneously self-varying but, more evidently, directly modulated by upstream dynamic pressure P_{sw}.

10.1.1.2 Spatial structure of the reconnection pattern inferred from mid-altitude and groundbased observations

Although there is evidence of intermittent reconnection, especially associated with changes in the solar wind, there is also evidence of spatial changes. Cluster observed structures in the cusp ion dispersion signatures that can be explained by a change of the reconnection location instead of a change in the reconnection rate. Trattner et al. (2003) reported such a case observed on July 25, 2001. The solar wind density and pressure were stable around 4 cm^{-3} and 600 km s^{-1}. The IMF-B_y and B_z were negative around -5 nT, starting at 23:00 UT, and then turned positive around 23:45 UT. Figure 10.4 shows the H^+ fluxes from C1, C3, and C4. C1 entered the cusp at about 23:15 UT, as indicated by a significant increase in ion energy and H^+ flux intensity (see white line labeled 1a). C1 subsequently observed the typical cusp ion energy dispersion for a southward interplanetary magnetic field, with lower energy ions arriving at higher latitudes. The characteristic ion energy decreased smoothly, indicating a constant magnetospheric reconnection rate at the magnetopause. At about 23:37 UT, C1 encountered a sudden increase in the ion energy dispersion (label 1c), consistent with a typical step-up ion signature that can occur if the satellite crossed onto magnetic field lines that were reconnected more recently. C3 crossed into the cusp at 23:54 UT, also indicated by a white line labelled 3a, in the colour spectrogram. C3 also observed a decreasing ion energy dispersion typical for a stable rate of reconnection with no further cusp structures later-on. The first Cluster satellite to enter the cusp on July 25, 2001, was C4 at 23:08 UT (label 4a). The cusp encounter was also followed by a decreasing ion energy dispersion which is reversed at about 23:13 UT. The precipitating ion energy reaches a new maximum at 23:15 UT (label 4b), the same time as C1 entered the cusp. C4 encountered the second sudden increase in ion energy at about 23:37 UT (label 4c), similar to the increase observed by C1 at about the same time.

Figure 10.5 shows a combination of the temporal and spatial separations of the Cluster spacecraft for the event. Shown are the Cluster magnetic foot points and a snapshot of the ionospheric convection streamlines for July 25, 2001, at 23:37 UT. The ionospheric convection streamlines, presented as contour lines, have been calculated using line-of-sight velocity data from the 8 operating northern hemisphere SuperDARN radars (Greenwald et al., 1995). Overlaid on the magnetic foot points are 14-minute intervals of the Cluster/CIS flux measurements presented in Figure 10.4, which are centered on the actual position of the Cluster satellites at 23:37 UT. This representation shows the actual Cluster measurements in time at their proper spatial location where they have been observed. At 23:37 UT the IMF shows a strong decrease in the value of B_z, which will subsequently result in a reversal from negative to positive B_z. An equatorward directed bulge in the convection pattern has moved rapidly equatorward, which in turn allowed the dawn convection cell (dashed black lines) to move equatorward as well. At 23:37 UT, C1 and C4 have progressed poleward far enough to be overtaken by the equatorward moving dawn convection cell. The transfer from one convection cell to another resulted in an almost simultaneous sudden increase of the ion energy dispersion (structures 1c and 4c) in Figure 10.4 on both satellites. The sudden increase in the ion energy dispersion coincides with a satellite moving into a neighboring spatially separated flux tube (or convection cell). The change in IMF conditions most probably caused a change in the location of the reconnection site, that in turn caused a shift in the positions of spatially separated flux tubes. This shift occurred without significant change in the reconnection rate.

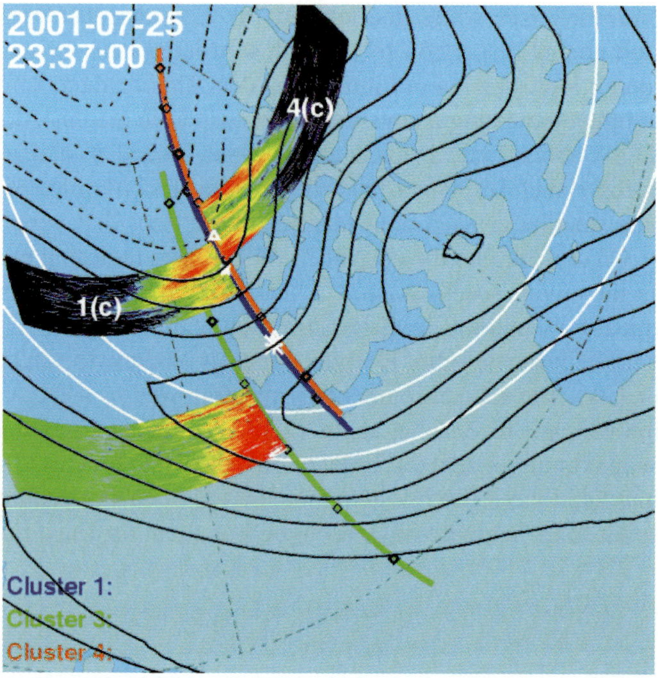

Figure 10.5. Composite plot of Cluster magnetic foot points and ionospheric convection streamlines for July 25, 2001, at 23:37 UT. Overlaid on the magnetic foot points are 14-minute intervals of the Cluster/CIS flux measurements presented in Figure 10.4, which are centered on the actual position of the Cluster satellites at 23:37 UT. The original entry points of C1 and C4 into the cusp are marked with a star and a triangle, respectively, along the tracks of their magnetic foot points. C1 and C4 are deep inside the cusp and have just entered the dawn convection cell (dashed lines) resulting in an almost simultaneous sudden increase of the ion energy dispersion on both satellites, as expected from a spatial interpretation of cusp structures. During that time C3 was still on closed field lines. (From Trattner et al., 2003).

10.1.1.3 Continuous reconnection

The Cluster mid-altitude cusp observations have revealed that structures in the ion dispersions can be caused by temporal variations of the reconnection rate as well as movements of the reconnection site due to solar wind variations. The fundamental question is whether reconnection is intrinsically intermittent, even when the solar wind plasma and magnetic field orientation are steady. We describe two studies, one for southward and the other for northward IMF, which provide evidence that reconnection at the magnetopause is continuous when the solar wind conditions are steady. When the IMF changes direction, a joint IMAGE-Cluster study reveals that the reconnection site moves, but reconnection never stops. Thus reconnection is still continuous in a global sense but may appear intermittent (or transient) at the local magnetopause.

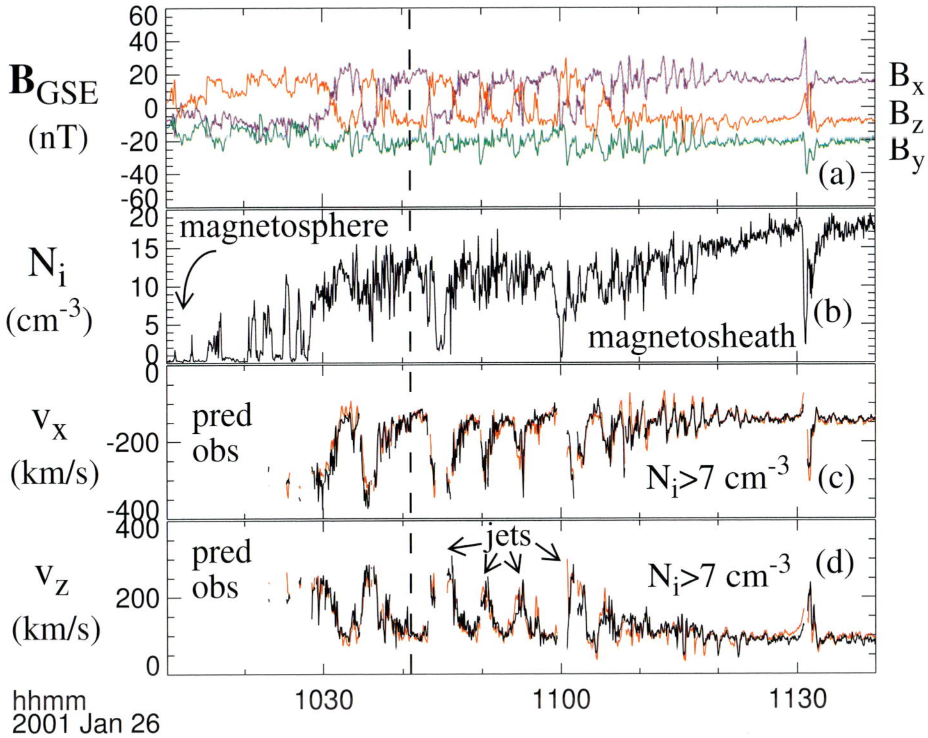

Figure 10.6. Observations by Cluster 1 of continuous reconnection at the equatorward of the cusp magnetopause for southward IMF. (a) magnetic field components in GSE, (b) ion number density, (c-d) x and z components of the predicted (red) and observed ion velocity for ion density > 7 cm^{-3}. The density threshold ensures that the comparison with theory is limited to the magnetopause (and excludes samples in the magnetosphere). The agreement between the predicted and observed Alfvénic flows is seen at all magnetopause crossings. The magnetopause is recognised by the rotation of the magnetosheath from negative B_z in the magnetosheath to positive B_z in the magnetosphere. Reconnection flows detected at all magnetopause crossings indicate continuous reconnection. (Figure adapted from Phan et al., 2004).

Continuous reconnection at dayside magnetopause during southward IMF. On January 26, 2001, the Cluster spacecraft detected reconnection jets at multiple crossings of the high-latitude duskside magnetopause over a period of more than 2 hours. The 4 spacecraft combined spent more than half of this time in the MP/BL and reconnection flows were observed whenever a spacecraft was in the magnetopause. These observations led to the interpretation that reconnection was continuous for over 2 hours (Phan et al., 2004). These observations were made under steady southward and dawnward interplanetary magnetic field (IMF) conditions. The magnetic shear across the local magnetopause was $\sim 100°$ and $\beta \sim 1$ in the adjacent magnetosheath.

According to MHD models of reconnection at the dayside magnetopause involving asymmetric plasma conditions on the two sides of the magnetopause, the magnetopause can be described as a rotational discontinuity (RD) (Levy et al., 1964). Across this type of discontinuity, the magnetosheath plasma is accelerated by the magnetic field tension force associated with the field line kink resulting from the linkage between magnetosheath and magnetospheric field lines. Figure 10.6 illustrates the quantitative agreement between predicted and observed reconnection flows for the 10:10–11:40 UT interval - which encompasses more than 10 partial or complete magnetopause crossings and an FTE (at 11:30 UT). The predicted velocities are based on a single reference magnetosheath time (10:41 UT) marked by the dashed line. The flow prediction is obtained by inserting the local magnetic field measurements and the reference velocity and field values into the jump relation for a 1-D rotational discontinuity (Hudson, 1970; Paschmann et al., 1986):

$$\Delta \mathbf{v}_{pred} = \mathbf{v}_{2t} - \mathbf{v}_{1t} = +\left(\frac{1-\alpha_1}{\mu_0 \rho_1}\right)^{1/2} \left[\left(\frac{1-\alpha_2}{1-\alpha_1}\right)\mathbf{B}_{2t} - \mathbf{B}_{1t}\right] \tag{10.1}$$

Subscript '1' denotes the reference time and '2' denotes the prediction for all other times. The + sign indicates that the spacecraft were always northward of the reconnection line. The pressure anisotropy factor is $\alpha = (p_{\parallel} - p_{\perp})\mu_0/B^2$, where $p_{\parallel}$ and $p_{\perp}$ are the plasma pressures parallel and perpendicular to B. This jump relation is closely related to the Walén test, performed in the deHoffmann-Teller (HT) frame (Sonnerup et al., 1987).

Figure 10.6 shows the (roughly) tangential v_x and v_z components of the predicted (red) and observed (black) flows. The comparison is restricted to measurements that have density > 7 cm^{-3} (roughly half of the magnetosheath density). This condition ensures that the comparison is made only across the RD, and does not extend into the lower density slow expansion fan region further inward from the magnetopause (Phan et al., 2001). The agreement between predicted and observed flows is remarkable, even for crossings that are far (more than 30 minutes) from the reference time. It is immediately clear from this comparison that all magnetopause crossings in this interval observed reconnection jets at, or very close to, the predicted Alfvén velocity, including the FTE at 11:30 UT.

The repeated detection of reconnection flows at all magnetopause crossings during the 2-hour period suggests that reconnection was operating continuously over this time period. However, because a single Cluster spacecraft spent only a brief amount of time in the magnetopause, one could not rule out the possibility that reconnection ceases when the spacecraft is not in the magnetopause. In other words, the onset of reconnection could conceivably cause the magnetopause to move, resulting in a crossing. With a single spacecraft one could never be sure. But with the multiple spacecraft following each other through the magnetopause, one could investigate whether reconnection signatures were detected by all spacecraft crossing the magnetopause consecutively and therefore rule out the possibility that recon-

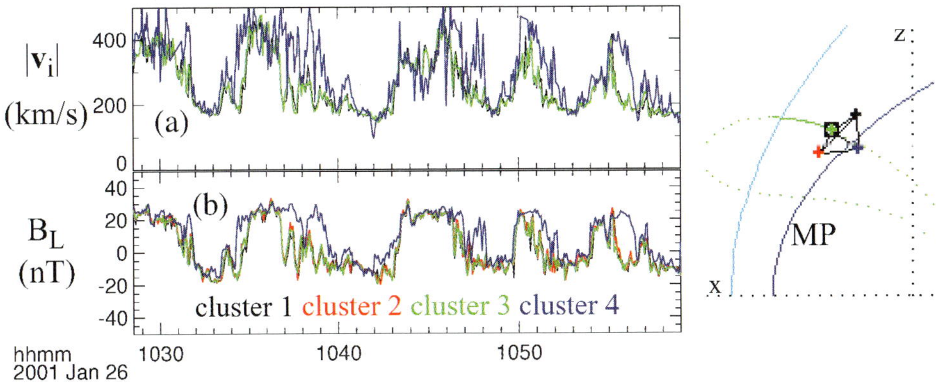

Figure 10.7. Reconnection jets detected by all spacecraft crossing the magnetopause, implying continuous reconnection. (a) Overlay of Spacecraft 1, 3, and 4 ion flow speed, (b) Overlay of Spacecraft 1-4 magnetic field BL component in the boundary normal coordinate system with L pointing approximately in the GSM z direction. Spacecraft 1, 2, and 3 are in the same plane tangential to the magnetopause and spacecraft 4 is closer to the Earth. (Figure adapted from Phan et al., 2004).

nection stops when a spacecraft leaves the magnetopause. The multi-spacecraft measurements also extend the total spacecraft residence time in the magnetopause.

Figure 10.7 shows the velocity traces from Cluster 1, 3, and 4 (not available from Cluster 2) and magnetic field traces from all 4 spacecraft. The configuration of the spacecraft on this day is such that spacecraft 1, 2, and 3 are in the same plane tangential to the magnetopause and spacecraft 4 is closer to the Earth. Thus, traces from spacecraft 1, 2, and 3 are nearly identical while spacecraft 4 provides the separation along the magnetopause normal. Jets are detected by the spacecraft residing in the magnetopause even when the others are outside (see for example 10:39, 10:48 and 10:56 UT). If one combines the jet observations from all 3 spacecraft where velocities have been measured, one finds that for more than half the time during this 2-hour interval at least one spacecraft was in the magnetopause and detected jets. Ideally one would like to have at least one spacecraft in the magnetopause at any given time so that the magnetopause can be monitored continuously, but this is not possible with a small number of spacecraft. Thus the most conservative conclusion from the observations presented here is that reconnection occurs at least 50% of the time in this 2-hour interval. In reality the percentage is likely to be closer to 100% since the jets are seen at every magnetopause crossing by all spacecraft.

It should be emphasised that the continuous detection of reconnection jets does not imply that reconnection proceeds at a steady rate. The flow acceleration across the magnetopause is independent of the reconnection rate (as long as the rate does not drop to zero). Phan et al. (2004) concluded, based on the evidence that the magnetopause crossings are caused by plasma blobs travelling along the magnetopause, that the reconnection rate may be modulated.

In a more recent study, Retino et al. (2004) have reported the detection of reconnection flows at all Cluster magnetopause crossings over a period of 4 hours, which again suggests that reconnection was continuous over that time span.

Continuous reconnection at high-latitude magnetopause during northward IMF. The example described in the previous section pertains to dayside (equatorward-of-the-cusp) reconnection during southward IMF. The question is whether tailward-of-the-cusp reconnection during northward IMF can also be continuous. Theories suggest that large (super-Alfvénic) shear flows can prevent steady reconnection (Cowley and Owen, 1989), thus it was unclear whether high-latitude reconnection, where the external magnetosheath flow is fast, could persist.

The Cluster-IMAGE conjunction occurred on March 18, 2002 (Figure 10.8). For a ~4-hour interval, the IMF was persistently northward directed ($B_z > 0$) while its y- (east-west) component fluctuated between positive and negative. During a brief interval (5 minutes), the Cluster spacecraft crossed the high latitude magnetopause and observed proton jets accelerated by reconnection at the magnetopause (Phan et al., 2003). The reconnection jets were observed on field lines that are linked to a proton auroral spot observed by the IMAGE spacecraft in the ionosphere. The energy fluxes of the precipitating protons observed by Cluster are consistent with the spot brightness. These simultaneous observations provide direct evidence that the spot represents the remote signature of high-latitude magnetopause reconnection. Thus one could use the proton aurora images to monitor reconnection on a global scale.

Although Cluster could only observe reconnection for 5 minutes, IMAGE observed the uninterrupted presence of the proton auroral spot over 4 hours (shown in Figure 10.8), which implies continuous reconnection over this entire time interval (Frey et al., 2003). The peak and average brightness of the spot remained high over the whole observation period with some fluctuations. The 5-second images were obtained on a two-minute cadence.

The magnetic local time (MLT) location of the spot follows very closely the changes in the east-west (y) component of the IMF (indicated by the arrows in the green inserts, with positive z pointing up and positive y to the left), with spot locations at pre-noon MLT for negative B_y and with post-noon locations for positive B_y. The correlation implies that reconnection is a directly driven process at the magnetopause and that the anti-parallel reconnection site produces the most intense proton aurora. As the IMF changes, the anti-parallel reconnection site moves on the magnetopause. This is seen in the ionosphere as motion in MLT but not in latitude. This motion, combined with the near constant brightness of the spot, indicates that the reconnection site is continuously rather than intermittently active. Thus, viewed globally, the reconnection process never stops at the high latitude magnetopause.

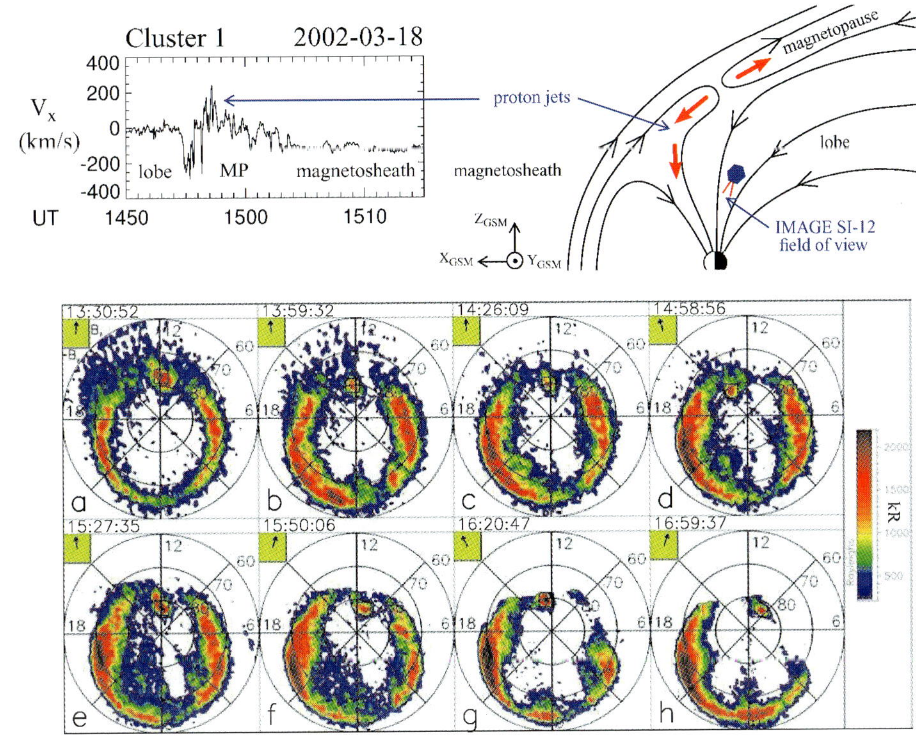

Figure 10.8. IMAGE-Cluster observations of continuous reconnection at poleward-of-the-cusp magnetopause during northward IMF. The observations of reconnection jets at the high-latitude magnetopause on March 18, 2002 by Cluster (top left panel) and a proton auroral spot by IMAGE on the same field lines provide evidence that the proton spot represents the remote signatures of reconnection. While Cluster observed the magnetopause for only 5 minutes, IMAGE detected the proton spot uninterruptedly for ~4 hours, implying continuous reconnection. The spot moves in local time, indicating a shift of the reconnection site, in response to variations in the IMF B_y. (Figure adapted from Phan et al., 2003; Frey et al., 2003).

Finally, as the spot moves in MLT in response to the solar wind B_y, it does not leave a long trail in latitude nor in MLT. Such a feature would indicate that proton aurorae are still created long (more than 4-5 minutes) after the cessation of reconnection.

The observations have been interpreted as follows. The continuous presence of the proton auroral spot even as the solar wind conditions changed implies continuous magnetic reconnection. In a global sense, it is significant that reconnection at the high-latitude magnetopause never stops for any period of time longer than 4-5 minutes. For more stable solar wind conditions, reconnection is continuous not only in a global sense, but is continuous at the local magnetopause as well. Note that this finding is consistent with the results of a survey by Pitout et al. (2004) (see mid-altitude cusp section) who found that smooth ion dispersion signatures observed in the mid-altitude cusp correspond to steady IMF whereas complicated dispersion signatures tend to correspond to variable IMF.

The fact that high-latitude reconnection can persist seems to be related to the presence of a plasma depletion layer during northward IMF which renders the magnetosheath flow adjacent to the high-latitude reconnection site sub-Alfvénic (Fuselier et al., 2002). This effect will be discussed further in Section 10.1.3.

10.1.1.4 Summary of observations of intermittent and continuous reconnection

It is clear from the examples in the previous sections that Cluster, together with ground based radar and IMAGE proton aurora observations, is ideal for the investigation of the temporal nature of the reconnection process.

Both intermittent and continuous (and even quasi-steady) reconnection have been convincingly observed. The accumulated evidence suggests that the steadiness of the solar wind determines the steadiness of reconnection. When the IMF and solar wind pressure are stable, reconnection at the local magnetopause can be continuous. If the IMF direction changes, the reconnection site moves and reconnection at the local magnetopause can become intermittent, but viewed globally, reconnection at the magnetopause never ceases.

10.1.2 Component vs. anti-parallel reconnection

This section describes direct evidence for component reconnection (or merging) by the detection of (1) a strong guide field along an X-line, and (2) a strong guide field in a flux transfer event.

One of the controversies related to reconnection at the magnetopause is whether the global configuration follows the ‘component merging’ (Sonnerup, 1974; Gonzalez and Mozer, 1974) or ‘anti-parallel merging’ (Crooker, 1979; Luhmann et al., 1984) models. Component merging proceeds with a finite guide field, i.e., a finite magnetic field component along the X-line, whereas the guide field is small or vanishes in the anti-parallel merging model. According to current understanding of processes in the reconnection diffusion region (e.g., Shay et al., 2001; Hesse et al., 2001; Pritchett, 2001), reconnection can proceed with a guide field, but at a slower rate than in the absence of a guide field (e.g., Sonnerup, 1974) (assuming that the total field strength is the same for the finite-guide-field and no-guide-field situations). Thus, unless future studies find that reconnection does not initiate with a guide field, one is really addressing a geophysical question of whether anti-parallel merging dominates to the extent of preventing component merging sites to occur on the magnetopause.

There are many ways of addressing this problem. One can test the anti-parallel and component merging models indirectly by examining the consequences of these models, or one can search directly for the presence or absence of a guide field at the reconnection X-line. It should be emphasised that anti-parallel merging is a subset of component merging. Thus finding the presence of anti-parallel reconnection in itself does not contradict the component merging model. But finding exclusively anti-parallel merging sites would favour the anti-parallel merging model. On the other hand, frequent detections of component merging signatures would rule out the anti-parallel merging model.

Indirect evidence favouring both models has been reported in a number of studies. Frequent detections of the reversal of the dawn-dusk components of the plasma flow across the low-latitude magnetopause in the presence of a finite IMF B_y were taken as evidence for component merging (Gosling et al., 1990a). Opposite sense of the FTE bipolar normal magnetopause detected north and south of the subsolar point for a variety of IMF B_y (Berchem and Russell, 1984) were interpreted as evidence for an X-line hinged at the subsolar point and directed according to the predictions of component merging. In another study, Trattner et al. (2004) deduced the location of the X-line based on time-of-flight effects in ion distributions observed by the Polar spacecraft and concluded that many of the reconnection ion distributions originated from component merging sites. On the other hand, ground-based radar observations of a split of the projection of the dayside reconnection X-line were taken as evidence for anti-parallel merging (Chisham et al., 2002). Recently, IMAGE observations of the motion of the dayside proton auroral spot in MLT in response to changes of the IMF B_y indicate that the site of maximum particle energisation at the magnetopause is in the vicinity of where the magnetosheath and magnetospheric field lines are anti-parallel. Although this finding indicates that reconnection does take place at the anti-parallel merging sites, it does not preclude the possibility that the X-line extends to lower magnetic shear regions where the plasma acceleration is not large enough for detection by the IMAGE/FUV proton aurora imager (Phan et al., 2003).

In terms of sampling the X-line directly, reports of encounters with the reconnection diffusion region at the magnetopause are rare. However, the presence of a guide field in the diffusion region would provide the most direct evidence for component merging. Scudder et al. (2002) and Vaivads et al. (2004) reported two different cases where the guide field is nearly zero at the high-latitude magnetopause. The diffusion region encounter reported by Mozer et al. (2002) at the subsolar magnetopause, on the other hand, has a finite guide field (20% of the anti-parallel field). In this section, we describe two different Cluster studies that provide evidence for component merging. The first event is a direct detection of a low magnetic shear X-line. The second is a flux transfer event with a strong guide field which is also indicative of component merging.

10.1.2.1 Direct detection of a guide field at an X-line

It is possible to distinguish between antiparallel and component merging models when spacecraft are close to the X-line and one can estimate the local magnetic shear across the magnetopause. Reconnection ion jet reversals have been inter-

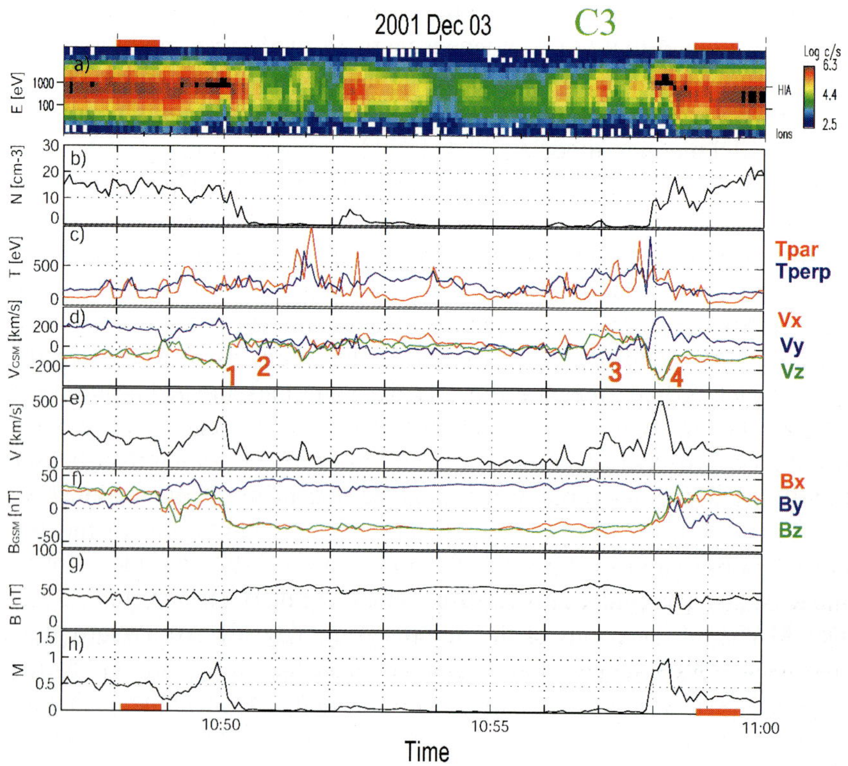

Figure 10.9. Two reconnection flow reversals (implying X-line crossings) observed by Cluster on December 3, 2001, 10:47-11:00 UT. The first jet reversal shows small magnetic shear across the magnetopause consistent with the component merging model. (a) ion energy spectrogram, (b) ion number density, (c) parallel and perpendicular ion temperatures, (d) ion velocity, (e) ion flow speed, (f) magnetic field, (g) magnetic field and (h) Alfvénic Mach number. Reconnection flows are indicated by red labels 1-4. (From Retino et al., 2004).

preted as a possible indication of the spacecraft passage close to the X-line (Gosling et al., 1991; Avanov et al., 2001; Phan et al., 2003). Retino et al. (2004) studied in detail two such ion jet reversals, one of which strongly supports the component merging model. On December 3, 2001 the Cluster spacecraft were skimming the high latitude duskside magnetopause, detecting a few ion jet reversals.

Figure 10.9 shows two ion jet reversals observed by spacecraft 3 within ~10 minutes. The first jet reversal at ~10:50 UT shows a passage from a tailward jet (indicated by the label 1) to a sunward jet (label 2) while the second jet reversal around ~10:58 UT shows a passage from a sunward jet (label 3) to a tailward jet (label 4). All four observed flows are consistent with the occurrence of magnetic reconnection: both fluid and kinetic evidence (D-shaped distribution functions) have been found. Reconnection is occurring tailward of the cusp as expected for northward IMF conditions. As one can see from the magnetic field changes in panel 6 of the figure, the first magnetopause crossing at $\sim$10:50 UT has low magnetic shear of $\sim 100^\circ$ degrees while the second one has a higher shear of $\sim 160^\circ$. For the first jet reversal (jets 1 and 2) the observations are consistent with component merging because of the small shear measured close to the reconnection site. The velocities of jets (both tailward and sunward) during the first jet reversal are smaller than those observed during the second jet reversal. This observation is also consistent with component merging because component merging predicts a weaker 'kick' experienced by the injected ions for smaller values of the magnetic shear (the velocity change is proportional to the change in **B** across the magnetopause which becomes smaller with decreasing shear angle). Furthermore, according to the anti-parallel merging model, no X-line is expected on the duskside magnetopause tailward of the cusp for IMF $B_y > 0$, $B_z > 0$. The observation close to the X-line of low magnetic shear on the dusk side in the southern hemisphere is therefore not consistent with antiparallel merging, but it is the direct evidence for component merging.

10.1.2.2 Guide field detected at the core of an FTE

Sonnerup et al. (2004) reported 4-spacecraft observations of a flux transfer event (see Section 'Flux Transfer Events' in De Keyser et al., 2005, for more details). The presence of a strong core field in the FTE was interpreted as evidence that the flux transfer event was generated by component merging.

10.1.3 Tailward-of-the-cusp reconnection

In this section we describe a statistical study of the occurrence and IMF dependence of reconnection at the magnetopause tailward of the cusp. The main findings are:

- Tailward-of-the-cusp reconnection has been observed only when the IMF has a northward component (i.e., only when the local magnetic shear $> 90^\circ$), in contrast to subsolar reconnection, which has been observed for shear $<90^\circ$.

- The occurrence rate of cusp reconnection is near 100% when the IMF is northward. This finding suggests that simultaneous cusp reconnection in the northern and southern hemispheres must be common.

- For northward IMF, the plasma depletion layer extends to high latitudes, where it reduces the plasma β and renders the magnetosheath flow sub-Alfvénic. These conditions allow the establishment of a stable X-line at the high-latitude magnetopause.

10.1.3.1 IMF dependence

In-situ observations have established the occurrence of reconnection jets at the low-latitude magnetopause (see Section 10.1.1) as well as the high-latitude magne-

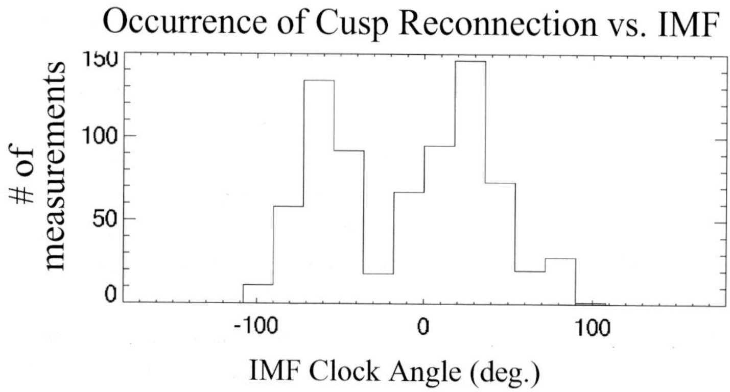

Figure 10.10. Occurrence of tailward-of-cusp reconnection (as detected at the magnetopause and in the high-altitude cusp) as a function of the IMF clock angle ($= \tan^{-1}(B_y/B_z)$). The occurrence is found to be confined almost exclusively to positive IMF B_z. (From Twitty et al., 2004).

topause tailward-of-the-cusp (e.g., Gosling et al., 1991; Kessel et al., 1996; Avanov et al., 2001).

At the low-latitude magnetopause, reconnection signatures have been detected for a large range of IMF orientations (e.g., Scurry et al., 1994) and local magnetic shear angles (e.g., Paschmann et al., 1986; Fuselier et al., 2000b). The Scurry et al. study found nearly equal occurrence of low-latitude enhanced flows for the IMF clock angle range of 60°-180°. Similarly, Phan et al. (1996) found nearly equal probability of detecting reconnection flows for the local shear range of 60°-180°.

At the high-latitude magnetopause tailward of the cusp, reconnection signatures have been reported when the IMF is northward in case studies (Kessel et al., 1996). However the occurrence of cusp reconnection for the full range of IMF clock angle was not known from in-situ observations before the launch of Cluster. Cluster's polar orbit results in crossings of tailward-of-the-cusp magnetopause and high-altitude cusp that are ideal for the survey of the occurrence and IMF dependence of tailward-of-the-cusp reconnection. Such a survey could shed light on the conditions for the occurrence of reconnection.

Twitty et al. (2004) reported the results of a systematic survey of high-altitude cusp and high-latitude magnetopause crossings by the Cluster spacecraft to investigate the IMF range in which reconnection flows originating from tailward-of-the-cusp reconnection are detected either at the magnetopause or in the high-altitude cusp.

The survey covers 3 years (2001-2003) of cusp and magnetopause crossings and is restricted to periods of relatively stable IMF. With this restriction, 12 magnetopause and 19 cusp crossings were found where reconnection flows were detected. The results (Figure 10.10) indicate that tailward-of-the-cusp reconnection occurs almost exclusively when the IMF has a northward component (or the clock angle is within ~90° of the GSM +z direction). When the IMF has a southward component, the high-altitude cusp typically reveals normal ion dispersion consistent with dayside reconnection instead of reversed dispersion. In addition, the transition from the cusp to the mantle/lobe is gradual (with no sharp boundaries or magnetic shear), and no reconnection flow signatures are seen near the lobe/mantle or at the tailward-of-the-cusp magnetopause.

These findings at first seem inconsistent with the component merging model which should allow tailward-of-cusp reconnection to occur for larger than 90° clock angle (or less than 90° in magnetic shear angle). However, it is possible that the finding indicates a geophysical effect rather than the failure of the component merging model. When the IMF has a southward component (clock angle > 90°), a high rate of equatorward-of-the-cusp reconnection could prevent the formation of a plasma depletion layer (PDL) adjacent to the low- and high-latitude magnetopause. With no PDL, the magnetosheath flow could be super-Alfvénic tailward of the cusp which, according to theory (Cowley and Owen, 1989), would prevent the establishment of a stable X-line. For northward IMF, the magnetosheath flow could be sub-Alfvénic due to the presence of a PDL (Zwan and Wolf, 1976) which extends from subsolar region to high latitudes (Fuselier et al., 2000a, 2002; Avanov et al., 2001; Petrinec et al., 2003). Indeed, Twitty et al. (2004) found that the magnetosheath flows in all tailward-of-the-cusp magnetopause crossings by Cluster under northward IMF were sub-Alfvénic.

10.1.3.2 Occurrence rate of tailward-of-the-cusp reconnection

The first part of the Twitty et al. (2004) survey involves only cases where reconnection flows are detected. It was found that these flows only occurred when the IMF has a northward component. One could turn the question around and ask how often tailward-of-the-cusp reconnection occurs when the IMF does have a northward component. It was found that during all tailward-of-the-cusp magnetopause crossings that occurred when the IMF had a northward component, reconnection flows were detected, implying an occurrence rate of 100%. For the high-altitude cusp crossings during stable northward IMF, in addition to the 19 cases with reconnection flows, 3 cases were found without clear reconnection jet signatures. Thus the overall reconnection occurrence rate (combining the magnetopause and the cusp cases) is better than 90%.

This occurrence rate is substantially higher than the 50% occurrence rate of reconnection flow signatures observed equatorward of the cusp for southward IMF (Phan et al., 1996). This difference at first seems surprising since the reconnection occurrence rate is often thought to be highest at the dayside magnetopause where the solar wind compression is strongest. The reason may be related to the dependence on the IMF of the magnetosheath plasma β adjacent to the magnetopause. The magnetosheath β is strongly dependent on the presence or absence of a depletion layer in the magnetosheath adjacent to the magnetopause. Previous studies found that reconnection events tend to correspond to low magnetosheath β (< 2) (e.g., Paschmann et al., 1986). The plasma β in all the magnetopause cases in the Twitty et al. (2004) survey was indeed low, varying between 0.1 and 2.7, with an average of 0.9. This low value is due to the presence of the PDL for northward IMF. On the other hand, at low-latitude during southward IMF, the PDL is usually weak or absent (Phan et al., 1994; Anderson et al., 1997). In that case, one may assume that the plasma β next to the magnetopause is simply the unmodified upstream magnetosheath β convected to the magnetopause. The magnetosheath β in that case could be large (unfavourable for reconnection) or small which could be the reason for a larger percentage of cases without reconnection.

An interesting implication of the high occurrence rate of cusp reconnection is that simultaneous reconnection in the northern and southern cusps must be common during northward IMF. However it remains unclear, from the survey performed so far, how often double-cusp reconnection involves the same flux tubes to create a boundary layer on closed field lines, as suggested by Song and Russell (1992).

10.1.4 Quantitative tests of reconnection occurrence

Cluster multi-spacecraft measurements and high resolution particle distribution samplings of the magnetopause provide tests of the occurrence of reconnection with unprecedented accuracy. The direct determination of the magnetopause current (from $\nabla\times \mathbf{B}$) allows the examination of the effect of the use of ions versus electrons in tests of fluid prediction of reconnection flows. The ion measurements however reveal an intriguing fact: that the kinetic signatures of reconnection, the D-shaped distributions, are often absent even when fluid signatures, Alfvénic jets, are clearly present.

10.1.4.1 Fluid signatures: Effects of magnetopause current and heavy ions

In previous studies, it had been found that sometimes the rotational discontinuity test (also called the 'Walén relation') fails at the dayside magnetopause (e.g., Paschmann et al., 1986; Sonnerup et al., 1990; Phan et al., 1996). This could mean that the magnetopause is not an RD, i.e., reconnection is not occurring in any location where it causes signatures to be seen at the observation site. However, there are cases where the observed flow direction agrees with the prediction but the flow magnitude is well off (e.g., less than 50% of the predicted value). For such cases some explanations have been provided. One is the effect on the Walén test when heavy ions (O^+ or He^{++}) may be present. These ions affect the ion moments if the moment computation assumes all ions to be protons (Paschmann et al., 1986; Puhl-Quinn and Scudder, 2000) and they reduce the Alfvén speed. The other is the use of ions instead of electrons in most previous Walén analyses. The electrons are the species that should be used because they are better tied to the magnetic field and because of the differential ion-electron motion associated with the magnetopause current (Scudder et al., 1999). Cluster has composition measurements as well as current measurements (from $\nabla \times \mathbf{B}$ measured by the 4 spacecraft) which permit direct tests of this hypothesis.

In the case on January 26, 2001 presented in Section 10.1.1.3 (Figure 10.6), the magnetopause current density is of the order of 50 nA m^{-2} (Phan et al., 2004). With a magnetosheath plasma density of 15 cm^{-3}, this implies an ion-electron differential motion of $\sim$20 km s^{-1}, which is only 10% of the 200 km s^{-1} flow-enhancements at the magnetopause. The current has to be substantially larger to affect the test of the RD condition. A statistical survey of 96 Cluster magnetopause crossings indicates that the average ion-electron differential motion associated with the magnetopause current is of the order of 30 km s^{-1}(Paschmann et al., 2004) (see also Dunlop et al., 2002). Thus the use of ions instead of electrons is unlikely to be the explanation for previous sub-Alfvénic flow cases.

In terms of the possible presence of heavy ions (O^+ or He^{++}) that could affect the HIA ion moments and therefore the Walén test (Paschmann et al., 1986; Puhl-Quinn and Scudder, 2000) the observed O^+ number density on January 26, 2001 is $\sim$2% of the total density. At this low O^+ density, the effect does not significantly alter the results (Paschmann et al., 1986). This explains the successful Walén test for all events on January 26, 2001. What needs to be determined in future studies is whether there are cases of abundant O^+ that could affect tests of the RD relation.

10.1.4.2 Kinetic signatures: Absence of D-shaped ion distributions in some reconnection jets

The existence of the D-shaped ion distributions at the magnetopause, a predicted kinetic effect of reconnection (Cowley, 1982), had been reported in a number of studies (e.g., Gosling et al., 1990b; Fuselier et al., 1991; Smith and Rodgers, 1991; Bauer et al., 1998; Phan et al., 2001). These distributions have the predicted low-energy cutoff along a constant $v_{\parallel}$ at precisely the deHoffmann-Teller (HT) velocity. Their detections provide quantitative evidence for reconnection which complement the observations of the fluid signatures of reconnection, namely the Alfvénic jets.

On occasion both fluid and kinetic signatures of reconnection have been seen in the same events. However, in a systematic survey of these signatures at the magnetopause, Bauer et al. (1998) found that more often than not, fluid signatures (Alfvénic flows) are observed without particle signatures ("D-shaped" ion distributions), and vice versa. This is a surprising finding since one would expect all fluid and kinetic signatures to be present in any reconnection events.

Cluster routinely provides full 3-D ion distributions (not just moments of 3-D distributions) at high-time resolution (4-12s), which permits examination of the occurrence of D-shaped ion distributions throughout the magnetopause current sheet. In the continuous reconnection event (January 26, 2001) reported by Phan et al. (2004), Alfvénic flows were detected at all magnetopause crossings (see Section 10.1.1.3). One thus expects to detect D-shaped ion distributions in the jets. Phan

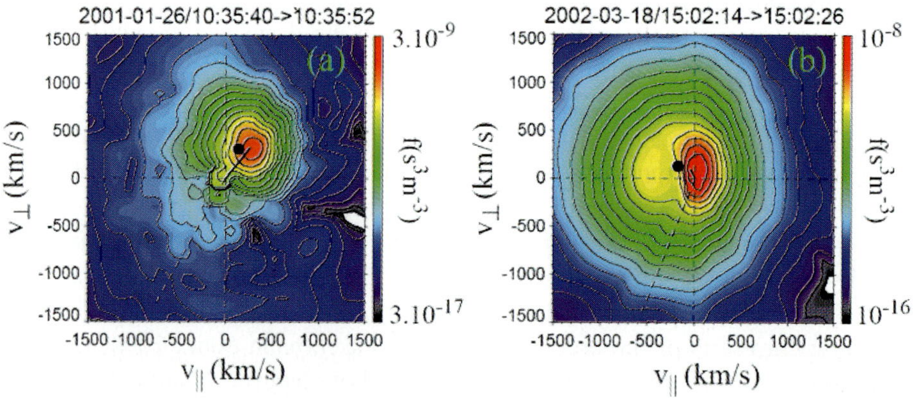

Figure 10.11. Absence and presence of D-shaped distributions in reconnecting magnetopause. Two-dimensional cuts of the 3D ion distributions through velocity space that contains the directions of the magnetic field (to the right) and E×B (upward). Black points indicate the deHoffmann-Teller (HT) velocity. (a) Example of non-D-shaped ion distribution in the January 6, 2001 reconnection jets, (b) D-shape ion distribution, with the low-energy cutoff along a constant $v_{||}$ at the HT velocity, near an X-line in a tailward of cusp reconnection event on March 18, 2002. (Adapted from Phan et al., 2004).

et al. examined every ion distribution measured in the jets and found that, contrary to the expectation, none of the distributions displays the expected D-shape with low-energy cutoff at the HT velocity. An example is shown in Figure 10.11 (left) during a magnetopause crossing from the magnetosheath into the magnetopause. It is seen that the distribution in the magnetopause jet is nearly isotropic in the bulk velocity frame, with no low-energy cutoffs at the HT velocity (the black dot). This finding does not depend on the location within the MP/BL where the distributions are sampled. Phan et al. (2001) had suggested that the D-shaped distributions are only present on the magnetosheath edge of the magnetopause because the plasma there had crossed the magnetopause locally. However, in this event the D-shape distributions are absent across the entire magnetopause. The high-resolution sampling of the magnetopause by more than 1 spacecraft rules out the possibility that the D-shaped distributions are confined to a thin layer and missed by the measurements.

The presence of the D-shaped distributions in some reconnection events (from previous reports) but not in others is presently not understood. Figure 10.11 (right) displays a clear D-shaped distribution detected on March 18, 2002 by Cluster in a reconnection event poleward of the cusp during northward IMF (Phan et al., 2003). One noted difference between the January 26, 2001, and March 18, 2002 events is that for the latter event, the jets (and the D-shaped distributions) were detected relatively close to the X-line (since the X-line drifted past the spacecraft), whereas in the January 26, 2001 case, the X-line is likely to be more than 9 R_E

away. Whether the distance to the X-line is a determining factor remains to be investigated.

10.1.5 Summary

In the first 3 years of the Cluster mission, due to its high-resolution multi-point observations as well as its orbit, Cluster has made significant contributions to the understanding of the large-scale properties of reconnection at the magnetopause. Highlights of the Cluster findings include:

- Convincing evidence for both intermittent and continuous reconnection at the magnetopause has been obtained from multi-point observations. Reconnection at the local magnetopause is continuous when the IMF is stable. When the IMF orientation changes, the reconnection site moves but reconnection never ceases. Thus reconnection may be intermittent at the local magnetopause but continuous on a global scale. Changes in the solar wind dynamic pressure can also change the reconnection rate and produce flux transfer events.
- A strong 'guide field' detected at a reconnection X-line, i.e., a finite magnetic field along the X-line, has provided direct evidence for component merging.
- Tailward-of-the-cusp reconnection has been found to occur only when the IMF has a northward component. The occurrence rate of cusp reconnection is nearly 100% when the IMF has a northward component, implying that cusp reconnection in the northern and southern hemispheres must be common. The high occurrence rate (in contrast to a rate of ~50% at the subsolar magnetopause) is thought to be due to the presence of a plasma depletion layer. In this layer, the plasma β is reduced, rendering the magnetosheath flow sub-Alfvénic and allowing the establishment of a stable X-line at the high-latitude magnetopause.

10.2 Kelvin-Helmholtz Instability at the Flank Magnetopause

In this section we describe Cluster observations of large-scale Kelvin-Helmholtz (KH) rolled-up vortices at the flank magnetopause. According to theory, the non-linear phase of the KH instability is a necessary condition for the transport of solar wind into the magnetosphere via this process.

While reconnection between solar wind and terrestrial magnetic fields can readily account for the formation of the LLBL during southward IMF conditions, there is presently no consensus as to how the solar wind plasma populates the low-latitude boundary layer when the IMF is oriented northward and parallel to the geomagnetic field. In this case, reconnection is less efficient or absent at the low-latitude magnetopause although, as already mentioned earlier, there is a suggestion that simultaneous northern and southern cusp reconnection (i.e., at remote sites) could result in the formation of the LLBL (Song and Russell, 1992). Fuselier et al.

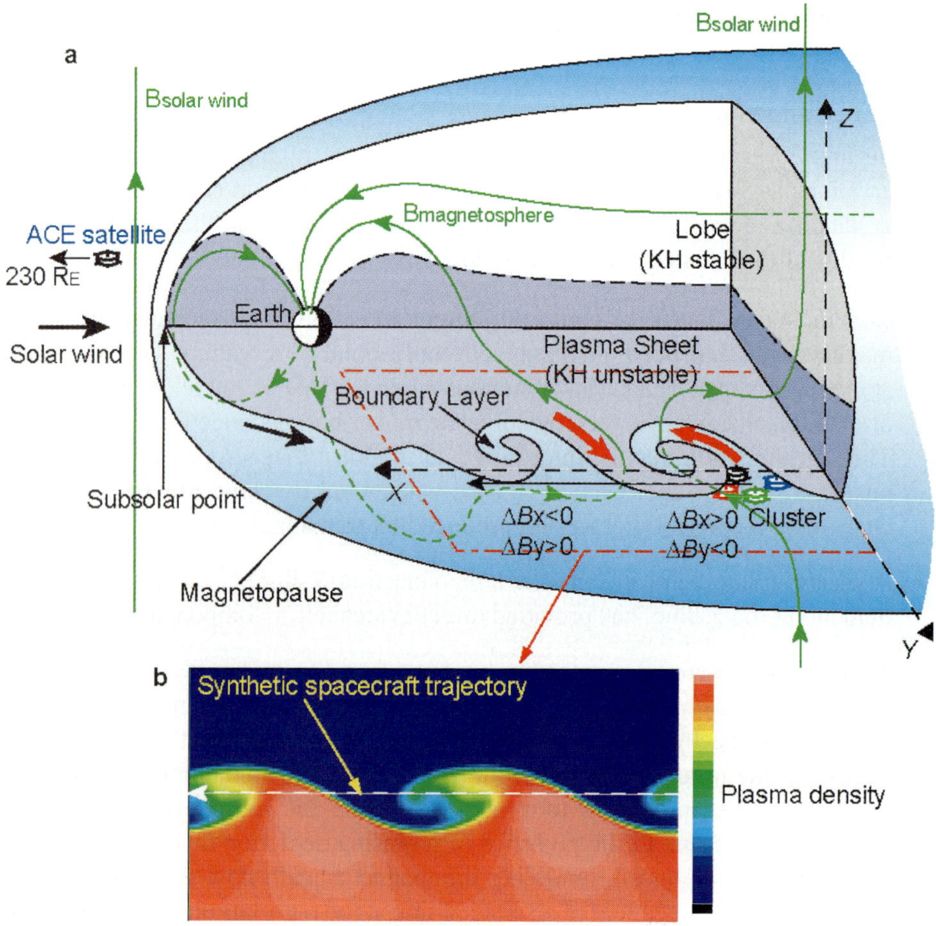

Figure 10.12. 3-D cutaway view of Earth's magnetosphere showing signatures of Kelvin-Helmholtz instability (KHI). (a) The KHI occurs at the interface between the magnetosheath and the plasma sheet because the plasma energy dominates in both regions, whereas it does not occur at the surface of the lobes where the magnetic energy dominates and the magnetic tension prevents it from deforming the magnetopause. Consequently, the KH vortices evolve only along the low-latitude magnetopause and only low-latitude portions of the magnetospheric and magnetosheath field lines are entrained into the vortices, inducing characteristic field perturbations in regions off the equatorial plane where the Cluster spacecraft were located. (b) Vortex structure resulting from a 3-D numerical simulation of the KHI. Colour-coded is the plasma density in an x-y cross section cut below the equatorial plane. The density, velocity, and magnetic field variations predicted when a synthetic satellite passes through the centre of the KH vortices are shown in Figure 10.13. (From Hasegawa et al., 2004).

(2002) pointed out that non-simultaneous reconnection (with reconnection in one cusp first and in the other hemisphere much later) can also give rise to an LLBL on closed field lines on the flanks.

Several candidate local-entry mechanisms unrelated to reconnection have also been proposed, one being the Kelvin-Helmholtz instability (KHI). This instability could occur along the flanks of the magnetosphere where the magnetosheath flow is fast relative to the stagnant magnetospheric plasma (Miura, 1984), as illustrated in Figure 10.12. Numerical simulation models (Thomas and Winske, 1993; Fujimoto and Terasawa, 1994; Huba, 1996; Nykyri and Otto, 2001; Matsumoto and Hoshino, 2004; Nakamura et al., 2004) suggest that fast plasma transport across the magnetopause can be accomplished by the KHI only when the KHI has grown sufficiently to form rolled-up vortices which can engulf plasmas from both sides of the magnetopause. In these models, the collapse of, or reconnection within, such a vortex (in the nonlinear phase of the KHI) is responsible for the plasma transport.

Multiple and quasi-periodic magnetopause encounters by single spacecraft and vortex-like flow perturbations near the boundary have been reported and are often interpreted as representing surface waves or vortices excited by the KHI (Ogilvie and Fitzenreiter, 1989; Kivelson and Chen, 1995; Hones et al., 1981; Fairfield et al., 2000; Fujimoto et al., 2003). But as long as these signatures are observed only by a single spacecraft one cannot tell unambiguously whether the KHI has reached its nonlinear stage to generate rolled-up vortices, which are crucial ingredients for plasma transport, or they are just ripples or small-amplitude KH vortices on the magnetopause surface. The KH rolled-up vortices, expected to form at the magnetopause, have complex structures, such as vortical plasma flow and a filament-like high density region intruding into the low density (magnetospheric) region (see Figure 10.12). To resolve such complex structures in the KH vortices, multipoint in situ measurements as carried out by the Cluster mission are essential, as are comparisons with 3D simulations.

Using Cluster multi-spacecraft measurements, Gustafsson et al. (2001) and Owen et al. (2004) reported the steepening of waves along the flank magnetopause. These observations could be interpreted as evidence for the KH instability in the process of becoming nonlinear at the magnetopause (see Section 'Surface Waves' in De Keyser et al., 2005).

More recently, Hasegawa et al. (2004) reported evidence for KH instability reaching its nonlinear phase along the flank magnetopause. The four Cluster spacecraft forming a tetrahedron made a fortuitous direct encounter with the rolled-up vortices (Figure 10.13) on November 20, 2001 when the upstream solar-wind magnetic field, observed by the ACE spacecraft, pointed northward, i.e., when reconnection was less efficient equatorward of the cusp, but the condition was favourable for the KHI at the low-latitude magnetopause. The magnetosheath and magnetospheric magnetic fields on the dusk flank magnetopause were approximately parallel to each other throughout the 16-minute interval (Figure 10.13 f-g). The period of the vortices encounter was embedded in a more than 13-hour interval of quasi-periodic plasma and magnetic field perturbations related to deformations of the magnetopause (09:30-23:30 UT). Rolled-up vortices were clearly identified using

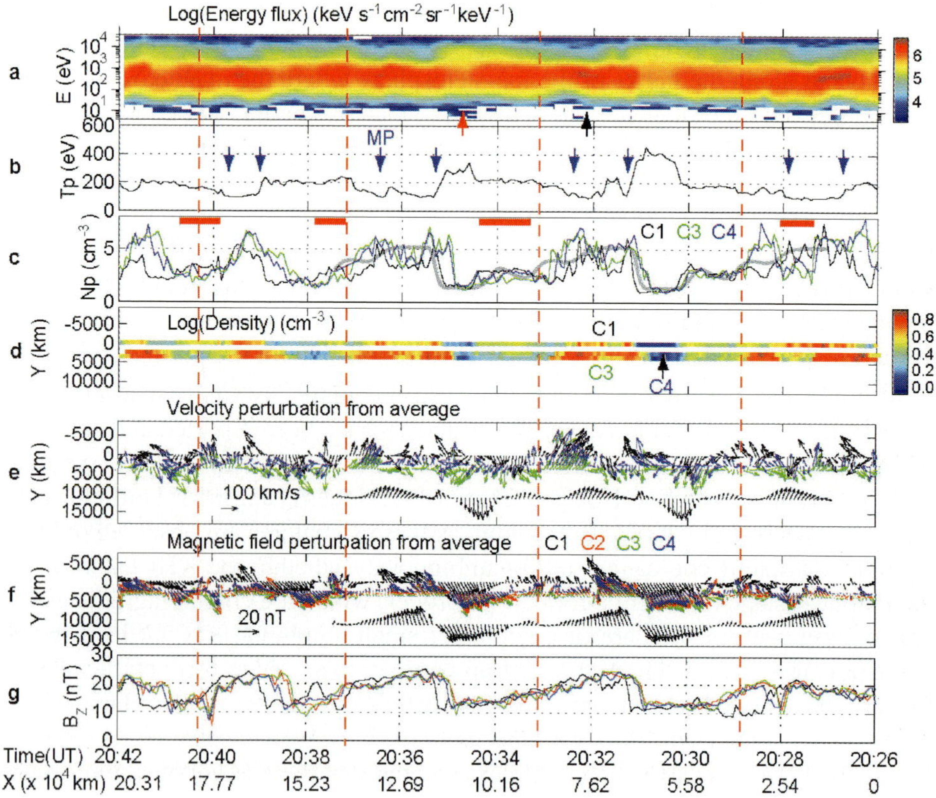

Figure 10.13. Detection by Cluster of large-scale plasma vortices on November 20, 2001 (20:26-20:42 UT). The satellites were 3 Earth radii behind the terminator and were separated by ~2000 km from each other. Time progresses to the left and has been translated into the position of the spacecraft. a, the omni-directional energy spectrogram of ions observed by the Cluster 1 spacecraft (C1); b, the C1 ion temperature. The blue arrows mark approximate locations of the magnetopause; c, the plasma density variations showing their similarity to that predicted by the numerical simulation (thick grey curve); d, the plasma density colour-coded and projected along the spacecraft trajectories. The y-axis is orthogonal to both x and the direction of the averaged magnetic field, B_{mean}, and points outward along the magnetopause normal; e, f, the x-y projection of the velocity and magnetic field deviations from the average velocity and magnetic field, respectively (C1:black, C2:red, C3:green, and C4:blue), along with the behaviour predicted by the simulation (black arrows shown in the lower part of panels e and f). The red vertical lines mark the approximate centres of the vortices. (From Hasegawa et al., 2004).

the multi-spacecraft information (Figure 10.13e). The flow vectors, transformed into the frame of the vortices and viewed from the north, rotate anticlockwise around the center of the vortices (marked by the red vertical lines in Fig. 10.13) as expected at the duskside magnetopause.

Near these vortices, the density observed by the Cluster 1 spacecraft (C1) located on the most magnetosphere side of the magnetopause (marked by the blue arrows in Figure 10.13 b was often higher than that observed by C3 or C4 (Figure 10.13 c-d). These instances of higher density at C1 are marked by red bars in Figure 10.13 c. Such high density regions appear to be connected to the dense solar wind on the anti-sunward side, for example at 20:27:30 UT in Figure 10.13 d, rather than detached from the solar-wind region. This feature is consistent with the simulation result (Figure 10.12 b) and indicates that the dense regions result from the roll-up of the solar-wind plasma associated with the growth of the KHI, not from the impulsive penetration process (Lemaire, 1977). The density variation observed by C1 (black curve in Figure 10.13 c) is similar to that expected when a synthetic spacecraft moves through the central portion of the simulated vortex (Figure 10.12 b) and crosses the magnetopause back and forth (as shown by the thick grey curve in Figure 10.13 c), suggesting that C1 was in the vicinity of the vortices center.

In addition to the density and flow signatures of the rolled-up KH vortices, a unique magnetic field perturbation pattern was identified associated with these vortices. This pattern should appear only in a 3-D configuration of the magnetosphere where the KH-unstable plasma sheet is sandwiched between the KH-stable northern and southern lobes (Figure 10.12 a). A numerical simulation of the KHI which considers this 3-D magnetosphere-like geometry predicted that this field perturbation manifests itself in the boundary regions between the plasma sheet and the lobes. This is because only low-latitude portions of the field lines surrounding the magnetopause are engulfed by the vortices while those at high latitudes are unaffected (see Figure 10.12 legend). Fig 10.13 f shows that magnetic field perturbations seen in the high density (solar-wind) region ($\Delta B_x > 0$ & $\Delta B_y < 0$) and in the low density (magnetospheric) region ($\Delta B_x < 0$ & $\Delta B_y > 0$) have polarities in precise agreement with the 3-D KHI effect on the magnetic field on the southward side of the equatorial plane where Cluster resided (Figure 10.12 a). The combined plasma and magnetic field observations provide unambiguous identification of rolled-up vortices at the magnetopause, which, according to theory, is a necessary ingredient for plasma transport via KHI.

Evidence for plasma transport across the magnetopause was indeed observed in this event. Cold solar-wind and hot magnetospheric ion populations are found to coexist in the vortices. Significant amounts of the solar-wind (< 2 keV) and magnetospheric (> 5 keV) ions were detected simultaneously in the same region on the magnetosphere side of the magnetopause (Hasegawa et al., 2004). The appearance of rolled-up vortices in the vicinity of the boundary layer suggests that the KHI mechanism could be responsible for plasma transport across the boundary, although the exact microphysical process that leads to plasma transport within the KH vortices could not be identified based on the available measurements. However, it was possible to rule out local reconnection since during the interval of the vortex observations, signatures of plasma acceleration due to magnetic stresses

(Paschmann et al., 1979; Sonnerup et al., 1981) and D-shaped ion distributions characteristic of reconnection (Cowley, 1982; Fuselier et al., 1991) were not observed. To exclude, conclusively, the possibility of remote (high-latitude) reconnection supplying the plasma observed at low latitude is more difficult. It was noted, however, that the boundary layer ions, detected off the equator, were flowing poleward in precisely the same direction as the magnetosheath flow. This seems inconsistent with the simple idea that high-latitude magnetopause reconnection, which would result in an equatorward flow at the observation point, produced the observed LLBL. These observations seem to imply that reconnection did not occur locally and high-latitude reconnection did not populate the plasma in the KH vortices. However, one cannot rule out the possibility that reconnection occurred in the past (but is no longer active) to produce a boundary layer which is now rolled-up by the KHI. Proving the actual occurrence of such a scenario is, however, even more difficult.

Based on the velocity measurements during the vortices interval, Hasegawa et al. estimated the length scale of one vortex to be 40,000-55,000 km. According to theory (Miura and Pritchett, 1982), the wavelength of the fastest growing KH mode is approximately 8 times the initial total thickness of the velocity shear layer. Thus, the initial thickness of the velocity shear layer is inferred to have been roughly 5,000-7,000 km. Furthermore, according to numerical simulations, the width of the sufficiently developed KH vortices equivalent to that of the plasma boundary layer reaches about 4 times the initial thickness. It is therefore inferred that a boundary layer with thickness in the range 20,000-28,000 km had formed in the most KH-unstable low-latitude regions near, or further downstream of, the observation site.

The Cluster results indicate that the KHI occurs at the flank magnetopause and that it may lead to solar wind entry, perhaps via non-reconnection associated processes, for northward solar-wind magnetic field. But the microphysical processes that cause the plasma transport in the KH vortices, which would control the rate at which mass and energy from the solar wind are transferred, remain to be understood.

10.3 Microphysics of Magnetopause Processes

Investigation of the microphysics of the magnetopause involves the resolution of ion and electron inertial scales. On these scales one expects kinetic processes to become important and plasma waves to play a role in the dynamics of plasma and magnetic field. Microphysical processes at the magnetopause and in the magnetopause boundary layer are believed to be responsible for collisionless magnetic reconnection and the generation of diffusive plasma entry into the magnetosphere.

Reconnection is an important mechanism for plasma transport that results in changes of magnetic topology. The process is initiated in a narrow 'diffusion region'. Though macroscopically the occurrence of reconnection is well established, the microphysical processes that lead to reconnection are still not well understood.

In addition to reconnection, general diffusive entry at the magnetopause may also contribute to local transport across the magnetopause. Here one refers not to collisional diffusion, which is small at the magnetopause, but rather diffusion from wave-particle interactions, driven by electromagnetic field fluctuations. Observations of wave activity are crucial for inferring microscopic mechanisms involved in the reconnection diffusion region and for diffusive entry. In principle, it may be relevant even for low diffusivity in the presence of eddy turbulent motion (e.g., Kelvin-Helmholtz turbulence). For KH turbulence, the turbulent eddies transport plasma reversibly over mesoscales and diffusive processes break the frozen-in condition only over the microscales found within the eddies.

In the following we discuss recent observations of the Cluster mission on microscales. At these scales, the processes can be described in a fluid picture by a generalised Ohm's law, as is briefly discussed in the next paragraphs. We then describe four-spacecraft observations of an ion diffusion region, followed by observations of high-frequency, lower hybrid, and low-frequency waves. The section concludes with an estimate of the diffusion coefficient near the magnetopause, based on the unique plasma gradient determination provided by Cluster, in conjunction with the observation of lower-hybrid waves.

10.3.1 Collisionless generalised Ohm's law

Breaking the frozen-in condition requires violation of the ideal-MHD Ohm's law $\mathbf{E} = -\mathbf{V} \times \mathbf{B}$. This can be done by adding terms to the right-hand side of the expression:

Two-fluid theory provides a simple generalisation of Ohm's law

$$\mathbf{E} + \mathbf{V} \times \mathbf{B} = \eta \mathbf{J} + \frac{1}{\varepsilon_0 \omega_{pe}^2} \hat{D}\mathbf{J} + \frac{1}{en} (\mathbf{J} \times \mathbf{B} - \nabla \cdot \mathsf{p}_e), \qquad (10.2)$$

where η is an effective resistivity, $\hat{D}$ is a modified convective time derivative, $\mathbf{V}$ is the bulk flow velocity, n is the quasineutral plasma density, $\mathbf{J} = \nabla \times \mathbf{B}/\mu_0$ is the electric current density, and p_e is the electron pressure tensor. The terms on the left describe the frozen-in nature of the magnetic field in an ideal plasma. The terms on the right describe deviations from the frozen-in state by resistive diffusion (Joule heating), electron inertia, the Hall effect, and electron pressure gradients, respectively.

In the collisionless plasma of the Earth's environment $\eta \approx 0$ and the resistive term can be dropped. However, in the presence of short wavelength electrostatic and electromagnetic waves 'anomalous' resistivity $\eta_{an} = \nu_{an}/\varepsilon_0 \omega_{pe}^2$ can be produced by scattering electrons off wave fronts (e.g., Sagdeev, 1979) and yielding an anomalous diffusivity $D_{an} = \eta_{an}/\mu_0$. The above two-fluid equations are then understood as 'averaged over the wave correlation length'. The 'anomalous collision frequency' ν_{an} depends on the nature of the waves. Expressions for different types of wave modes have been given in the literature (Sagdeev, 1979; LaBelle and

Treumann, 1988; Sibeck et al., 1999; Treumann and Sckopke, 1999). Cluster observations related to the effect of low frequency magnetic and electrostatic waves will be discussed below in more detail.

One of the most important terms accessible to Cluster is the Hall term $\mathbf{J} \times \mathbf{B}$. It becomes significant on scales comparable to, or smaller than, the ion inertial-scale length $\lambda_i = c/\omega_{pi}$. This happens in the reconnection diffusion region which is a region close to the reconnection X-line where electron and ion flows decouple from each other. The Hall term introduces the dynamics of whistler and kinetic Alfvén waves (Rogers et al., 2001). In observations of reconnection in the magnetospheric tail (Nagai et al., 2001; Øieroset et al., 2001; Runov et al., 2003) and at the magnetopause (Mozer et al., 2003; Vaivads et al., 2004) the effect of the Hall term has been detected in the signature of an out-of-plane quadrupolar magnetic field component confined to the reconnection ion diffusion region as predicted by theory (Sonnerup, 1979; Terasawa, 1983; Mandt et al., 1994; Birn et al., 2001).

In examining the validity of the generalised Ohm's law for the the magnetopause, it is of vital interest to assess the importance of its various terms. Cluster provides the following opportunities:

- to estimate anomalous diffusion coefficients based on wave measurements;
- to determine the balance of the electric field by the various terms in Ohm's law, in particular the Hall term;
- and to identify the closure of field-aligned currents emanating from the ion diffusion region.

Cluster spacecraft separations and plasma data cadence do not permit full investigation of processes on the electron inertial scale, $\lambda_e = c/\omega_{pe}$, or processes caused by the divergence of the electron pressure tensor. In other words, the relative contribution of the electron inertial term in Ohm's law and the spatial derivatives of the electron pressure terms cannot be assessed.

10.3.2 Ion diffusion region observations

One of the key issues in the physics of reconnection is the understanding of the microphysical processes in the region where reconnection is initiated – the small diffusion regions, where the magnetic flux is no longer frozen into the motion of the ions (ion diffusion region) and the electrons (electron diffusion region). Single-spacecraft observations have established the collisionless character of the ion-diffusion region (Nagai et al., 2001; Øieroset et al., 2001; Mozer et al., 2002). However, single-satellite measurements involve ambiguity in distinguishing spatial and temporal features of the diffusion region (Mozer et al., 2002). In this section we describe multi-spacecraft observations of the ion diffusion region that reveal stable Hall electric and magnetic field structures, bifurcation of the current sheet, and intense field-aligned currents along the separatrices.

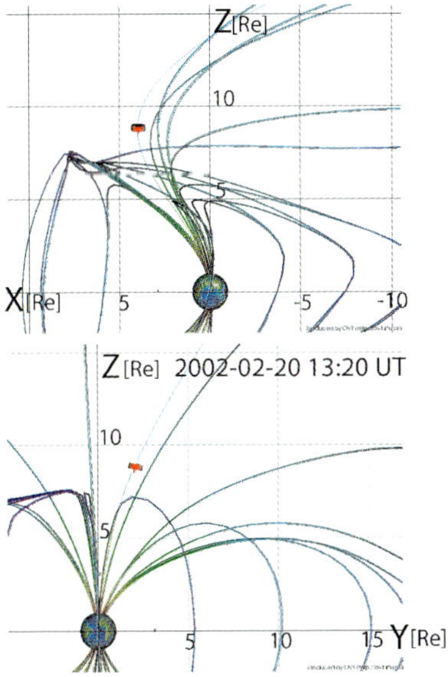

Figure 10.14. Location of the four Cluster spacecraft (marked in red) and the magnetic field lines of the Earth's magnetosphere (Tsyganenko 2001-model, interplanetary magnetic field IMF in GSE (B_x, B_y, B_z)=(0,7,-2) nT, and solar wind pressure 3.5 nPa are averages from ACE spacecraft). Assuming that reconnection occurs in regions with anti-parallel magnetic fields, the varying IMF B_z (from +3 nT to -7 nT within 10 min) is consistent with a diffusion region located dusk- and tailward of the cusp, as observed by Cluster (Figure taken from Vaivads et al., 2004).

Vaivads et al. (2004) recently reported Cluster observations during a passage of the diffusion region where the Hall effects were observed by all spacecraft crossing the ion diffusion region consecutively, indicating that the large amplitude Hall fields are stable spatial structures. The measurements were taken on February 20, 2002 when Cluster crossed the magnetopause poleward of the polar cusp several times while reconnection took place between the northward IMF and the southward magnetospheric field at high latitudes. Reconnection occurred between the plasma mantle and the magnetosheath. Figure 10.14 shows the Cluster configuration for this event. The observational results combined with a numerical simulation are given in Figure 10.15. A 2-D two-fluid MHD simulation of the diffusion region for plasma parameters similar to observation (Rogers et al., 2003) is given on the left of this figure. The colour coding is for the out-of-plane magnetic field component (so-called Hall field) showing its quadrupolar structure.

The relative times at which the Cluster spacecraft observe variations in the reconnecting magnetic field component provide an estimate of the magnetopause velocity $V_{mp} \sim -120\ (0.82, 0.54, -0.18)\ \mathrm{km\,s^{-1}}$ GSE. This velocity can be reliably estimated only using multi-spacecraft data; it establishes, with a high confidence, the magnetopause reference frame and the spatial scale of the magnetopause cur-

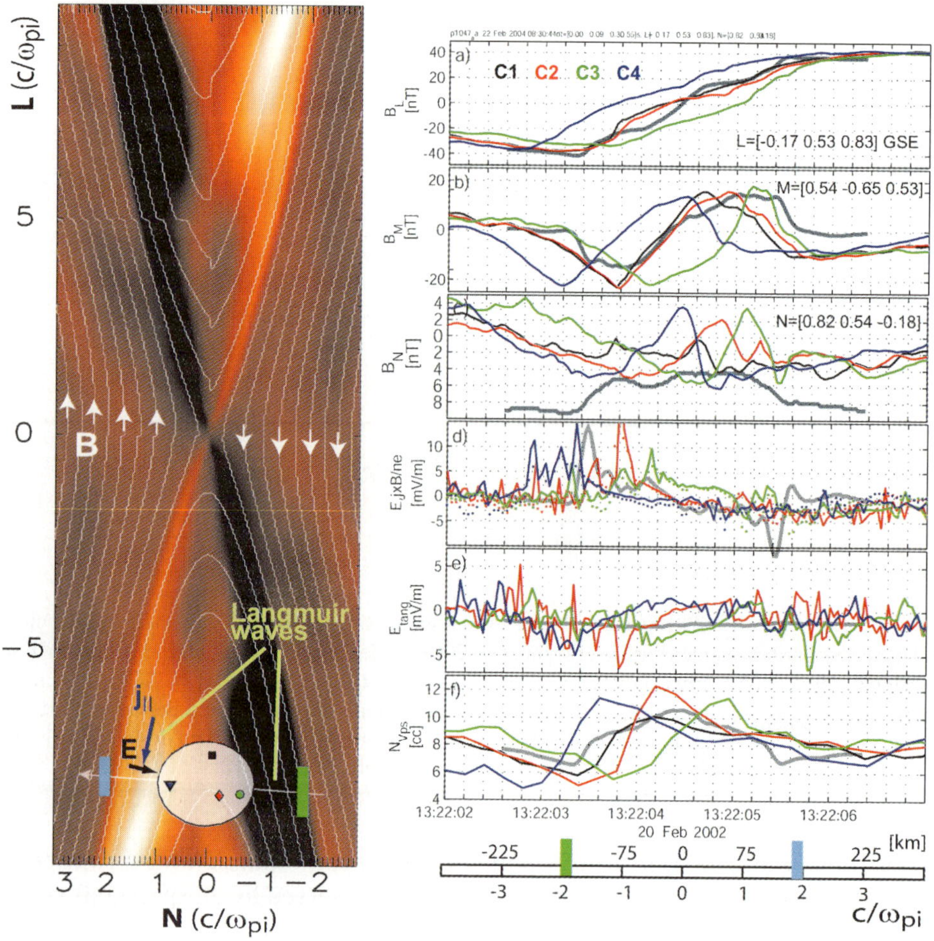

Figure 10.15. *Left*: Structure of the diffusion region from a numerical Hall fluid simulation by Rogers et al. (2003). The magnetic field lines are shown in white, the out-of-plane magnetic field is colour coded (white duskward, black dawnward). Also shown: projection of Cluster configuration with colour coded spacecraft, and approximate location relative to the diffusion region. *Right*: Cluster observations, simulation results are in grey lines. *a*: Reconnecting magnetic field component. *b*: Out-of-plane magnetic field component. *c*: Normal magnetic field component. *d*: Electric field normal to the magnetopause, directly observed (solid lines), $\mathbf{j} \times \mathbf{B}/ne$ (dotted lines). *e*: Tangential electric field with average value about -1 mV m^{-1}. *f*: plasma density from satellite potential. At the bottom the spatial scale obtained from the four-spacecraft magnetopause velocity estimate is given. The observations are consistent with fast collisionless reconnection. (From Vaivads et al., 2004).

rent layer (bottom of figure). The current sheet thickness is about 300 km, $\sim 4\lambda_i$. All 4 spacecraft observe a similar current structure (panel a), with the largest deviation seen by Cluster-3. Thus, the current sheet is planar on the scale of spacecraft separation ($\sim \lambda_i$) and is stable on the time scale of one second (approximately the ion gyroperiod). In addition, all 4 spacecraft observe that the current sheet is bifurcated, with strongest current (gradient in B_L) along the outer edges.

The Hall magnetic field B_M (panel b) exhibits a bipolar variation. There is no significant constant offset in B_M, indicating the absence of an extra (external) guide field in this case. The fact that all 4 spacecraft, crossing the magnetopause consecutively, observe Hall fields indicates that this is a stable spatial feature of the diffusion region rather than some brief temporal variation. This is an important conclusion that was not possible to confirm with the measurements from earlier single spacecraft missions. The negative sign of the normal component of the magnetic field $B_N \approx$-3 nT in combination with a negative B_M followed by a positive B_M is consistent with the Cluster spacecraft crossing the diffusion region south of the X-line.

For the local magnetopause being a rotational discontinuity, the $B_N \sim 10\%$ of B_L yields a reconnection rate of ~0.1, a value that is obtained in numerical simulations of fast collisionless reconnection (e.g., Hesse et al., 2001; Shay et al., 2001). It corresponds to an inflow velocity of $\sim 25\,\mathrm{km\,s^{-1}}$, or a tangential electric field in the magnetopause reference frame of $E_{\mathrm{tang}} \sim 1\,\mathrm{mV\,m^{-1}}$. The observed tangential electric field component fluctuates, but is on average of this value (panel e).

Panel f shows that in the center of the current sheet (plasma outflow region) the density increases significantly (by ~50%), and there is a similarly large density dip when entering the current sheet. The density dips had been seen in simulations (e.g., Shay et al., 2001) along the separatrices as a result of enhancements of the magnetic field strength associated with the Hall current loops. The absence of a significant density gradient across the magnetopause shows that reconnection is almost symmetric at this position where the magnetospheric mantle and magnetosheath fields reconnect.

In addition to the Hall magnetic field, Cluster also observed the Hall electric field (panel d). The electric field component normal to the magnetopause, E_n changes sign from positive to negative in the center of the current sheet. The dotted lines show the corresponding $\mathbf{j} \times \mathbf{B}/ne$ terms of the generalised Ohm's law, Eq. 10.2. E_n and $\mathbf{j} \times \mathbf{B}/ne$ are of same order and exhibit very similar variations thus indicating that most of the electric field E_n is balanced by the $\mathbf{j} \times \mathbf{B}/ne$-Hall term in the generalised Ohm's law. This good agreement is best seen within the narrow region of strong E_n at ~13:22:04 UT. Thus the Cluster measurements confirm the dominant role of the Hall term on the scale of the ion diffusion region. In addition, the presence of a cross-magnetic field electric field on a spatial scale comparable to, or smaller than, the ion gyroradius implies that ions will be accelerated by this field as they move from the outside of current sheet to the center of current sheet. The cross-field potential drop across the region of the strongest electric field amounts to ~300 V. A proton accelerating through 300 V increases its speed by $\sim 250\,\mathrm{km\,s^{-1}}$. This is approximately the Alfvén velocity in the inflow region and comparable to the predicted outflow velocity of ions from the reconnection region.

Figure 10.16 shows the location of the parallel currents in the ion diffusion region of the reconnection site. The out-of-plane magnetic field component B_M in panel a exhibits a bipolar signature which is attributed to the Hall currents in the reconnection region. Panel b shows the parallel current obtained by using the four Cluster spacecraft to define the appropriate coordinate system and then using the single spacecraft magnetic field perturbations to calculate the current. As predicted by simulations, strong parallel currents in the direction away from the X-line occur all along the outer edge of the bipolar B_M structure. (Since $\nabla \cdot \mathbf{J} = 0$, there must also be return currents toward the X-line further inward from the separatrix.)

10.3.3 High-frequency waves

High frequency plasma waves at the magnetopause can be taken as indicators of electron beams which have been accelerated at the magnetopause. In the absence of very high temporal resolution (<4s) of the electron distribution measurements this is the best and in most cases the only way of inferring such beams. Since the main reasons for particle acceleration along the magnetic field can be sought in the process of reconnection, such beams provide information on the processes inside the diffusion region. Here we report on two such observations, one of which is the event described in the previous section. The key finding is that localised high frequency waves along the separatrices are directly related to field-aligned currents.

Figure 10.16 shows the relation between the passage of an ion diffusion region tailward of the cusp during reconnection as was discussed above. In this figure, panel c shows the electric field power integrated over a broad frequency range from 2-80 kHz which includes the plasma frequency f_{pe}. All four spacecraft show that the highest amplitudes of the waves are along the external edges of the bipolar disturbance in B_M. These regions coincide with the separatrix of the reconnection site which is located north of the Cluster quartet. As shown above, the regions of strong emissions coincide with narrow regions of strong parallel currents. For the first time the high-frequency waves in the vicinity of a reconnection site are directly related to the parallel currents flowing along the separatrices. While single spacecraft measurements have indicated a relation between high frequency waves and separatrices already earlier (Farrell et al., 2002), the multi-point measurements were crucial to establish the existence of a direct relation between the two.

The wave spectrum is broad-band, exhibiting a spectral peak near the plasma frequency, panel d. This suggests that these waves are a combination of Langmuir (or upper hybrid) waves and solitary structures or electron acoustic emissions. The presence of these waves indicates the presence of electron beams along the separatrix and thus serve as a diagnostic tool for the reconnection site. The waves could also play a role in electron thermalization. Neither the beam nor this thermalization is, however, observable due to the narrowness of the transition.

Figure 10.17 shows another example of high-frequency wave observations associated with reconnection, but obtained at some distance from the X-line and in

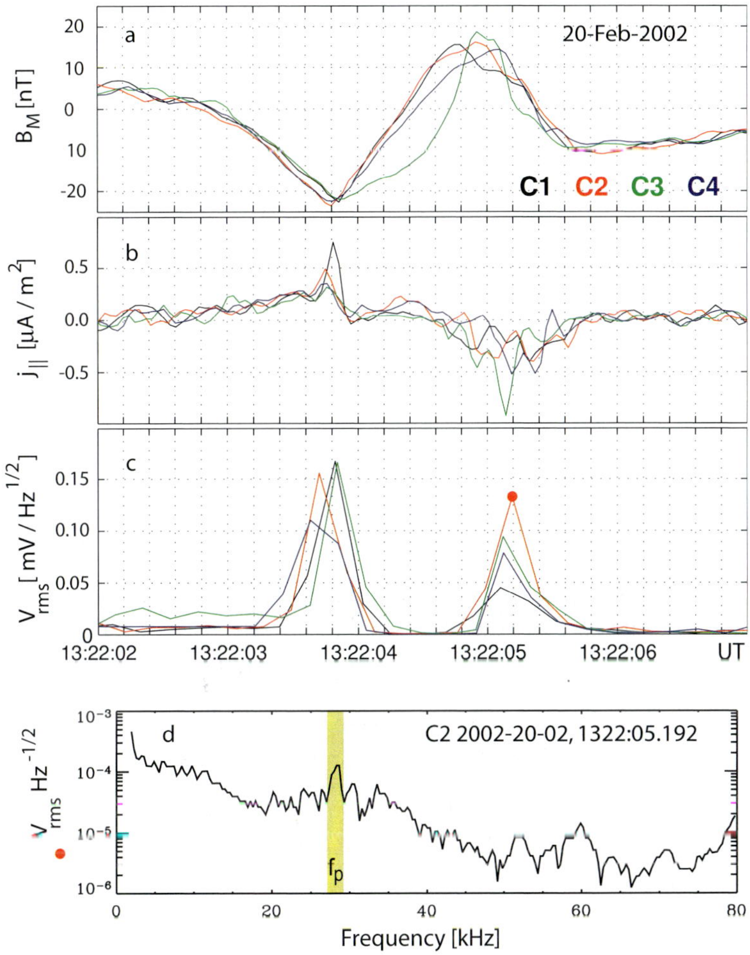

Figure 10.16. Observations of field-aligned currents and Langmuir/upper hybrid waves at the separatrix. For Cluster 2,3 and 4 the time series have been shifted -0.09 s, -0.3 s, and +0.55 s into the magnetopause frame. *a*: Out-of-plane magnetic field component; *b*: Field-aligned current with direction away from the X-line; *c*: Total spectral density in the frequency range 2-80 kHz; *d*: Wave spectrum exhibiting intense emissions at the plasma frequency, $f_{pe} \sim 28$ kHz, marked by the red dot in panel c. (From Vaivads et al., 2004).

a high plasma-β region. The figure shows a sequence of Cluster 3 measurements on March 4, 2002 (Khotyaintsev et al., 2004). The spacecraft was in the high-β regime at this time outside the magnetopause with the exception of the short time interval around 09:38:30 UT, when it crossed the center of an FTE (as indicated by the maximum in the total magnetic field and the fast tailward flow with negative v_y-component, and with low β).

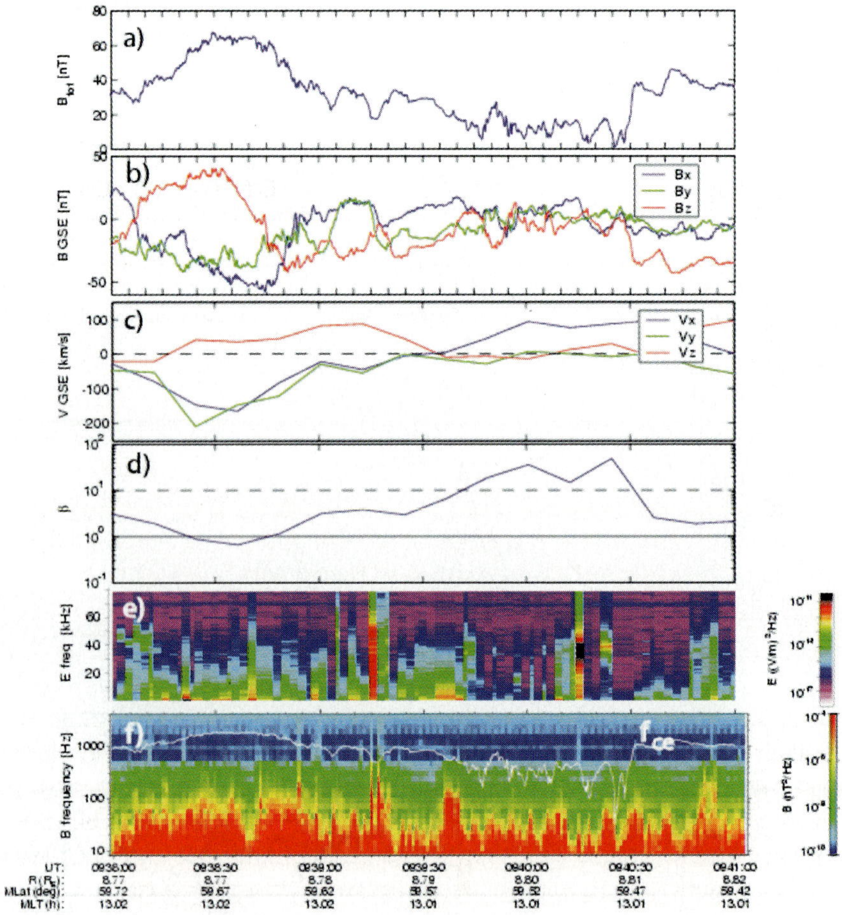

Figure 10.17. Cluster-3 measurements on March 4, 2002. *a*: total magnetic field magnitude, *b*: magnetic field components, *c*: plasma flow velocity, *d*: plasma β, *e*: electric field spectrogram between 2-40 kHz, *f*: magnetic field spectrogram, between 8 Hz and 4 kHz. (From Khotyaintsev et al., 2004).

The interesting times are at ~09:39:20 UT and ~09:40:15 UT when the electric spectrum exhibited intense high frequency emissions. The first of these events is a broadband event, while the second shows a spectral peak around the local plasma frequency at f_{pe} ~40 kHz. Both events are close to narrow strong current sheets, with currents up to 0.5 μA m^{-2} (determined by using the multi-spacecraft curlometer technique, but not shown here) of which the first coincides with the end of the passage across the FTE. The second is much closer to the magnetopause crossing at ~0940:30 UT.

Figure 10.18 shows the integrated electric wave power in the frequency interval 2-80 kHz (top panel) and the magnetic component B_x-GSE (maximum variance)

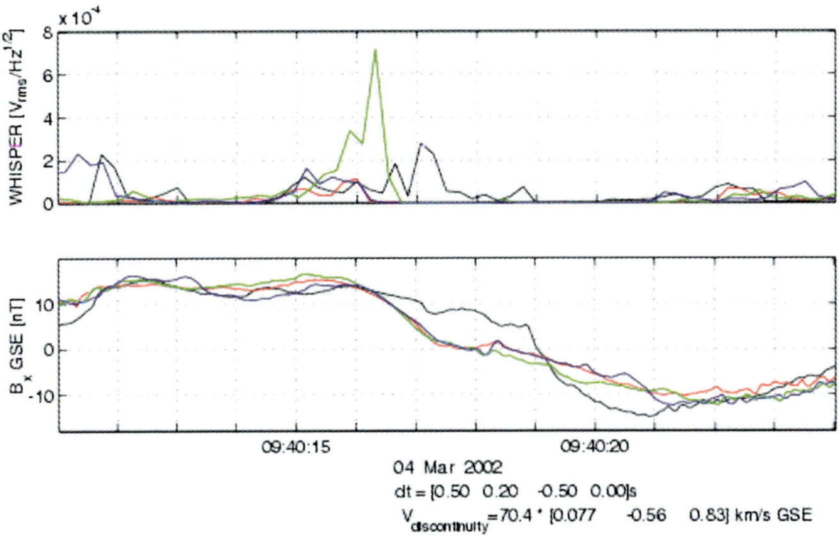

Figure 10.18. Integrated wave power. *Top*: Integrated electric wave power integrated in the frequency range 2-80 kHz. *Bottom*: The B_x-GSE component. (From Vaivads et al., 2004).

for the second high frequency event for all 4 spacecraft (the spacecraft at this time are at their closest separation, $\sim$100 km). All spacecraft see the highest electric wave power at the edges of the current sheet (indicated by the change in B_x) but at different intensities. Since the whole crossing of the layer took not more than 2 s, it was not possible for the PEACE instrument to provide a sufficiently fast measurement of the electron distribution. The observation of high frequency waves is the only and strongest indication for the presence of electron beams flowing along the magnetic field. Presumably these beams are emanating from the reconnection site at the cusp magnetopause along the magnetic separatrices.

10.3.4 Lower-hybrid waves

In this section we describe Cluster observations of intense lower-hybrid wave activity along the separatrix at the inner edge of the magnetopause boundary layer. Electron beams are the probable energy source for these waves.

Satellite measurements show that the magnetopause is almost always subject to enhanced wave activity, often of broad-band character, at frequencies close to the local lower hybrid frequency (Gurnett et al., 1979; Anderson et al., 1982; Tsurutani and Thorne, 1982; LaBelle and Treumann, 1988; André et al., 2004). Since the magnetopause is characterised by density gradients and by electron motion relative to the ions that is directed transverse to the magnetic field, some of the instabilities that may occur are drift instabilities, in particular the lower-hybrid drift instability (Davidson, 1978), electron beam instabilities and/or the electron-shear flow instability (Drake et al., 1994). These waves may be involved in particle acceleration, in diffusive plasma entry across the magnetopause from the magnetosheath into the magnetospheric boundary layer, and in reconnection.

The current understanding of the role of lower hybrid waves in reconnection remains controversial. Particle-in-cell (PIC) full particle simulations in 2-D and 3-D demonstrate that lower hybrid waves evolve in the current sheet and contribute to the onset of reconnection (Shinohara et al., 2001; Scholer et al., 2003). It seems that once reconnection has developed, it is no longer sensitive to the presence of lower hybrid waves. The reason for this insensitivity is that the reconnection rates obtained in different simulations either allowing or inhibiting lower hybrid waves are similar. From a crossing of the Polar spacecraft close to the reconnection site at the magnetopause, Bale et al. (2002) showed that lower-hybrid wave amplitudes were too small to support reconnection. Similar results have been obtained in laboratory experiments (Carter et al., 2002). More elaborate observations of such waves are therefore very helpful and can improve our understanding of their role in reconnection.

André et al. (2004) and Vaivads et al. (2004) have recently used Cluster wave-field measurements at frequencies near f_{LH}, taken some distance from the reconnection site at the magnetopause to infer the nature and intensity of the waves. Vaivads et al. (2004) studied a high latitude, northern hemisphere, dayside magnetopause crossing at MLT $\sim$14 on February 6, 2002 at 08:11:57 UT (see Figure 10.19). Cluster was at minimum spacecraft separation of $\sim$100 km, The magnetopause normal $\mathbf{n}_{\mathrm{MP}}$ was obtained from times when the spacecraft detected the magnetopause density gradient. Cluster 1 and 4 were closest to being on the same flux tube, with only $\sim$20 km separation transverse to **B**. Cluster 3 was the last to cross the magnetopause while Cluster 1, 2, and 4 crossed it in rapid sequence.

The overall structure of this dayside magnetopause region equatorward of the cusp has been described by André et al. (2004). These authors, using high time resolved PEACE electron data from Cluster sampled at a time resolution below the spacecraft spin period, showed that the largest amplitude electric fields were found on scales between c/ω_{pe} and c/ω_{pi} in the region marked in yellow in Figure 8.24 of De Keyser et al. (2005). These coincide with a narrow density depletion layer inside the magnetopause with steep density gradients and strong currents. Field-aligned currents flowing in this region are associated with field-aligned electron beams (André et al., 2004). This layer might be associated with the separatrix emanating from a distant reconnection site. It also coincides with the inner boundary of the magnetopause separating regions with and without high energy plasma sheet electrons.

The lower hybrid waves are expected to have wavelengths much shorter than the spacecraft separation and thus they are not expected to be correlated between different spacecraft. We therefore show data from only Cluster- 4, which had the

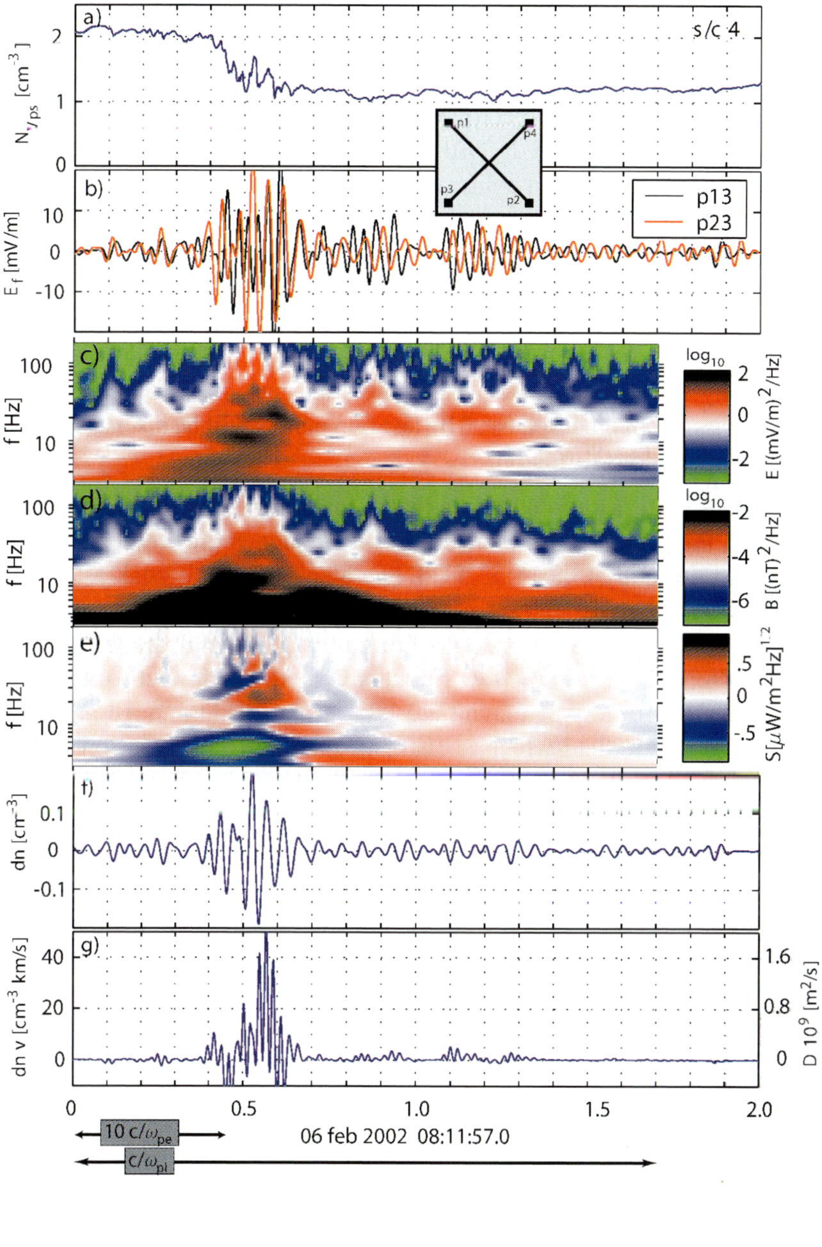

Figure 10.19. Lower hybrid drift wave observations near the magnetopause on February 6, 2002. *a*: plasma density n_{Vps} derived from satellite potential, *b*: electric field between p1 and p3, p2 and p3 band-pass filtered 15-35 Hz (inset showing geometry of probes), *c*: E-spectrogram, *d*: B-spectrogram, *e*: field-aligned Poynting flux spectrogram, *f*: band-pass filtered density fluctuations dn_{Vps} at 15–35 Hz, *g*: electron flux across magnetopause; vertical axis on the right gives an estimate of diffusion coefficient D for a density gradient of $25\,\mathrm{m}^{-4}$. (From Vaivads et al., 2004).

highest time resolution as the EFW instrument was in the internal burst mode. The importance of having multiple spacecraft will be discussed later.

Figure 10.19a shows the density as seen by Cluster-4. Figure 10.19b shows the band-pass filtered (15-35 Hz) electric wave field measured between the probes p13 and p23 (see inset). The lower-hybrid frequency is ~ 30 Hz. This frequency range shows a strong wave packet in coincidence with the density gradient. The plasma-β during the event was in the range of 0.3 to 0.5. The wave amplitude is mainly perpendicular to the local magnetic field indicating that these are lower-hybrid drift waves, excited by the density gradient. Panels c and d of the same figure show the respective electric and magnetic wavelet spectra of the measured signals between 0-180 Hz. Panel e gives the corresponding Poynting flux, panel f the band-pass filtered density fluctuations, and panel g the band-pass filtered 'diffusive' flux of electrons across the magnetopause. The last two panels will be discussed later when we consider anomalous diffusion.

The wavelet spectra show the broadband character of the localised emissions. Furthermore, the peaks near the lower-hybrid frequency are well expressed in the electric spectrum, and at 08:11:57.5 UT one also observes a peak at ~ 100 Hz in the whistler band in both electric and magnetic fluctuations. It is interesting to consider the colour-coded field-aligned Poynting flux (green antiparallel, red parallel). The energy flux is largest in the low frequency range and lowest for the whistlers, where it is about 2 orders of magnitude less. The whistlers move antiparallel to the ambient magnetic field which is also anti-parallel to the electron beam at the density gradient. On the other hand, the field aligned velocity of the lower hybrid-frequency waves is positive (parallel to the electrons). Both these observations and the fact that electron beams carry much more energy than the waves suggest that the electrons are involved in the interaction, possibly generating the lower hybrid and whistler waves. Moreover, the lower hybrid waves propagate obliquely possessing a magnetic component, i.e., they are not purely electrostatic. In this frequency range the presence of the magnetic component suggests that the lower hybrid waves are partially in the whistler mode and probably propagate near the resonance cone.

Estimates of the phase velocity across the magnetic field, applying interferometric methods (Vaivads et al., 2004), yield large though strongly fluctuating lower hybrid phase velocities $(\omega/k)_{LH} > 100\,\mathrm{km\,s^{-1}}$. Typical wave lengths lie between 3-10 km, substantially shorter than the spacecraft separation, which in retrospect explains the de-correlation of the waves between the spacecraft even for these short spacecraft separations. These observations do not resolve the importance of lower-hybrid waves in reconnection. They do, however, show that in certain regions (e.g., along the separatrices), the lower-hybrid wave activity can be intense.

The example above demonstrates that Cluster can provide very detailed information on the dispersive properties of the lower-hybrid waves at the magnetopause. However, more events will be needed to understand the role of these waves in the reconnection process. Particularly useful would be cases where Cluster is located closer to the X-line.

10.3.5 Low-frequency waves

This section describes Cluster observations of low-frequency waves at, and in the vicinity of, the magnetopause. The main points are:

- Determination of the dispersion relation and the identification of low-frequency wave modes in the magnetopause and magnetosheath region.
- Observations of turbulence in the magnetopause over a wide range of frequencies.
- Conclusion that electromagnetic broadband waves might be the result of weak turbulence processes

Identification of low frequency wave modes in the magnetopause current and boundary layers, respectively, during reconnection and in the absence of reconnection, is of importance for the understanding of the internal dynamics of the magnetopause, plasma transport across the magnetopause, and the onset, maintenance and spatio-temporal evolution of reconnection. Theoretically, the magnetopause and boundary layer can be subject to different instabilities, fluid and kinetic, driven by the inhomogeneities and currents present in the plasma and fields encountered in the transition region or convected into the magnetopause enabling wave transformation. In all cases the magnetopause current layer is subject to enhanced wave activity at low and high frequencies.

In this section we focus on low frequency waves in the frequency range well below f_{ce}. Near the magnetopause the plasma is usually overdense, i.e., the electron cyclotron frequency is much less than the plasma frequency, $f_{ce} \ll f_{pe}$.

10.3.5.1 Spectral properties

It is well known (Gurnett et al., 1979; Perraut et al., 1979; Anderson et al., 1982) that low-frequency waves observed at the magnetopause appear in both the electromagnetic and electrostatic polarisations. They are most intense at the magnetopause boundary itself. For illustration we show in Figure 10.20 an example of such waves measured by Cluster 2 during a particular crossing of the magnetopause. These spectra have been obtained by combining FGM data and STAFF data for the magnetic field measurements, EFW for the electric field. The low frequency spectra are obtained with the waveform unit, the high frequency parts from the onboard Spectrum Analyser (STAFF-SA). The spectra of the electric and magnetic fluctuations in the frequency range from $\sim$0 Hz to near the lower hybrid frequency $f_{LH} \simeq \sqrt{f_{ce} f_c}$ are quite different.

For frequencies below the proton gyrofrequency (f_{ci}) the slopes of the electric and magnetic spectra are similar, with the magnetic intensity exceeding the elec-

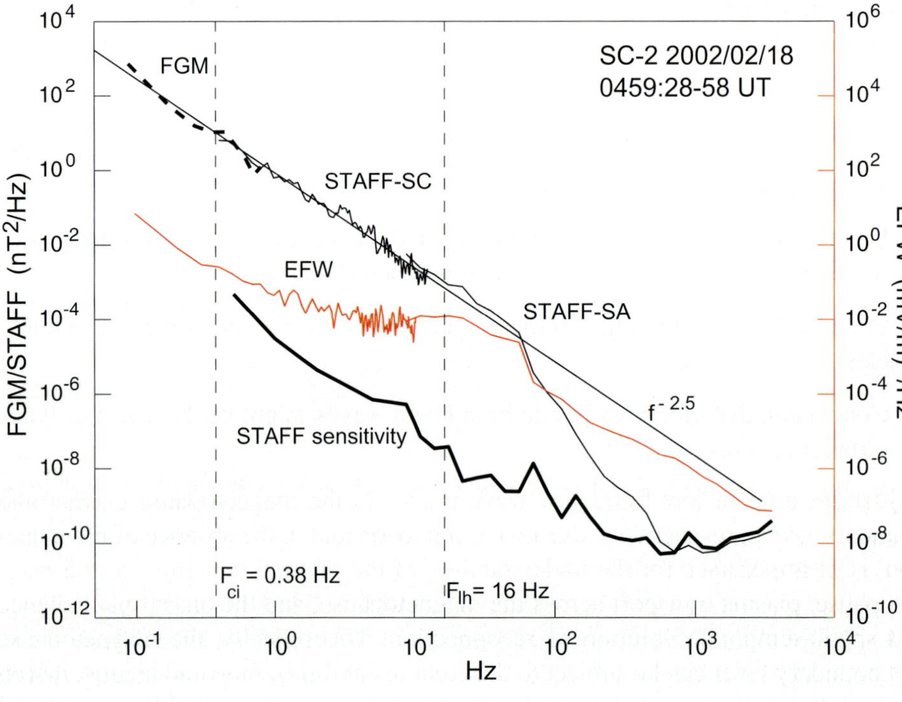

Figure 10.20. Low-frequency electric and magnetic wave power spectra measured by the various wave instrumentation on Cluster 2 at a crossing of the dayside magnetopause. Red: electric wave spectrum (right ordinate). Black: magnetic power spectrum (left ordinate). The curve labeled 'STAFF sensitivity' is the magnetic instrumental noise level. (Figure provided by D. Attié).

tric by two orders of magnitude as is expected for magnetohydrodynamic waves. Stepping up in frequency, the magnetic spectrum retains its slope of $\sim f^{-2.5}$ up to frequencies above the lower hybrid frequency f_{LH} in the whistler range. In contrast, the electric spectrum flattens out with increasing frequency into the ion cyclotron harmonic range, exhibiting a significant hump around the lower hybrid frequency. This hump ends at ~50 Hz, in coincidence with the decay in the magnetic wave power. At higher frequencies the electric wave power starts exceeding the magnetic power. It decreases roughly like $\sim f^{-1}$, while close to the electron gyrofrequency the magnetic power drops to the instrumental noise level. These spectra are interpreted as broadband magnetic noise with its electric counterpart, ion-cyclotron waves and superposition of electrostatic waves around the lower hybrid frequency (lower hybrid waves). Waves of the latter kind were discussed in Section 10.3.4.

The low-frequency, broadband magnetic wave spectra resemble the neighboring magnetosheath ULF waves. Their intensity is very low on the magnetospheric side where they decay with frequency as $\sim f^{-2.7}$. In the magnetosheath the waves are much more intense having a spectral slope of ~ -2.3. In the magnetopause they

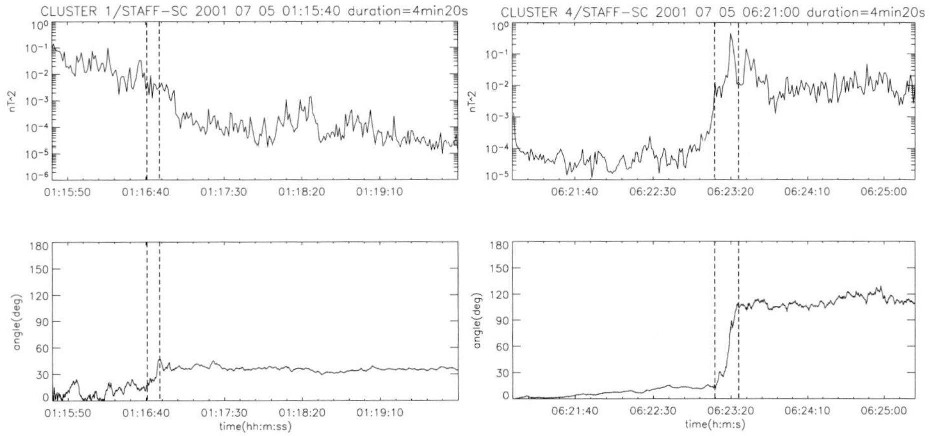

Figure 10.21. *Top*: Power of ULF fluctuations in the range from 2-10 Hz for two different crossings. *Bottom*: Rotation of the magnetic field in the plane of the magnetopause. The current layers are between pairs of dashed lines. (Figure provided by D. Attié).

exhibit a maximum that is 10 times the level in the magnetosheath (Rezeau and Belmont, 2001). The spectrum in the magnetopause is thus more intense but not as 'hard''as in the magnetosheath.

10.3.5.2 Generation of low frequency waves

The source of at least some of these magnetopause waves may lie in the magnetosheath. Exploration of the more homogeneous magnetosheath turbulence by exploiting the special multi-spacecraft properties of Cluster will then help understanding them. In particular, application of the *k*-filtering method to magnetosheath observations of low-frequency electromagnetic waves has provided evidence for the role of the Doppler shift (Sahraoui et al., 2003) in generating broadband low-frequency frequency spectra of the kind shown in Figure 10.20.

As described in Chapter 3, application of the *k*-filtering method is uniquely suited for the Cluster data. Accounting for the Doppler shift of the power spectrum in the fast streaming (bulk velocity $\mathbf{v}_{\mathrm{MSH}}$) magnetosheath showed that the frequency of these waves in the plasma frame is $f \ll f_{ci}$, i.e., far below the proton gyro-frequency. In this frequency range, the *k*-filtering technique can identify the wave modes. The wave spectra measured in the magnetosheath and close to the magnetopause resolve into a mixture of mirror modes, Alfvén modes, and magnetosonic modes of different wave vectors. Since the Doppler shift $\mathbf{k} \cdot \mathbf{v}_{\mathrm{MSH}}$ of the different modes depends on the wave number, mixing of different wave modes occurs at a given frequency in the spacecraft-frame. Without this identification, the earlier interpretation (Rezeau et al., 1999) of the phases of the apparently turbulent low-frequency wave spectra is in doubt.

The importance of low frequency magnetic waves from the magnetosheath for processes in the magnetopause was discussed at the beginning of this section. These waves can be resonantly converted at the magnetopause into shear-Alfvén waves which propagate backward and amplify resonantly by becoming trapped in the current layer where the wave vector becomes perpendicular to the magnetic field $\mathbf{k}\cdot\mathbf{B} \approx 0$ (Belmont et al., 1995; Johnson and Cheng, 1997; De Keyser et al., 1999; Belmont and Rezeau, 2001). When the wave vectors have no preferred direction, wave trapping should simply be stronger for large magnetic field shear angle across the magnetopause. This is one mechanism of creating a high intensity of low-frequency magnetic waves in the magnetopause current layer that can be directly checked using the Cluster observations.

A comparison of two Cluster magnetopause crossings is shown in Figure 10.21. For magnetic field shear angles $< 40°$, the Cluster 1 measurements lack any maximum in the fluctuating power, while the Cluster 4 observations at magnetic field shear angles $\sim 110°$ exhibit a pronounced maximum in the magnetopause current layer. This result suggests that resonant conversion does occur in the large-shear crossings. Such resonant conversion of the magnetosonic magnetosheath modes into Alfvén waves at the magnetopause (Belmont and Rezeau, 2001) has been suggested as a trigger of small scale magnetopause reconnection.

A more realistic model of the low frequency wave activity and its generation in the magnetopause transition should not only consider magnetosonic wave modes but should also refer to other incident waves, e.g., mirror modes. These are particularly interesting for the ignition of reconnection as they consist of localised regions with a transverse size of λ_i (Constantinescu et al., 2003; Treumann et al., 2004) containing strongly reduced magnetic fields encountering the magnetopause from the side of the magnetosheath. In addition, streaming instabilities at the magnetopause such as the Kelvin-Helmholtz instability (Sckopke et al., 1981; Scholer and Treumann, 1997) will contribute as well, either directly or via cascading into other modes.

10.3.6 Anomalous diffusion: Estimates

Cluster measurements of particles and fields are used to estimate the anomalous diffusion at the magnetopause. The key findings are:

- Flux and density-gradient based estimates of the diffusivity at the inner separatrix of the ion diffusion region yield a value of the the order of the Bohm diffusion.

- The order of magnitude agreement with the diffusivity determined independently from the lower hybrid wave power suggests that lower hybrid turbulence contributes substantially to diffusion, but only within a narrow layer. The role of ULF (MHD) waves is also likely to be important since the diffusion they imply is similar in magnitude to that of lower-hybrid waves.

Both the electric field fluctuations in the lower hybrid waves and the electromagnetic fluctuations can be used in an estimate of the relevant diffusion coefficients at the magnetopause. It should once more be noted in this context that the concept of diffusive entry is not well defined in application to the magnetopause. What is meant is a sufficiently strong violation of the frozen-in condition by anomalous processes, described by an anomalous collision frequency ν_{an} or an anomalous diffusion coefficient D_{an}, as defined in the paragraph following Eq. 10.2. On the other hand, the diffusive entry can also be determined directly from the measured plasma flux across a magnetic boundary. Both methods will be applied here to the case of the observations on February 6, 2002 already discussed above (Vaivads et al., 2004).

When using low frequency magnetic fluctuations, one possibility is to assume that such fluctuations will scatter the particles in pitch angle. Since the waves have a low frequency, pitch-angle scattering will affect the ions while the electrons will then be carried along by ambipolar electric fields in order to maintain quasi-neutrality. The magnetic diffusion coefficient $D_{B,an}$ can then be written approximately as (Tsurutani and Thorne, 1982)

$$D_{B,an} \approx \frac{k_B T_\perp}{2eB_0} \left\langle \left(\frac{\delta B}{B_0} \right)^2 \right\rangle . \tag{10.3}$$

The factor in front depends on the perpendicular temperature and is essentially the Bohm diffusion coefficient. Measured values are $T_i \approx 10\,\mathrm{MK}$, $B_0 = 48\,\mathrm{nT}$, and $\langle (\delta B/B_0)^2 \rangle \approx 0.15$ (for frequencies up to 0.1 Hz), which yields a diffusivity of $D_{B,an} \sim 1.4 \times 10^9\,\mathrm{m^2\,s^{-1}}$. This estimate is to be taken with caution because of the intrinsic uncertainties.

For the case under consideration we can use the Cluster observations for cross checking (Vaivads et al., 2004) whether diffusive entry is (occasionally) important. The occurrence of lower-hybrid drift waves in the density gradient of the ion diffusion region (Mozer et al., 2002; Bale et al., 2002), described above, suggests that lower hybrid waves should be involved in reconnection as is also suggested by numerical 2D and 3D particle-in-cell simulations of reconnection (Shinohara et al., 2001; Scholer et al., 2003; Pritchett and Coroniti, 2004).

We estimate the diffusion coefficient due to lower hybrid waves in a narrow current layer that can be a separatrix emanating from a distant X-line (Vaivads et al., 2004) (see Figure 10.19). The diffusion coefficient can be estimated in two ways. First, one can use theoretical expression for the anomalous collision frequency generated by lower hybrid drift wave turbulence (e.g., Treumann and Baumjohann, 1997, p. 332), which in the regime of high density yields a lower-hybrid particle scattering rate

$$\nu_{LH} \approx \omega_{LH} \frac{\omega_{pe}^2}{\omega_{LH}^2} \frac{W_{LH}}{nk_B T}, \tag{10.4}$$

with W_{LH} being the average lower-hybrid wave energy density. Inserting the relevant values taken from Vaivads et al. (2004) one finds that $\nu_{LH} \sim \omega_{LH}$. The corresponding value of the transverse diffusion coefficient is

$$D_{LH\perp} = \frac{v_{e,th}^2 \nu_{LH}}{\omega_{ce}^2} \sim 0.5 \times 10^9\,\mathrm{m^2 s^{-1}} \tag{10.5}$$

A second way to estimate the diffusion coefficient has been given by Vaivads et al. (2004). If one assumes that electron transport across the narrow layer is a diffusion process, then the diffusion coefficient can be estimated using the values of electron flux due to lower hybrid waves and the values of the density gradient. Electron flux and density gradient cannot be directly measured at the narrow scale of the current layer. Instead, Vaivads et al. (2004) use the variations in satellite potential as a proxy for density fluctuations and in addition assume that electron flux is dominated by the $\mathbf{E} \times \mathbf{B}$ drift. The fluctuating diffusive electron flux $\langle dn\mathbf{v}_{fn}\rangle$ is shown in panel g of Figure 10.19, where the index n indicates the component of the flux normal to the magnetopause. The diffusion coefficient is obtained from the interval ~0.35-0.5 s containing the steepest density gradient:

$$D_{\perp} \sim \frac{\langle dn\mathbf{v}_{fn}\rangle}{\nabla \langle n_{Vps}\rangle} \sim 10^9 \quad \mathrm{m^2 s^{-1}} \tag{10.6}$$

where the average density gradient $\nabla \langle n_{Vps}\rangle$ ~25 $\mathrm{m^{-4}}$. The two methods of estimating the diffusion coefficient are different but give values that are, by order of magnitude, in a good agreement. The resulting value of the diffusion coefficient is of the order of Bohm diffusivity.

The agreement between the above two estimates demonstrates that, along the separatrices, bursts of localised lower hybrid drift fluctuations can induce strong enough scattering of plasma across the magnetic field to account for the observed fluxes and gradients. It also demonstrates that the separatrix is a dynamical transition region of width of the order of λ_i, where the frozen-in property of the plasma is broken locally and kinetic processes induced by gradients, field-aligned currents and electron beams dominate the plasma dynamics on microscopic scales.

The cross-check given in this section has been made possible by the availability of the Cluster multi-spacecraft data in combination with very high-time resolution field data. The multi-spacecraft data allowed estimating the geometry and speed of the narrow current layer while the high resolution field data allowed analysing the spectral wave properties. Finally, the interferometric capability of the electric field instrument made possible the analysis of the dispersive wave properties. The Cluster data-based consistency test of the diffusivity gives confidence to the diffusivity determined from lower-hybrid wave observations. It also demonstrates that substantial diffusion can locally be generated by large amplitude lower hybrid turbulence.

10.3.7 Conclusions

The microphysics in the collisionless environment of the magnetopause contributes to dissipation, heating and transport as well as wave generation and transformation, and to turbulent interactions. The following are highlights of the Cluster achievements in the first three years of its mission:

- The Hall region of the reconnection has been spatially resolved and it has been shown that the normal (Hall) electric field is approximately balanced by the Hall-Lorentz force term in the generalised Ohm's law for conditions without a guide magnetic field.
- Localised, large amplitude electrostatic waves have been detected near the lower hybrid frequency in connection with passages through the reconnection related separatrices emanating from the ion diffusion (Hall) region.
- Flux and density-gradient based estimates of the diffusivity in the ion diffusion region at the inner separatrix yield a diffusivity of the order of Bohm diffusion.
- The order of magnitude agreement of the above estimates with the diffusivity determined independently from the lower hybrid wave power suggests that lower hybrid turbulence contributes substantially to dissipation and diffusive transport at the separatrices.
- High-frequency localised waves along the separatrices are directly related to field-aligned currents.
- The power-law, low-frequency, magnetic wave turbulence spectra near the magnetopause consist of magnetosonic, mirror and Alfvén wave modes.

Needless to say, there remains a large number of unresolved problems. Among them are:

- Are the spectra observed at the magnetopause generated locally, or are they the result of the superposition of wave modes converted from the magnetosheath?
- What is the role of low-frequency turbulence in diffusive entry, reconnection, plasma heating, and transport?
- Under what conditions can anomalous resistivity and diffusion become important? Earlier studies inferred low levels of lower hybrid turbulence (Bale et al., 2002) in the diffusion region at the magnetopause. Does the high diffusion coefficient found by Cluster in intense, highly sporadic turbulence possibly apply only to separatrix crossings?
- What is the relation between reconnection and generation of field-aligned currents and kinetic Alfvén waves? Where do these currents and waves emanate?

Are they the closure-currents of the ion diffusion-region Hall currents? Are these currents carried by kinetic Alfvén waves?

- Do fluctuations such as lower-hybrid drift and MHD waves cause spontaneous onset of reconnection or is magnetopause reconnection driven by inflow? Simulations suggest that lower hybrid waves can be important at onset but unimportant once reconnection is going on. Is the reported low intensity of lower hybrid waves in the diffusion region therefore due to the fact that observations are rarely made at the initial state of reconnection?
- What is the role of external conditions (guide fields, velocity shear, external drive, density gradients etc.) in the microphysics of reconnection?

Acknowledgements

We thank Masaki Fujimoto for his helpful comments. The work of TDP and BUÖS was supported by NASA, that of CPE by ESA. Cluster research in France is funded by CNES. Research of AV was supported by the Swedish Research Council. BR was supported in part by an NSF grant.

References

Anderson, B. J., T.-D. Phan, and S. A. Fuselier: 1997, 'Relationships between plasma depletion and subsolar reconnection'. *J. Geophys. Res.* **102**(11), 9531–9542.

Anderson, R. R., T. E. Eastman, C. C. Harvey, M. M. Hoppe, B. T. Tsurutani, and J. Etcheto: 1982, 'Plasma waves near the magnetopause'. *J. Geophys. Res.* **87**(16), 2087–2107.

André, M., A. Vaivads, S. C. Buchert, A. N. Fazakerley, and A. Lahiff: 2004, 'Thin electron-scale layers at the magnetopause'. *Geophys. Res. Lett.* **31**, L03803, doi:10.1029/2003GL018137.

Avanov, L. A., S. A. Fuselier, and O. L. Vaisberg: 2001, 'High-latitude magnetic reconnection in sub-Alfvénic flow: Interball tail observations on May 29, 1996'. *J. Geophys. Res.* **106**(15), 29491–29502.

Bale, S. D., F. S. Mozer, and T. Phan: 2002, 'Observation of lower hybrid drift instability in the diffusion region at a reconnecting magnetopause'. *Geophys. Res. Lett.* **29**, 2180, doi:10.1029/2002GL016113.

Bauer, T., G. Paschmann, N. Schopke, W. Baumjohann, R. Treumann, and T. Phan: 1998, 'AMPTE/IRM observations of particles and fields at the dayside low-latitude magnetopause'. In: *Geospace Mass and Energy Flow: Results From the International Solar-Terrestrial Physics Program*, Vol. 104 of *Geophysical Monograph Series*. Washington, D. C., USA: American Geophysical Union, p. 51.

Belmont, G., F. Reberac, and L. Rezeau: 1995, 'Resonant amplification of magnetosheath MHD fluctuations at the magnetopause'. *Geophys. Res. Lett.* **22**, 295–298.

Belmont, G. and L. Rezeau: 2001, 'Magnetopause reconnection induced by magnetosheath Hall-MHD fluctuations'. *J. Geophys. Res.* **106**, 10751–10760.

Berchem, J. and C. T. Russell: 1984, 'Flux transfer events on the magnetopause - Spatial distribution and controlling factors'. *J. Geophys. Res.* **89**(18), 6689–6703.

Birn, J., J. F. Drake, M. A. Shay, B. N. Rogers, R. E. Denton, M. Hesse, M. Kuznetsova, Z. W. Ma, A. Bhattacharjee, A. Otto, and P. L. Pritchett: 2001, 'Geospace Environmental Modeling (GEM) magnetic reconnection challenge'. *J. Geophys. Res.* **106**, 3715–3720.

Bosqued, J. M., C. P. Escoubet, H. Frey, M. Dunlop, J. Berchem, A. Marchaudon, J. C. Cerisier, A. Fazerkerley, B. Lavraud, H. Rème, H. Laakso, , and A. Balogh: 2004, 'Multipoint observations of transient reconnection signatures in the dayside cusp precipitation: A CLUSTER-IMAGE detailed case study'. *J. Geophys. Res.* p. submitted.

Bosqued, J. M., T. D. Phan, I. Dandouras, C. P. Escoubet, H. Rème, A. Balogh, M. W. Dunlop, D. Alcaydé, E. Amata, M. B. Bavassano-Cattaneo, R. Bruno, C. Carlson, A. M. DiLellis, L. Eliasson, V. Formisano, L. M. Kistler, B. Klecker, A. Korth, H. Kucharek, R. Lundin, M. McCarthy, J. P. McFadden, E. Moebius, G. K. Parks, and J.-A. Sauvaud: 2001, 'CLUSTER observations of the high-latitude magnetopause and cusp: initial results from the CIS ion instruments'. *Ann. Geophys.* **19**, 1545–1566.

Carter, T. A., M. Yamada, H. Ji, R. M. Kulsrud, and F. Trintchouk: 2002, 'Experimental study of lower-hybrid drift turbulence in a reconnecting current sheet'. *Phys. Plasmas* **9**, 3272.

Chisham, G., I. J. Coleman, M. P. Freeman, M. Pinnock, and M. Lester: 2002, 'Ionospheric signatures of split reconnection X-lines during conditions of positive IMF Bz and By = Bz: Evidence for the antiparallel merging hypothesis'. *J. Geophys. Res.* **107**(A10), 1323, doi:10.1029/2001JA009124.

Constantinescu, O. D., K.-H. Glassmeier, R. Treumann, and K.-H. Fornaçon: 2003, 'Magnetic mirror structures observed by Cluster in the magnetosheath'. *Geophys. Res. Lett.* **30**, 1802, doi:10.1029/2003GL017313.

Cowley, S. W. H.: 1982, 'The causes of convection in the earth's magnetosphere - A review of developments during the IMS'. *Rev. Geophys. Space Phys.* **20**, 531–565.

Cowley, S. W. H. and C. J. Owen: 1989, 'A simple illustrative model of open flux tube motion over the dayside magnetopause'. *Planet. Space Sci.* **37**, 1461–1475.

Crooker, N. U.: 1979, 'Dayside merging and cusp geometry'. *J. Geophys. Res.* **84**(13), 951–959.

Davidson, R. C.: 1978, 'Quasi-linear stabilization of lower-hybrid-drift instability'. *Physics of Fluids* **21**, 1375–1380.

De Keyser, J., M. W. Dunlop, C. J. Owen, B. U. Ö. Sonnerup, S. Haaland, A. Vaivads, R. Lundin, L. Rezeau, and G. Paschmann: 2005, 'Magnetopause and Boundary Layer'. *Space Sci. Rev.* **this issue**.

De Keyser, J., M. Roth, F. Reberac, L. Rezeau, and G. Belmont: 1999, 'Resonant amplification of MHD waves in realistic subsolar magnetopause configurations'. *J. Geophys. Res.* **104**(13), 2399–2410.

Drake, J. F., J. Gerber, and R. G. Kleva: 1994, 'Turbulence and transport in the magnetopause current layer'. *J. Geophys. Res.* **99**(18), 11211–11224.

Dunlop, M. W., A. Balogh, K.-H. Glassmeier, and P. Robert: 2002, 'Four-point Cluster application of magnetic field analysis tools: The Curlometer'. *J. Geophys. Res.* **107**(A11), 1384, doi:10.1029/2001JA005088.

Escoubet, C. P., M. F. Smith, S. F. Fung, P. C. Anderson, R. A. Hoffman, E. M. Basinska, and J. M. Bosqued: 1992, 'Staircase ion signature in the polar cusp - A case study'. *Geophys. Res. Lett.* **19**, 1735–1738.

Fairfield, D. H., A. Otto, T. Mukai, S. Kokubun, R. P. Lepping, J. T. Steinberg, A. J. Lazarus, and T. Yamamoto: 2000, 'Geotail observations of the Kelvin-Helmholtz instability at the equatorial magnetotail boundary for parallel northward fields'. *J. Geophys. Res.* **105**(14), 21159–21174.

Farrell, W. M., M. D. Desch, M. L. Kaiser, and K. Goetz: 2002, 'The dominance of electron plasma waves near a reconnection X-line region'. *Geophys. Res. Lett.* **29**, 1902, doi:10.1029/2002GL014662.

Farrugia, C. J., P. E. Sandholt, W. F. Denig, and R. B. Torbert: 1998, 'Observation of a correspondence between poleward moving auroral forms and stepped cusp ion precipitation'. *J. Geophys. Res.* **103**(12), 9309–9316.

Frey, H. U., T. D. Phan, S. A. Fuselier, and S. B. Mende: 2003, 'Continuous magnetic reconnection at Earth's magnetopause'. *Nature* **426**, 533–537.

Fujimoto, M. and T. Terasawa: 1994, 'Anomalous ion mixing within an MHD scale Kelvin-Helmholtz vortex'. *J. Geophys. Res.* **99**(18), 8601–8613.

Fujimoto, M., T. Tonooka, and T. Mukai: 2003, 'Vortex-like fluctuations in the magnetotail flanks and their possible roles in plasma transport'. In: P. T. Newell and T. Onsager (eds.): *Earth's Low-Latitude Boundary Layer*, Vol. 133 of *Geophysical Monograph Series*. Washington, D. C., USA: American Geophysical Union, pp. 241–251.

Fuselier, S. A., H. U. Frey, K. J. Trattner, S. B. Mende, and J. L. Burch: 2002, 'Cusp aurora dependence on interplanetary magnetic field Bz'. *J. Geophys. Res.* **107**(A7), 1111, doi:10.1029/2001JA900165.

Fuselier, S. A., D. M. Klumpar, and E. G. Shelley: 1991, 'Ion reflection and transmission during reconnection at the earth's subsolar magnetopause'. *Geophys. Res. Lett.* **18**, 139–142.

Fuselier, S. A., S. M. Petrinec, and K. J. Trattner: 2000a, 'Stability of the high-latitude reconnection site for steady northward IMF'. *Geophys. Res. Lett.* **27**, 473–476.

Fuselier, S. A., K. J. Trattner, and S. M. Petrinec: 2000b, 'Cusp observations of high- and low-latitude reconnection for northward interplanetary magnetic field'. *J. Geophys. Res.* **105**(14), 253–266.

Gonzalez, W. D. and F. S. Mozer: 1974, 'A quantitative model for the potential resulting from reconnection with an arbitrary interplanetary magnetic field'. *J. Geophys. Res.* **79**(18), 4186–4194.

Gosling, J. T., J. R. Asbridge, S. J. Bame, W. C. Feldman, G. Paschmann, N. Sckopke, and C. T. Russell: 1982, 'Evidence for quasi-stationary reconnection at the dayside magnetopause'. *J. Geophys. Res.* **87**(16), 2147–2158.

Gosling, J. T., M. F. Thomsen, S. J. Bame, R. C. Elphic, and C. T. Russell: 1990a, 'Plasma flow reversals at the dayside magnetopause and the origin of asymmetric polar cap convection'. *J. Geophys. Res.* **95**(14), 8073–8084.

Gosling, J. T., M. F. Thomsen, S. J. Bame, R. C. Elphic, and C. T. Russell: 1991, 'Observations of reconnection of interplanetary and lobe magnetic field lines at the high-latitude magnetopause'. *J. Geophys. Res.* **96**(15), 14097.

Gosling, J. T., M. F. Thomsen, S. J. Bame, T. G. Onsager, and C. T. Russell: 1990b, 'The electron edge of the low latitude boundary layer during accelerated flow events'. *Geophys. Res. Lett.* **17**, 1833–1836.

Greenwald, R. A., W. A. Bristow, G. J. Sofko, C. Senior, J.-C. Cerisier, and A. Szabo: 1995, 'Super dual auroral radar network radar imaging of dayside high-latitude convection under northward interplanetary magnetic field: Toward resolving the distorted two-cell versus multicell controversy'. *J. Geophys. Res.* **100**(9), 19661–19674.

Gurnett, D. A., R. R. Anderson, B. T. Tsurutani, E. J. Smith, G. Paschmann, G. Haerendel, S. J. Bame, and C. T. Russell: 1979, 'Plasma wave turbulence at the magnetopause - Observations from ISEE 1 and 2'. *J. Geophys. Res.* **84**(13), 7043–7058.

Gustafsson, G., A. M., T. Carozzi, A. I. Eriksson, C.-G. Fthammar, R. Grard, G. Holmgren, J. A. Holtet, N. Ivchenko, T. Karlsson, Y. Khotyaintsev, S. Klimov, H. Laakso, P.-A. Lindqvist, B. Lybekk, G. Marklund, F. Mozer, K. Mursula, A. Pedersen, B. Popielawska, S. Savin, K. Stasiewicz,

P. Tanskanen, A. Vaivads, and J.-E. Wahlund: 2001, 'First results of electric field and density observations by Cluster EFW based on initial months of operation'. *Ann. Geophys.* **19**, 1219–1240.

Hasegawa, H., M. Fujimoto, T.-D. Phan, H. Rème, A. Balogh, M. W. Dunlop, C. Hashimoto, and R. TanDokoro: 2004, 'Transport of solar wind into Earth's magnetosphere through rolled-up Kelvin-Helmholtz vortices'. *Nature* **430**, 755–758.

Hesse, M., J. Birn, and M. Kuznetsova: 2001, 'Collisionless magnetic reconnection: Electron processes and transport modeling'. *J. Geophys. Res.* **106**(15), 3721–3736.

Hones, E. W., S. J. Bame, J. R. Asbridge, J. Birn, G. Paschmann, N. Sckopke, and G. Haerendel: 1981, 'Further determination of the characteristics of magnetospheric plasma vortices with Isee 1 and 2'. *J. Geophys. Res.* **86**(15), 814–820.

Huba, J. D.: 1996, 'The Kelvin-Helmholtz instability: Finite Larmor radius magnetohydrodynamics'. *Geophys. Res. Lett.* **23**, 2907–2910.

Hudson, P. D.: 1970, 'Discontinuities in an anisotropic plasma and their identification in the solar wind'. *Planet. Space Sci.* **18**, 1611–1622.

Johnson, J. R. and C. Z. Cheng: 1997, 'Kinetic Alfvén waves and plasma transport at the magnetopause'. *Geophys. Res. Lett.* **24**, 1423–1426.

Kessel, R. L., S.-H. Chen, J. L. Green, S. F. Fung, S. A. Boardsen, L. C. Tan, T. E. Eastman, J. D. Craven, and L. A. Frank: 1996, 'Evidence of high-latitude reconnecting during northward IMF: Hawkeye observations'. *Geophys. Res. Lett.* **23**, 583–586.

Khotyaintsev, Y., A. Vaivads, Y. Ogawa, B. Popielawska, M. André, S. Buchert, P. Décréau, B. Lavraud, and H. Rème: 2004, 'Cluster observations of high-frequency waves in the exterior cusp'. *Ann. Geophys.* **22**, 2403–2411.

Kivelson, G. K. and S.-H. Chen: 1995, 'The magnetopause: Surface waves and instabilities and their possible dynamical consequences'. In: P. Song, B. U. Ö. Sonnerup, and M. F. Thomsen (eds.): *Physics of the Magnetopause*, Vol. 90 of *Geophysical Monograph Series*. Washington, D. C., USA: American Geophysical Union, pp. 257–268.

LaBelle, J. and R. A. Treumann: 1988, 'Plasma waves at the dayside magnetopause'. *Space Science Reviews* **47**, 175–202.

Lemaire, J.: 1977, 'Impulsive penetration of filamentary plasma elements into the magnetospheres of the earth and Jupiter'. *Planet. Space Sci.* **25**, 887–890.

Levy, R. H., H. E. Petschek, and G. L. Siscoe: 1964, 'Aerodynamic aspects of the magnetospheric flow'. *AIAA Journal* **2**, 2065.

Lockwood, M., A. Fazakerley, H. J. Opgenoorth, J. Moen, A. P. van Eyken, M. Dunlop, J.-M. Bosqued, W. Denig, J. A. Wild, G. Lu, C. Cully, P. Eglitis, I. W. McCrea, M. A. Hapgood, M. N. Wild, R. R. Stamper, W. Denig, M. Taylor, J. A. Wild, G. Provan, O. Amm, K. Kauristie, T. Pulkkinen, A. Strømme, P. Prikryl, F. Pitout, A. Balogh, H. Rème, R. Behlke, T. Hansen, R. Greenwald, H. Frey, S. K. Morley, D. Alcaydé, P.-L. Blelly, E. Donovan, M. Engebretson, M. Lester, J. Watermann, and M. M. F.: 2001a, 'Coordinated Cluster and ground-based instrument observations of transient changes in the magnetopause boundary layer during an interval of predominantly northward IMF: relation to reconnection pulses and FTE signatures'. *Ann. Geophys.* **19**, 1613–1640.

Lockwood, M., B. S. Lanchester, H. U. Frey, K. Throp, S. K. Morley, S. E. Milan, and M. Lester: 2003, 'IMF control of cusp proton emission intensity and dayside convection: implications for component and anti-parallel reconnection'. *Ann. Geophys.* **21**, 955–982.

Lockwood, M., H. J. Opgenoorth, A. P. van Eyken, A. Fazakerley, J.-M. Bosqued, W. Denig, J. A. Wild, C. Cully, R. Greenwald, G. Lu, O. Amm, H. Frey, A. Strømme, P. Prikryl, M. A. Hapgood, M. N. Wild, R. Stamper, M. Taylor, I. McCrea, K. Kauristie, T. Pulkkinen, F. Pitout, A. Balogh,

M. Dunlop, H. Rème, R. Behlke, T. Hansen, G. Provan, P. Eglitis, S. K. Morley, D. Alcaydé, P.-L. Blelly, J. Moen, E. Donovan, M. Engebretson, M. Lester, J. Watermann, and M. M. F.: 2001b, 'Coordinated Cluster, ground-based instrumentation and low-altitude satellite observations of transient poleward-moving events in the ionosphere and in the tail lobe'. *Ann. Geophys.* **19**, 1589–1612.

Lockwood, M. and M. F. Smith: 1992, 'The variation of reconnection rate at the dayside magnetopause and cusp ion precipitation'. *J. Geophys. Res.* **97**(16), 14841.

Luhmann, J. G., R. J. Walker, C. T. Russell, N. U. Crooker, J. R. Spreiter, and S. S. Stahara: 1984, 'Patterns of potential magnetic field merging sites on the dayside magnetopause'. *J. Geophys. Res.* **89**(18), 1741–1744.

Mandt, A., R. E. Denton, and J. F. Drake: 1994, 'Transition to whistler mediated magnetic reconnection'. *Geophys. Res. Lett.* **21**, 73–76.

Marchaudon, A., J. Cerisier, J. Bosqued, M. Dunlop, J. Wild, P. Décréau, M. Förster, D. Fontaine, and H. Laakso: 2004, 'Transient plasma injections in the dayside magnetosphere: one-to-one correlated observations by Cluster and SuperDARN'. *Ann. Geophys.* **22**, 141–158.

Matsumoto, Y. and M. Hoshino: 2004, 'Onset of turbulence induced by a Kelvin-Helmholtz vortex'. *Geophys. Res. Lett.* **31**, L02807, doi:10.1029/2003GL018195.

Miura, A.: 1984, 'Anomalous transport by magnetohydrodynamic Kelvin-Helmholtz instabilities in the solar wind-magnetosphere interaction'. *J. Geophys. Res.* **89**(18), 801–818.

Miura, A. and P. L. Pritchett: 1982, 'Nonlocal stability analysis of the MHD Kelvin-Helmholtz instability in a compressible plasma'. *J. Geophys. Res.* **87**(16), 7431–7444.

Moen, J., J. Holtet, A. Pedersen, B. Lybekk, K. Svenes, K. Oksavik, W. Denig, E. Lucek, F. Sraas, and A. M: 2001, 'Cluster boundary layer measurements and optical observations at magnetically conjugate sites'. *Ann. Geophys.* **19**, 1655.

Mozer, F. S., S. D. Bale, and T. D. Phan: 2002, 'Evidence of Diffusion Regions at a Subsolar Magnetopause Crossing'. *Physical Review Letters* **89**(1), 015002.

Mozer, F. S., S. D. Bale, T. D. Phan, and J. A. Osborne: 2003, 'Observations of Electron Diffusion Regions at the Subsolar Magnetopause'. *Physical Review Letters* **91**(24), 245002.

Nagai, T., I. Shinohara, M. Fujimoto, M. Hoshino, Y. Saito, S. Machida, and T. Mukai: 2001, 'Geotail observations of the Hall current system: Evidence of magnetic reconnection in the magnetotail'. *J. Geophys. Res.* **106**, 25929–25950.

Nakamura, T. K., D. Hayashi, M. Fujimoto, and I. Shinohara: 2004, 'Decay of MHD-Scale Kelvin-Helmholtz Vortices Mediated by Parasitic Electron Dynamics'. *Phys. Rev. Lett.* **92**(14), 145001.

Newell, P. T. and C. Meng: 1991, 'Ion acceleration at the equatorward edge of the cusp - Low altitude observations of patchy merging'. *Geophys. Res. Lett.* **18**, 1829–1832.

Nykyri, K. and A. Otto: 2001, 'Plasma transport at the magnetospheric boundary due to reconnection in Kelvin-Helmholtz vortices'. *Geophys. Res. Lett.* **28**, 3565–3568.

Ogilvie, K. W. and R. J. Fitzenreiter: 1989, 'The Kelvin-Helmholtz instability at the magnetopause and inner boundary layer surface'. *J. Geophys. Res.* **94**(13), 15113–15123.

Øieroset, M., T. D. Phan, M. Fujimoto, R. P. Lin, and R. P. Lepping: 2001, 'In situ detection of collisionless reconnection in the Earth's magnetotail'. *Nature* **412**, 414–417.

Onsager, T. G., C. A. Kletzing, J. B. Austin, and H. Mackiernan: 1993, 'Model of magnetosheath plasma in the magnetosphere - Cusp and mantle particles at low-altitudes'. *Geophys. Res. Lett.* **20**, 479–482.

Owen, C. J., A. N. Fazakerley, P. J. Carter, A. J. Coates, I. C. Krauklis, S. Szita, M. G. G. T. Taylor, P. Travnicek, G. Watson, R. J. Wilson, A. Balogh, and M. W. Dunlop: 2001, 'CLUSTER PEACE observations of electrons during magnetosheric flux transfer events'. *Ann. Geophys.* **19**, 1509–1522.

Owen, C. J., M. G. G. T. Taylor, I. C. Krauklis, A. N. Fazakerley, M. W. Dunlop, and J. M. Bosqued: 2004, 'Cluster observations of surface waves on the dawn flank magnetopause'. *Ann. Geophys.* **22**, 971–983.

Paschmann, G., W. Baumjohann, N. Sckopke, I. Papamastorakis, and C. W. Carlson: 1986, 'The magnetopause for large magnetic shear - AMPTE/IRM observations'. *J. Geophys. Res.* **91**(10), 11099–11115.

Paschmann, G., S. Haaland, B. U. Ö. Sonnerup, E. Georgescu, B. Klecker, A. Vaivads, H. Rème, T. D. Phan, and H. Hasegawa: 2004, 'Characteristics of the near-tail dawn magnetopause and boundary layer'. *Ann. Geophys.* submitted.

Paschmann, G., I. Papamastorakis, N. Sckopke, G. Haerendel, B. U. Ö. Sonnerup, S. J. Bame, J. R. Asbridge, J. T. Gosling, C. T. Russel, and R. C. Elphic: 1979, 'Plasma acceleration at the earth's magnetopause - Evidence for reconnection'. *Nature* **282**, 243–246.

Perraut, S., R. Gendrin, P. Robert, and A. Roux: 1979, 'Magnetic pulsations observed on board GEOS-2 in the ULF range during multiple magnetopause crossings'. In: *Alpbach Conf. Proc. ESA SP-148: Magnetospheric Boundary Layers,*. pp. 113–122.

Petrinec, S. M., K. J. Trattner, and S. A. Fuselier: 2003, 'Steady reconnection during intervals of northward IMF: Implications for magnetosheath properties'. *J. Geophys. Res.* **108**(A12), 1458, doi:10.1029/2003JA009979.

Phan, T., H. U. Frey, S. Frey, L. Peticolas, S. Fuselier, C. Carlson, H. Rème, J.-M. Bosqued, A. Balogh, M. Dunlop, L. Kistler, C. Mouikis, I. Dandouras, J.-A. Sauvaud, S. Mende, J. McFadden, G. Parks, E. Moebius, B. Klecker, G. Paschmann, M. Fujimoto, S. Petrinec, M. F. Marcucci, A. Korth, and R. Lundin: 2003, 'Simultaneous Cluster and IMAGE observations of cusp reconnection and auroral proton spot for northward IMF'. *Geophys. Res. Lett.* **30**, 1509, doi:10.1029/2003GL016885.

Phan, T. D., M. W. Dunlop, G. Paschmann, B. Klecker, J. M. Bosqued, H. Rème, A. Balogh, C. Twitty, F. S. Mozer, C. W. Carlson, C. Mouikis, and L. M. Kistler: 2004, 'Cluster observations of continuous reconnection at the magnetopause under steady interplanetary magnetic field conditions'. *Ann. Geophys.* **22**, 2355 – 2367.

Phan, T.-D., G. Paschmann, W. Baumjohann, N. Sckopke, and H. Lühr: 1994, 'The magnetosheath region adjacent to the dayside magnetopause: AMPTE/IRM observations'. *J. Geophys. Res.* **99**(18), 121–141.

Phan, T.-D., G. Paschmann, and B. U. Ö. Sonnerup: 1996, 'Low-latitude dayside magnetopause and boundary layer for high magnetic shear 2. Occurrence of magnetic reconnection'. *J. Geophys. Res.* **101**(10), 7817–7828.

Phan, T. D., B. U. Ö. Sonnerup, and R. P. Lin: 2001, 'Fluid and kinetics signatures of reconnection at the dawn tail magnetopause: Wind observations'. *J. Geophys. Res.* **106**(15), 25489–25502.

Pitout, F., C. P. Escoubet, H. Rème, and A. Balogh: 2004, 'Cluster survey of the mid-altitude cusp: morphology and plasma properties'. *Ann. Geophys.* submitted.

Pritchett, P. L.: 2001, 'Geospace Environment Modeling magnetic reconnection challenge: Simulations with a full particle electromagnetic code'. *J. Geophys. Res.* **106**(15), 3783–3798.

Pritchett, P. L. and F. V. Coroniti: 2004, 'Three-dimensional collisionless magnetic reconnection in the presence of a guide field'. *J. Geophys. Res.* **109**, A01220, doi:10.1029/2003JA009999.

Puhl-Quinn, P. A. and J. D. Scudder: 2000, 'Systematics of ion Walén analysis of rotational discontinuities using E/Z measurements'. *J. Geophys. Res.* **105**(14), 7617–7628.

Retino, A., M. B. Bavassano-Cattaneo, M. F. Marcucci, A. Vaivads, M. Andre, Y. Khotyaintsev, T. Phan, G. Pallocchia, H. Rème, E. Moebius, B. Klecker, C. Carlson, M. McCarthy, A. Korth, R. Lundin, and A. Balogh: 2004, 'Cluster multispacecraft observations at the high latitude duskside magnetopause: implications for continuous and component magnetic reconnection'. *Ann. Geophys.* p. in press.

Rezeau, L. and G. Belmont: 2001, 'Magnetic turbulence at the magnetopause, a key problem for understanding the solar wind/ magnetosphere exchanges'. *Space Science Reviews* **95**, 427–441.

Rezeau, L., G. Belmont, N. Cornilleau-Wehrlin, F. Reberac, and C. Briand: 1999, 'Spectral law and polarization properties of the low-frequency waves at the magnetopause'. *Geophys. Res. Lett.* **26**, 651–654.

Rogers, B. N., R. E. Denton, and J. F. Drake: 2003, 'Signatures of collisionless magnetic reconnection'. *J. Geophys. Res.* **108**(A3), 1111, doi:10.1029/2002JA009699.

Rogers, B. N., R. E. Denton, J. F. Drake, and M. A. Shay: 2001, 'Role of Dispersive Waves in Collisionless Magnetic Reconnection'. *Phys. Rev. Lett.* **87**(19), 195004;doi:10.1103/PhysRevLett.87.195004.

Runov, A., R. Nakamura, W. Baumjohann, R. A. Treumann, T. L. Zhang, M. Volwerk, Z. Vörös, A. Balogh, K.-H. Glassmeier, B. Klecker, H. Rème, and L. Kistler: 2003, 'Current sheet structure near magnetic X-line observed by Cluster'. *Geophys. Res. Lett.* **30**, 1579, doi:10.1029/2002GL016730.

Russell, C. T. and R. C. Elphic: 1978, 'Initial ISEE magnetometer results - Magnetopause observations'. *Space Sci. Rev.* **22**, 681–715.

Sagdeev, R. Z.: 1979, 'The 1976 Oppenheimer lectures: Critical problems in plasma astrophysics. II. Singular layers and reconnection'. *Reviews of Modern Physics* **51**, 11–20.

Sahraoui, F., J. L. Pinçon, G. Belmont, L. Rezeau, N. Cornilleau-Wehrlin, P. Robert, L. Mellul, J. M. Bosqued, A. Balogh, P. Canu, and G. Chanteur: 2003, 'ULF wave identification in the magnetosheath: The k-filtering technique applied to Cluster II data'. *J. Geophys. Res.* **108**(A9), 1335, doi:10.1029/2002JA009587.

Scholer, M., I. Sidorenko, C. H. Jaroschek, R. A. Treumann, and A. Zeiler: 2003, 'Onset of collisionless magnetic reconnection in thin current sheets: Three-dimensional particle simulations'. *Physics of Plasmas* **10**, 3521.

Scholer, M. and R. A. Treumann: 1997, 'The Low-Latitude Boundary Layer at the Flanks of the Magnetopause'. *Space Sci. Rev.* **80**, 341–367.

Sckopke, N., G. Paschmann, G. Haerendel, B. U. Ö. Sonnerup, S. J. Bame, T. G. Forbes, E. W. Hones, and C. T. Russell: 1981, 'Structure of the low-latitude boundary layer'. *J. Geophys. Res.* **86**, 2099–2110.

Scudder, J. D., F. S. Mozer, N. C. Maynard, and C. T. Russell: 2002, 'Fingerprints of collisionless reconnection at the separator, I, Ambipolar-Hall signatures'. *J. Geophys. Res.* **107**(A10), 1294, doi:10.1029/2001JA000126.

Scudder, J. D., P. A. Puhl-Quinn, F. S. Mozer, K. W. Ogilvie, and C. T. Russell: 1999, 'Generalized Walén tests through Alfvén waves and rotational discontinuities using electron flow velocities'. *J. Geophys. Res.* **104**(13), 19817–19834.

Scurry, L., C. T. Russell, and J. T. Gosling: 1994, 'A statistical study of accelerated flow events at the dayside magnetopause'. *J. Geophys. Res.* **99**(18), 14815.

Shay, M. A., J. F. Drake, B. N. Rogers, and R. E. Denton: 2001, 'Alfvénic collisionless magnetic reconnection and the Hall term'. *J. Geophys. Res.* **99**(15), 3759–3772.

Shinohara, I., H. Suzuki, M. Fujimoto, and M. Hoshino: 2001, 'Rapid Large-Scale Magnetic-Field Dissipation in a Collisionless Current Sheet via Coupling between Kelvin-Helmholtz and Lower-Hybrid Drift Instabilities'. *Phys. Rev. Lett.* **87**(9), 095001.

Sibeck, D. G., G. Paschmann, R. A. Treumann, S. A. Fuselier, W. Lennartsson, M. Lockwood, R. Lundin, K. W. Ogilvie, T. G. Onsager, T.-D. Phan, M. Roth, M. Scholer, N. Sckopke, K. Stasiewicz, and M. Yamauchi: 1999, 'Plasma Transfer Processes at the Magnetopause'. *Space Sci. Rev.* **88**, 207–283.

Smith, M. F. and D. J. Rodgers: 1991, 'Ion distributions at the dayside magnetopause'. *J. Geophys. Res.* **96**(15), 11617.

Song, P. and C. T. Russell: 1992, 'Model of the formation of the low-latitude boundary layer for strongly northward interplanetary magnetic field'. *J. Geophys. Res.* **97**(16), 1411–1420.

Sonnerup, B. U. Ö.: 1974, 'Magnetopause reconnection rate'. *J. Geophys. Res.* **79**, 1546.

Sonnerup, B. U. Ö.: 1979, 'Magnetic field reconnection'. In: *Solar system plasma physics. Volume 3. (A79-53667 24-46) Amsterdam, North-Holland Publishing Co., 1979,* pp. 45–108.

Sonnerup, B. U. Ö., H. Hasegawa, and G. Paschmann: 2004, 'Anatomy of a flux transfer event seen by Cluster'. *Geophys. Res. Lett.* **31**, 11803, doi:10.1029/2004GL0201343.

Sonnerup, B. U. Ö., I. Papamastorakis, G. Paschmann, and H. Lühr: 1987, 'Magnetopause properties from AMPTE/IRM observations of the convection electric field - Method development'. *J. Geophys. Res.* **92**(11), 12137–12159.

Sonnerup, B. U. Ö., I. Papamastorakis, G. Paschmann, and H. Lühr: 1990, 'The magnetopause for large magnetic shear - Analysis of convection electric fields from AMPTE/IRM'. *J. Geophys. Res.* **95**(14), 10541–10557.

Sonnerup, B. U. Ö., G. Paschmann, I. Papamastorakis, N. Sckopke, G. Haerendel, S. J. Bame, J. R. Asbridge, J. T. Gosling, and C. T. Russell: 1981, 'Evidence for magnetic field reconnection at the earth's magnetopause'. *J. Geophys. Res.* **86**(15), 10049–10067.

Terasawa, T.: 1983, 'Hall current effect on tearing mode instability'. *Geophys. Res. Lett.* **10**, 475–478.

Thomas, V. A. and D. Winske: 1993, 'Kinetic simulations of the Kelvin-Helmholtz instability at the magnetopause'. *J. Geophys. Res.* **98**(17), 11425.

Trattner, K. J., S. A. Fuselier, W. K. Peterson, M. Boehm, D. Klumpar, C. W. Carlson, and T. K. Yeoman: 2002, 'Temporal versus spatial interpretation of cusp ion structures observed by two spacecraft'. *J. Geophys. Res.* **107**(A10), 1287, doi:10.1029/2001JA000181.

Trattner, K. J., S. A. Fuselier, and S. M. Petrinec: 2004, 'Location of the reconnection line for northward interplanetary magnetic field'. *J. Geophys. Res.* **109**(A18), A03219, doi:10.1029/2003JA009975.

Trattner, K. J., S. A. Fuselier, T. Yeoman, C. Carlson, W. Peterson, A. Korth, H. Rème, J. Sauvaud, and N. Dubouloz: 2005, 'Spatial and Temporal Cusp Structures Observed by Multiple Spacecraft and Ground Based Observations'. *Surveys in Geophysics.* submitted.

Trattner, K. J., S. A. Fuselier, T. K. Yeoman, A. Korth, M. Fraenz, C. Mouikis, H. Kucharek, L. M. Kistler, C. P. Escoubet, H. Rème, I. Dandouras, J. A. Sauvaud, J. M. Bosqued, B. Klecker, C. Carlson, T. Phan, J. P. McFadden, E. Amata, and L. Eliasson: 2003, 'Cusp structures: combining multi-spacecraft observations with ground-based observations'. *Ann. Geophys.* **21**, 2031–2041.

Treumann, R. A. and W. Baumjohann: 1997, *Advanced space plasma physics.* London: Imperial College Press, 1997.

Treumann, R. A., C. H. Jaroschek, K. Hallatschek, O. A. Pokhotelov, W. Baumjohann, and E. Georgescu: 2004, 'The strange physics of low frequency mirror mode turbulence in the high temperature plasma of the magnetosheath'. *Nonlin. Proc. Geophys.* **11**, in press.

Treumann, R. A. and N. Sckopke: 1999, 'Diffusion Processes'. *Space Sci. Rev.* **88**, 389–402.

Tsurutani, B. T. and R. M. Thorne: 1982, 'Diffusion processes in the magnetopause boundary layer'. *Geophys. Res. Lett.* **9**, 1247–1250.

Twitty, C., T. Phan, G. Paschmann, B. Lavraud, H. Rème, and M. Dunlop: 2004, 'Cluster survey of cusp reconnection and its IMF dependence'. *Geophys. Res. Lett.* **31**, L19808, doi:10.1029/2004GL020646.

Vaivads, A., M. André, S. Buchert, J.-E. Wahlund, A. Fazakerley, and N. Cornilleau-Wehrlin: 2004, 'Cluster observations of lower hybrid turbulence within thin layers at the magnetopause'. *Geophys. Res. Lett.* **31**, L03804, doi:10.1029/2003GL018142.

Vaivads, A., Y. Khotyaintsev, M. André, A. Retinò, S. C. Buchert, B. N. Rogers, P. Décréau, G. Paschmann, and T. D. Phan: 2004, 'Structure of the Magnetic Reconnection Diffusion Region from Four-Spacecraft Observations'. *Phys. Rev. Lett.* **93**(10), 105001.

Vontrat-Reberac, A., J. M. Bosqued, M. G. G. T. Taylor, B. Lavraud, D. Fontaine, M. W. Dunlop, H. Laakso, N. Cornilleau-Werhlin, P. Canu, and A. Fazakerley: 2003, 'Cluster observations of the high-altitude cusp for northward interplanetary magnetic field: A case study'. *J. Geophys. Res.* **108**(A9), 1346, doi:10.1029/2002JA009717.

Wild, J. A., S. W. H. Cowley, J. A. Davies, H. Khan, S. E. Milan, G. Provan, T. K. Yeoman, A. Balogh, M. W. Dunlop, K.-H. Fornacon, and E. Georgescu: 2001, 'First simultaneous observations of flux transfer events at the high-latitude magnetopause by the Cluster spacecraft and pulsed radar signatures in the conjugate ionosphere by the CUTLASS and EISCAT radars'. *Ann. Geophys.* **19**, 1491–1508.

Zwan, B. J. and R. A. Wolf: 1976, 'Depletion of solar wind plasma near a planetary boundary'. *J. Geophys. Res.* **81**(10), 1636–1648.

Abbreviations and Acronyms

ASPOC	Active Spacecraft Potential Control
CIS	Cluster Ion Spectrometry
CIS/HIA	CIS Hot Ion Analyzer
CIS/CODIF	CIS ion COmpostion and DIstribution Function analyzer
DWP	Digital Wave Processor
EDI	Electron Drift Instrument
EFW	Electric Fields and Waves
FGM	Fluxgate Magnetometer
PEACE	Plasma Electron and Current Experiment
PEACE/HEEA	PEACE High-Energy Electron Analzyer
PEACE/LEEA	PEACE Ligh-Energy Electron Analzyer
RAPID	Research with Adaptive Particle Imaging Detectors
STAFF	Spatio-Temporal Analysis of Field Fluctutations Experiment
STAFF-SA	STAFF Spectrum Analyzer
STAFF-SC	STAFF Search Coil
WBD	Wide Band Data
WHISPER	Waves of High Frequency and Sounder for Probing of Electron Density by Relaxation

AACGM	Altitude Adjusted Corrected Geomagnetic Coordinates
ACE	Advanced Composition Explorer
AIC	Alfvén ion cyclotron waves
AMPTE	Active Magnetospheric Particle Tracer Explorers
AMPTE-IRM	AMPTE Ion Release Module
AMPTE-UKS	AMPTE UK Satellite
AU	Astronomical Unit
BBELF	broadband extremely low frequency
BEN	broadband electrostatic noise
BL	boundary layer
BS	bow shock
CME	coronal mass ejection
CTA	constant thickness approach
CVA	constant velocity approach
DA	discontinuity analyzer
DMSP	Defense Meteorological Satellite Program
EISCAT	European Incoherent Scatter Radar
EMIC	electromagnetic ion cyclotron
ESR	EISCAT Svalbard Radar
ESW	electrostatic solitary waves
FTE	flux-transfer event
GS	Grad-Shafranov
GSE	geocentric solar ecliptic coordinates
GSM	geocentric solar magnetospheric coordinates

HCS	heliospheric current sheet
HF	high frequency
HFA	hot flow anomaly
HPS	heliospheric plasma sheet
HT	deHoffmann-Teller
ICME	interplanetary manifestations of CMEs
IES	isolated electrostatic structures
IMF	interplanetary magnetic field
IMP	Interplanetary Monitoring Platform
ISEE	International Sun-Earth Explorers
KH	Kelvin-Helmholtz
KHI	Kelvin-Helmholtz instability
LF	low frequency
LLBL	low-latitude boundary layer
MDS	magnetosheath dispersed signatures
MFA	mean field-aligned coordinates
MFED	magnetic wave field energy distribution
MFR	minimum Faraday residue
MHD	magnetohydrodynamics
MLT	magnetic local time
MMR	minimum mass-flux residue
MMS	mirror mode structure
MP	magnetopause
MTV	minimum thickness variation
MVA	minimum variance analysis
MVAB	minimum variance analysis on B
MVABC	MVAB with $\langle \mathbf{B} \rangle \cdot \mathbf{n} = 0$ constraint
MVAJ	minimum variance analysis of current density
OCB	open-closed field line boundary
PIC	Particle in Cell
PDL	plasma depletion layer
PM	plasma mantle
PMS	planar magnetic structure
RD	rotational discontinuity
R_E	Earth radius
TD	tangential discontinuity
TDS	time domain sample
ULF	ultra-low frequency
WHAMP	Waves in Homogeneous Anistropic Magnetized Plasma

Author Affiliations and Addresses

David Attié
CETP/IPSL
10-12 av. de l'Europe
78140 Vélizy
France
david.attie@cetp.ipsl.fr

Stuart Bale
Space Sciences Laboratory
University of California, Berkeley
Berkeley, CA 94720-7450
USA
bale@ssl.berkeley.edu

Michael Balikhin
University of Sheffield
Mappin Street
Sheffield S1 3JD
United Kingdom
balikhin@acse.shef.ac.uk

Andre Balogh
The Blackett Laboratory
Imperial College of Science and Technology
Prince Consort Rd.
London SW7 2BZ
United Kingdom
a.balogh@imperial.ac.uk

Yulia Bogdanova
Mullard Space Science Laboratory
University College London
Holmbury St. Mary
Dorking, Surrey, RH5 6NT
United Kingdom
jb@mssl.ucl.ac.uk

David Burgess
Astronomy Unit
Queen Mary, University of London
Mile End Rd.
London E1 4NS
United Kingdom
d.burgess@qmul.ac.uk

Peter Cargill
Space and Atmospheric Physics
The Blackett Laboratory
Imperial College
Prince Consort Rd.
London SW7 2BZ
United Kingdom
p.cargill@imperial.ac.uk

Dragos Constantinescu
Institut für Geophysik und
extraterrestrische Physik
TU-Braunschweig
Mendelssohnstr. 3
D-38106 Braunschweig
Germany
d.constantinescu@tu-bs.de

Nicole Cornilleau-Wehrlin
CETP/IPSL
10-12 av. de l'Europe
78140 Vélizy
France
nicole.cornilleau@cetp.ipsl.fr

Johan De Keyser
Belgian Institute for Space Aeronomy (BISA)
Ringlaan 3
B-1180 Brussels
Belgium
johan.dekeyser@bira-iasb.oma.be

Pierrette Décréau
LPCE/CNRS and Université d'Orléans
3A av. de la recherche scientifique
F-45071 Orléans
France
pdecreau@cnrs-orleans.fr

Malcolm Dunlop
SSTD
RAL
Chilton
Didcot, OX11 0QX
United Kingdom
m.dunlop@rl.ac.uk

Jonathan Eastwood
Goddard Space Flight Center
MC 696
Greenbelt, MD 20771
USA
jeastwood@lepvax.gsfc.nasa.gov

C. Philippe Escoubet
ESTEC
Keplerlaan 1
Postbus 299
NL-2200 AG Noordwijk
The Netherlands
philippe.escoubet@esa.int

Stephen Fuselier
Department L9-42
Lockheed-Martin Advanced Techn. Center
Bldg. 255
3251 Hanover Street
Palo Alto, CA 94304
USA
fuselier@spasci.com

Melvyn Goldstein
Goddard Space Flight Center
MC 692
NASA
Greenbelt, MD 20771
USA
melvyn.goldstein@gsfc.nasa.gov

Benjamin Grison
CETP/IPSL
10-12 av. de l'Europe
78140 Vélizy
France
Benjamin.Grison@cetp.ipsl.fr

Stein Haaland
Max-Planck-Institut für extraterrestrische Physik
Postfach 1312
D-85741 Garching
Germany
seh@mpe.mpg.de

Tim Horbury
The Blackett Laboratory
The Blackett Laboratory
Imperial College
Prince Consort Rd.
London SW7 2BZ
United Kingdom
t.horbury@imperial.ac.uk

Vladimir Krasnosselskikh
LPCE/CNRS
3A, av. de la Recherche Scientificque
F-45071 Orléans Cedex 2
France
vkrasnos@cnrs-orleans.fr

Harald Kucharek
Space Science Center
University of New Hampshire
39 College Road
Morse Hall
Durham, NH 03824
USA
harald.kucharek@unh.edu

Benoit Lavraud
Space Science and Application
Los Alamos National Laboratory
P.O. Box 1663, MS D466
Los Alamos, NM 87545
USA
lavraud@lanl.gov

Bertrand Lembège
CETP/IPSL 10-12, av. de l'Europe
F-78140 Vélizy
France
bertrand.lembege@cetp.ipsl.fr

Elizabeth Lucek
Space and Atmospheric Physics
The Blackett Laboratory
Imperial College
Prince Consort Rd.
London SW7 2BZ
United Kingdom
e.lucek@imperial.ac.uk

Rickard Lundin
Swedish Institute of Space Physics
P.O. Box 812
SE-98128 Kiruna
Sweden
rickard.lundin@irf.se

Christian Mazelle
CESR/CNRS
9, av. du Colonnel Roche
BP 4346
F-31028 Toulouse cedex 4
France
christian.mazelle@cesr.fr

Karim Meziane
Physics Department
University of New Brunswick
P.O. Box 4400
Fredericton, NB E3B5A3
Canada
karim@unb.ca

Eberhard Möbius
Space Science Center
and Departement of Physics
University of New Hampshire
Durham, NH 03824-3525
USA
eberhard.moebius@unh.edu

Yasuhito Narita
Institut für Geophysik und
extraterrestrische Physik
TU-Braunschweig
Mendelssohnstrasse 3
D-38106 Braunschweig
Germany
y.narita@tu-bs.de

Katariina Nykyri
Space and Atmospheric Physics Group
The Blackett Laboratory
Imperial College
Prince Consort Rd.
London SW7 2BZ
United Kingdom
k.nykyri@imperial.ac.uk

Christopher Owen
Mullard Space Science Laboratory
University College London
Holmbury St. Mary
Dorking, Surrey, RH5 6NT
United Kingdom
cjo@mssl.ucl.ac.uk

Götz Paschmann
International Space Science Institute
Hallerstr. 6
CH-3012 Bern
Switzerland
goetz.paschmann@issi.unibe.ch

Tai Phan
Space Sciences Lab.
University of California Berkeley
Berkeley, CA 94720
USA
phan@ssl.berkeley.edu

Jolene Pickett
Department of Physics and Astronomy
The University of Iowa
Iowa City, IA 52242
USA
pickett@uiowa.edu

Jean-Louis Pinçon
LPCE/CNRS and Université d'Orléans
3A av. de la recherche scientifique
F-45071 Orléans
France
jlpincon@cnrs-orleans.fr

Laurence Rezeau
CETP/IPSL/UPMC
10-12 av. de l'Europe
F-78140 Vélizy
France
laurence.rezeau@cetp.ipsl.fr

Barrett Rogers
Dartmouth College
HB 6127
Wilder Bldg.
Hanover, NH 03755
USA
barrett.rogers@dartmouth.edu

Fouad Sahraoui,
CETP/IPSL
10-12, av. de l'Europe
F-78140 Vélizy
France
fouad.sahraoui@cetp.ipsl.fr

Manfred Scholer
Max-Planck-Institut für extraterrestrische Physik
Postfach 1312
D-85741 Garching
Germany
mbs@mpe.mpg.de

Steve Schwartz
The Blackett Laboratory
Imperial College
Prince Consort Rd.
London SW7 2BZ
United Kingdom
s.schwartz@imperial.ac.uk

Bengt Sonnerup
Thayer School of Engineering
Dartmouth College
8000 Cummings Hall
Hanover, NH 03755-8000
USA
sonnerup@dartmouth.edu

Michelle Thomsen
Space and Atmospheric Sciences Group
Los Alamos National Laboratory
MS D466
Los Alamos, NM 87545
USA
mthomsen@lanl.gov

Rudolf Treumann
Max-Planck-Institut für extraterrestrische Physik
Postfach 1312
D-85741 Garching
Germany
tre@mpe.mpg.de

Andris Vaivads
Swedish Institute of Space Physics
Box 537
SE-75121 Uppsala
Sweden
andris@irfu.se

Simon Walker
Department of Automatic Control and Systems Engineering
University of Sheffield
Mappin Street
Sheffield S1 3JD
United Kingdom
simon.walker@sheffield.ac.uk

Space Science Series of ISSI

1. R. von Steiger, R. Lallement and M.A. Lee (eds.): *The Heliosphere in the Local Insterstellar Medium.* 1996 ISBN 0-7923-4320-4
2. B. Hultqvist and M. Øieroset (eds.): *Transport Across the Boundaries of the Magnetosphere.* 1997 ISBN 0-7923-4788-9
3. L.A. Fisk, J.R. Jokipii, G.M. Simnett, R. von Steiger and K.-P. Wenzel (eds.): *Cosmic Rays in the Heliospehre.* 1998 ISBN 0-7923-5069-3
4. N. Prantzos, M. Tosi and R. von Steiger (eds.): *Primordial Nuclei and Their Galactic Evolution.* 1998 ISBN 0-7923-5114-2
5. C. Fröhlich, M.C.E. Huber, S.K. Solanki and R. von Steiger (eds.): *Solar Composition and its Evolution – From Core to Corona.* 1998 ISBN 0-7923-5496-6
6. B. Hultqvist, M. Øieroset, Goetz Paschmann and R. Treumann (eds.): *Magnetospheric Plasma Sources and Losses.* 1999 ISBN 0-7923-5846-5
7. A. Balogh, J.T. Gosling, J.R. Jokipii, R. Kallenbach and H. Kunow (eds.): *Co-rotating Interaction Regions.* 1999 ISBN 0-7923-6080-X
8. K. Altwegg, P. Ehrenfreund, J. Geiss and W. Huebner (eds.): *Composition and Origin of Cometary Materials.* 1999 ISBN 0-7923-6154-7
9. W. Benz, R. Kallenbach and G.W. Lugmair (eds.): *From Dust to Terrestrial Planets.* 2000 ISBN 0-7923-6467-8
10. J.W. Bieber, E. Eroshenko, P. Evenson, E.O. Flückiger and R. Kallenbach (eds.): *Cosmic Rays and Earth.* 2000 ISBN 0-7923-6712-X
11. E. Friis-Christensen, C. Fröhlich, J.D. Haigh, M. Schüssler and R. von Steiger (eds.): *Solar Variability and Climate.* 2000 ISBN 0-7923-6741-3
12. R. Kallenbach, J. Geiss and W.K. Hartmann (eds.): *Chronology and Evolution of Mars.* 2001 ISBN 0-7923-7051-1
13. R. Diehl, E. Parizot, R. Kallenbach and R. von Steiger (eds.): *The Astrophysics of Galactic Cosmic Rays.* 2001 ISBN 0-7923-7051-1
14. Ph. Jetzer, K. Pretzl and R. von Steiger (eds.): *Matter in the Universe.* 2001 ISBN 1-4020-0666-7
15. G. Paschmann, S. Haaland and R. Treumann (eds.): *Auroral Plasma Physics.* 2002 ISBN 1-4020-0963-1
16. R. Kallenbach, T. Encrenaz, J. Geiss, K. Mauersberger, T.C. Owen and F. Robert (eds.): *Solar System History from Isotopic Signatures of Volatile Elements.* 2003 ISBN 1-4020-1177-6
17. G. Beutler, M.R. Drinkwater, R. Rummel and R. Von Steiger (eds.): *Earth Gravity Field from Space – from Sensors to Earth Sciences.* 2003 ISBN 1-4020-1408-2
18. D. Winterhalter, M. Acuña and A. Zakharov (eds.): *"Mars" Magnetism and its Interaction with the Solar Wind.* 2004 ISBN 1-4020-2048-1
19. T. Encrenaz, R. Kallenbach, T.C. Owen and C. Sotin: *The Outer Planets and their Moons.* ISBN 1-4020-3362-1
20. G. Paschmann, S.J. Schwartz, C.P. Escoubet and S. Haaland (eds.): *Outer Magnetospheric Boundaries: Cluster Results* ISBN 1-4020-3488-1

Springer – Dordrecht / Boston / London